Optimal Control

CONTEMPORARY SOVIET MATHEMATICS

Series Editor: Revaz Gamkrelidze, *Steklov Institute, Moscow, USSR*

COHOMOLOGY OF INFINITE-DIMENSIONAL LIE ALGEBRAS
D. B. Fuks

DIFFERENTIAL GEOMETRY AND TOPOLOGY
A. T. Fomenko

LINEAR DIFFERENTIAL EQUATIONS OF PRINCIPAL TYPE
Yu. V. Egorov

OPTIMAL CONTROL
V. M. Alekseev, V. M. Tikhomirov, and S. V. Fomin

THEORY OF SOLITONS: The Inverse Scattering Method
S. Novikov, S. V. Manakov, L. P. Pitaevskii, and V. E. Zakharov

TOPICS IN MODERN MATHEMATICS: Petrovskii Seminar No. 5
Edited by O. A. Oleinik

Optimal Control

V. M. Alekseev
V. M. Tikhomirov
and S. V. Fomin

Department of Mathematics and Mechanics
Moscow State University
Moscow, USSR

Translated from Russian by
V. M. Volosov

CONSULTANTS BUREAU • NEW YORK AND LONDON

Library of Congress Cataloging in Publication Data

Alekseev, V. M. (Vladimir Mikhailovich)
Optimal control.
(Contemporary Soviet mathematics)
1. Control theory. 2. Mathematical optimization. I. Tikhomirov, V. M. II. Fomin, S. V. (Sergei Vasil'evich III. Title. IV. Series.
QA402.3.A14 1987 001.53 87-6935
ISBN 0-306-10996-4

This translation is published under an agreement with the Copyright Agency of the USSR (VAAP).

A Division of Plenum Publishing Corporation
233 Spring Street, New York, N.Y. 10013

Printed in the United States of America

PREFACE

There is an ever-growing interest in control problems today, connected with the urgent problems of the effective use of natural resources, manpower, materials, and technology. When referring to the most important achievements of science and technology in the 20th Century, one usually mentions the splitting of the atom, the exploration of space, and computer engineering. Achievements in control theory seem less spectacular when viewed against this background, but the applications of control theory are playing an important role in the development of modern civilization, and there is every reason to believe that this role will be even more significant in the future.

Wherever there is active human participation, the problem arises of finding the best, or optimal, means of control. The demands of economics and technology have given birth to optimization problems which, in turn, have created new branches of mathematics.

In the Forties, the investigation of problems of economics gave rise to a new branch of mathematical analysis called linear and convex programming. At that time, problems of controlling flying vehicles and technological processes of complex structures became important. A mathematical theory was formulated in the mid-Fifties known as optimal control theory. Here the maximum principle of L. S. Pontryagin played a pivotal role. Optimal control theory synthesized the concepts and methods of investigation using the classical methods of the calculus of variations and the methods of contemporary mathematics, for which Soviet mathematicians made valuable contributions.

This book is intended to serve as a text in various courses in optimization that are offered in universities and colleges. A brief outline of the general concept and structure of the book is given below.

The history of the investigation of problems on extremum or, as we usually say, "extremal problems," did not begin in our time – such problems have attracted the attention of mathematicians for ages. Chapter 1, which is written for a wide audience with diverse mathematical backgrounds, follows the trend of development of this science. Although the mathematical apparatus used there is minimal, the style of presentation is that of a precise mathematical text, confined to the most impressive yet elementary periods in the history of the investigation of extremal problems. The aim is to unite the early concepts of J. Kepler and P. Fermat, the problems posed by C. Huygens, I. Newton, and J. Bernoulli, and the methods of

J. L. Lagrange, L. Euler, and K. T. W. Weierstrass with present-day theory which, in fact, is an extension of the works of these great scientists. Chapter 1 also describes some methods of solving concrete problems and presents examples for solving these problems based on a unified approach.

The remainder of the book is intended for mathematicians. The appearance of the new theory stimulated the development of both the old and new branches of mathematical analysis, yet not every one of these branches is adequately represented in modern mathematical education. That part of classical mathematical analysis centered on the topic of implicit functions plays an exceptional role at present in all aspects of finite-dimensional and infinite-dimensional analysis. The same applies to the principles of convex analysis. Finally, in the development of optimal control theory, the basic principles of the theory of differential equations remain valid for equations with discontinuous right-hand sides. The above three divisions of mathematical analysis and geometry are dealt with in Chapter 2.

Chapters 3 and 4 deal with the theory of extremal problems, a substantial portion of which forms the subject matter in the half-year courses given by the authors at the Department of Mathematics and Mechanics of Moscow State University. The text is so arranged that the proof of the main theorems takes no more than one lecture (two sessions of 45 minutes each in the Soviet Union). The presentation of the material is thorough, the authors avoiding omissions and references to the obvious.

Some standard notation from set theory, functional analysis, etc. is used in the text without further comment, and, to facilitate reading, a list of basic notation with brief explanations is given at the end of the book.

Sections in the text marked with an asterisk are intended to acquaint the reader with modern methods in the theory of extremal problems. The presentation of the material in these sections follows closely the style of a monograph although a few references are made to some theorems which have become classical but are not at present included in traditional mathematical programs. These sections are based on special courses known as "Convex Analysis," "Supplementary Chapters in the Theory of Extremal Problems," and others that have been given for a number of years at the Department of Mathematics and Mechanics of Moscow State University.

The book is intended for a broad audience: primarily for students of universities and colleges with comprehensive mathematical programs, and also for engineers, economists, and mathematicians involved in the solution of extremal problems. The Introduction (Chapter 1) is written for the last group of readers, and researchers intimately associated with the theory of extremal problems will find the Notes and Guide to the Literature at the end of the book useful.

Credit must be given to the late Professor S. V. Fomin in organizing and expanding the course in optimal control at the Department of Mathematics and Mechanics of Moscow State University. It was on his initiative that the present book was written. His untimely death in the prime of his life did not allow him to see this work realized. In the preparation of this book the authors used the original text of the lectures given by S. V. Fomin and V. M. Tikhomirov. Professor V. M. Alekseev, one of the co-authors of this book, passed away on December 1, 1980. This book is his last major work.

I express my gratitude to the Faculty of General Control Problems of the Department of Mathematics and Mechanics of Moscow State University for their participation in the discussion of teaching methods for the course in optimal control and the method of presentation of the subjects touched upon in this book.

I am indebted to A. A. Milyutin for his creative work in the formulation of the mathematical concepts on which this book is based. Thanks are also due to A. I. Markushevich for his valuable advice and to A. P. Buslaev and G. G. Magaril-Il'yaev for reading the manuscript and making a number of valuable comments.

The English translation incorporates suggestions made by me and V. M. Alekseev. It includes two supplementary sections: Section 3.5 of Chapter 3 and the whole of Chapter 5. I express my deep gratitude to Professor V.M. Volosov for translating the book and for the various refinements that he made in the translation.

V. M. Tikhomirov

CONTENTS

Chapter 1
INTRODUCTION

Chapter 2
MATHEMATICAL METHODS OF THE THEORY OF EXTREMAL PROBLEMS

Chapter 3
THE LAGRANGE PRINCIPLE FOR SMOOTH PROBLEMS WITH CONSTRAINTS

Chapter 4
THE LAGRANGE PRINCIPLE FOR PROBLEMS OF THE CLASSICAL CALCULUS OF VARIATIONS AND OPTIMAL CONTROL THEORY

Chapter 1

INTRODUCTION

1.1. How Do Extremal Problems Appear?

Striving to improve oneself is a characteristic of human beings, and when one has to choose from several possibilities one naturally tries to find the *optimal* possibility among them.

The word "optimal" originates from the Latin "optimus," meaning the best. To choose the optimal possibility one has to solve a problem of finding a *maximum* or a *minimum*, i.e., the greatest or the smallest values of some quantities. A maximum or a minimum is called an *extremum* (from the Latin "extremus," meaning furthest off). Therefore the problems of finding a maximum or a minimum are called *extremal problems*.

The methods of solving and analyzing all kinds of extremal problems constitute special branches of mathematical analysis. The term *optimization problems* is used in almost the same sense; this term indicates more distinctly its connection with practical applications of mathematics.

The aim of this book is to present to the reader the theory of extremal problems and the methods of their solution. However, before proceeding to the formal and logically consistent presentation of this branch of mathematics, we turn to the past to gain a better understanding of the reasons motivating scientists to pose and solve extremal problems in their applied aspect, i.e., as optimization problems.

> Mercatique solum, facti
> de nomine Byrsam
> Taurino quantum possent
> circumdare tergo
> *P. Vergilius Maro*, "The Aeneid"*

1.1.1. The Classical Isoperimetric Problem. Dido's Problem. Problems of finding maximum and minimum values were first formulated in antiuity. The classical isoperimetric problem is perhaps the most ancient of the extremal problems known to us. It is difficult to say when the idea was first stated that the circle and the sphere have the greatest "capacity" among all the closed curves of the same length and all the surfaces of the same area. Simplicius (6th century A.D.), one of the last pupils

*They bought a space of ground, which (Byrsa called, from the bull's hide) they first enclosed..." [The Works of Virgil, Oxford University Press (1961), p. 144, translated by J. Dryden.]

of the school of Neoplatonism at Athens who compiled extensive commentaries on Aristotle's work (4th century B.C.), writes: "It was proved before Aristotle (since the latter uses it as a known fact) and later proved by Archimedes and Zenodoros more completely that, among the isoperimetric figures, the circle, and among the solids of constant surface area, the ball, have the greatest capacity." These words indicate the statement of the following extremal problems: among the plane closed curves of a given length find the curve enclosing the greatest area, and among the closed spatial surfaces of a given area find the surface enclosing the greatest volume.

Such a formulation of the problem is natural for a Platonist and is connected with the search for perfect forms. As is known, the circle and the ball were symbols of perfection in ancient times.

A more matter-of-fact motivation of the same isoperimetric problem and a number of similar problems are found in a sufficiently distinct though naive form in the legend of Dido. Let us recall the plot of the legend as it is narrated in "The Aeneid" by the Roman poet Virgil (see the epigraph at the beginning of this subsection).

The Phoenician princess Dido with a small group of inhabitants of the city of Tyre fled from her brother, the tyrant Pygmalion, and set sail for the West along the Mediterranean coast. Dido and her faithful companions chose a good place on the north coast of Africa (at present the shore of the Gulf of Tunis) and decided to found a settlement there. It seems that among the natives there was not much enthusiasm for this idea. However, Dido managed to persuade their chieftain Hiarbas to give her "as much land as she could enclose with the hide of a bull." Only later did the simple-hearted Hiarbas understand how cunning and artful Dido was: she then cut the bull's hide into thin strips, tied them together to form an extremely long thin thong, and surrounded with it a large extent of territory and founded the city of Carthage* there. In commemoration of this event the citadel of Carthage was called Byrsa.† According to the legend, all these events occurred in 825 (or 814) B.C.

Analyzing the situation, we see that here there are several possibilities of stating an optimization problem.

A) It is required to find the optimal form of a lot of land of the maximum area S for a given perimeter L.

Clearly, this is that same classical isoperimetric problem.‡ Its solution is a circle.

Exercise. Assuming that the hide of a bull is a rectangle of 1×2 m and that the width of the thong is 2 mm, determine L and the maximum S.

[The authors could not find information on the exact dimensions of Byrsa. Situated on a high hill (63 m above sea level) it was unlikely to be very large. Compare the length of the wall of the Kremlin in Moscow — 2235 m.]

---

*From the Phoenician name "Quarthadasht" ("new town").

†Byrsa is a word Punic in origin ("bull's hide"). The name "Punic" originates from the Latin "Punicus" pertaining to Carthage or its inhabitants.

‡Another real situation leading to the same problem was described in the story "How much land does a man need?" by L. N. Tolstoy. For the analysis of this story from the geometrical viewpoint, see the book "Zanimatel'naya geometriya" (Geometry for Entertainment) by J. A. Perelman, Moscow, Gostekhizdat (1950), Chapter 12 (in Russian).

The solution of the isoperimetric problem is contained in the following assertion:

If a rectifiable curve of length L encloses a plane figure of area S, then

$$L^2 \geqslant 4\pi S, \tag{1}$$

and, moreover, the equality takes place here if and only if the curve is a circle.

Relation (1) is called the *isoperimetric inequality*. For its proof, see [21].

B) Other statements of the problems are obtained if we suppose (which is quite natural) that Dido wanted to preserve the access to the sea. To distinguish between these problems and the classical isoperimetric problem, we shall call the former Dido's problems. For the sake of simplicity, we shall first consider the case of a rectilinear coastline (Fig. 1).

Dido's First Problem. Among all arcs of length L, lying within a half-plane bounded by a straight line l, with end points A, B $\in l$ find the arc which together with the line segment [AB] encloses a figure of the maximum area S.

Solution. Let ACB be an arbitrary admissible arc with end points A, B $\in l$, enclosing a figure of area S (Fig. 1). Together this arc and its symmetric reflection in l form a closed curve of length 2L enclosing a figure of area 2S. According to the isoperimetric inequality,

$$(2L)^2 \geqslant 4\pi 2S, \tag{2}$$

whence

$$S \leqslant L^2/(2\pi). \tag{3}$$

Consequently the maximum value of S can only be equal to $L^2/(2\pi)$, and this value is in fact attained if ACB is a semicircle subtending the diameter [AB]. The problem possesses a single solution to within a shift along the straight line l (why?).

C) In the foregoing problem the end points A and B of the sought-for arc could occupy arbitrary positions on the straight line l. Now, what occurs when the end points are fixed?

Dido's Second Problem. Among all arcs of length L, lying in a half-plane bounded by a straight line l, with fixed end points A, B $\in l$, find the arc which together with the line segment [AB] encloses a figure of the maximum area.

Solution. It is evident that the problem makes sense only for L > |AB| (otherwise, then either there is no arc satisfying the conditions of the problem or there is only one such arc, namely (for L = |AB|) the line segment [AB] itself. By analogy with the first problem, it is natural to expect that the solution is an arc of a circle for which [AB] is a chord. Such an arc $\hat{ACB}$ is determined uniquely. Let us complete it to the whole circle by adding the arc ADB (Fig. 2). Let us denote the length of the arc ADB by λ and the area of the segment of the circle bounded by this arc and the line segment [AB] by σ.

Now let ACB be an arbitrary arc satisfying the conditions of the problem which together with [AB] encloses an area S. The length of the closed curve ACBD is equal to L + λ and the area bounded by it is equal to S + σ. According to (1), we have $4\pi(S + \sigma) \leqslant (L + \lambda)^2$, whence

$$S \leqslant \frac{1}{4\pi}(L+\lambda)^2 - \sigma.$$

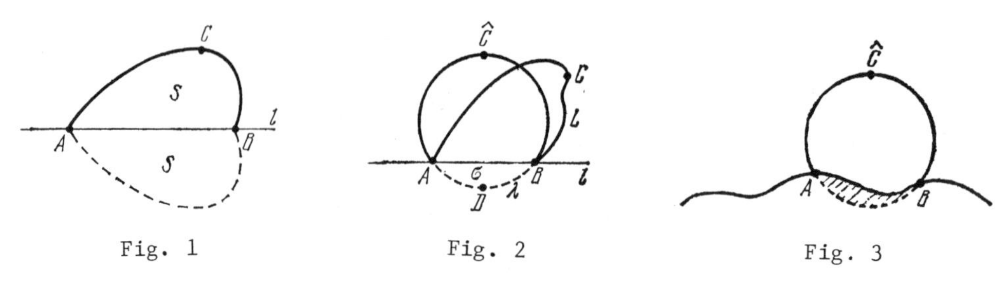

Fig. 1 Fig. 2 Fig. 3

As in (1), the equality in this relation and, consequently, the maximum S are attained if and only if the curve ACBD is a circle, i.e., if the two arcs are equal: $ACB = A\hat{C}B$.

Attention should be paid to the following distinction between the two problems we have considered. In Dido's first problem the set of the admissible curves is broader since the positions of the points A and B are not prescribed. By the way, without loss of generality, one of them, say A, can be regarded as fixed. Then the position of the point B is determined by an additional condition: $A\hat{C}B$ is not simply an arc of a circle as in Dido's second problem but is a semicircle. In the equivalent form this condition is stated thus: the sought-for arc approaches the straight line l at angles of 90° at its end points. Later we shall see that this demonstrates a general principle: when the end points of a sought-for curve possess some freedom we must require that at these points certain conditions called *transversality conditions* be fulfilled. As to the shape of the sought-for curve, it is the same in both the problems and is determined by a certain equation (the *Euler equation*) which must hold along the curve. In the case under consideration the sought-for curve must have one and the same curvature at all its points.

D) Now let us consider the case when the coastline is not a straight line. We shall confine ourselves to fixed end points. It can easily be understood that if the coastline between A and B deviates slightly from a straight line, the solution will be again the same circular arc $A\hat{C}B$ as before.

The above proof is completely applicable if we denote by σ the area shaded in Fig. 3. Figure 4 shows what occurs when there is a big gulf between A and B. For instance, let a canal DC be dug perpendicularly to AB. Assuming that the border of the city must go along the coastline we see that the solution of the problem is the same arc $A\hat{C}B$ when the point D remains inside $A\hat{C}B$ (Fig. 4a). In case the canal DC intersects the arc $A\hat{C}B$ but the inequality $|AC| + |CB| < L$ holds, the solution is the curve composed of the two circular arcs AC and CB (Fig. 4b). In the limiting case $|AC| + |CB| = L$ the solution is the broken line ACB, and for $|AC| + |CB| > L$ there exists no solution.

This version of the problem could be called *Dido's problem with phase constraints*.

E) Finally, let us dwell on one more version of Dido's problem. Suppose that for some reason (for instance, because of the ban imposed by the priests of Eshmun, a god whose temple was built in Byrsa) the walls of the city must not be inclined to the coastline (which we again assume to be rectilinear) at an angle greater than 45°. Such a problem is an *optimal control problem*. Its solution can be found with the aid of the *Pontryagin maximum principle* which will be presented later. In a typical case the solution is as shown in Fig. 5. The line segments AC and BE form angles of 45° with the coastline, and CDE is an arc of a circle.

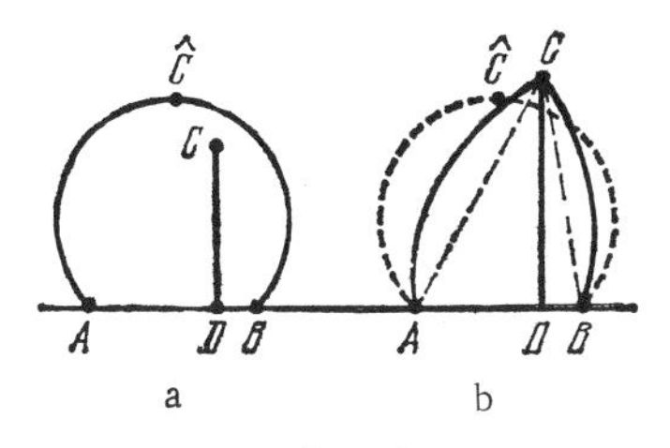

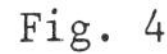
Fig. 4

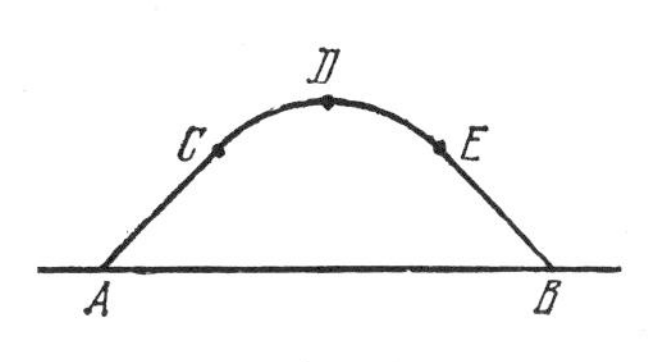

Fig. 5

1.1.2. Some Other Ancient Extremal Problems in Geometry. The isoperimetric problem has been discussed sufficiently thoroughly, and in this section we shall dwell on some other extremal probelms of geometrical character which were considered by mathematicians of different centuries. In particular, such problems are found in the works of Euclid, Archimedes, and Apollonius, the great mathematicians of antiquity.

In Euclid's "Elements" (4th century B.C.) only one maximum problem is found (Book 6, Proposition 27). In terms of modern language it is stated thus:

The Problem of Euclid. In a given triangle ABC inscribe a parallelogram ADEF (Fig. 6) with the greatest area.

Solution. The vertices $\hat{D}$, $\hat{E}$, and $\hat{F}$ of the sought-for parallelogram are the midpoints of the corresponding sides of the given triangle. This can be proved in various ways. For instance, it can easily be shown that the parallelograms $\hat{D}DG\hat{E}$ and $FH\hat{E}\hat{F}$ have equal areas. It follows that the area of the parallelogram ADEF is less than that of the parallelogram $A\hat{D}\hat{E}\hat{F}$ since the latter is equal to the area of the figure $ADG\hat{E}HF$ containing the parallelogram ADEF. ■

In Archimedes' works (3rd century B.C.) known to us the isoperimetric problem is not mentioned. It is still unknown what was done by Archimedes in this field, and therefore the words of Simplicius quoted in Section 1.1.1 still remain mysterious. However, in the work "On a Ball and a Cylinder" by Archimedes there is a solution of a problem on figures of a given area. Namely, in this work Archimedes stated and solved *the problem on the maximum volume that spherical segments with a given lateral surface area may have*.

The solution of this problem is a hemisphere (just as a semicircle is the solution of Dido's second problem).

"Conics" is the name of the greatest work of Apollonius (3rd-2nd centuries B.C.). The fifth book of "Conics" is related to our topic. Van der Waerden* writes: Apollonius states the "problem of how to draw the shortest and the longest line segments from a point O (see Fig. 7) to a conic section. However, he does even more than he promises: he finds all the straight lines passing through O which intersect the conic section at right angles (at present they are called normals) and analyzes for what position of O the problem has two, three, or four solutions." Moving the point O he "determines the ordinates of the boundary points G_1 and G_2 where the number of the normals through O passes at once from 2 to 4 and vice versa."

After the downfall of the ancient civilization the scientific activity in Europe had been at a standstill until the 15th century. In the 16th century the foundations of algebra were laid down and the first extremal problem of algebraic character appeared.

*B. L. Van der Waerden, Ontwakende Wetenschap, Groningen (1950).

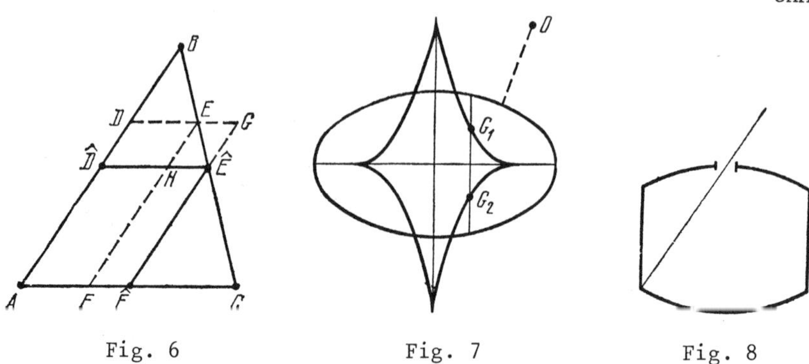

Fig. 6 Fig. 7 Fig. 8

For example, N. Tartaglia (16th century) posed the following problem: *divide the number 8 into two parts such that the product of their product by their difference is maximal.*

Until the 17th century no general methods of solving extremal problems were elaborated, and each problem was solved by means of a special method. In 1615 the book "New Stereometry of Wine Barrels"* by J. Kepler was published. Kepler begins his book thus: "The year when I got married the grape harvest was rich and wine was cheap, and therefore as a thrifty man I decided to store some wine. I bought several barrels of wine. Some time later the wine merchant came to measure the capacity of the barrels in order to set the price for the wine. To this end he put an iron into every barrel and immediately announced how much wine there was in the barrel without any calculations." (This "method" is demonstrated in Fig. 8.)

Kepler was very much surprised. He found it strange that it was possible to determine the capacity of barrels of different shape by means of a single measurement. Kepler writes: "I thought it would be appropriate for me to take a new subject for my mathematical studies: to investigate the geometrical laws of this convenient measurement and to find its foundations." To solve the stated problem Kepler laid the foundation of differential and integral calculus and at the same time formulated the first rules for solving extremal problems. He writes: "Under the influence of their good genius who was undoubtedly a good geometer the coopers began shaping barrels in such a way that from the measurement of the given length of a line it was possible to judge upon the maximum capacity of the barrel, and since changes are insensible in the vicinity of every maximum, small random deviations do not affect the capacity appreciably."

The words we have underlined contain the basic algorithm of finding extrema which was later stated as an exact theorem first by Fermat in 1629 (for polynomials) and then by Newton and Leibniz and was called the *theorem of Fermat.*

It should also be noted that Kepler solved a number of concrete extremal problems and, in particular, the problem on a *cylinder of the greatest volume inscribed in a ball.*

To conclude this section, we shall present one more geometrical problem in which many mathematicians of the 17th century were interested (Cavalieri, Viviani, Torricelli, Fermat, etc.). It was studied in the 19th century by the Swiss–German geometer J. Steiner and therefore is called *Steiner's problem.*†

Steiner's Problem. Find a point on a plane such that the sum of the distances from that point to three given points on the plane is smallest.

*J. Kepler, Nova Stereometria Doliorum Vinariorum, Linz (1615).

†It should be noted that later problems of this kind appeared in highway engineering, oil pipeline engineering, and laying urban service lines.

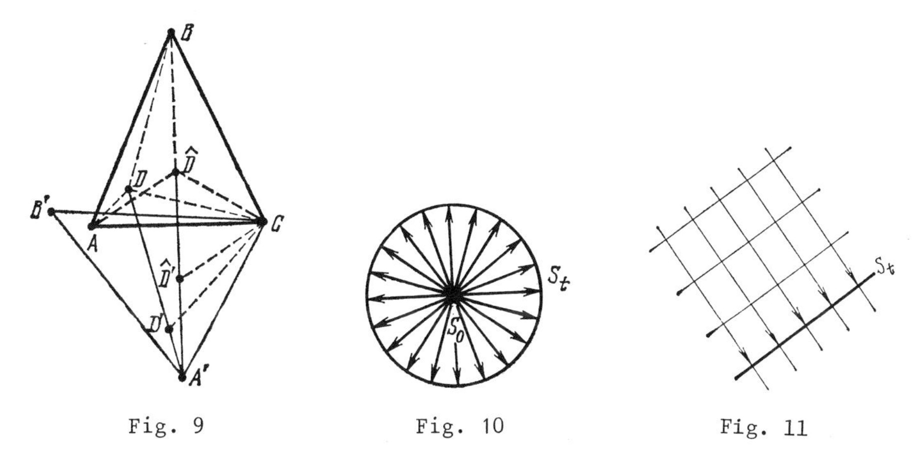

Fig. 9 Fig. 10 Fig. 11

Solution. We shall present an elegant geometrical solution of the problem for the case when the three given points are the vertices of an acute triangle. Let the angle at the vertex C of the triangle ABC (Fig. 9) be not smaller than 60°. Let us turn the triangle ABC through an angle of 60° about the point C. This results in the triangle A'B'C. Now we take an arbitrary point D within the triangle ABC and denote by D' the image of D under this rotation through 60°. Then the sum of the lengths $|AD| + |BD| + |CD|$ is equal to the length of the broken line $|BD| + |DD'| + |D'A'|$ because the triangle CDD' is equilateral and $|D'A'| = |DA|$.

Now let $\hat{D}$ be the *Torricelli point*, i.e., the point at which the angles of vision of the three sides of the triangle are equal to 120°, and let $\hat{D}'$ be the image of $\hat{D}$ under the rotation. Now it can easily be understood that the points B, $\hat{D}$, $\hat{D}'$, and A' lie in one straight line, and hence the Torricelli point is the solution of the problem. ■

We have acquainted the reader with a number of problems of geometrical character posed and solved at different times. Their statement can only be partly justified from the viewpoint of practical demands, and perhaps they were mainly stimulated by the desire to demonstrate the elegance of geometry itself.

1.1.3. Fermat's Variational Principle and Huygens' Principle. The Light Refraction Problem. The name of Pierre Fermat is connected with the formulation of the first variational principle for a physical problem. We mean *Fermat's variational principle in geometrical optics*. The law of light refraction was established experimentally by W. Snell. Soon after that R. Descartes suggested a theoretical explanation of Snell's law. However, according to Descartes, the speed of light in a denser medium (say, in water) is greater than in a rarer medium (e.g., in the air). This seemed strange to many scientists.

Fermat gave another explanation of the phenomenon. His basic idea was that a ray of light "chooses" a trajectory so that the time of travel from one point to another along this trajectory be minimal (in comparison with any other trajectories joining the same points). In a homogeneous medium where the speed of light propagation is the same at all points and in all directions the time of travel along a trajectory is proportional to its length. Therefore, the minimum time trajectory connecting two points A and B is simply the line segment [AB]: light propagates along straight lines in a homogeneous medium.

At present the derivation of Snell's law from Fermat's variational principle is included in textbooks for schools (also see Section 1.6.1).

Fermat's principle is based on the assumption that light propagates along certain lines. C. Huygens (1629-1695) suggested another explanation of the laws of propagation and refraction of light based on the concept of light as a wave whose front moves with time.

Abstracting from the discussion of the physical foundations of this idea and, in particular, the question of whether light is a wave or a flux of particles, we shall give the following definition which is visual rather than rigorous.

By a *wave front* S_t is meant the set of points that are reached in time t by light propagating from a source S_0.

For instance, if S_0 is a point source (Fig. 10) and the medium is homogeneous, then S_t is a sphere (a "spherical wave") of radius vt with center at S_0. As t increases the wave front expands uniformly in all directions with velocity v. As to the lines of light propagation, they form a pencil of rays (which are the radii) orthogonal to S_t at each instant of time t. As the spherical wave moves off the source, the wave front becomes flatter and still flatter, and if we imagine that the source recedes to infinity, in the limit the wave front becomes a plane moving uniformly with velocity v and remaining perpendicular to the beam of rays of light which are parallel to one another (Fig. 11).

To determine the motion of a wave front in more complicated situations Huygens uses the following rule (*Huygens' principle*). Every point of a wave front S_t becomes a secondary source of light, and in time Δt we obtain the family of the wave fronts of all these secondary sources, and the real wave front $S_{t+\Delta t}$ at the instant of time $t + \Delta t$ is the envelope of this family (Fig. 12).

It can easily be seen that the propagation of light in a homogeneous medium and the limiting case of a plane wave satisfy Huygens' principle. As an example of the application of this principle for the simplest homogeneous medium, we shall present the derivation of Snell's law "according to Huygens."

Let a parallel beam of rays of light be incident on a plane interface Σ between two homogeneous media; for simplicity, we shall assume that Σ is horizontal and the light is incident from above (Fig. 13). By v_1 and v_2 we denote the speed of light above and below Σ, and by α_1 and α_2 we denote the angle of incidence and the angle of refraction (reckoned from the normal N to Σ). The wave front A_1A_2A propagates with velocity v_1, and at some time t the light issued from the point A reaches the interface Σ at the point B. After that the point B becomes a secondary source of spherical waves propagating in the lower medium with velocity v_2. The light reaches the point C_1 at time $t_1 = t + \frac{|B_1C_1|}{v_1} = t + |BC_1|\frac{\sin\alpha_1}{v_1}$, and the intermediate point $D \in [BC_1]$ at time $t_2 = t + |BD|\frac{\sin\alpha_1}{v_1}$. At time t_1 the radius of the spherical wave issued from the secondary source B becomes equal to $r_1 = v_2(t_1 - t) = |BC_1|\frac{v_2}{v_1}\sin\alpha_1$, and the radius of the spherical wave from D becomes equal to $r_2 = v_2(t_1 - t_2) = |DC_1|\frac{v_2}{v_1}\sin\alpha_1$. The tangents C_1C and C_1C_2 to these spheres coincide because

$$\sin\widehat{BC_1C} = \frac{r_1}{|BC_1|} = \frac{v_2}{v_1}\sin\alpha_1 = \frac{r_2}{|DC_1|} = \sin\widehat{DC_1C_2}.$$

The point D was taken on BC_1 quite arbitrarily, and consequently the envelope of the secondary waves at time t_1 is the straight line CC_1 forming an angle α_2 with Σ such that $\sin\alpha_2 = \frac{v_2}{v_1}\sin\alpha_1$. But this is nothing other than

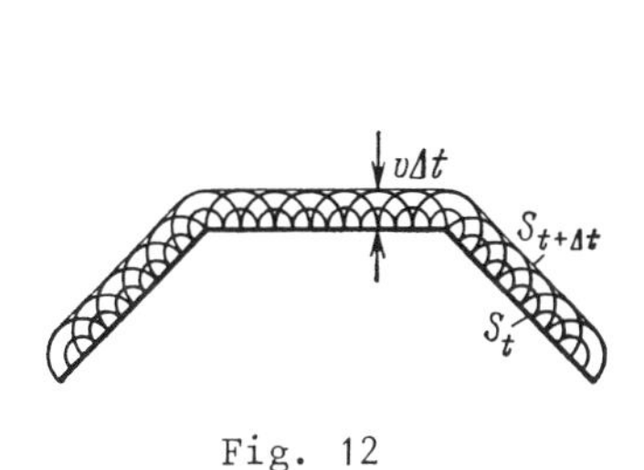

Fig. 12

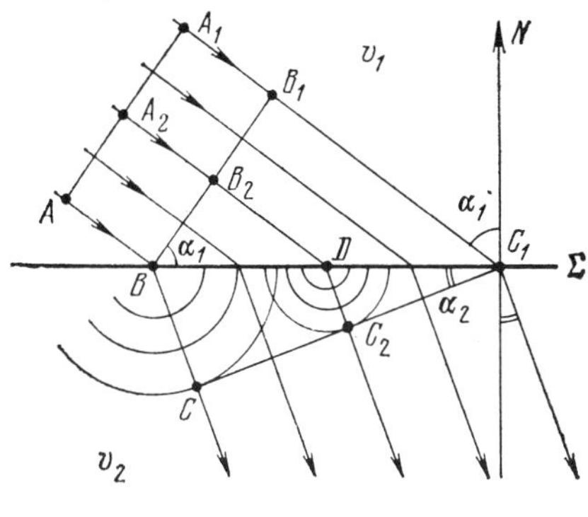

Fig. 13

Snell's law:

$$\frac{\sin\alpha_2}{\sin\alpha_1}=\frac{v_2}{v_1}.$$

The principles of Huygens and Fermat are closely interrelated. Indeed, let the position of a wave front S_t at an instant of time t be known. Where will it be in some time Δt? Let us take a point $C\in S_{t+\Delta t}$ (Fig. 14). By definition, there exist $A\in S_0$ and a path AC traveled by the light during time $t + \Delta t$, and, according to Fermat's principle, it takes the light more time to travel from A to C along any other path. By the continuity, there is a point B on the arc AC such that it takes the light time t to travel from A to B and time Δt to travel from B to C. Since the arc AC possesses the minimality property, the arcs AB and BC possess the same property. Indeed, if, for instance, there existed a path along which the light could travel from A to B in shorter time than t, then adding the arc BC to this path we would obtain a path from A to C along which the light propagates during time shorter than $t + \Delta t$, which is impossible. It follows that, firstly, $B\in S_t$, and, secondly, the point C belongs to the wave front corresponding to the point source located at B and to time Δt. This is in complete agreement with Huygens' principle: the point B became a secondary source, and the wave propagating from it reaches the point C in time Δt. The idea of a wave front, Huygens' principle, and the way of reasoning we have sketched served later as a basis for the *Hamilton–Jacobi theory* and in the middle of our century for the so-called *dynamic programming*, which is an important tool for solving various applied extremal problems.

* * *

After Fermat's variational principle many other variational principles were discovered: first in mechanics and then in physics. Gradually most of the scientists began to believe that nature always "chooses" a motion as if an extremal problem were solved. Here it is expedient to quote Euler: "In everything that occurs in the world the meaning of a maximum or a minimum can be seen." In our time C. L. Siegel said jokingly: "According to Leibniz, our world is the best of all the possible worlds, and therefore its laws can be described by extremal principles."

1.1.4. The Brachistochrone Problem. The Origin of the Calculus of Variations. In 1696 an article by Johann Bernoulli was published with an intriguing title: "Problema novum, ad cujus solutionem mathematici invitantur" ("A New Problem to Whose Solution Mathematicians Are Invited"). The following problem was stated there: "Suppose that we are given two points A and B in a vertical plane (Fig. 15). Determine the path AMB along which a body M which starts moving from the point A under the action of its own

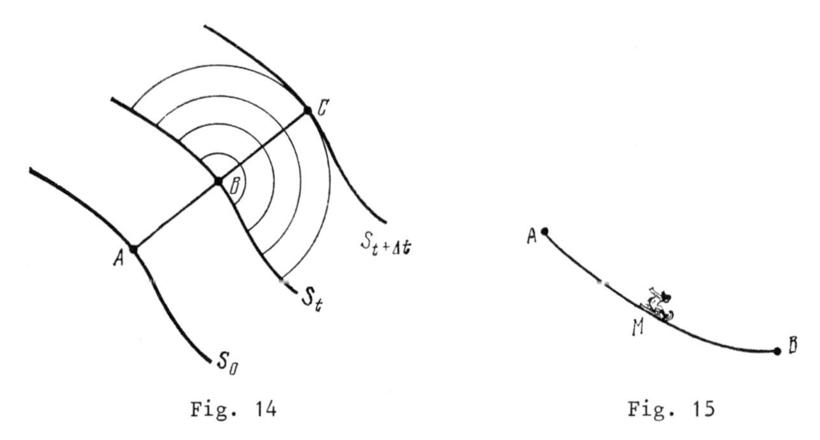

Fig. 14 Fig. 15

gravity reaches the point B in the shortest time."*

The solution of this problem ("so beautiful and until recently unknown," according to Leibniz) was given by Johann Bernoulli himself and also by Leibniz, Jacob Bernoulli, and an anonymous author. Experts immediately guessed that it was Newton ("ex unge leonem,"† as Johann Bernoulli put it). The curve of shortest descent or the brachistochrone turned out to be a cycloid. Leibniz' solution was based on the approximation of curves with broken lines. The idea of broken lines was later developed by Euler and served as a basis of the so-called direct methods of the calculus of variations. The remarkable solution given by Jacob Bernoulli was based on Huygens' principle and the concept of a "wave front." However, the most popular solution is the one given by Johann Bernoulli himself, and we shall present it here.

Let us introduce a coordinate system (x, y) in the plane so that the x axis is horizontal and the y axis is directed downward (Fig. 16). According to Galileo's law, the velocity of the body M at the point with coordinates (x, y(x)) is independent of the shape of the curve $y(\cdot)$ between the points A and (x, y(x)) [provided that the body moves downward without friction along the curve $y(\cdot)$] and is dependent on the ordinate y(x) solely and is equal to $\sqrt{2gy(x)}$, where g is the acceleration of gravity. It is required to find the shortest time it takes the body to travel along the path from A to B, i.e., the integral

$$T=\int\limits_{\widehat{AB}}\frac{ds}{v}=\int\limits_{\widehat{AB}}\frac{ds}{\sqrt{2gy(x)}}$$

(here ds is the differential of arc length) must be minimized.

However (by virtue of Fermat's principle stated in Section 1.1.3), we obtain exactly the same problem if we investigate the trajectories of light in a nonhomogeneous (two-dimensional) medium where the speed of light at a point (x, y) is equal to $\sqrt{2gy}$. Further, Johann Bernoulli "splits" the medium into parallel layers within each of which the speed of light is assumed to be constant and equal to v_i; i = 1, 2,... (Fig. 17). By virtue of Snell's law, we obtain

$$\frac{\sin\alpha_1}{v_1}=\frac{\sin\alpha_2}{v_2}=\ldots\Leftrightarrow\frac{\sin\alpha_i}{v_i}=\text{const},$$

*It should be noted that in Galileo's "Dialogues" there is a slight hint to the statement of the brachistochrone problem: Galileo proves that a body which moves along a chord reaches the terminal point later than if it moved along the circular arc subtended by the chord.
†"Painting a lion from the claw" (*Plutarch*, On the Cessation of Oracles).

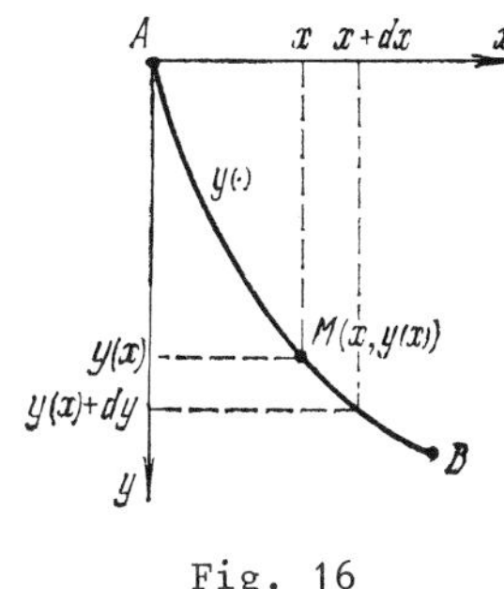

Fig. 16

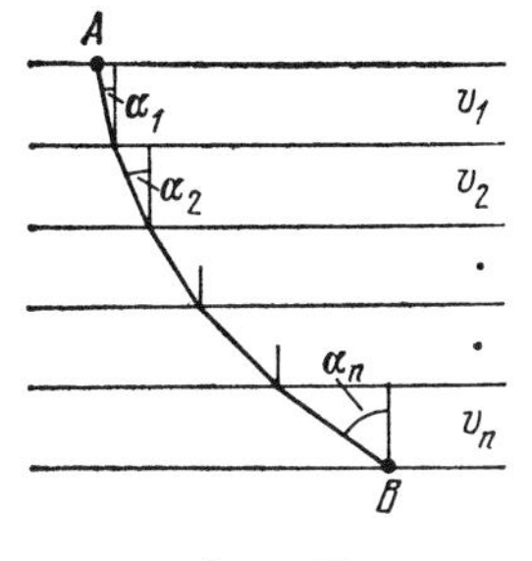

Fig. 17

where α_i are the angles of incidence of the ray. Performing the refinement of the layers and passing to the limit (Johann Bernoulli did not, of course, dwell on the justification of the validity of this procedure), we conclude that

$$\frac{\sin\alpha(x)}{v(x)} = \text{const},$$

where $v(x) = \sqrt{2gy(x)}$ and $\alpha(x)$ is the angle between the tangent to the curve $y(\cdot)$ at the point $(x, y(x))$ and the axis Oy, i.e., $\sin\alpha(x) = 1/\sqrt{1+(y'(x))^2}$. Thus, the equation of the brachistochrone is written in the following form:

$$\sqrt{1+(y')^2}\sqrt{y} = C \Leftrightarrow y' = \sqrt{\frac{C-y}{y}} \Leftrightarrow \frac{dy\sqrt{y}}{\sqrt{C-y}} = dx.$$

Integrating this equation (with the aid of the substitution $y = C\sin^2\frac{t}{2}$, $dx = C\sin^2\frac{t}{2}\,dt$), we arrive at the equation of a cycloid:

$$x = C_1 + \frac{C}{2}(t - \sin t), \quad y = \frac{C}{2}(1 - \cos t).$$

We would like to stress an important distinction between the problem of Euclid and Kepler's problem (on an inscribed cylinder), on the one hand, and, say, the brachistochrone problem, on the other hand. The latter is that the set of all parallelograms inscribed in a triangle and the set of all cylinders inscribed in a ball depend only on *one* parameter. Hence in these problems it is required to find an extremum of a *function of one variable*. As to the brachistochrone problem, it requires one to find an extremum of a *function of infinitely many variables*. The history of mathematics made a sudden jump: from the dimension one it passed at once to the infinite dimension or from the theory of extrema of functions of one variable to the theory of problems of the type of the brachistochrone problem, i.e., to the *calculus of variations*, as this branch of mathematics was called in the 18th century.

Soon after the work by Johann Bernoulli many other problems similar to the brachistochrone problem were solved: on the shortest lines on a surface, on the equilibrium of a heavy thread, etc.

Traditionally the year 1696, the year of the brachistochrone problem, is considered to be the starting point of the calculus of variations. However, this is not exactly so, and we shall discuss it in the next section.

1.1.5. Newton's Aerodynamic Problem. In 1687 the book "Mathematical Principles of Natural Philosophy" by I. Newton was published. In Section VII titled "The Motion of Fluids, and the Resistance Made to Projected Body" Newton considers the problem on the resistance encountered by a ball and a cylinder in a "rare" medium, and further in the "Scholium" he investigates the question of the resistance to a frustum of a cone moving in

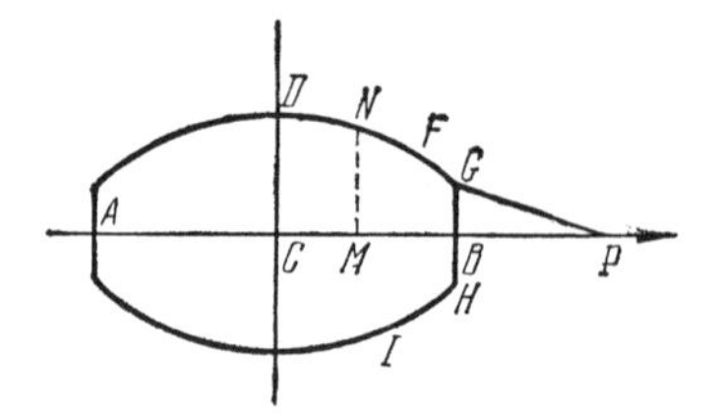

Fig. 18

the same "rare" medium.* In particular, he finds that among all the cones of given width and altitude the cone encountering the least resistance has an angle of 135°. He makes the remark that this result "may be of some use in shipbuilding" and then writes: "Quod si figura DNFG ejusmodi sit ut, si ab ejus puncto quovis N ad axem AB demittatur perpendiculum NM, et dicatur recta GP quae parellela sit rectae figuram tangenti in N, et axem productam sicet in P, fuerit MN ad GP ut GP^{cub} ad $4BP \times GB^{q}$, solidum quod figurae hujus revolutione circa axem AB describitur resistetur minime omnium ejusdem longitudinis & latitudinis."

Newton's words can be translated thus: "When the curve DNFG is such that if a perpendicular is dropped from its arbitrary point N to the axis AB and a straight line GP parallel to the tangent to the curve at the point N is drawn [from the given point G] which intersects the axis at the point P, then [there holds the proportion] $MN:GP = GP^3:(4BP \times GB^2)$, the solid obtained by the revolution of this curve about the axis AB encounters the least resistance in the above-mentioned rare medium among all the other solids of the same length and width" (see Fig. 18 drawn by Newton). Newton did not explain how he had obtained his solution. Later he gave his commentators a sketch of the derivation, but it was published only in 1727-1729 when the first stage of the calculus of variations was completed. Newton's preparatory material (published only in our time) shows that he had had a good command of the elements of many mathematical constructions which were realized later by Euler and Lagrange. As we shall see, Newton's problem belongs to the optimal control theory whose elaboration began only in the Fifties of our century rather than to the calculus of variations.

1.1.6. The Diet Problem and the Transportation Problem. Let us suppose that an amount of a certain product is kept in several stores and that the product must be delivered to several shops. Let the cost of the transportation of the unit of the product from each store to each shop be known, and let it be known how much product must be delivered to each shop. The *transportation problem* consists in working out a transportation plan optimal in the given situation, i.e., it is required to indicate what amount of the product must be transported from each store to each of the shops for the total cost of the transportation to be minimal. The *diet problem* is similar to the transportation problem; it consists in working out a diet with minimum cost satisfying the consumption requirements for a given assortment of products on condition that the content of nutrient substances in each of the products and the cost of the unit of each product are known.

A great number of problems of this kind appear in practical economics. The two above-mentioned problems belong to the branch of mathematics called *linear programming*. The linear programming theory was created quite recently – in the Forties-Fifties of our century.

*For example, see I. Newton, Mathematical Principles of Natural Philosophy, Cambridge (1934), pp. 327 and 333 (translated into English by Andrew Motte in 1729).

1.1.7. The Time-Optimal Problem. We shall present the simplest example of an extremal problem with a "technical" content. Suppose that there is a cart moving rectilinearly without friction along horizontal rails. The cart is controlled by an external force which can be varied within some given limits. It is required that the cart be stopped at a certain position in the shortest time. In what follows this problem will be called the *simplest time-optimal problem.*

A characteristic peculiarity of extremal problems in engineering is that the acting forces are divided into two groups. One of them consists of natural forces (such as the gravitational force) and the other includes forces controlled by a man (e.g., the thrust force). Here there naturally appear constraints on the controlled forces connected with limitations of technical means.

The theory of the solution of such problems was constructed still later — at the end of the Fifties. It is called the *optimal control theory.*

* * *

So, where do extremal problems appear from? We tried to demonstrate by the examples presented here that there are many answers to this question. Extremal problems appear both from natural science, economics, and technology and from mathematics itself due to its own demands. That is why the theory of extremal problems and its practical aspect — the optimization theory — have won a wide popularity in our time.

1.2. How Are Extremal Problems Formalized?

1.2.1. Basic Definitions. Each of the problems of Section 1.1 was formulated descriptively in terms of the specific field of knowledge where the problem appeared. This is the way in which extremal problems are usually posed, and, generally speaking, it is not necessary to solve every problem analytically. For example, Euclid's and Steiner's problems were solved in Section 1.1.2 purely geometrically. However, if we want to make use of the advantages of the analytical approach, then the first thing to do is to translate the problem from the "descriptive" language into the formal language of mathematical analysis. Such a translation is called *formalization.*

The exact statement of an extremal problem includes the following components: a *functional*† $f: X \to \overline{\mathbf{R}}$ defined on a set X, and a *constraint,* i.e., a subset $C \subseteq X$. (By $\overline{\mathbf{R}}$ we denote the "extended real line," i.e., the collection of all real numbers together with the values $-\infty$ and $+\infty$.) The set X is sometimes called the *class of underlying elements,* and the points $x \in C$ are said to be *admissible with respect to the constraint.* The problem itself is formulated thus: *find the extremum* (i.e., the infimum or the supremum) *of the functional* f *over all* $x \in C$. For this problem we shall use the following standard notation:

$$f(x) \to \inf (\sup); \quad x \in C. \tag{1}$$

Thus for the exact statement of the problem we must describe X, f, and C.

If X = C, then problem (1) is called a *problem without constraints.* A point $\hat{x}$ is called a *solution* of problem (1), namely, it is said to yield an *absolute minimum (maximum)* in the problem (1), if $f(x) \geqslant f(\hat{x})$ $[f(x) \leqslant f(\hat{x})]$ for all $x \in C$. As a rule, we shall write all the problems as minimization problems replacing, when necessary, a problem of the type $f(x) \to \sup$, $x \in C$, by the problem $\tilde{f}(x) \to \inf$, $x \in C$, where $\tilde{f}(x) = -f(x)$. In those

†In the theory of extremal problems scalar functions are often called *functionals.*

cases when we want to stress that it does not make any difference to us whether a minimization or a maximization problem is considered we write $f(x) \to \text{extr}$.

Further, the set X is usually endowed with a topology, i.e., the notion of closeness of elements makes sense in X. To this end we can, for instance, specify a set of neighborhoods in X (as it is done in the standard manner in $\mathbf{R}^n$ or in a normed space). If X is a topological space, then $\hat{x}$ is called a *local minimum* if there exists a neighborhood U of the point $\hat{x}$ such that $\hat{x}$ is a solution of the problem $f(x) \to \inf$, $x \in C \cap U$. A *local maximum* is defined in a similar way.

1.2.2. The Simplest Examples of Formalization of Extremal Problems. We shall present the formalization of some of the problems mentioned in Section 1.1. Let us begin with the *problem of Euclid* (see Section 1.1.2, Fig. 6). From the similarity of the triangles DBE and ABC, we obtain $h(x)/H = x/b$. Here x is the side |AF| of the parallelogram ADEF, H is the altitude of ΔABC, h(x) is the altitude of ΔBDE, and $b = |AC|$ is the length of the side AC. The area of the parallelogram ADEF is equal to $(H - h(x))x = H(b - x)x/b$. Now we obtain the following formalization of the problem of Euclid:

$$\frac{H(b-x)x}{b} \to \sup, \ 0 \leqslant x \leqslant b, \Leftrightarrow x(x-b) \to \inf, \ x \in [0, b]. \tag{1}$$

Here the three components every formalization consists of are:

$$X = \mathbf{R}, \ f = H(b-x)x/b, \ C = [0, b].$$

Let us formalize the *problem of Archimedes* on spherical segments with a given surface area (see Section 1.1.2). Let h be the altitude of a spherical segment, and let R be the radius of the ball. As is known from geometry, the volume of a spherical segment is equal to $\pi h^2(R - h/3)$ while its surface area is equal to $2\pi Rh$. We see that the problem of Archimedes can be formalized in two ways:

$$\pi h^2\left(R - \frac{h}{3}\right) \to \sup, \ 2\pi Rh = a, \quad R \geqslant 0, \ 2R \geqslant h \geqslant 0, \tag{2}$$

or, eliminating R from the functional in (2),

$$\frac{ha}{2} - \frac{\pi h^3}{3} \to \sup, \ 0 \leqslant h \leqslant \sqrt{\frac{a}{\pi}} \tag{2'}$$

(the last inequality is a consequence of $h \leqslant 2R \Rightarrow a \geqslant \pi h^2$). In the first case we have $X = \mathbf{R}^2_+$, $f = \pi h^2(R - h/3)$, $C = \{(R, h) \mid 2\pi Rh = a, \ 2R \geqslant h\}$; in the second case, putting $X = [0, \sqrt{a/\pi}]$, we obtain a problem without constraints with the functional $f = ha/2 - \pi h^3/3$.

Kepler's problem on the cylinder of maximum volume inscribed in a ball (see Section 1.1.2) admits of the following evident formalization:

$$2\pi x(1 - x^2) \to \sup; \ 0 \leqslant x \leqslant 1 \ (X = \mathbf{R}, \ C = [0, 1]). \tag{3}$$

Here the radius of the ball is equal to unity, and x is half the altitude of the cylinder.

The *problem of light refraction* at the interface between two homogeneous media which is solved with the aid of Fermat's variational principle (see Section 1.1.3 and Fig. 19) is reducible to the following problem. Let the interface between two homogeneous media be the plane $z = 0$, the speed of light in the upper and in the lower half-spaces being v_1 and v_2, respectively. We must find the trajectory of the ray of light propagating from the point $A = (0, 0, \alpha)$, $\alpha > 0$, to the point $B = (\xi, 0, -\beta)$, $\beta > 0$. Due to the symmetry, the ray must lie in the plane $y = 0$. Let $C = (x, 0, 0)$ be the point of refraction of the ray. Then the time of propagation of the ray from A to B is equal to

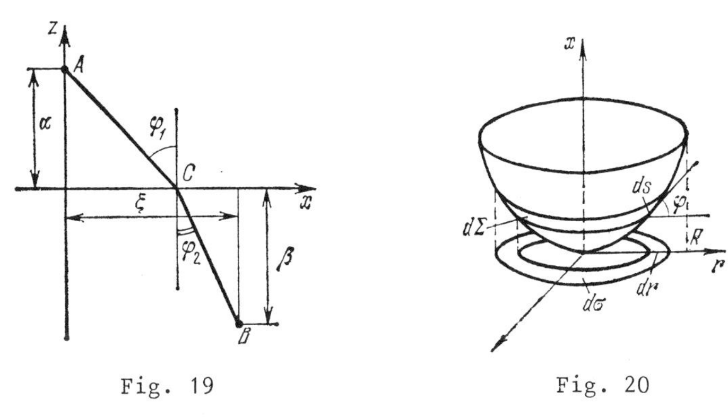

Fig. 19 Fig. 20

$$T(x)=\frac{\sqrt{\alpha^2+x^2}}{v_1}+\frac{\sqrt{\beta^2+(\xi-x)^2}}{v_2}.$$

According to Fermat's principle, the coordinate $\hat{x}$ of the point of refraction is found from the solution of the problem

$$T(x)=\frac{\sqrt{\alpha^2+x^2}}{v_1}+\frac{\sqrt{\beta^2+(\xi-x)^2}}{v_2}\to\inf \qquad (X=C=\mathbf{R}). \tag{4}$$

It should be noted that we have obtained a problem without constraints.

Similarly, the problem without constraints

$$|x-\xi_1|+|x-\xi_2|+|x-\xi_3|\to\inf, \tag{5}$$

where $X = C = \mathbf{R}^2$, ξ_1, ξ_2, and ξ_3 are three given points in the plane $\mathbf{R}^2$, and $|x| = \sqrt{x_1^2 + x_2^2}$ is a formalization of *Steiner's problem* (see Section 1.1.2). Let us stress an important peculiarity of the functional of the problem (5): it is a convex function which, however, is not everywhere differentiable.

The *problem of Apollonius* on the shortest distance from a point $\xi = (\xi_1, \xi_2)$ to an ellipse specified by an equation of the form $x_1^2/a_1^2 + x_2^2/a_2^2 = 1$ obviously admits of the following formalization:

$$(x_1-\xi_1)^2+(x_2-\xi_2)^2\to\inf,\quad \frac{x_1^2}{a_1^2}+\frac{x_2^2}{a_2^2}=1$$
$$\left(X=\mathbf{R}^2,\ C=\left\{x\,\middle|\,\frac{x_1^2}{a_1^2}+\frac{x_2^2}{a_2^2}=1\right\}\right). \tag{6}$$

Thus, we have learned the formalization of the simplest problems. The formalization of more complex problems arising in natural sciences, technology, and economics is a special and nontrivial problem. In these problems the formalization itself depends on physical or some other hypotheses. In the next section this will be demonstrated by the example of Newton's problem.

<u>1.2.3. The Formalization of Newton's Problem</u>. Naturally, the formalization of this problem depends on the resistance law of the medium. Newton thought of the medium (he called it "rare") as consisting of fixed particles with a constant mass m possessing the properties of perfectly elastic balls. We shall also assume this hypothesis.

Let a solid of revolution about the x axis (Fig. 20) move in the direction opposite to that of the x axis ("downward") in Newton's "rare" medium with a velocity v. In its rotation about the x axis an element dr on the r axis describes an annulus of area $d\sigma = 2\pi dr$; to this annulus there corresponds a surface element $d\Sigma$ on the body of revolution. During time dt this surface element "displaces" a volume $dV = 2\pi r dr v dt$. Let ρ be the density of the medium. Then the number of the particles which have hit the surface element is $N=\frac{\rho}{m}dV=\frac{\rho\, 2\pi r\, dr\, v}{m}dt$, where m is the mass of a particle. Let us calculate the force dF acting on the surface element dΣ during time dt. Let the element ds be inclined to the r axis at an angle φ. When a particle is reflected from dΣ, it gains an increment of momentum equal to $m(\mathbf{v}_2-\mathbf{v}_1)=-2mv\cos\varphi\cdot\mathbf{n}$, where $v = |\mathbf{v}_1| = |\mathbf{v}_2|$, $\mathbf{n}$ is the unit normal vector to dΣ, and $\varphi=\arctan\frac{dx}{dy}$ is the angle between ds and the horizontal direction. By virtue of Newton's third law, the body gains the opposite momentum increment $m\,2v\cos\varphi\cdot\mathbf{n}$, and during time dt there are N such increments; by virtue of symmetry, the sum of the components of the momentum orthogonal to the axis of revolution is equal to zero while the resultant axial component of the increment of the momentum is equal to

$$Nm\,2v\cos\varphi\cos\varphi=\frac{2\rho\pi r\, dr\, v\, dt}{m}m2v\cos^2\varphi=4\rho\pi v^2 r\, dr\, dt\cos^2\varphi.$$

By Newton's second law, this expression is equal to dFdt, whence dF = $kr\,dr\cos^2\varphi$, $k=4\rho\pi v^2$, and the resultant resistance force is

$$F=k\int_0^R\frac{r\,dr}{1+(dx/dr)^2}. \quad (1)$$

Hence, replacing r and t by R and T we arrive at the extremal problem

$$\int_0^T\frac{t\,dt}{1+\dot{x}^2}\to\inf,\quad x(0)=0,\quad x(T)=\xi. \quad (2)$$

It can easily be understood without solving the problem (Legendre was the first to note this fact in 1788) that the infimum in the problem is equal to zero. Indeed, if we choose a broken line x(·) with a very large absolute value of the derivative (Fig. 21), then integral (2) will be very small. On the other hand, the integral in (2) is nonnegative for any function x(·). Thus the infimum of the values of the integral is equal to zero.

For this Newton was many times subjected to criticism. For example, in the book* by the distinguished mathematician L. C. Young it is written: "Newton formulated a variational problem of solid of least resistance, in which the law of resistance assumed is physically absurd and ensures that the problem has no solution – the more jagged the profile, the less the assumed resistance.... If this problem had been even approximately correct... there would be no need today for costly wind tunnel experiments." But is all this really so in the situation we are discussing? First of all, it should be noted that Newton himself did not formalize the problem. This was done (and not quite adequately) by other people. For the correct formalization one must take into account the monotonicity of the profile, which was tacitly implied (when the profile is jagged the particles undergo multiple reflections, which leads to distortion of the whole phenomenon). The requirement of monotonicity makes the problem physically consistent. If this fact is taken into account, it is seen that the solution given by Newton himself is not only "approximately correct" but marvellously correct in every detail, which will be discussed later. And what is more, the physical hypotheses Newton had put forward and the solution of the aerodynamic problem he had given turned out to be very important for modern supersonic aerodynamics when it became necessary to construct high-speed and high-altitude flying vehicles.

The assumption on the monotonicity leads to the following correct formalization of Newton's problem:

*L. C. Young, Lectures on the Calculus of Variations and Optimal Control Theory, W. B. Saunders, Philadelphia–London–Toronto (1969), p. 23.

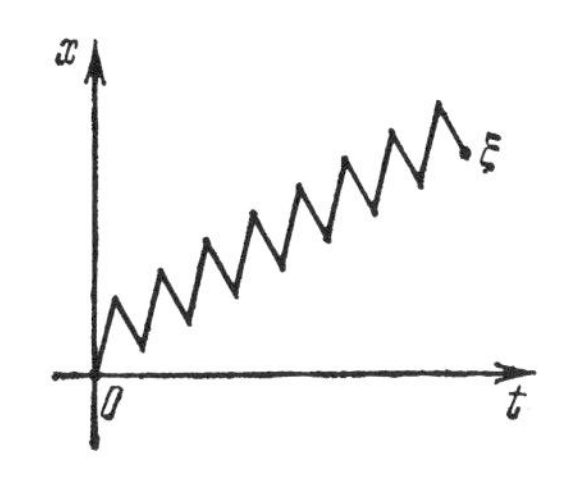

Fig. 21

$$\int_0^T \frac{t\,dt}{1+\dot{x}^2} \to \inf, \quad x(0)=0, \quad x(T)=\xi, \quad \dot{x}\in \mathbf{R}_+. \tag{3}$$

1.2.4. Various Formalizations of the Classical Isoperimetric Problem and the Brachistochrone Problem. The Simplest Time-Optimal Problem. The first two problems mentioned in the title of this section belong to the most widely known ones, but they turn out to be perhaps the most difficult for a complete analysis. We shall formalize each of them twice: first in the traditional and well-known manner and then by means of a method which is not so widely known. In this way we want to stress that, in principle, the formalization procedure is not unique. An adequate choice of a formalization is in fact an independent problem whose successful solution is to a great extent dependent on one's skill.

We shall begin with the classical isoperimetric problem. Let the length of the curve be L, and let the curve itself be represented parametrically by means of functions $x(\cdot)$ and $y(\cdot)$, the parameter being the arc length s reckoned along the curve from one of its points. Then the relation $\dot{x}^2(s) + \dot{y}^2(s) = 1$ holds at each point, and, moreover, $x(0) = x(L)$ and $y(0) = y(L)$ since the curve is closed.

For definiteness, let the sought-for curve be located so that its center of gravity is at the origin, i.e., the equalities $\int_0^L x(s)\,ds = \int_0^L y(s)\,ds = 0$ take place. The area S bounded by the curve $(x(\cdot), y(\cdot))$ is equal to $\int_0^L x\dot{y}\,ds$. We thus obtain the following formalization:

$$S=\int_0^L x\dot{y}\,ds \to \sup; \quad \dot{x}^2(s)+\dot{y}^2(s)=1,$$
$$\int_0^L x(s)\,ds=\int_0^L y(s)\,ds=0, \quad x(0)=x(L), \quad y(0)=y(L). \tag{1}$$

However, the same problem can be formalized in a different way. Suppose that an airplane is required to circumnavigate the greatest area in a given time and return to the airfield. If the maximum velocity of the airplane is independent of the direction of the flight, the natural formalization of the given problem is the following:

The area must be the greatest $\Leftrightarrow \frac{1}{2}\int_0^T (x(t)v(t)-y(t)u(t))\,dt \to \sup$, where $\dot{x}(t) = u(t)$, $\dot{y}(t) = v(t)$.

The maximum velocity of the airplane is equal to $V \Leftrightarrow u^2+v^2 \leqslant V^2$.

The airplane must return to its airfield $\Leftrightarrow x(0)=x(T)$, $y(0)=y(T)$.

The problem also admits of a more general formulation when the maximum velocity depends on the direction of the flight (e.g., in the presence of wind). Then we have a more general problem:

$$\frac{1}{2}\int_0^T (xv - yu)\,dt \to \sup, \quad \dot{x} = u, \; \dot{y} = v, \\ x(0) = x(T), \; y(0) = y(T), \; (u, v) \in A, \tag{2}$$

where A is the set of all admissible values of the velocity of the airplane.

If A is a circle, we obviously arrive at the classical isoperimetric problem, and if A is a "shifted circle" (which corresponds to constant wind), we obtain the well-known *Chaplygin problem*.

Now we shall present the most traditional formalization of the *brachistochrone problem*. As in Section 1.1.4, let us introduce a coordinate system (x, y) in the plane (Fig. 16) so that the x axis is horizontal and the y axis is directed downward. Without loss of generality, we can assume that the point A coincides with the origin. Let the coordinates of the point B be (x_1, y_1), $x_1 > 0$, $y_1 > 0$ (see Fig. 16), and let $y(\cdot)$ be the function specifying the equation of the curve joining the points A and B. We remind the reader that, according to Galileo's law, the velocity of the body M at the point $(x, y(x))$ is independent of the shape of the curve $y(\cdot)$ in the interval $(0, x)$ and is only dependent on the ordinate $y(x)$ itself, the value of this velocity being equal to $\sqrt{2gy(x)}$, where g is the acceleration of gravity. Consequently, the time T it takes the body to travel along an element of the curve $ds = \sqrt{dx^2 + dy^2}$ from the point $(x, y(x))$ to the point $(x + dx, y(x) + dy)$ is equal to $ds/\sqrt{2gy(x)}$, whence the following formalization of the brachistochrone problem is obtained:

$$\mathcal{J}(y(\cdot)) = \int_0^{x_1} \frac{\sqrt{1 + (y')^2}}{\sqrt{2gy}}\,dx \to \inf, \quad y(0) = 0, \quad y(x_1) = y_1. \tag{3}$$

Let us present another formalization of the brachistochrone problem whose idea is similar to that of the second formalization of the classical isoperimetric problem; here we shall follow the article by Johann Bernoulli (1969) mentioned earlier in which he proceeded from Fermat's variational principle.

Suppose that there is a nonhomogeneous medium in which the speed of light propagation depends solely on the "depth" y according to the law $v^2 = 2gy$. Then, by virtue of Fermat's variational principle, the ray of light will travel along the path from A to B in the shortest time. In this way we obtain a formalization of the brachistochrone problem in the form of a *time-optimal problem*:

$$T \to \inf, \; \dot{x} = \sqrt{y}\,u, \; \dot{y} = \sqrt{y}\,v, \; u^2 + v^2 = 2g, \\ x(0) = y(0) = 0, \quad x(T) = x_1, \quad y(T) = y_1. \tag{4}$$

The formalization of the *simplest time-optimal problem* (Section 1.1.7) is also similar to the above. Let the mass of the cart be m, let its initial coordinate be x_0, and let the initial velocity be v_0. The external force (the thrust force) and the variable coordinate of the cart will be denoted by u and x(t), respectively. Then, by Newton's second law, $m\ddot{x} = u$. The constraint on the thrust force will be set in the form $u \in [u_1, u_2]$. We thus have

$$T \to \inf, \; m\ddot{x} = u, \; u \in [u_1\; u_2], \\ x(0) = x_0, \quad \dot{x}(0) = v_0, \quad x(T) = \dot{x}(T) = 0. \tag{5}$$

The statement of the problem we have obtained is very similar to (2) and (4).

Now we note an important fact: we have in fact "underformalized" the problems. We mean that, for instance, in the formalization (3) the domain of the functional $\mathcal{J}$ is not indicated precisely, and consequently it is yet unknown for what class of curves the problem should be considered, i.e., the set X (see Section 1.2.1) is not defined. The same applies to the other formalizations of the present section. By the way, the "classics" often did not pay attention at all to the precise formalizations of the problems and simply solved them in the "underformalized" form. However, in what follows we are going to be precise, and therefore we shall have to do this rather tedious job: to stipulate every time within what class of objects the solution is being sought (or was found).

1.2.5. Formalization of the Transportation and the Diet Problems. We shall begin with the *transportation problem*. Let us introduce the following notation:

a_i is the amount of the product (measured in certain units) kept in the i-th store, $1 \leqslant i \leqslant m$;

b_j is the demand for the product (expressed in the same units) of the j-th shop;

c_{ij} is the cost of the transportation of the unit of the product from the i-th store to the j-th shop;

x_{ij} is the amount of the product (in the same units) to be transported from the i-th store to the j-th shop.

Then the total cost of the transportation is equal to $\sum_{i=1}^{m}\sum_{j=1}^{n} c_{ij}x_{ij}$, and it must be minimized. Here we have the following constraints:

a) $x_{ij} \in \mathbf{R}_+$ (this is an obvious constraint on the amount of the product to be transported);

b) $\sum_{j=1}^{n} x_{ij} \leqslant a_i$ (this constraint means that it is impossible to transport more product from a store than the amount kept there);

c) $\sum_{i=1}^{m} x_{ij} = b_j$ (this means that it is necessary to transport exactly the amount of the product which is demanded).

As a result, we arrive at the following formalization:

$$\sum_{i=1}^{m}\sum_{j=1}^{n} c_{ij}x_{ij} \to \inf, \quad \sum_{j=1}^{n} x_{ij} \leqslant a_i, \quad \sum_{i=1}^{m} x_{ij} = b_j, \quad x_{ij} \in \mathbf{R}_+. \tag{1}$$

The *diet problem* is formalized as simply as the transportation problem. Let there be n products (corn, milk, etc.) and m components (fat, protein, carbohydrates, etc.). Let us suppose that for the full value diet b_j units of the j-th component are needed, and let us denote by a_{ij} the amount of the j-th component contained in a unit amount of the i-th product and by c_i the cost of a unit of the i-th product.

Denoting by x_i the amount of the i-th product in the diet, we obtain the problem

$$\sum_{i=1}^{n} c_i x_i \to \inf, \quad \sum_{i=1}^{n} a_{ij}x_i \geqslant b_j, \quad x_i \geqslant 0. \tag{2}$$

1.2.6. The Basic Classes of Extremal Problems. In Section 1.1 we already mentioned briefly that in the theory of extremal problems a number of sufficiently well-defined classes of problems are separated out. Before describing them we shall present a brief review of the methods by means of which the constraints were specified in the problems formalized above.

In the first place, we encountered formalizations in which there were no constraints at all (say, in the light refraction problem or in Steiner's problem). In the second place, constraints were sometimes specified by a system of equalities [for example, in the problem of Apollonius or in the formalization (3) of the brachistochrone problem in Section 1.2.4 where the boundary conditions were specified by equalities]. In the third place, constraints were sometimes expressed by inequalities (e.g., in the transportation problem). Finally, in the fourth place, some constraints were written in the form of inclusions [for instance, the constraint $\dot{x} \in \mathbf{R}_+$ in Newton's problem, and the constraint $(u, v) \in A$, where $A = \{(u, v);\ u^2 + v^2 \leqslant 1\}$ in the formalization (2) of the classical isoperimetric problem in Section 1.2.4].

It should be stressed that such a division is somewhat conditional. For example, the constraint $\dot{x} \in \mathbf{R}_+$ in Newton's problem could have been written in the form of the inequality $\dot{x} \geqslant 0$ while the constraint $(u, v) \in A$ in the classical isoperimetric problem could have been written in the form of the inequality $u^2 + v^2 \leqslant 1$. Conversely, every inequality $f(x) \leqslant 0$ can be replaced by the equality $f(x) + u = 0$ and the inclusion $u \in \mathbf{R}_+$, etc.

Nevertheless, from the viewpoint accepted in the present book the division of the constraints into equalities and inequalities, on the one hand, and inclusions, on the other hand, is in a sense meaningful. In courses of mathematical analysis the Lagrange multiplier rule is presented for the solution of "conditional extremum" problems (this will be discussed in detail in Section 1.3). As is known, at the initial stage of the application of this rule the "Lagrange function" is formed which involves both the functional under investigation and the functions specifying the constraints. It may happen that for some reason or other it is convenient not to include some of the constraints in the Lagrange function. So, these are the constraints which are not involved in the Lagrange function in the solution of the corresponding problem that are separated out in the form of inclusions. Here, as in the formalization of a problem (which can usually be carried out in many ways and whose adequate choice among the possible formalizations depends on one's skill), there is no uniqueness in the division of the constraints. Now we pass to the description of the basic classes of extremal problems.

In what follows, we shall consider, from a sufficiently general viewpoint, the following four classes.

I. Smooth Problems with Equality and Inequality Constraints. Here the class of underlying elements X is usually a normed space,* and the constraint C is specified by an equality $F(x) = 0$, where F is a mapping from X into another normed space Y and by a finite number of inequalities $f_i(x) \leqslant 0$, $i = 0, 1, \ldots, m$. As a result, we obtain the class of the problems

$$f_0(x) \to \inf, \quad F(x) = 0, \quad f_i(x) \leqslant 0, \quad i = 1, \ldots, m. \tag{1}$$

It is supposed here that the functions f_i, $i = 1, 2, \ldots, m$, and the mapping F possess some smoothness properties. The smooth problem $f_0(x) \to \inf$ will be called an *elementary smooth problem*.†

II. The Classical Calculus of Variations. Here the traditional class of underlying elements is a Banach space $X = C^1([t_0, t_1], \mathbf{R}^n)$ of continuously differentiable n-dimensional vector functions $x(\cdot) = (x_1(\cdot), \ldots, x_n(\cdot))$ in

*Here and below the terms "normed space" and "Banach space" are used only for the sake of correctness of the statement of the problem. The reader will find the exact definitions in Section 2.1.1. For simplicity, the reader may assume that in problem (1) the element x is a vector $x = (x_1, \ldots, x_n)$ in the n-dimensional arithmetic space $\mathbf{R}^n$.

†Maximum problems or those involving the opposite inequalities ($\geqslant$) can easily be brought to the form (1) (see Section 3.2).

which the norm is determined by the formulas

$$\|x(\cdot)\|_1 = \max(\|x(\cdot)\|_0,\ \|\dot{x}(\cdot)\|_0),$$
$$\|x(\cdot)\|_0 = \max_{1 \leqslant i \leqslant n} \left(\max_{t \in [t_0, t_1]} |x_i(t)| \right).$$

The functionals in the problems of the classical calculus of variations are usually of the following types:

integral functionals, i.e., functionals of the form

$$\mathcal{I}(x(\cdot)) = \int_{t_0}^{t_1} L(t, x, \dot{x})\, dt = \int_{t_0}^{t_1} L(t, x_1(t), \ldots, x_n(t), \dot{x}_1(t), \ldots, \dot{x}_n(t))\, dt; \tag{2}$$

endpoint functionals, i.e., functionals of the form

$$\mathcal{T}(x(\cdot)) = l(x(t_0), x(t_1)) = l(x_1(t_0), \ldots, x_n(t_0), x_1(t_1), \ldots, x_n(t_1)); \tag{3}$$

functionals of *mixed form*, i.e.,

$$\mathcal{B}(x(\cdot)) = \mathcal{I}(x(\cdot)) + \mathcal{T}(x(\cdot)). \tag{4}$$

[In what follows functionals of the form (4) will also be called Bolza functionals.]

The constraints in the problems of the classical calculus of variations usually split into two groups:

differential constraints of the form

$$M(t, x(t), \dot{x}(t)) = 0 \Leftrightarrow M_i(t, x_1(t), \ldots, x_n(t), \dot{x}_1(t), \ldots, \dot{x}_n(t)) = 0,\ i = 1, 2, \ldots, p; \tag{5}$$

boundary conditions of the form

$$\psi(x(t_0), x(t_1)) = 0 \Leftrightarrow \psi_j(x_1(t_0), \ldots, x_n(t_0), x_1(t_1), \ldots, x_n(t_1)), \quad j = 1, 2, \ldots, s. \tag{6}$$

The functions L and M_i in (2) and (5) and the functions l and ψ_j in (3) and (6) depend on 2n + 1 and 2n variables, respectively, and are assumed to be smooth. The problem

$$\mathcal{I}(x(\cdot)) \to \inf,\quad M(t, x, \dot{x}) = 0,\quad \psi(x(t_0), x(t_1)) = 0 \tag{7}$$

is called the *Lagrange problem*. The problem

$$\mathcal{B}(x(\cdot)) \to \inf,\quad M(t, x, \dot{x}) = 0,\ \psi(x(t_0), x(t_1)) = 0$$

is called the *Bolza problem*.

The problem

$$\mathcal{T}(x(\cdot) \to \inf,\quad M(t, x, \dot{x}) = 0,\quad \psi(x(t_0), x(t_1)) = 0$$

is called the *Mayer problem*.

The problem without constraints

$$\mathcal{B}(x(\cdot)) = \int_{t_0}^{t_1} L(t, x(t), \dot{x}(t))\, dt + l(x(t_0), x(t_1)) \to \inf \tag{8}$$

will be called the *elementary Bolza problem*.

The problem

$$\mathcal{I}(x(\cdot)) = \int_{t_0}^{t_1} L(t, x(t), \dot{x}(t))\, dt \to \inf,\quad x(t_0) = x_0,\ x(t_1) = x_1 \tag{9}$$

will be called the *simplest vector problem of the classical calculus of variations;* in the case n = 1 it will be called the *simplest problem of the classical calculus of variations*. For the sake of simplicity, here we shall confine ourselves to problems with fixed time.

In Chapter 4 the reader will find a more general statement in which the functional and the constraints also depend on the variable instants of time t_0 and t_1.

III. Convex Programming Problems. Here the class of underlying elements X is a linear space, and the constraint C is determined by a system of equalities $F(x) = 0$ ($F: X \to Y$, where Y is another linear space), inequalities $f_i(x) \leq 0$, $i = 1, 2, \ldots, m$, and inclusions $x \in A$. As a result, the class of problems

$$f_0(x) \to \inf, \; F(x) = 0, \; f_i(x) \leqslant 0, \; i = 1, \ldots, m, \; x \in A, \tag{10}$$

is obtained.

It is supposed that the functions f_i, $i = 0, 1, \ldots, m$, are convex, the mapping F is affine [i.e., $F(x) = \Lambda x + \eta$, where η is a fixed vector and Λ is a linear operator from X into Y], and A is a convex set. If all the functions f_i in (10) are linear and A is a certain standard cone, then the problem (10) is called a *linear programming problem*. If there are no constraints in (10), the problem $f_0(x) \to \inf$ with a convex function f_0 will be called an *elementary convex problem without constraints*.

IV. Optimal Control Problems. In this book we shall consider the following class of optimal control problems:

$$J_0(x(\cdot), u(\cdot), t_0, t_1) = \int_{t_0}^{t_1} f_0(t, x(t), u(t))\, dt + \psi_0(t_0, x(t_0), t_1, x(t_1)) \to \inf,$$

$$\dot{x} = \varphi(t, x, u), J_i(x(\cdot), u(\cdot), t_0, t_1) = \int_{t_0}^{t_1} f_i(t, x(t), u(t))\, dt + \psi_i(t_0, x(t_0), t_1, x(t_1)) \lesseqgtr 0, \qquad i = 1, \ldots, m, \tag{11}$$

where $x \in \mathbf{R}^n$ and $u \in \mathbf{R}^r$. The symbol $J_i \lesseqgtr 0$ means that the i-th constraint is of the form $J_i = 0$ or $J_i \leqslant 0$ or $J_i \geqslant 0$.

In the general case t_0 and t_1 in (11) are not fixed, all the functions $f: \mathbf{R} \times \mathbf{R}^n \times \mathbf{R}^r \to \mathbf{R}$, $\varphi: \mathbf{R} \times \mathbf{R}^n \times \mathbf{R}^r \to \mathbf{R}^n$, $\psi: \mathbf{R} \times \mathbf{R}^n \times \mathbf{R} \times \mathbf{R}^n \to \mathbf{R}^s$, and $g = \mathbf{R} \times \mathbf{R}^n \times \mathbf{R} \times \mathbf{R}^n \to \mathbf{R}$ are assumed to be continuous jointly with respect to their arguments and continuously differentiable with respect to the variables t and x; the set $\mathfrak{U}$ is an arbitrary subset of $\mathbf{R}^r$.

For the complete description of the problem it only remains to explain what objects form the class of underlying elements. To begin with, we shall consider the collection of vector functions $(x(\cdot), u(\cdot))$ where $u(\cdot)$ is defined and piecewise continuous on $[t_0, t_1]$ and, moreover, the inclusion $u(t) \in \mathfrak{U}$ holds for all t, and $x(\cdot)$ is continuous on $[t_0, t_1]$ and differentiable at all the points except those where $u(\cdot)$ suffers discontinuities; besides, it is supposed that the equality $\dot{x}(t) = \varphi(t, x(t), u(t))$ holds at all the points where $x(\cdot)$ is differentiable.

The problems in which t_1 plays the role of the functional are called *time-optimal problems*.

* * *

Now let us see to what classes the problems of Sections 1.2.1-1.2.5 belong.

The *problem of Euclid* [see (1) in Section 1.2.2] can be considered both as a smooth problem and as a convex programming problem. The formalizations (2) and (2') of the *problem of Archimedes* in Section 1.2.2 and *Kepler's problem* (3) of Section 1.2.2 belong to the class of smooth problems. The *light refraction problem* [see (4) in Section 1.2.2] is both an elementary smooth problem and an elementary convex problem. *Steiner's problem* [see (5) in Section 1.2.2] is an elementary convex problem. The *problem of Apollonius* [see (6) in Section 1.2.2] is a smooth problem with equality constraints. *Newton's problem* [see (3) in Section 1.2.3] is an optimal control problem. The formalization (1) in Section 1.2.4 of the *classical isoperimetric problem* belongs to the classical calculus of variations while the formalization (2) of the same problem belongs to the optimal control theory. The *brachistochrone problem* [see (3) in Section 1.2.4] is the simplest problem of the classical calculus of variations; the same problem in the formalization (4) of Section 1.2.4 is a time-optimal problem of the optimal control theory. The *transportation problem* and the *diet problem* [see (1) and (2) in Section 1.2.5] are linear programming problems; the *simplest time-optimal problem* is a problem of the optimal control theory.

Thus the problems have been formalized and classified. Now we shall see what methods the apparatus of mathematical analysis provides for the solution of the problems.

1.3. Lagrange Multiplier Rule and the Kuhn–Tucker Theorem

1.3.1. The Theorem of Fermat.

The first general analytical method of solving extremal problems was elaborated by Pierre Fermat. It must have been discovered in 1629, but for the first time it was presented sufficiently completely in a letter of 1636 to G. P. Roberval.* In terms of modern mathematical language Fermat's approach (Fermat himself developed it only for polynomials) reduces to the fact that at a point $\hat{x}$ which yields an extremum in the problem without constraints $f(x) \to \text{extr}$ the equality $f'(\hat{x}) = 0$ must be fulfilled. As we remember, the first hint to this result can be found in Kepler's words in his "New Stereometry of Wine Barrels."

The exact meaning of Fermat's reasoning was elucidated 46 years later in 1684 when Leibniz' work appeared, in which the foundations of mathematical analysis were laid down. Even the title of this work itself, which begins with the words "Nova methodus pro maximis et minimis..." ("A New Method of Finding the Greatest and the Smallest Values..."), shows the important role of the problem of finding extrema in the development of modern mathematics. In this article Leibniz not only obtained the relation $f'(\hat{x}) = 0$ as a necessary condition for an extremum (at present this result is called the *theorem of Fermat*) but also used the second differential to distinguish between a maximum and a minimum. However, we should remind the reader that most of the results presented by Leibniz had also been known to Newton by that time. But his work "The Method of Fluxions," which had been basically completed by 1671, was published only in 1736.

Now we shall remind the reader of some facts of mathematical analysis and then pass to Fermat's theorem itself. To begin with, we shall consider the one-dimensional case when the functions are defined on the real line $\mathbf{R}$. A function of one variable $f\colon \mathbf{R} \to \mathbf{R}$ is said to be *differentiable at a point* $\hat{x}$ if there is a number α such that

$$f(\hat{x}+\lambda) = f(\hat{x}) + \alpha\lambda + r(\lambda),$$

where $r(\lambda) = o(|\lambda|)$, i.e., for any $\varepsilon > 0$ there is $\delta > 0$ such that $|\lambda| \leqslant \delta$ implies $|r(\lambda)| \leqslant \varepsilon|\lambda|$. The number α is called the *derivative of* f *at the point* $\hat{x}$ and denoted by $f'(\hat{x})$. Thus, $f'(\hat{x}) = \lim_{\lambda \to 0} (f(\hat{x}+\lambda) - f(\hat{x}))/\lambda$.

Theorem of Fermat. Let f be a function of one variable differentiable at a point $\hat{x}$. If $\hat{x}$ is a point of local extremum, then

$$f'(\hat{x}) = 0. \tag{1}$$

*Oevres de Fermat, Vol. 1, Paris (1891), p. 71.

The points x at which the relation (1) holds are said to be *stationary*.

According to the general definitions of Section 1.2.1, a point $\hat{x}$ yields a local minimum (maximum) for the function f if there is $\varepsilon > 0$ such that the inequality $|x - \hat{x}| < \varepsilon$ implies the inequality $f(x) \geqslant f(\hat{x})$ ($f(x) \leqslant f(\hat{x})$). By Fermat's theorem, the points of local extremum (maximum or minimum) are stationary. The converse is not of course true: for example, $f(x) = x^3$, $\hat{x} = 0$.

Proof. Let us suppose that $\hat{x}$ is a point of local minimum of the function f but $f'(\hat{x}) = \alpha \neq 0$. For definiteness, let $\alpha < 0$. On setting $\varepsilon = |\alpha|/2$ we find $\delta > 0$ from the definition of the derivative such that $|\lambda| < \delta$ implies $|r(\lambda)| < |\alpha||\lambda|/2$. Then for $0 < \lambda < \delta$ we obtain

$$f(\hat{x}+\lambda)=f(\hat{x})+\alpha\lambda+r(\lambda) \leqslant f(\hat{x})+\alpha\lambda+\frac{|\alpha|\lambda}{2}=f(\hat{x})-\frac{|\alpha|\lambda}{2}<f(\hat{x}),$$

i.e., $\hat{x}$ is not a point of local minimum of f. This contradiction proves the theorem. ■

There are a number of different definitions of derivatives for the case of several variables (and all the more so for "infinitely many" variables). In Chapter 2 (see Section 2.2.1) this will be discussed in greater detail. Here we shall only remind the reader of the basic definition (whose infinite-dimensional version was stated by Fréchet).

A function $f: \mathbf{R}^n \to \mathbf{R}$ of n variables is said to be *differentiable at a point* $\hat{x}$ if there are numbers $(\alpha_1,\ldots,\alpha_n)$ (we briefly denote them by α) such that

$$f(\hat{x}+h)=f(\hat{x})+\sum_{i=1}^{n}\alpha_i h_i+r(h),$$

where $|r(h)| = o(|h|)$, i.e., for any $\varepsilon > 0$ there is $\delta > 0$ such that $|h| = \sqrt{h_1^2 + \ldots h_n^2} < \delta$ implies $|r(h)| \leqslant \varepsilon h$. The collection of numbers $\alpha = (\alpha_1,\ldots,\alpha_n)$ is called the *derivative of* f *at the point* $\hat{x}$ and is also denoted by $f'(\hat{x})$. We stress that $f'(\hat{x})$ is a set of a numbers. The numbers $\alpha_i = \lim_{\lambda \to 0} [f(\hat{x}+\lambda e_i)-f(\hat{x})]/\lambda$, $i = 1,\ldots,n$, where $e_i = (0,\ldots,1,\ldots,0)$ (the unity occupies the i-th place), are called *partial derivatives of* f and denoted by $f_{x_i}(\hat{x})$ or $\partial f(\hat{x})/\partial x_i$. Thus, $f'(\hat{x}) = (f_{x_1}(\hat{x}),\ldots,f_{x_n}(\hat{x}))$. The relation $f'(\hat{x}) = 0$ means that $f_{x_1}(\hat{x}) = \ldots = f_{x_n}(\hat{x}) = 0$.

The "one-dimensional" Fermat theorem implies an obvious corollary:

COROLLARY (Fermat's Theorem for Functions of n Variables). Let f be a function of n variables differentiable at a point $\hat{x}$. If $\hat{x}$ is a point of local extremum of the function f, then

$$f'(\hat{x})=0 \Leftrightarrow f_{x_i}(\hat{x})=0, \quad i=1,\ldots,n. \tag{2}$$

The points $\hat{x} \in \mathbf{R}^n$ at which equalities (2) take place are also called *stationary*.

The Proof of the Corollary. If the function f has an extremum at a point $\hat{x}$, then the point zero must be a point of extremum of the function $\varphi_i(\lambda)=f(\hat{x}+\lambda e_i)$. By Fermat's theorem, $\varphi_i'(0)=0$, and we have $\varphi_i'(0)\overset{\text{def}}{=}f_{x_i}(\hat{x})$.

1.3.2. The Lagrange Multiplier Rule. This section is devoted to smooth finite-dimensional problems. We once again remind the reader of the whimsical historical development of the calculus of variations. After the methods of solving one-dimensional extremal problems had been elaborated there came the time of the classical calculus of variations. In his "Analytical mechanics"* Lagrange stated the famous multiplier rule for variational

*J. L. Lagrange, Mécanique Analytique, Paris (1788).

problems with constraints, i.e., for the infinite-dimensional case, and it was only about a decade later that he applied this rule to finite-dimensional problems (in 1797).

By a *finite-dimensional smooth extremal problem with equality constraints* (or a *conditional extremum problem*) is meant the problem

$$f_0(x) \to \text{extr}, \quad f_1(x) = \ldots = f_m(x) = 0. \tag{1}$$

Here the functions $f_k\colon \mathbf{R}^n \to \mathbf{R}$, $k = 0, 1, \ldots, m$, must possess some smoothness (i.e., differentiability) properties. In the present section we shall assume that all the functions f_k are continuously differentiable in some neighborhood U in the space $\mathbf{R}^n$ (in the sense that all the partial derivatives $\partial f_k/\partial x_i$ exist and are continuous in U).

A point $\hat{x} \in U$, $f_k(\hat{x}) = 0$, $k = 1, 2, \ldots, m$, yields a local minimum (maximum) in the problem (1) if there is $\varepsilon > 0$ such that if a point $x \in U$ satisfies all the constraints ($f_i(x) = 0$, $i = 1, \ldots, m$) and $|x - \hat{x}| < \varepsilon$, then $f_0(x) \geqslant f_0(\hat{x})$ ($f_0(x) \leqslant f_0(\hat{x})$).

The function

$$\mathscr{L} = \mathscr{L}(x, \lambda, \lambda_0) = \sum_{k=0}^{m} \lambda_k f_k(x), \tag{2}$$

where $\lambda = (\lambda_1, \ldots, \lambda_m)$, is called the *Lagrange function** of the problem (1), and the numbers $\lambda_0, \lambda_1, \ldots, \lambda_m$ are called the *Lagrange multipliers*.

The Lagrange Multiplier Rule. 1. Let all the functions $f_0, f_1, \ldots, f_m$ in problem (1) be continuously differentiable in a neighborhood of a point $\hat{x}$. If $\hat{x}$ is a point of local extremum in problem (1), then there are Lagrange multipliers $\hat{\lambda} = (\hat{\lambda}_1, \ldots, \hat{\lambda}_m)$ and $\hat{\lambda}_0$, not all zero, such that there hold the stationarity conditions for the Lagrange function with respect to x:

$$\mathscr{L}_x(\hat{x}, \hat{\lambda}, \hat{\lambda}_0) = 0 \Leftrightarrow \mathscr{L}_{x_i}(\hat{x}, \hat{\lambda}_1, \ldots, \hat{\lambda}_m, \hat{\lambda}_0) = 0, \quad i = 1, 2, \ldots, n. \tag{3}$$

2. For $\hat{\lambda}_0$ to be nonzero ($\hat{\lambda}_0 \neq 0$) it is sufficient that the vectors $f_1'(\hat{x}), \ldots, f_m'(\hat{x})$ be linearly independent.

Thus, we have n + m equations with n + m + 1 unknowns for determining $\hat{x}$, $\hat{\lambda}$, and $\hat{\lambda}_0$:

$$\frac{\partial}{\partial x_i}\left(\sum_{k=0}^{m} \lambda_k f_k(x)\right) = 0, \quad i = 1, \ldots, n, \quad f_1(x) = \ldots = f_m(x) = 0. \tag{4}$$

It should be taken into consideration that the Lagrange multipliers are determined to within proportionality. If it is known that $\hat{\lambda}_0 \neq 0$ (this is the most important case because if $\hat{\lambda}_0 = 0$, then the relations (3) only express the degeneracy of the constraints and are not connected with the functional), we can obtain the equality $\hat{\lambda}_0 = 1$ by multiplying all $\hat{\lambda}_i$ by a constant. Then the number of the equations becomes equal to that of the unknowns.

Equations (4) can also be written in the following more symmetric form:

$$\mathscr{L}_x = 0, \quad \mathscr{L}_\lambda = 0. \tag{4'}$$

*In English mathematical literature L is called the Lagrangian. In the present translation we follow the authors' distinction between the "Lagrange function" and the "Lagrangian" (Sections 1.3.3, 1.4.1, 1.5.1, 1.5.3, 3.1.5, 3.3.1, 4.1.1, and 4.2.2) — Translator's note.

The solutions of these equations are called *stationary points* of the problem (1).

From the times of Lagrange for almost a whole century the multiplier rule was stated with $\hat{\lambda}_0 = 1$ (Lagrange's formulation will be presented a little later), although in this form it is incorrect without some additional assumptions, for instance, without the assumption on the linear independence of $f_i'(x)$, $i = 1,\dots,m$. To confirm what has been said let us consider the problem

$$x_1 \to \inf, \quad x_1^2 + x_2^2 = 0.$$

Its evident and single solution is $\hat{x} = (0, 0)$ because this is the only admissible point. Now let us try to form the Lagrange function with $\hat{\lambda}_0 = 1$ and then apply the algorithm (3). We have $\mathcal{L} = x_1 + \lambda(x_1^2 + x_2^2)$, whence

$$\frac{\partial \mathcal{L}}{\partial x_1} = 0 \Rightarrow 2\lambda x_1 + 1 = 0, \quad \frac{\partial \mathcal{L}}{\partial x_2} = 0 \Rightarrow 2\lambda x_2 = 0,$$

and we see that the first of these equations contradicts the equation $x_1^2 + x_2^2 = 0$.

Usually as a regularity condition, guaranteeing that $\hat{\lambda}_0 \neq 0$, the linear independence of the derivatives $f_1'(\hat{x}),\dots,f_m'(\hat{x})$ is used (see the assertion 2 of the above theorem). However, as a rule, the verification of this condition is more complicated than the direct verification, with the aid of Eq. (3), of the fact that $\hat{\lambda}_0$ cannot be equal to zero (see the solutions of the problems in Section 1.6). Therefore, the above formulation of the theorem, which involves no additional assumptions other than the smoothness conditions, is very convenient.

The proof of the theorem is based on one of the most important theorems of finite-dimensional differential calculus, namely the inverse function theorem (see [14, Vol. 1, p. 455] and [9, Vol. 1]).

<u>The Inverse Function Theorem.</u> Let $\psi_1(x_1,\dots,x_s),\dots,\psi_s(x_1,\dots,x_s)$ be s functions of s variables continuously differentiable in a neighborhood of a point $\hat{x}$. Moreover, let the Jacobian

$$\mathcal{J} = \det\left(\frac{\partial \psi_i(\hat{x})}{\partial x_j}\right)$$

be nonzero. Then there exist $\varepsilon_0 > 0$ and $\delta_0 > 0$ such that for any $\eta = (\eta_0,\dots,\eta_s)$, $|\eta| \leqslant \varepsilon_0$, there is ξ, $|\xi| \leqslant \delta_0$, such that $\psi(\hat{x} + \xi) = \psi(\hat{x}) + \eta$, and, moreover, $\xi \to 0$ when $\eta \to 0$.

More general versions of this theorem will be proved in Section 2.3.

<u>Proof of the Lagrange Multiplier Rule.</u> Let a point $\hat{x}$ yield a local minimum in problem (1). Here there are two possibilities: either the vectors $f_0'(\hat{x}),\dots,f_m'(\hat{x})$ are linearly independent or these vectors are linearly dependent.

A) Let the vectors be linearly dependent. Then, by the definition of linear dependence, there are $\hat{\lambda}_k$, $k = 0, 1,\dots,m$, not all zero, for which the condition (3) holds.

B) Let the vectors be linearly independent. We shall show that this assumption contradicts the fact that $\hat{x}$ yields a local minimum in the problem (1).

Let us consider the mapping $\Phi(x) = (f_0(x) - f_0(\hat{x}), f_1(x),\dots,f_m(x))$. By assumption, the vectors $f_0'(\hat{x}),\dots,f_m'(\hat{x})$ are linearly independent, i.e., the rank of the matrix

$$A=\begin{pmatrix} \frac{\partial f_0(\hat{x})}{\partial x_1}, & \dots, & \frac{\partial f_0(\hat{x})}{\partial x_n} \\ \dots & \dots & \dots \\ \frac{\partial f_m(\hat{x})}{\partial x_1}, & \dots, & \frac{\partial f_m(\hat{x})}{\partial x_n} \end{pmatrix}$$

is equal to m + 1. For definiteness, let us suppose that the first m + 1 columns of the matrix A are linearly independent, i.e.,

$$\det\begin{pmatrix} \frac{\partial f_0(\hat{x})}{\partial x_1}, & \dots, & \frac{\partial f_0(\hat{x})}{\partial x_{m+1}} \\ \dots & \dots & \dots \\ \frac{\partial f_m(\hat{x})}{\partial x_1}, & \dots, & \frac{\partial f_m(\hat{x})}{\partial x_{m+1}} \end{pmatrix}\neq 0.$$

Then the functions

$$\psi_1(x_1, \dots, x_{m+1})=f_0(x_1, \dots, x_{m+1}, \hat{x}_{m+2}, \dots, \hat{x}_n)-f_0(\hat{x}_1, \dots, \hat{x}_n),$$
$$\psi_2(x_1, \dots, x_{m+1})=f_1(x_1, \dots, x_{m+1}, \hat{x}_{m+2}, \dots, \hat{x}_n), \dots,$$
$$\psi_{m+1}(x_1, \dots, x_{m+1})=f_m(x_1, \dots, x_{m+1}, \hat{x}_{m+2}, \dots, \hat{x}_n)$$

satisfy the conditions of the inverse function theorem. By that theorem, for any ε, $0 < \varepsilon < \varepsilon_0$, there are $x_1(\varepsilon), \dots, x_{m+1}(\varepsilon)$ such that

$$\begin{gathered} f_0(x_1(\varepsilon), \dots, x_{m+1}(\varepsilon), \hat{x}_{m+2}, \dots, \hat{x}_n)-f_0(\hat{x})=-\varepsilon, \\ f_1(x_1(\varepsilon), \dots, x_{m+1}(\varepsilon), \hat{x}_{m+2}, \dots, \hat{x}_n)=0, \\ \dots\dots\dots \\ f_m(x_1(\varepsilon), \dots, x_{m+1}(\varepsilon), \hat{x}_{m+2}, \dots, \hat{x}_n)=0, \end{gathered} \tag{5}$$

and, moreover, $x_i(\varepsilon) \to \hat{x}_i$ for $\varepsilon \to 0$; $i = 1, 2, \dots, m + 1$. Now (5) directly implies that $\hat{x}$ is not a point of local minimum. We have thus arrived at the desired contradiction. The assertion 2 of the theorem is obvious. ■

The generalization of the Lagrange multiplier rule to infinite-dimensional problems with equality and inequality constraints will be presented in Chapter 3.

Now we shall make some more historical remarks. The multiplier rule was first mentioned in Euler's works on isoperimetric problems (in 1744). Later (in 1788) it was put forward by Lagrange in his "Analytical mechanics" for a wide class of variational problems (the so-called *Lagrange problems*; this class of problems will be discussed later). In 1797 in his book "The Theory of Analytical Functions"* Lagrange also touches upon the question of finite-dimensional extremal problems.

He writes: "On peut les réduire à ce principe générale. Lors qu'une fonction de plusieurs variables doit être un maximum ou minimum, et qu'il y a entre ces variables une ou plusieurs équations, il suffira d'ajouter à la fonction proposée les fonctions qui doivent être nulles, multipliées chacune par une quantité indéterminée, et là chercher ensuite le maximum ou minimum comme si les variables étaient indépendantes; les équations qu'on trouvées, serviront à déterminer toutes les inconnues."

"The following general principle can be stated. If a maximum or a minimum of a function of several variables is being sought under the condition that there is a constraint connecting the variables which is determined by one or several equations, one must add the functions determining the equations of the constraint multiplied by indeterminate multipliers to the function to be minimized and then seek the maximum or the minimum of the constructed sum as if the variables were independent. The equations

*J. L. Lagrange, Théorie des Fonctions Analytiques, Paris (1813).

obtained, when added to the constraint equations, will serve for determining all the unknowns."

We can say without exaggeration that the major part of the present book is devoted to the explanation and the interpretation of Lagrange's idea in its application to problems of different nature. Let us again present here the main idea of Lagrange. Let it be required to find an extremum in the problem (1). Then we should form the Lagrange function (2) and consider the problem $\mathscr{L}(x, \hat{\lambda}, \hat{\lambda}_0) \to$ extr without constraints. According to the theorem of Fermat for functions of n variables, the necessary conditions for this problem without constraints yield the desired equations $\mathscr{L}_x = 0$. Thus, in accordance with the Lagrange principle, the equations for an extremum in the problem with constraints coincide with the equations for the problem $\mathscr{L}(x, \hat{\lambda}, \hat{\lambda}_0) \to$ extr without constraints for an adequate choice of the Lagrange multipliers $\hat{\lambda}$, $\hat{\lambda}_0$.

Further, we shall see that the general idea of Lagrange proves to be correct for a great number of problems.

1.3.3. The Kuhn–Tucker Theorem. In this section we shall consider convex programming problems for which the idea of Lagrange assumes the most complete form. The investigation of this class of problems has been going on for a comparatively short time. The fundamentals of linear programming (this is a special case of convex programming) were laid down in the work by L. V. Kantorovich [55] in 1939. The Kuhn–Tucker theorem (which is the basic result in the present section) was proved in 1951.

Let us consider the following extremal problem (a *convex programming problem*):

$$f_0(x) \to \inf, \quad f_i(x) \leqslant 0, \quad i = 1, \ldots, m, \; x \in A, \tag{1}$$

where X is a linear space (not necessarily finite-dimensional), f_i are convex functions on X, and A is a convex subset of X. Here it is important that (1) is a minimization problem and not a maximization problem.

We remind the reader that a set C lying in a linear space is said to be *convex* if, along with any two of its points x and y, it contains the whole line segment $[x, y] = \{z \mid z = \alpha x + (1 - \alpha)y, \; 0 \leqslant \alpha \leqslant 1\}$. A function $f: X \to \mathbf{R}$ is said to be *convex* if *Jensen's inequality*

$$f(\alpha x + (1-\alpha) y) \leqslant \alpha f(x) + (1-\alpha) f(y), \quad \forall \alpha, \quad 0 \leqslant \alpha \leqslant 1,$$

holds for any $x, y \in X$ or, which is the same, if the "epigraph" of f, i.e., the set

$$\operatorname{epi} f = \{(\alpha, x) \in \mathbf{R} \times X \mid \alpha \geqslant f(x)\},$$

is a convex set in the product space $\mathbf{R} \times X$.

The function

$$\mathscr{L} = \mathscr{L}(x, \lambda, \lambda_0) = \sum_{k=0}^{m} \lambda_k f_k(x), \tag{2}$$

where $\lambda = (\lambda_1, \ldots, \lambda_m)$, is called the *Lagrange function* of the problem (1), and the numbers $\lambda_0, \ldots, \lambda_m$ are called the *Lagrange multipliers*. Here the constraint $x \in A$ is not involved in the expression (2).

The Kuhn–Tucker Theorem. 1. Let X be a linear space, let $f_i: X \to \mathbf{R}$, $i = 0, 1, \ldots, m$, be convex functions on X, and let A be a convex subset of X.

If $\hat{x}$ is a solution of problem (1), then there are Lagrange multipliers $\hat{\lambda}_0$, $\hat{\lambda}$, not all zero, such that

a) $\min_{x \in A} \mathscr{L}(x, \hat{\lambda}, \hat{\lambda}_0) = \mathscr{L}(\hat{x}, \hat{\lambda}, \hat{\lambda}_0)$ (this is the minimum principle);

b) $\hat{\lambda}_i \geqslant 0$, $i = 0, 1, \ldots, m$ (the nonnegativity condition);

c) $\hat{\lambda}_i f_i(\hat{x}) = 0$, $i = 1, 2, \ldots, m$ (the condition of complementary slackness).

2. If $\lambda_0 \neq 0$, then the conditions a)-c) are sufficient for the admissible point x to be a solution of problem (1).

3. For $\hat{\lambda}_0$ to be nonzero ($\hat{\lambda}_0 \neq 0$) it is sufficient that there exist a point $\bar{x} \in A$, for which the Slater condition holds: $f_i(\bar{x}) < 0$, $i = 1, \ldots, m$.

Hence, if the Slater condition is fulfilled, we can assume that $\hat{\lambda}_0 = 1$.

The relation a) expresses Lagrange's idea in the most complete form: if $\hat{x}$ yields a minimum in a problem with constraints (inequality constraints) then the same point is a point of minimum of the Lagrange function (in the problem not involving those constraints which are included in the Lagrange function). The relations b) and c) are characteristic of problems with inequality constraints (for more detail, see Section 3.2).

The proof of the Kuhn–Tucker theorem is based on one of the most important theorems of convex analysis, namely, the *separation theorem*. However, here it suffices to prove the simplest finite-dimensional version of this theorem which we state below and prove in the next section.

The Finite-Dimensional Separation Theorem. Let C be a convex subset of $\mathbf{R}^N$ not containing the point 0. Then there are numbers $a_1, \ldots, a_N$ such that the inequality

$$\sum_{i=1}^{N} a_i x_i \geqslant 0 \qquad (3)$$

holds for any $x = (x_1, \ldots, x_N) \in C$ (in other words, the hyperplane $\sum_{i=1}^{N} a_i x_i = 0$, passing through the point 0, divides $\mathbf{R}^N$ into two parts within one of which the set C is entirely contained).

Proof of the Kuhn–Tucker Theorem. Let $\hat{x}$ be a solution of the problem. Without loss of generality, we can assume that $f_0(\hat{x}) = 0$ [if otherwise, we introduce the new function $\widetilde{f_0(x)} = f_0(x) - f_0(\hat{x})$]. Let us put

$$C = \{\mu \in \mathbf{R}^{m+1},\ \mu = (\mu_0, \ldots, \mu_m) \mid \exists x \in A\colon\ f_0(x) < \mu_0,\quad f_i(x) \leqslant \mu_i,\ i \geqslant 1\}. \qquad (4)$$

The further part of the proof is split into several stages.

A) The Set C is Nonempty and Convex. Indeed, any vector $\mu \in \mathbf{R}^{m+1}$ with positive components belongs to C because we can put $x = \hat{x}$ in (4). Consequently, the set C is nonempty. Let us prove its convexity. Let $\mu = (\mu_0, \ldots, \mu_m)$ and $\mu' = (\mu_0', \ldots, \mu_m')$ belong to C, $0 \leqslant \alpha \leqslant 1$, and let x and x' be elements of A such that $f_0(x) < \mu_0$, $f_0(x') < \mu_0'$, $f_i(x) \leqslant \mu_i$, and $f_i(x') \leqslant \mu_i'$, $i \geqslant 1$ [according to (4), such elements x and x' exist]. Let us put $x_\alpha = \alpha x + (1 - \alpha)x'$. Then $x_\alpha \in A$ since A is convex, and, by the convexity of f_i,

$$f_i(x_\alpha) = f_i(\alpha x + (1-\alpha)x') \leqslant \alpha f_i(x) + (1-\alpha) f_i(x') \begin{cases} < \alpha\mu_0 + (1-\alpha)\mu_0', & i = 0, \\ \leqslant \alpha\mu_i + (1-\alpha)\mu_i', & i \geqslant 1, \end{cases}$$

i.e., the point $\alpha\mu + (1 - \alpha)\mu'$ belongs to C.

B) The Point $0 \in \mathbf{R}^{m+1}$ Does Not Belong to C. Indeed, if the point 0 belonged to C, then, by virtue of definition (4), this would imply the existence of an element $(\tilde{x}) \in A$, for which the inequalities $f_0(\tilde{x}) < 0$, $f_i(\tilde{x}) \leqslant 0$, $i \geqslant 1$, hold. However, from these inequalities it follows that

$\hat{x}$ is not a solution of the problem. Consequently, $0 \notin C$.

Since C is convex and $0 \notin C$, we can apply the separation theorem according to which there are $\hat{\lambda}_0, \ldots, \hat{\lambda}_m$ such that

$$\sum_{i=0}^{m} \hat{\lambda}_i \mu_i \geqslant 0 \tag{5}$$

for any $\mu \in C$.

C) The Multipliers $\hat{\lambda}_i$, $i \geqslant 0$, in (5) Are Nonnegative. Indeed, in Section A) it was already said that any vector with positive components belongs to C; in particular, the vector $(\varepsilon, \ldots, \varepsilon, 1, \varepsilon, \ldots, \varepsilon)$ belongs to C, where $\varepsilon > 0$ and 1 occupies the i_0-th place, $i_0 \geqslant 1$. Substituting this point into (5), we see that $\hat{\lambda}_{i_0} \geqslant -\varepsilon \sum_{i \neq i_0} \hat{\lambda}_i$, whence, by the arbitrariness of $\varepsilon > 0$, it follows that $\hat{\lambda}_{i_0} \geqslant 0$.

D) The Multipliers $\hat{\lambda}_i$, $i \geqslant 1$, Satisfy the Conditions of Complementary Slackness. Indeed, if $f_{j_0}(\hat{x}) = 0$, $j_0 \geqslant 1$, then the equality $\hat{\lambda}_{j_0} f_{j_0}(\hat{x}) = 0$ is trivial. Let $f_{j_0}(\hat{x}) < 0$. Then the point $(\delta, 0, \ldots, 0, f_{j_0}(\hat{x}), 0, \ldots, 0)$, where the number $f_{j_0}(\hat{x})$ occupies the j_0-th place and $\delta > 0$, belongs to C [to show this it suffices to substitute the point $\hat{x}$ for the point x in (4)].

Substituting this point $(\delta, 0, \ldots, 0, f_{j_0}(\hat{x}), 0, \ldots, 0)$ into (5), we see that $\hat{\lambda}_{j_0} f_{j_0}(\hat{x}) \geqslant -\hat{\lambda}_0 \delta$, whence, by the arbitrariness of $\delta > 0$, the inequality $\hat{\lambda}_{j_0} \leqslant 0$ follows. However, as was proved in Section B), we have $\hat{\lambda}_{j_0} \geqslant \geqslant 0$, and hence $\hat{\lambda}_{j_0} = 0$ and $\hat{\lambda}_{j_0} f_{j_0}(\hat{x}) = 0$.

E) The Minimum Principle Holds at the Point $\hat{x}$. Indeed, let $x \in A$. Then the point $(f_0(x) + \delta, f_1(x), \ldots, f_m(x))$ belongs to C for any $\delta > 0$ [see definition (4)]. Therefore, taking into account (5), we obtain $\hat{\lambda}_0 f_0(x) + \sum_{i=1}^{m} \hat{\lambda}_i f_i(x) \geqslant -\hat{\lambda}_0 \delta$, whence (by virtue of the arbitrariness of $\delta > 0$) the inequality $\mathscr{L}(x, \hat{\lambda}, \hat{\lambda}_0) \geqslant 0$ follows.

Now, taking into account the equality $f_0(\hat{x}) = 0$ and the conditions of complementary slackness, we obtain

$$\mathscr{L}(x, \hat{\lambda}, \hat{\lambda}_0) \geqslant 0 = \sum_{i=0}^{m} \hat{\lambda}_i f_i(\hat{x}) = \mathscr{L}(\hat{x}, \hat{\lambda}, \hat{\lambda}_0)$$

for any $x \in A$. Assertion 1 of the theorem has been proved.

F) Let Us Prove Assertion 2. If we suppose that $\lambda_0 \neq 0$, then we can put $\lambda_0 = 1$ [because the Lagrange multipliers satisfying the relations a)-c) of the theorem preserve their properties when multiplied by an arbitrary positive factor]. Then for any admissible $x\,(x \in A,\ f_i(x) \leqslant 0,\ i \geqslant 1)$, we obtain

$$f_0(x) \overset{\text{b)}}{\geqslant} f_0(x) + \sum_{i=1}^{m} \hat{\lambda}_i f_i(x) = \mathscr{L}(x, \hat{\lambda}, 1) \overset{\text{a)}}{\geqslant} \mathscr{L}(\hat{x}, \hat{\lambda}, 1) = f_0(\hat{x}) + \sum_{i=1}^{m} \hat{\lambda}_i f_i(\hat{x}) \overset{\text{c)}}{=} f_0(\hat{x}),$$

i.e., x is a solution of the problem.

G) Let Us Prove Assertion 3. Let the Slater condition be fulfilled [i.e., let for some $\bar{x} \in A$ the inequalities $f_i(\bar{x}) < 0$, $i \geqslant 1$, hold] and, moreover, let $\hat{\lambda}_0 = 0$ in assertion 1. Then we immediately arrive at a contradiction:

$$\mathscr{L}(\bar{x}, \hat{\lambda}, 0) = \sum_{i=0}^{m} \hat{\lambda}_i f_i(\bar{x}) < 0 = \mathscr{L}(\hat{x}, \hat{\lambda}, 0)$$

(in the calculation we have used the fact that $\hat{\lambda}_i$, $i \geqslant 1$, are not all zero), and, at the same time, by virtue of a), we have $\mathscr{L}(\bar{x}, \hat{\lambda}, 0) \geqslant \mathscr{L}(\hat{x}, \hat{\lambda}, 0)$. ■

We shall also present another version of the Kuhn–Tucker theorem. Since $\hat{\lambda}_0 > 0$ when the Slater conditions are fulfilled, and since the Lagrange multipliers are determined to within a positive factor, we can assume that $\hat{\lambda}_0 = 1$. Now the Lagrange function

$$\mathscr{L}(x, \lambda, 1) = f_0(x) + \sum_{i=1}^{m} \lambda_i f_i(x)$$

is defined on the set

$$A \times \mathbf{R}_+^m = \{(x, \lambda) \mid x = (x_1, \ldots, x_m) \in A; \lambda = (\lambda_1, \ldots, \lambda_m), \lambda_i \geqslant 0\},$$

and the relations a)-c) are equivalent to the fact that $(\hat{x}, \hat{\lambda})$ is a *saddle point* of the Lagrange function, i.e.,

$$\min_{x \in A} \mathscr{L}(x, \hat{\lambda}, 1) = \mathscr{L}(\hat{x}, \hat{\lambda}, 1) = \max_{\lambda \geqslant 0} \mathscr{L}(\hat{x}, \lambda, 1). \tag{6}$$

Indeed, the left equality (6) coincides with a), and the right one is a consequence of c) and b):

$$\mathscr{L}(\hat{x}, \hat{\lambda}, 1) = f_0(\hat{x}) + \sum_{i=1}^{m} \hat{\lambda}_i f_i(\hat{x}) = f_0(\hat{x}) \geqslant f_0(\hat{x}) + \sum_{i=1}^{m} \lambda_i f_i(\hat{x}) = \mathscr{L}(\hat{x}, \lambda, 1).$$

In Sections 2.6, 3.1, 3.3, and 4.3 we shall present some other theorems of convex analysis and convex programming.

Remark. In "convex analysis" it turns out to be convenient to consider functions whose values belong to the extended real line $\overline{\mathbf{R}} = \mathbf{R} \cup \{-\infty, +\infty\}$; that is, it is convenient to admit that the functions may assume the infinite values $-\infty$, $+\infty$. The rules of arithmetic are applicable, with some limitations, to the extended real line (this will be discussed in greater detail in the footnote to Section 2.6.1). It can easily be seen that the above proof of the Kuhn–Tucker theorem remains valid, without changes, for such functions as well.

1.3.4. Proof of the Finite-Dimensional Separation Theorem. The statement of the theorem was given earlier. We remind the reader that C is a convex set in $\mathbf{R}^N$ and $0 \notin C$.

A) Let lin C be the *linear hull* of the set C, i.e., the smallest linear subspace containing C. Here there are two possible cases: either lin C ≠ $\mathbf{R}^N$ or lin C = $\mathbf{R}^N$. In the first case lin C is a proper subspace of $\mathbf{R}^N$, and therefore there exists a hyperplane $\sum_{i=1}^{N} a_i x_i = 0$, containing lin C and passing through zero. This is the sought-for hyperplane.

B) If lin C = $\mathbf{R}^N$, then among the vectors belonging to C there are N linearly independent ones, and, due to the linear independence, they form a basis in $\mathbf{R}^N$. Let us denote these vectors by $\tilde{e}_1, \ldots, \tilde{e}_N$; $\tilde{e}_i \in C$, $i = 1, \ldots, N$.

Let us consider two convex cones in $\mathbf{R}^N$: the "negative orthant"

$$\mathscr{K}_1 = \left\{ x \in \mathbf{R}^N \,\middle|\, x = \sum_{i=1}^{N} \beta_i \tilde{e}_i,\ \beta_i < 0,\ i = 1, \ldots, N \right\}$$

and the conic hull of the set C:

$$\mathscr{K}_2 = \left\{ x \in \mathbf{R}^N \,\middle|\, x = \sum_{i=1}^{s} \alpha_i \xi_i,\ \xi_i \in C,\ \alpha_i \geqslant 0,\ i = 1, \ldots, s,\ s \text{ is arbitrary} \right\}.$$

These cones do not intersect. Indeed, if a vector $\tilde{x}=-\sum_{i=1}^{N}\tilde{\gamma}_i\tilde{e}_i$, $\tilde{\gamma}_i>0$, belonged to $\mathscr{K}_2$, then there would be s, $\tilde{\alpha}_i \geqslant 0$ and $\tilde{\xi}_i \in C$, such that $\tilde{x}=\sum_{i=1}^{s}\tilde{\alpha}_i\tilde{\xi}_i$. But then the point 0 would belong to C because this point could be represented in the form of a convex combination

$$0=\sum_{i=1}^{s}\tilde{\alpha}_i\tilde{\xi}_i-\tilde{x}=\frac{\sum_{i=1}^{s}\tilde{\alpha}_i\tilde{\xi}_i+\sum_{i=1}^{N}\tilde{\gamma}_i\tilde{e}_i}{\sum_{i=1}^{s}\tilde{\alpha}_i+\sum_{i=1}^{N}\tilde{\gamma}_i}$$

of points belonging to C.

C) Since $\mathscr{K}_1$ is open, what is proved in B) implies that none of the points of $\mathscr{K}_1$ can belong to the closure $\overline{\mathscr{K}}_2$ of the set $\mathscr{K}_2$. (It should be noted that $\overline{\mathscr{K}}_2$ is closed and convex. Why?) Let us take an arbitrary point belonging to $\mathscr{K}_1$, say $x_0=-\sum_{i=1}^{N}\tilde{e}_i$, and find the nearest point $\xi_0\in\overline{\mathscr{K}}_2$ to x_0. There exists such a point; namely, this is the point belonging to the compact set $\overline{\mathscr{K}}_2\cap B(x_0,|x_0|)$, at which the continuous function $f(x) = |x - x_0|$ attains its minimum.

D) Let us draw a hyperplane H through ξ_0 perpendicular to $x_0 - \xi_0$ and show that it is the desired one, i.e., $0\in H$ and the set C is entirely contained in one of the two closed half-spaces bounded by that hyperplane. We shall prove even more; namely, we shall show that if $\mathring{H}$ is the interior of one of the half-spaces, the one which contains the point x_0, then $\mathring{H} \cap \overline{\mathscr{K}}_2=\varnothing$. Since C is a subset of $\overline{\mathscr{K}}_2$, it is entirely contained in the closed half-space which is the complement to $\mathring{H}$.

Let us assume the contrary, and let $\xi_1\in\mathring{H}\cap\overline{\mathscr{K}}_2$. Then the angle $\widehat{x_0\xi_0\xi_1}$ is acute. Moreover, $[\xi_0,\xi_1]\in\overline{\mathscr{K}}_2$, since $\overline{\mathscr{K}}_2$ is convex. Let us drop a perpendicular (x_0,ξ_2), $\xi_2\in(\xi_0,\xi_1)$, from x_0 to the straight line (ξ_0, ξ_1) and show that ξ_0 is not the point of $\overline{\mathscr{K}}_2$ nearest to x_0. Indeed, the points ξ_0, ξ_1, and ξ_2 lie in one straight line and $\xi_2\in\mathring{H}$ (why?). If $\xi_2\in[\xi_0,\xi_1]$, then $\xi_2\in\overline{\mathscr{K}}_2$ and $|x_0 - \xi_2| < |x_0 - \xi_0|$ (the perpendicular is shorter than an inclined line). If ξ_1 lies between ξ_0 and ξ_2, then $|x_0 - \xi_1| < |x_0 - \xi_0|$ (of two inclined lines the shorter is the one whose foot is nearer to the foot of the perpendicular).

On the other hand, H passes through the point 0 because, if otherwise, the ray issued from 0 toward the point ξ_0 which is entirely contained in $\overline{\mathscr{K}}_2$ (why?) would have common points with $\mathring{H}$.

1.4. Simplest Problem of the Classical Calculus of Variations and Its Generalizations

1.4.1. Euler Equation. Soon after the work by Johann Bernoulli on the brachistochrone had been published, there appeared (and were solved) many problems of this type. Johann Bernoulli suggested that his disciple L. Euler should find the general method of solving such problems.

In 1744 was published Euler's work "Methodus inveniendi lineas curvas maximi minimive proprietate gaudentes sive solutio problematis isoperimetrici latissimo sensu accepti": "A method of finding curves possessing the properties of a maximum or a minimum or the solution of the isoperimetric problem taken in the broadest sense." In this work the theoretical foundations of a new branch of mathematical analysis were laid down. In particular, approximating curves with broken lines, Euler derived a second-order differential equation for the extremals. Later Lagrange called it the *Euler equation*. In 1759 the first work by Lagrange on the calculus of variations appeared in which new investigation methods were elaborated. Lagrange "varied" the curve suspected to be an extremal, separated out the *principal linear parts* from the increments of the functionals which he called *variations*, and used the fact that the variations must vanish at a point of extremum. Later on the Lagrange method became generally accepted. This is the method that we shall use to derive the Euler equation. It should also be mentioned that after Lagrange's works had appeared Euler suggested that the whole branch of mathematics in which the Lagrange method was used should be called the *calculus of variations*.

Now we proceed to the derivation of the Euler equation for the simplest problem of the classical calculus of variations. In Section 1.2.6 this name was given to the extremal problem

$$\mathcal{J}(x(\cdot))=\int_{t_0}^{t_1} L(t, x, \dot{x})\,dt \to \text{extr}, \quad x(t_0)=x_0, \quad x(t_1)=x_1, \tag{1}$$

considered in the space of continuously differentiable functions $C^1([t_0, t_1])$, $-\infty < t_0 < t_1 < \infty$. $C^1([t_0, t_1])$ is a Banach space, i.e., a normed space complete with respect to the norm

$$\|x(\cdot)\|_1 = \max\left(\max_{t\in[t_0, t_1]} |x(t)|, \max_{t\in[t_0, t_1]} |\dot{x}(t)|\right).$$

We shall suppose that the function L (we call it the *integrand* or the *Lagrangian* of the problem) together with its partial derivatives L_x and $L_{\dot{x}}$ is continuous jointly with respect to the arguments t, x, $\dot{x}$.

<u>THEOREM (a Necessary Condition for an Extremum in the Simplest Problem of the Classical Calculus of Variations).</u> Let a function $\hat{x}(\cdot)\in C^1([t_0, t_1])$ yield a local extremum in the problem (1). Then it satisfies the equation

$$-\frac{d}{dt}L_{\dot{x}}(t, \hat{x}(t), \dot{\hat{x}}(t))+L_x(t, \hat{x}(t), \dot{\hat{x}}(t))=0. \tag{2}$$

Equation (2) is called the *Euler equation*. An admissible function $\hat{x}(\cdot)$ for which the equation holds is called a *stationary point* of the problem (1) or an *extremal*. Thus, the admissible elements which yield local extrema in the problem (1) are extremals; generally speaking, the converse is not true.

In the calculus of variations a local extremum in the space $C^1([t_0, t_1])$ is said to be *weak*. According to the general definition, a function $\hat{x}(\cdot)$ yields a local minimum (maximum) in the problem (1) in the space $C^1([t_0, t_1])$ if there is $\varepsilon > 0$ such that for any function $x(\cdot)\in C^1([t_0, t_1])$ satisfying the conditions $x(t_0) = x(t_1) = 0$ and $\|x(\cdot)\|_1 \leqslant \varepsilon$ there holds the inequality

$$\mathcal{J}(\hat{x}(\cdot)+x(\cdot))\geqslant \mathcal{J}(\hat{x}(\cdot)) \qquad (\leqslant \mathcal{J}(\hat{x}(\cdot))).$$

We shall carry out the proof of the theorem twice. First we shall make use of Lagrange's reasoning. However, to this end we shall have to assume additionally that the function $t \to L_{\dot{x}}(t, \hat{x}(t), \dot{\hat{x}}(t))$ is continuously differentiable. After that we shall prove a very important lemma of DuBois–Reymond from which the theorem under consideration follows without that additional assumption as well. A generalization of the construction elaborated by DuBois–Reymond will play an essential role in Chapter 4.

The *proof* of the theorem consists of three stages.

<u>A) The Definition of the First Lagrange Variation.</u> Let $x(\cdot)\in C^1([t_0, t_1])$. Let us consider the function $\lambda \mapsto \varphi(\lambda)=\mathcal{J}(\hat{x}(\cdot)+\lambda x(\cdot))$. We have

$$\varphi(\lambda)=\int_{t_0}^{t_1} F(t,\lambda)\,dt=\int_{t_0}^{t_1} L(t,\hat{x}(t)+\lambda x(t),\dot{\hat{x}}(t)+\lambda\dot{x}(t))\,dt.$$

The conditions imposed on L allow us to differentiate the function F under the integral sign (to this end it is sufficient that the functions $(t,\lambda)\to F(t,\lambda)$ and $(t,\lambda)\to \partial F(t,\lambda)/\partial\lambda$ be continuous (see [14, Vol. 2, p. 661] and [9, Vol. 2]). Differentiating and substituting $\lambda = 0$, we obtain

$$\varphi'(0)-\int_{t_0}^{t_1}(q(t)x(t)+p(t)\dot{x}(t))\,dt, \tag{3}$$

where

$$p(t)=L_{\dot{x}}(t,\hat{x}(t),\dot{\hat{x}}(t)),\quad q(t)=L_x(t,\hat{x}(t),\dot{\hat{x}}(t)). \tag{3a}$$

Thus, $\lim\limits_{\lambda\to 0}(\mathcal{J}(\hat{x}(\cdot)+\lambda x(\cdot))-\mathcal{J}(\hat{x}(\cdot))/\lambda)$ exists for any $x(\cdot)\in C^1([t_0,t_1])$. Let us denote this limit by $\delta\mathcal{J}(\hat{x}(\cdot),x(\cdot))$. The function $x(\cdot)\mapsto\delta\mathcal{J}(\hat{x}(\cdot),x(\cdot))$ is called the *first Lagrange variation of the functional* $\mathcal{J}$.

B) The Transformation of the First Variation with the Aid of Integration by Parts. Let $x(t_0) = x(t_1) = 0$, and let the function $p(\cdot)$ be continuously differentiable. Following Lagrange's method, we have

$$\delta\mathcal{J}(\hat{x}(\cdot),x(\cdot))=\int_{t_0}^{t_1}(q(t)x(t)+p(t)\dot{x}(t))\,dt=\int_{t_0}^{t_1}a(t)x(t)\,dt, \tag{4}$$

where $a(t) = -\dot{p}(t) + q(t)$ [the product $p(t)x(t)$ vanished at the end points of the interval of integration].

We know that $x(\cdot)$ yields a local extremum (for definiteness, let it be a minimum) in problem (1). It follows that the function $\lambda\mapsto\varphi(\lambda)$ has a local minimum at zero. By virtue of Fermat's theorem (see Section 1.3.1), $\varphi'(0)=\delta\mathcal{J}(\hat{x}(\cdot),x(\cdot))=0$. Comparing this with (4), we conclude that for an arbitrary function $x(\cdot)\in C^1([t_0,t_1])$ such that $x(t_0) = x(t_1) = 0$ the equality $\int_{t_0}^{t_1}a(t)x(t)\,dt=0$ takes place.

C) The Fundamental Lemma of the Classical Calculus of Variations (Lagrange Lemma). Let a continuous function $t\mapsto a(t)$, $t\in[t_0,t_1]$, possess the property that $\int_{t_0}^{t_1}a(t)x(t)\,dt=0$ for any continuously differentiable function $x(\cdot)$ satisfying the condition $x(t_0) = x(t_1) = 0$. Then $a(t) \equiv 0$.

Proof. Let us suppose that $a(\tau) \neq 0$ at a point $\tau\in[t_0,t_1]$. Then the continuity of $a(\cdot)$ implies that there is a closed interval $\Delta=[\tau_1,\tau_2]\subset(t_0,t_1)$, on which $a(\cdot)$ does not vanish. For definiteness, let $a(t)\geqslant m>0$, $t\in\Delta$. Let us construct the function

$$\tilde{x}(t)=\begin{cases}(t-\tau_1)^2(t-\tau_2)^2, & t\in\Delta,\\ 0, & t\notin\Delta.\end{cases}$$

It can easily be verified that $\tilde{x}(\cdot)\in C^1([t_0,t_1])$; moreover, $\tilde{x}(t_0) = \tilde{x}(t_1) = 0$. By the condition of the lemma, $\int_{t_0}^{t_1}a(t)\tilde{x}(t)\,dt=0$. On the other hand, by the mean value theorem (see [14, Vol. 2, p. 115] and [9, Vol. 1]), we have $\int_{t_0}^{t_1}a(t)\tilde{x}(t)\,dt=a(\xi)\int_{t_0}^{t_1}\tilde{x}(t)\,dt>0$, and we arrive at a contradiction which proves the lemma. ■

Comparing what was established in A) and B), we see that $-\dot{p}(t) + q(t) \equiv 0$, which is what we had to prove.

D) DuBois–Reymond's Lemma. Let functions $a(\cdot)$ and $b(\cdot)$ be continuous on $[t_0, t_1]$, and let

$$\int_{t_0}^{t_1} (a(t)\dot{x}(t) + b(t)x(t))\,dt = 0 \tag{5}$$

for any function $x(\cdot)\in C^1([t_0, t_1])$ such that $x(t_0) = x(t_1) = 0$. Then the function $a(\cdot)$ is continuously differentiable and $da/dt = b(t)$.

Proof. Integrating by parts in the second summand of the integrand in (5), we obtain

$$0 = \int_{t_0}^{t_1} \left(a(t) - \int_{t_0}^{t} b(s)\,ds\right)\dot{x}(t)\,dt + \int_{t_0}^{t} b(s)\,ds\, x(t)\Big|_{t_0}^{t_1} = \int_{t_0}^{t_1} \left(a(t) - \int_{t_0}^{t} b(s)\,ds\right)\dot{x}(t)\,dt. \tag{6}$$

Let us prove that $\varphi(t) = a(t) - \int_{t_0}^{t} b(s)\,ds = \text{const}$. Indeed, if this is not so, then there are τ_1 and τ_2 such that $\varphi(\tau_1) \neq \varphi(\tau_2)$. Without loss of generality, we can assume that τ_1 and τ_2 are interior points of the closed interval $[t_0, t_1]$ and that $\varphi(\tau_1) > \varphi(\tau_2)$. Let us choose $\delta > 0$ so that $[\tau_i - \delta, \tau_i + \delta] \subset [t_0, t_1]$; $i = 1, 2$, and so that $\varphi(s+\tau_1) - \varphi(s+\tau_2) \geqslant \alpha > 0$ for $|s| \leqslant \delta$.

Now let us take a function $x(\cdot)$ whose derivative has a special form:

$$\dot{x}(t) = \begin{cases} -(t-\tau_1+\delta)(t-\tau_1-\delta), & t\in[\tau_1-\delta, \tau_1+\delta], \\ +(t-\tau_2+\delta)(t-\tau_2-\delta), & t\in[\tau_2-\delta, \tau_2+\delta], \\ 0 \text{ for all the other } t. \end{cases} \tag{7}$$

If we put $x(t_0) = 0$ then, by virtue of (7), we have $x(t_1) = \int_{t_0}^{t_1} \dot{x}(t)\,dt = 0$, and hence (6) must hold for such a function. However,

$$\int_{t_0}^{t_1} \left[a(t) - \int_{t_0}^{t} b(s)\,ds\right]\dot{x}(t)\,dt = \int_{\tau_1-\delta}^{\tau_1+\delta} \varphi(t)\dot{x}(t)\,dt + \int_{\tau_2-\delta}^{\tau_2+\delta} \varphi(t)\dot{x}(t)\,dt =$$

$$= \int_{-\delta}^{\delta} (\varphi(\tau_1+s) - \varphi(\tau_2+s))(\delta-s)(s+\delta)\,ds \geqslant \alpha\int_{-\delta}^{\delta} (\delta-s)(s+\delta)\,ds > 0.$$

We thus arrive at a contradiction, which shows that $\varphi(t) = a(t) - \int_{t_0}^{t} b(s)\,ds = \text{const}$, and this implies that $a(\cdot)$ is differentiable and $da/dt = b$. ■

Applying DuBois–Reymond's lemma to the first variation (3), we conclude that $p(\cdot)\in C^1([t_0, t_1])$ and $\dot{p}(t) = q(t)$.

The argument we have just used contains the so-called *method of variations* in its incipiency with the aid of which various necessary conditions for extremum are derived. The essence of the method consists in the following. Let $\hat{x}$ be a point suspected to yield a minimum in the problem $f(x) \to \inf$, $x\in C$. Then one can try to construct a continuous mapping $\lambda \to x(\lambda)$, $\lambda\in \mathbf{R}_+$, such that $x(0) = \hat{x}$ and $x(\lambda)\in C$, $0\leqslant\lambda\leqslant\lambda_0$. It is natural to call this curve a variation of the argument. Let us put $\varphi(\lambda) = f(x(\lambda))$. Further, let us suppose that the function φ is differentiable on the right with respect to λ. If $\hat{x}$ does in fact yield a minimum, then the inequality $\varphi'(+0) = \lim_{\lambda\downarrow 0}(\varphi(\lambda) - \varphi(0))/\lambda \geqslant 0$ must hold because it is evident that $\varphi(\lambda) \geqslant \varphi(0) = f(\hat{x})$. If one manages to construct a sufficiently broad set of variations of the argument, the set of the inequalities $\varphi'(+0)\geqslant 0$ corresponding to all these variations forms a necessary condition for a minimum. When deriving the Euler equation, we used "directional variations": $x(\lambda) = \hat{x} + \lambda x$. In the derivation of Weierstrass' necessary condition and the maximum principle we shall use variations of another form, namely, the so-called "needlelike variations" (see Sections 1.4.4 and 1.5.4).

In conclusion, we indicate two first integrals of the Euler equation:

1) If the Lagrangian L does not depend on x, the Euler equation possesses the obvious integral

$$p(t)=L_{\dot{x}}(t, \dot{x}(t))=\text{const.}$$

It is called the *momentum integral*.

2) If the Lagrangian L does not depend on t, the Euler equation possesses the first integral

$$H(t)=L_{\dot{x}}(\hat{x}(t), \dot{\hat{x}}(t))\dot{\hat{x}}(t)-L(\hat{x}(t), \dot{\hat{x}}(t))=\text{const.}$$

It is called the *energy integral* (both the names originate from classical mechanics; we shall have a chance to come back to this question; see Sections 4.4.6 and 4.4.7). To prove this we calculate the derivative:

$$\frac{dH}{dt}=L_{\dot{x}}(\hat{x}(t), \dot{\hat{x}}(t))\ddot{\hat{x}}(t)+\frac{d}{dt}L_{\dot{x}}(\hat{x}(t), \dot{\hat{x}}(t))\dot{\hat{x}}(t)-$$
$$-L_x(\hat{x}(t), \dot{\hat{x}}(t))\dot{\hat{x}}(t)-L_{\dot{x}}(\hat{x}(t), \dot{\hat{x}}(t))\ddot{\hat{x}}(t)=\left(\frac{d}{dt}L_{\dot{x}}(\hat{x}(t), \dot{\hat{x}}(t))-L_x(\hat{x}(t), \dot{\hat{x}}(t))\right)\dot{\hat{x}}(t)=0$$

(by virtue of the Euler equations). Consequently, H(t) = const.

Remark. In the course of the calculation we have made the additional assumption on the existence of the second derivative $\ddot{\hat{x}}(t)$. This assumption may be weakened, but it is impossible to do without it completely.

It should also be noted that the Euler equation is a second-order differential equation. Its general integral depends on two arbitrary constants. Their values are found from the boundary conditions.

1.4.2. Necessary Conditions in the Bolza Problem. Transversality Conditions. The simplest problem considered in the foregoing section is a problem with constraints: the boundary conditions $x(t_0) = x_0$ and $x(t_1) = x_1$ form two equality constraints. The problem considered in the present section (the so-called Bolza problem) is a problem without constraints. Let L satisfy the same conditions as in Section 1.4.1; let $l = l(x_0, x_1)$ be a continuously differentiable function of two variables. We shall consider the following extremum problem in the space $C^1([t_0, t_1])$:

$$\mathscr{B}(x(\cdot))=\int_{t_0}^{t_1} L(t, x, \dot{x})\,dt+l(x(t_0), x(t_1))\to\text{extr.} \tag{1}$$

This is the problem that is called the *Bolza problem*.

THEOREM (a Necessary Condition for an Extremum in the Bolza Problem). Let a function $\hat{x}(\cdot)\in C^1([t_0, t_1])$ yield a local minimum in the problem (1). Then there hold the Euler equation

$$-\frac{d}{dt}L_{\dot{x}}(t, \hat{x}(t), \dot{\hat{x}}(t))+L_x(t, \hat{x}(t), \dot{\hat{x}}(t))=0 \tag{2}$$

and the transversality conditions

$$L_{\dot{x}}(t_0, \hat{x}(t_0), \dot{\hat{x}}(t_0))=\frac{\partial l(\hat{x}(t_0), \hat{x}(t_1))}{\partial x_0},$$
$$L_{\dot{x}}(t_1, \hat{x}(t_1), \dot{\hat{x}}(t_1))=-\frac{\partial l(\hat{x}(t_0), \hat{x}(t_1))}{\partial x_1}. \tag{3}$$

Proof. In exactly the same way as at the first stage of the proof of the theorem of the foregoing section we obtain the following expression for the first variation of the functional $\mathscr{B}$:

$$\delta\mathscr{B}(\hat{x}(\cdot),\ x(\cdot))=\int_{t_0}^{t_1}(q(t)x(t)+p(t)\dot{x}(t))\,dt+\alpha_0 x(t_0)+\alpha_1 x(t_1), \tag{4}$$

where p(·) and q(·) are determined by relations (3a) of the foregoing section and $\alpha_i=\frac{\partial l(\hat{x}(t_0),\ \hat{x}(t_1))}{\partial x_i}$, i = 0, 1. Since $\hat{x}(\cdot)$ is a point of local minimum, $\delta\mathscr{B}(\hat{x}(\cdot);\ x(\cdot))=0$ for any point $x(\cdot)\in C^1([t_0,\ t_1])$. In particular, $\delta\mathscr{B}(\hat{x}(\cdot);\ \hat{x}(\cdot))=0$ for any point $x(\cdot)\in C^1([t_0,\ t_1])$ such that $x(t_0)=x(t_1)=0$. By DuBois–Reymond's lemma, the function p(·) is continuously differentiable, and, moreover, $dp(t)/dt = q(t)$. Integrating by parts in (4) and using the equality $\dot{p}(t) = q(t)$, we arrive at the following expression for the first Lagrange variation of the Bolza functional:

$$0=\delta\mathscr{B}(\hat{x}(\cdot),\ x(\cdot))=\int_{t_0}^{t_1}(q(t)-\dot{p}(t))\,x(t)\,dt+(\alpha_0-p(t_0))\,x(t_0)+$$
$$+(\alpha_1+p(t_1))x(t_1)=(\alpha_0-p(t_0))x(t_0)+(\alpha_1+p(t_1))\,x(t_1). \tag{5}$$

Putting, in succession, $x(t) = (t - t_0)$ and then $x(t) = t - t_1$ in (5) we conclude that $\alpha_0 = p(t_0)$, $-\alpha_1 = p(t_1)$. ■

Thus, we have again obtained a second-order differential equation and two boundary conditions (the transversality conditions).

We have considered the "one-dimensional" Bolza problem. The vector problem is stated quite analogously:

$$\mathscr{B}(x(\cdot))=\int_{t_0}^{t_1}L(t,\ x_1,\ \ldots,\ x_n,\ \dot{x}_1,\ \ldots,\ \dot{x}_n)\,dt+$$
$$+l(x_1(t_0),\ \ldots,\ x_n(t_0),\ x_1(t_1),\ \ldots,\ x_n(t_1))\to\text{extr}, \tag{1'}$$

where

$$L:\ \mathbf{R}\times\mathbf{R}^n\times\mathbf{R}^n\to\mathbf{R},\qquad l:\ \mathbf{R}^n\times\mathbf{R}^n\to\mathbf{R}.$$

In this case the necessary conditions for an extremum are also of the form (2), (3), but the corresponding equation should be understood in the vector sense:

$$-\frac{d}{dt}L_{\dot{x}_j}(t_1,\hat{x}_1(t),\ \ldots,\ \hat{x}_n(t),\ \dot{\hat{x}}_1(t),\ \ldots,\ \dot{\hat{x}}_n(t))+$$
$$+L_{x_j}(t,\ \hat{x}_1(t),\ \ldots,\ \hat{x}_n(t),\ \dot{\hat{x}}_1(t),\ \ldots,\ \dot{\hat{x}}_n(t))=0, \tag{2'}$$

$$L_{\dot{x}_j}(t_i,\ \hat{x}_1(t_i),\ \ldots,\ \hat{x}_n(t_i),\ \dot{\hat{x}}_1(t_i),\ \ldots,\ \dot{\hat{x}}_n(t_i))=$$
$$=(-1)^i\frac{\partial l(\hat{x}_1(t_0),\ \ldots,\ \hat{x}_n(t_0),\ \hat{x}_1(t_1),\ \ldots,\ \hat{x}_n(t_1))}{\partial x_{ij}}, \tag{3'}$$
$$i=0,1;\quad j=1,\ \ldots,\ n.$$

1.4.3. The Extension of the Simplest Problem. In Section 1.4.1 we considered the simplest problem of the classical calculus of variations in the space $C^1([t_0,\ t_1])$, which is traditional. The space $C^1([t_0,\ t_1])$ is of course convenient, but it is by far not always natural for a given problem. In particular, under the conditions imposed on the integrand in that space [the assumption that L, L_x, and $L_{\dot{x}}$ are continuous jointly with respect to the arguments (t, x, $\dot{x}$)] the existence of the solution in $C^1([t_0,\ t_1])$ cannot always be guaranteed. Below we give one of the simplest examples.

Hilbert's Example. Let us consider the problem

$$\mathscr{I}(x(\cdot))=\int_0^1 t^{2/3}\dot{x}^2\,dt\to\inf,\quad x(0)=0,\quad x(1)=1.$$

Here the Euler equation possesses the momentum integral (see Section 1.4.1) $t^{2/3}\dot{x} = C$, whence it follows that the only admissible extremal is $\hat{x}(t) = t^{1/3}$. This extremal does not belong to the space $C^1([t_0, t_1])$ (why?). At the same time, it yields an absolute minimum in the problem. Indeed, let $x(\cdot)$ be an arbitrary absolutely continuous function* for which the integral $\mathcal{J}(x(\cdot))$ is finite and $x(0) = x(1) = 0$. Then

$$\mathcal{J}(\hat{x}(\cdot)+x(\cdot))=\int_0^1 t^{2/3}(\dot{\hat{x}}^2(t)+2\dot{\hat{x}}(t)\dot{x}(t)+\dot{x}^2(t))\,dt=$$

$$=\mathcal{J}(\hat{x}(\cdot))+\frac{2}{3}\int_0^1 \dot{x}(t)\,dt+\mathcal{J}(x(\cdot))=\mathcal{J}(\hat{x}(\cdot))+\mathcal{J}(x(\cdot))\geqslant\mathcal{J}(\hat{x}(\cdot)).$$

Hence, the solution of the problem exists but does not belong to $C^1([t_0, t_1])$.

Among the famous "Hilbert problems" there are some devoted to the calculus of variations. In particular, the twentieth problem concerns the existence of the solution. Hilbert† writes: "I am sure that it will be possible to prove the existence with the aid of some general basic principle indicated by the Dirichlet principle and which will probably enable us to come nearer to the question whether every regular variational problem admits of a solution... if, when necessary, the very notion of a solution is given an extended interpretation." Hilbert's optimism is confirmed by a great deal of experience in the classical mathematical analysis. Here we shall demonstrate the problem of an "extended interpretation" of a solution by the example of the simplest variational problem.

One of the constructions of an extension consists in adding new elements to the original class of underlying elements and extending the functional to the new elements. Here we shall make a small step in this direction: namely, we shall extend the simplest problem to the class of piecewise smooth functions.

We remind the reader that by a piecewise smooth function $x(\cdot)$ on $[t_0, t_1]$ is meant a function possessing the property that the function itself is continuous while its derivative is piecewise continuous, i.e., continuous everywhere on $[t_0, t_1]$ except at a finite number of points $t_0 < \tau_1 < \ldots < \tau_m < t_1$, and, moreover, the derivative $\dot{x}(\cdot)$ has discontinuities of the first kind at the points τ_i. For example, the function

$$x(t)=\begin{cases} 0, & -1\leqslant t\leqslant 0,\\ t^2\sin\frac{1}{t}, & 0<t,\end{cases}$$

possesses the derivative which is not piecewise continuous. The collection of all piecewise smooth functions on $[t_0, t_1]$ will be denoted by $KC^1([t_0, t_1])$.

The integral functional

$$\mathcal{J}(x(\cdot))=\int_{t_0}^{t_1} L(t, x(t), \dot{x}(t))\,dt \tag{1}$$

(cf. Section 1.4.1) can be extended in a natural way to the elements $x(\cdot)\in KC^1([t_0, t_1])$ since the integrand is piecewise continuous for these $x(\cdot)$ and the integral exists.

*For absolutely continuous functions, see Section 2.1.8. If the reader is not familiar with this notion, it is allowable to assume that $x(\cdot)$ is continuously differentiable in the half-open interval $(0, 1]$, is integrable in the improper sense on the closed interval $[0, 1]$, possesses the finite integral $\mathcal{J}(x(\cdot))$, and $x(0) = x(1) = 0$.

†D. Hilbert, Gesammelte Abhandlungen, Vol. 3, Springer-Verlag, Berlin (1935).

Now let us consider the simplest problem

$$\mathcal{J}(x(\cdot)) \to \text{extr}, \quad x(t_0) = x_0, \quad x(t_1) = x_1 \tag{2}$$

in the space $KC^1([t_0, t_1])$. At the end of this section we shall give an example of a function L satisfying the standard conditions for which, however, a problem of the form (2) possesses no solution in the space $C^1([t_0, t_1])$ while its solution in $KC^1([t_0, t_1])$ exists, which shows that this extension of the problem is reasonable (although in Hilbert's example the minimum is not attained in KC^1 either).

Here the notion of a local solution needs a somewhat more precise definition. If, as before, we include in a neighborhood of an element $\hat{x}(\cdot) \in KC^1([t_0, t_1])$ those $x(\cdot)$ for which both the differences $x(t) - \hat{x}(t)$ themselves and their derivatives are small (cf. the definition of a weak extremum in Section 1.4.1), then the newly added elements will lie far from the old ones. Indeed, if the derivative $\dot{\hat{x}}(\cdot)$ has a jump of magnitude δ at some point then none of the functions $x(\cdot) \in C^1([t_0, t_1])$ can satisfy the inequality $|\dot{x}(t) - \dot{\hat{x}}(t)| < \delta/2$ for all t [for which $\dot{\hat{x}}(t)$ exists]. Generally, this is inconvenient because it is usually desirable that the old space be everywhere dense in the extended space. Therefore, the closeness in $KC^1([t_0, t_1])$ will be defined by comparing only the functions themselves but not their derivatives. This leads to the replacement of the notion of a weak extremum (Section 1.4.1) by the notion of a strong extremum.

Definition. A function $\hat{x}(\cdot) \in KC^1([t_0, t_1])$ is said to yield a strong minimum (maximum) in the problem (2) if there exists $\varepsilon > 0$ such that for any $x(\cdot) \in KC^1([t_0, t_1])$, satisfying the conditions $x(t_0) = x_0$, $x(t_1) = x_1$, and

$$\|x(\cdot) - \hat{x}(\cdot)\|_0 = \sup_{t \in [t_0, t_1]} |x(t) - \hat{x}(t)| < \varepsilon, \tag{3}$$

the inequality $\mathcal{J}(x(\cdot)) \geqslant \mathcal{J}(\hat{x}(\cdot))$ ($\mathcal{J}(x(\cdot)) \leqslant \mathcal{J}(\hat{x}(\cdot))$) holds.

The Lemma on "Rounding the Corners." 1) If a function $L = L(t, x, \dot{x})$ is continuous jointly with respect to its arguments, then

$$\inf_{\substack{x(\cdot) \in KC^1([t_0, t_1]) \\ x(t_0) = x_0,\ x(t_1) = x_1}} \mathcal{J}(x(\cdot)) = \inf_{\substack{x(\cdot) \in C^1([t_0, t_1]) \\ x(t_0) = x_0,\ x(t_1) = x_1}} \mathcal{J}(x(\cdot)). \tag{4}$$

2) Equality (4) is retained when only those $x(\cdot) \in KC^1([t_0, t_1])$ are taken which satisfy inequalities (3) for given $\hat{x}(\cdot)$ and $\varepsilon > 0$.

3) The assertion also remains true when inf is replaced by sup.

Proof. Since $KC^1([t_0, t_1]) \supset C^1([t_0, t_1])$, the left-hand member of (4) does not exceed the right-hand member, and it only remains to prove the opposite inequality. Let us assume the contrary. If $\inf_{KC^1} \mathcal{J}(x(\cdot)) < \inf_{C^1} \mathcal{J}(x(\cdot))$, then there is a piecewise-smooth function $\tilde{x}(\cdot)$ and $\eta > 0$ such that $\mathcal{J}(\tilde{x}(\cdot)) < \inf_{C_1} \mathcal{J}(x(\cdot)) - \eta$. Let τ_i, $i = 1, 2, \ldots, m$, be the points of discontinuity of the derivative $\dot{\tilde{x}}$, and let $\Delta_i = \dot{\tilde{x}}(\tau_i + 0) - \dot{\tilde{x}}(\tau_i - 0)$ be the jumps of the derivative at these points.

The continuous function L is bounded on the closure of the bounded set

$$\mathcal{K} = \{(t, x, \dot{x}) \mid t_0 \leqslant t \leqslant t_1,\ |x - \tilde{x}(t)| \leqslant \max|\Delta_i|\delta_0/4,\ |\dot{x} - \dot{\tilde{x}}(t)| \leqslant \max|\Delta_i|/2\}$$

i.e., $|L(t, x, \dot{x})| \leqslant M$.

The function

$$a(t) = \begin{cases} \dfrac{(1-|t|)^2}{4}, & |t| \leqslant 1, \\ 0, & |t| > 1 \end{cases} \tag{5}$$

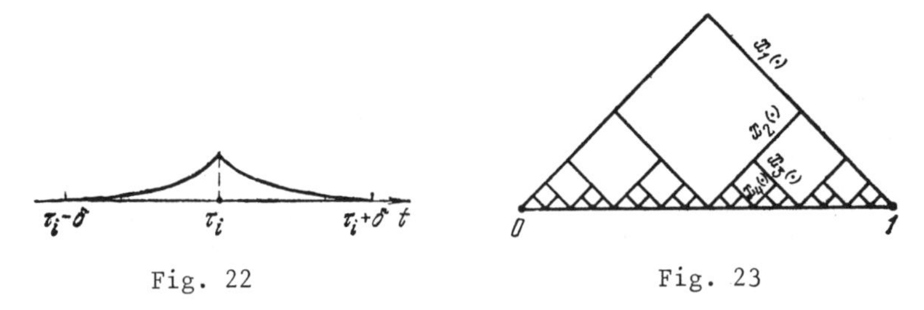

Fig. 22 Fig. 23

is continuous, and its derivative has a jump of magnitude -1 at $t = 0$. Therefore, the function $\delta a\left(\frac{t-\tau_i}{\delta}\right)$, whose graph is obtained from the graph of $a(\cdot)$ by a similitude transformation and a shift (Fig. 22), is also continuous everywhere except at the point a_i where, as before, it has a jump of magnitude -1.

Now it can easily be verified that the function

$$x_\delta(t) = \tilde{x}(t) + \sum_{i=1}^{m} \Delta_i \delta a\left(\frac{t-\tau_i}{\delta}\right)$$

is continuous on $[t_0, t_1]$ together with its derivative, and, moreover, $x_\delta(t) \equiv \tilde{x}(t)$ outside the closed intervals $[\tau_i - \delta, \tau_i + \delta]$. In particular, these intervals do not overlap for sufficiently small δ, $x_\delta(t_i) = \tilde{x}(t_i) = x_i$, $i = 0, 1$, $|x_\delta(t) - \tilde{x}(t)| \leqslant \max|\Delta_i|\delta/4$, and $|\dot{x}_\delta(t) - \dot{\tilde{x}}(t)| \leqslant \max|\Delta_i|/2$, because, according to (5), we have $|a(t)| \leqslant 1/4$ and $|\dot{a}(t)| \leqslant 1/2$. Therefore, for $0 < \delta < \delta_0$ we have

$$\mathcal{J}(x_\delta(\cdot)) - \mathcal{J}(\tilde{x}(\cdot)) = \int_{t_0}^{t_1} L(t, x_\delta(t), \dot{x}_\delta(t))\,dt - \int_{t_0}^{t_1} L(t, \tilde{x}(t), \dot{\tilde{x}}(t))\,dt =$$

$$= \sum_{i=1}^{m} \int_{\tau_i-\delta}^{\tau_i+\delta} (L(t, x_\delta(t), \dot{x}_\delta(t)) - L(t, \tilde{x}(t), \dot{\tilde{x}}(t)))\,dt,$$

and consequently

$$\mathcal{J}(x_\delta(\cdot)) \leqslant \mathcal{J}(\tilde{x}(\cdot)) + 4m\delta M < \inf_{C^1} \mathcal{J}(x(\cdot)) - \eta + 4m\delta M < \inf_{C^1} \mathcal{J}(x(\cdot))$$

if δ is sufficiently small. Since $x_\delta(\cdot) \in C^1([t_0, t_1])$, we have arrived at a contradiction. For small δ the function $x_\delta(\cdot)$ satisfies the inequalities (3) provided that $\tilde{x}(\cdot)$ satisfies them. This implies the second assertion of the lemma. The third assertion is obvious. ■

COROLLARY. If a function $\hat{x}(\cdot) \in C^1([t_0, t_1])$ yields an absolute or a strong minimum (maximum) in problem (2) in the space $C^1([t_0, t_1])$, then it possesses the same property in the space $KC^1([t_0, t_1])$ as well.

The extension we have constructed is not in fact always sufficient. Besides, the space $KC^1([t_0, t_1])$ is not complete [with respect to the metric determined by the left-hand side of (3)]. It would be more natural to extend the problem to the class $W^1_\infty([t_0, t_1])$ of functions satisfying the Lipschitz condition.* But this would require some facts of the theory of functions. There is another important construction of an extension of the simplest problem which will be demonstrated by the example below.

*It should be noted, however, that in this extension we would not obtain the existence in Hilbert's problem because the function $t \to t^{1/3}$ does not satisfy the Lipschitz condition.

<u>The Bolza Example.</u> Let us consider the problem

$$\mathcal{J}(x(\cdot)) = \int_0^1 ((1-\dot{x}^2)^2 + x^2)\,dt \to \inf, \quad x(0) = 0, \quad x(1) = \xi.$$

Let us suppose that $\xi = 0$. Then it is clear that there exists no solution of the problem. Indeed, here the infimum of the functional is equal to zero. To show this it sufficies to consider the following minimizing sequence of functions belonging to $KC^1([0, 1])$:

$$x_n(t) = \int_0^t \operatorname{sgn} \sin 2\pi 2^n \tau\, d\tau, \quad n = 1, 2, \ldots \text{ (Fig. 23)}.$$

The sequence of the functions $x_n(\cdot)$ is uniformly convergent to zero, and $\dot{x}_n^2(t) \equiv 1$ except at a finite number of points, i.e., $\mathcal{J}(x_n(\cdot)) = \int_0^1 x_n^2(t)\,dt \to 0$. On the other hand, if $x_0(\cdot) \equiv 0$, then $\mathcal{J}(x_0(\cdot)) = 1$, and if $x(\cdot) \not\equiv 0$, then $\mathcal{J}(x(\cdot)) \geqslant \int_0^1 x^2 dt > 0$.

The matter is that the functional $\mathcal{J}$ is not lower semicontinuous: as was established, in any arbitrarily small neighborhood of the function $x_0(\cdot) \equiv 0$ on which the functional assumes the value equal to unity there are functions for which the value of the functional is much smaller [since $\mathcal{J}(x_n(\cdot) \to 0$]. It turns out that it is possible to construct a "lower semicontinuous" extension of the problem in which the functional $\mathcal{J}$ is replaced by the functional

$$\mathcal{Y}(x(\cdot)) = \varliminf_{y(\cdot) \to x(\cdot)} \mathcal{J}(y(\cdot)),$$

while the original set of functions (say $C^1([t_0, t_1])$ or $KC^1([t_0, t_1])$) remains the same. The passage to the limit for $y(\cdot) \to x(\cdot)$ is understood here in the sense of the space $C([t_0, t_1])$ (i.e., in the sense of the uniform convergence). It turns out that the functional $\mathcal{Y}$ admits of a very simple description; namely, in the Bolza example we have

$$\mathcal{Y}(x(\cdot)) = \int_0^1 (((\dot{x}^2 - 1)_+)^2 + x^2)\,dt,$$

where

$$(\dot{x}^2 - 1)_+ = \begin{cases} 0, & |\dot{x}| \leqslant 1, \\ \dot{x}^2 - 1, & |\dot{x}| > 1. \end{cases}$$

In the general case when

$$\mathcal{J}(x(\cdot)) = \int_{t_0}^{t_1} L(t, x, \dot{x})\,dt,$$

we have

$$\mathcal{Y}(x(\cdot)) = \int_{t_0}^{t_1} \tilde{L}(t, x, \dot{x})\,dt,$$

where $\tilde{L}$ is the "convexification" of L with respect to $\dot{x}$, i.e., the function $\dot{x} \to \tilde{L}(t, x, \dot{x})$ is the greatest convex function not exceeding the function $\dot{x} \to L(t, x, \dot{x})$. This assertion is known as *Bogolyubov's theorem.*

The same Bolza example can be used to construct a problem of the type (2) in which the minimum is attained on a curve with a corner. However, for purely technical reasons, we shall consider a somewhat different problem:

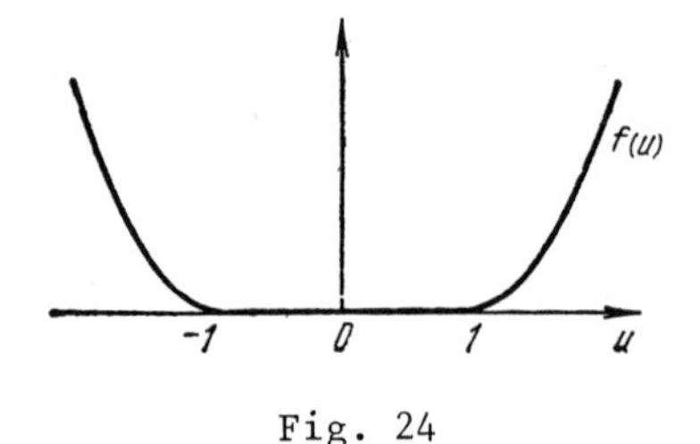

Fig. 24

$$\mathcal{J}(x(\cdot))=\int_{t_0}^{t_1}(f(\dot{x})+x^2)\,dt\to\inf,\quad x(t_0)=x_0,\quad x(t_1)=x_1, \tag{6}$$

where

$$f(u)=((|u|-1)_+)^2=\begin{cases}(u-1)^2, & u\geqslant 1,\\ 0, & |u|\leqslant 1,\\ (u+1)^2, & u\leqslant -1.\end{cases}$$

The function f(u) is continuously differentiable and convex (Fig. 24). In particular, the graph of f(u) does not lie lower than any of its tangents; i.e., we always have

$$f(v)\geqslant f(u)+f'(u)(v-u).$$

Therefore, for any two functions $x(\cdot), \hat{x}(\cdot)\in KC^1([t_0,t_1])$ satisfying the given boundary conditions $x(t_i) = \hat{x}(t_i) = x_i$, i = 1, 2, we have

$$\mathcal{J}(x(\cdot))-\mathcal{J}(\hat{x}(\cdot))=\int_{t_0}^{t_1}(f(\dot{x}(t))-f(\dot{\hat{x}}(t))+x^2(t)-\hat{x}^2(t))\,dt\geqslant$$

$$\geqslant\int_{t_0}^{t_1}(f'(\dot{\hat{x}}(t))(\dot{x}(t)-\dot{\hat{x}}(t))+2\hat{x}(t)(x(t)-\hat{x}(t))+(x(t)-\hat{x}(t))^2)\,dt\geqslant$$

$$\geqslant\int_{t_0}^{t_1}(f'(\dot{\hat{x}}(t))(\dot{x}(t)-\dot{\hat{x}}(t))+2\hat{x}(t)(x(t)-\hat{x}(t)))\,dt.$$

Now let us suppose that $\hat{x}(\cdot)$ satisfies the Euler equation

$$\frac{d}{dt}f'(\dot{\hat{x}}(t))=2\hat{x}(t) \tag{7}$$

at all points of its differentiability. Integrating by parts on each of the intervals of continuity of $\dot{\hat{x}}(\cdot)$, we obtain

$$\mathcal{J}(x(\cdot))-\mathcal{J}(\hat{x}(\cdot))\geqslant\int_{t_0}^{t_1}\left(-\frac{d}{dt}f'(\dot{\hat{x}}(t))+2\hat{x}(t)\right)(x(t)-\hat{x}(t))\,dt+$$

$$+\sum_i[f'(\dot{\hat{x}}(\tau_i+0)-f'(\dot{\hat{x}}(\tau_i-0))(x(\tau_i)-\hat{x}(\tau_i))=0$$

provided that, in addition to (7), the function $p(t) = f'(\dot{\hat{x}}(t))$ (the momentum) is continuous. Hence the function $\hat{x}(\cdot)$ [satisfying the Euler equation, the condition of continuity of $p(\cdot)$, and the boundary conditions] is a solution of the problem (6). For instance, the function $\hat{x}(t) = e^{|t|}$ having a corner at the point t = 0 is a solution of this problem for $t_0 = 1$, $t_1 = 1$, $x_0 = x_1 = e$. Indeed,

$$\ddot{\hat{x}}(t)=\begin{cases}e^t, & t\geqslant 0,\\ e^{-t}, & t<0,\end{cases}$$

whence $|\dot{\hat{x}}(t)| \geqslant 1$, and consequently the function

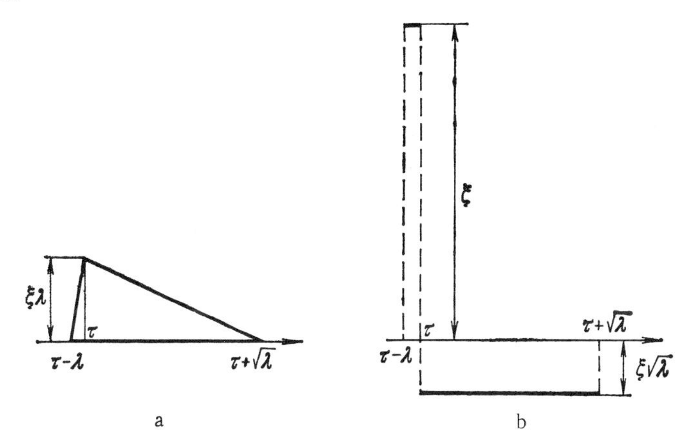

a b

Fig. 25

$$p(t)=f'(\dot{\hat{x}}(t))=\begin{cases} 2(e^t-1), & t\geqslant 0,\\ 2(-e^{-t}+1), & t<0,\end{cases}$$

has no jump at the point $t = 0$: $p(+0) = p(-0) = 0$. Moreover, $\frac{d}{dt}p(t)=2e^{|t|}=2\hat{x}(t)$, and hence the Euler equation is satisfied.

1.4.4. Needlelike Variations. Weierstrass' Condition. The notion of a strong extremum was introduced into the calculus of variations by Weierstrass. To prove a necessary condition for a strong minimum Weierstrass used special variations of the following form:

$$h_\lambda(t)=h_\lambda(t;\tau,\xi)=\begin{cases} \xi\lambda+(t-\tau)\xi, & t\in[\tau-\lambda,\tau],\\ \xi\lambda-(t-\tau)\xi\sqrt{\lambda}, & t\in[\tau,\tau+\sqrt{\lambda}],\end{cases}$$

$$x_\lambda(t)=\hat{x}(t)+h_\lambda(t) \tag{1}$$

(see Fig. 25a).

The derivative of the variation $h_\lambda(\cdot)$ has the form shown in Fig. 25b. The graph of the derivative resembles a needle to some extent, which accounts for the name "needlelike." The needlelike variations are convenient for investigating problems on a strong extremum.

Now we proceed to the derivation of Weierstrass' necessary condition. Let us consider the simplest problem of the classical calculus of variations:

$$\mathcal{J}(x(\cdot))=\int_{t_0}^{t_1} L(t,x,\dot{x})\,dt\to\inf,\quad x(t_0)=x_0,\quad x(t_1)=x_1, \tag{2}$$

on the class $KC^1([t_0, t_1])$ of piecewise-smooth functions. Let $\hat{x}(\cdot)$ be an extremal suspected to yield a strong minimum; for definiteness, we shall suppose that $\hat{x}(\cdot)$ is smooth.

Following the general idea of the method of variations, we shall consider the function

$$\varphi(\lambda)=\mathcal{J}(x_\lambda(\cdot))=\mathcal{J}(\hat{x}(\cdot)+h_\lambda(\cdot)), \tag{3}$$

where h_λ is determined by the formulas (1), τ is an interior point of $[t_0, t_1]$, and ξ is an arbitrary number.

For sufficiently small $\lambda \geqslant 0$ the function $x_\lambda(\cdot)$ is admissible in problem (2), i.e., $x_\lambda(t_i) = \hat{x}(t_i) + h_\lambda(t_i) = x_i$, $i = 0, 1$.

The function $\varphi(\cdot)$ is defined for nonnegative λ. Let us prove that it is differentiable on the right at the point zero. From definition (1) it is readily seen that

$$\begin{aligned} &\text{a) } \|h_\lambda(\cdot)\|_0 = \max_{t\in[t_0,t_1]} |h_\lambda(t)| = O(\lambda), \\ &\text{b) } |\dot{h}_\lambda(t)| = \xi\sqrt{\lambda} = O(\sqrt{\lambda}), \quad t\in(\tau, \tau+\sqrt{\lambda}). \end{aligned} \tag{4}$$

It follows that

$$\varphi(\lambda)-\varphi(0) = \int_{\tau-\lambda}^{\tau} (L(t, x_\lambda(t), \dot{\hat{x}}(t)+\xi) - L(t, \hat{x}(t), \dot{\hat{x}}(t)))\,dt + \\ + \int_{\tau}^{\tau+\sqrt{\lambda}} (L(t, x_\lambda(t), \dot{x}_\lambda(t)) - L(t, \hat{x}(t), \dot{\hat{x}}(t)))\,dt = \mathscr{Y}_1 + \mathscr{Y}_2. \tag{5}$$

The integral $\mathscr{Y}_1$ in (5) can be estimated thus:

$$\mathscr{Y}_1 = \lambda(L(\tau, \hat{x}(\tau), \dot{\hat{x}}(\tau)+\xi) - L(\tau, \hat{x}(\tau), \dot{\hat{x}}(\tau))) + o(\lambda) \tag{6}$$

[to this end one should make use of the mean value theorem of differential calculus (see Section 2.2.3) and the estimate (4a)].

To estimate the second integral $\mathscr{Y}_2$ we represent the difference

$$\Delta = L(t, \hat{x}(t)+h_\lambda(t), \dot{\hat{x}}(t)+\dot{h}_\lambda(t)) - L(t, \hat{x}(t), \dot{\hat{x}}(t))$$

in the form

$$\Delta = L_x(t, \hat{x}(t), \dot{\hat{x}}(t))\,h_\lambda(t) + L_{\dot{x}}(t, \hat{x}(t), \dot{\hat{x}}(t))\,\dot{h}_\lambda(t) + o(\sqrt{\lambda})$$

[here again one must apply the mean value theorem and the estimate (4b)], perform integration by parts in the second term and use the fact that

$$-\frac{d}{dt}L_{\dot{x}} + L_x \Big|_{\hat{x}(t)} = 0$$

[because $\hat{x}(\cdot)$ is an extremal]. As a result, we obtain

$$\mathscr{Y}_2 = -\xi\lambda L_{\dot{x}}(\tau, \hat{x}(\tau), \dot{\hat{x}}(\tau)) + \int_{\tau}^{\tau+\sqrt{\lambda}} o(\sqrt{\lambda})\,dt = -\xi\lambda L_{\dot{x}}(\tau, \hat{x}(\tau), \dot{\hat{x}}(\tau)) + o(\lambda). \tag{7}$$

Comparing (6) and (7), we find

$$\varphi(\lambda)-\varphi(0) = \lambda(L(\tau, \hat{x}(\tau), \dot{\hat{x}}(\tau)+\xi) - L(\tau, \hat{x}(\tau), \dot{\hat{x}}(\tau)) - \xi L_{\dot{x}}(\tau, \hat{x}(\tau), \dot{\hat{x}}(\tau))) + o(\lambda). \tag{8}$$

Hence, the function $\varphi(\cdot)$ possesses the right-hand derivative at the point $\lambda = 0$:

$$\varphi'(+0) = \lim_{\lambda\downarrow 0} \frac{\varphi(\lambda)-\varphi(0)}{\lambda} = L(\tau, \hat{x}(\tau), \dot{\hat{x}}(\tau)+\xi) - L(\tau, \hat{x}(\tau), \dot{\hat{x}}(\tau)) - \xi L_{\dot{x}}(\tau, \hat{x}(\tau), \dot{\hat{x}}(\tau)).$$

However, if $\hat{x}(\cdot)$ yields a strong minimum, then $\mathscr{I}(x_\lambda(\cdot)) \geqslant \mathscr{I}(\hat{x}(\cdot))$, and consequently

$$\varphi'(+0) = \lim_{\lambda\downarrow 0} (\varphi(\lambda)-\varphi(0))/\lambda \geqslant 0,$$

i.e., the relation

$$\mathscr{E}(\tau, \hat{x}(\tau), \dot{\hat{x}}(\tau), \dot{\hat{x}}(\tau)+\xi) = \\ = L(\tau, \hat{x}(\tau), \dot{\hat{x}}(\tau)+\xi) - L(\tau, \hat{x}(\tau), \dot{\hat{x}}(\tau)) - \xi L_{\dot{x}}(\tau, \hat{x}(\tau), \dot{\hat{x}}(\tau)) \geqslant 0 \tag{9}$$

holds for any $\xi\in\mathbb{R}$. The function

$$\mathscr{E}(t, x, y, z) = L(t, x, y, z) - L(t, x, y) - (z-y)L_y(t, x, y)$$

is called the *Weierstrass function*. We have thus proved the following:

THEOREM (Weierstrass' Necessary Condition for a Strong Minimum). For an extremal $\hat{x}(\cdot)\in C^1([t_0, t_1])$ of the simplest problem of the classical calculus of variations (2) to yield a strong minimum it is necessary that the equality

$$\mathscr{E}(\tau, \hat{x}(\tau), \dot{\hat{x}}(\tau), \dot{\hat{x}}(\tau)+\xi)=L(\tau, \hat{x}(\tau), \dot{\hat{x}}(\tau)+\xi)-L(\tau, \hat{x}(\tau), \dot{\hat{x}}(\tau))-\xi L_{\dot{x}}(\tau, \hat{x}(\tau), \dot{\hat{x}}(\tau))\geqslant 0$$

hold for any $\tau\in[t_0, t_1]$ and any $\xi\in\mathbf{R}$.

1.4.5. Isoperimetric Problem and the Problem with Higher Derivatives. By the *isoperimetric problem* (with fixed end points) in the calculus of variations is usually meant the following problem:

$$\mathscr{I}_0(x(\cdot))=\int_{t_0}^{t_1} f_0(t, x, \dot{x})dt\rightarrow \text{extr},$$

$$\mathscr{I}_i(x(\cdot))=\int_{t_0}^{t_1} f_i(t, x, \dot{x})\,dt=\alpha_i,\ i=1, \ldots, m, \tag{1}$$

$$x(t_0)=x_0,\ x(t_1)=x_1.$$

It is supposed that the functions f_i satisfy the same conditions as those for L in Section 1.4.1: the functions themselves and their partial derivatives with respect to x and $\dot{x}$ are continuous jointly with respect to their arguments. For the sake of simplicity, we shall confine ourselves to the case m = 1.

THEOREM (a Necessary Condition for an Extremum in the Isoperimetric Problem). Let a function $\hat{x}(\cdot)$ yield a local extremum (relative to the space $C^1([t_0, t_1])$) in the problem

$$\mathscr{I}_0(x(\cdot))=\int_{t_0}^{t_1} f_0(t, x, \dot{x})\,dt\rightarrow \text{extr},$$

$$\mathscr{I}_1(x(\cdot))=\int_{t_0}^{t_1} f_1(t, x, \dot{x})\,dt=\alpha_1,\quad x(t_i)=x_i,\ i=0, 1, \tag{1'}$$

and, moreover, let the functions $t\to f_{0\dot{x}}(t, \hat{x}(t), \dot{\hat{x}}(t))$ and $t\to f_{1\dot{x}}(t, \hat{x}(t), \dot{\hat{x}}(t))$ belong to $C^1([t_0, t_1])$. Then there are numbers, $\hat{\lambda}_0$ and $\hat{\lambda}_1$, not vanishing simultaneously, such that for the integrand $L = \hat{\lambda}_0 f_0 + \hat{\lambda}_1 f_1$ the Euler equation holds:

$$-\frac{d}{dt}L_{\dot{x}}(t, \hat{x}(t), \dot{\hat{x}}(t))+L_x(t, \hat{x}(t), \dot{\hat{x}}(t))=0. \tag{2}$$

Proof. A) By analogy with what was done at the beginning of the proof of the theorem of Section 1.4.1, we calculate the first Lagrange variations of the functionals $\mathscr{I}_0$ and $\mathscr{I}_i$:

$$\delta\mathscr{I}_i(\hat{x}(\cdot), x(\cdot))=\int_{t_0}^{t_1}(p_i(t)\dot{x}(t)+q_i(t)x(t))\,dt,\ i=0,1, \tag{3}$$

where

$$p_i(t)=f_{i\dot{x}}(t, \hat{x}(t), \dot{\hat{x}}(t)),\ q_i(t)=f_{ix}(t, \hat{x}(t), \dot{\hat{x}}(t)).$$

Here there are two possible cases: either $\delta\mathscr{I}_1\equiv 0$, $\forall x(\cdot)\in C^1([t_0, t_1])$, $x(t_0) = x(t_1) = 0$, or $\delta\mathscr{I}_1\not\equiv 0$. In the first case, by the theorem of Section 1.4.1, we have

$$-\frac{d}{dt}f_{1\dot{x}}(t, \hat{x}(t), \dot{\hat{x}}(t))+f_{1x}(t, \hat{x}(t), \dot{\hat{x}}(t))=0.$$

and putting $\lambda_0 = 0$ and $\lambda_1 = 1$ we immediately obtain (2).

B) Let $\delta\mathcal{J}_1 \neq 0$; then there exists a function $y(\cdot) \in C^1([t_0, t_1])$, $y(t_0) = y(t_1) = 0$, for which $\delta\mathcal{J}_1(\hat{x}(\cdot), y(\cdot)) \neq 0$. Let us consider the functions of two variables

$$\varphi_i(\alpha, \beta) = \mathcal{J}_i(\hat{x}(\cdot) + \alpha x(\cdot) + \beta y(\cdot)), \quad i = 0, 1.$$

We leave it to the reader to verify that they are continuously differentiable in a neighborhood of zero and that

$$\frac{\partial \varphi_i(0,0)}{\partial \alpha} = \delta\mathcal{J}_i(\hat{x}(\cdot), x(\cdot)), \quad \frac{\partial \varphi_i(0,0)}{\partial \beta} = \delta\mathcal{J}_i(\hat{x}(\cdot), y(\cdot)).$$

LEMMA. For any function $x(\cdot) \in C^1([t_0, t_1])$ such that $x(t_0) = x(t_1) = 0$ there holds the relation

$$\frac{\partial(\varphi_0, \varphi_1)}{\partial(\alpha, \beta)}(0, 0) = \begin{vmatrix} \dfrac{\partial \varphi_0}{\partial \alpha}, & \dfrac{\partial \varphi_0}{\partial \beta} \\ \dfrac{\partial \varphi_1}{\partial \beta} & \dfrac{\partial \varphi_1}{\partial \beta} \end{vmatrix}_{(0,\,0)} = \begin{vmatrix} \delta\mathcal{J}_0(\hat{x}(\cdot), x(\cdot)), & \delta\mathcal{J}_0(\hat{x}(\cdot), y(\cdot)) \\ \delta\mathcal{J}_1(\hat{x}(\cdot), x(\cdot)), & \delta\mathcal{J}_1(\hat{x}(\cdot), y(\cdot)) \end{vmatrix} = 0. \tag{4}$$

Proof. If determinant (4) is nonzero, then there is a neighborhood of the point (0, 0) which goes into a neighborhood of the point $\varphi_1(0, 0))$, $(\varphi_0(0, 0)$, under the mapping $(\alpha, \beta) \mapsto (\varphi_0(\alpha, \beta), \varphi_1(\alpha, \beta))$ (the corresponding theorem on the inverse mapping was already used in Section 1.3.2). In particular, there are α and β, and, consequently, an admissible function $\hat{x}(\cdot) + \alpha x(\cdot) + \beta y(\cdot)$, such that $\varphi_0(\alpha, \beta) = \mathcal{J}_0(\hat{x}(\cdot) + \alpha x(\cdot) + \beta y(\cdot)) = \mathcal{J}_0(\hat{x}(\cdot)) - \varepsilon$, where $\varepsilon > 0$ and $\varphi_1(\alpha, \beta) = \varphi_1(0, 0) = \mathcal{J}_1(\hat{x}(\cdot)) = \alpha_1$. This contradicts the fact that $\hat{x}(\cdot)$ yields a local minimum in the problem (1). ■

C) From equality (4), we obtain

$$\delta\mathcal{J}_0(\hat{x}(\cdot), x(\cdot)) - \frac{\delta\mathcal{J}_0(\hat{x}(\cdot), y(\cdot))}{\delta\mathcal{J}_1(\hat{x}(\cdot), y(\cdot))}\delta\mathcal{J}_1(\hat{x}(\cdot), x(\cdot)) = 0$$

(by assumption, the denominator is nonzero here). Putting $\hat{\lambda}_0 = 1$, $\hat{\lambda}_1 = -\delta\mathcal{J}_0(\hat{x}(\cdot), y(\cdot)) / \delta\mathcal{J}_1(\hat{x}(\cdot), y(\cdot))$, we see that

$$\hat{\lambda}_0 \delta\mathcal{J}_0(\hat{x}(\cdot), x(\cdot)) + \hat{\lambda}_1 \delta\mathcal{J}_1(\hat{x}(\cdot), x(\cdot)) \equiv 0 \tag{5}$$

for any function $x(\cdot) \in C^1([t_0, t_1])$ satisfying the condition $x(t_0) = x(t_1) = 0$.

It can easily be seen that the left-hand side of (5) has the form

$$\int_{t_0}^{t_1} ((\hat{\lambda}_0 p_0(t) + \hat{\lambda}_1 p_1(t))\dot{x}(t) + (\hat{\lambda}_0 q_0(t) + \hat{\lambda}_1 q_1(t))x(t))\,dt.$$

Applying the fundamental lemma of the classical calculus of variations, we obtain the equation

$$-\frac{d}{dt}(\hat{\lambda}_0 p_0(t) + \hat{\lambda}_1 p_1(t)) + (\hat{\lambda}_0 q_0(t) + \hat{\lambda}_1 q_1(t)) = 0,$$

coinciding with (2). ■

Problem (1') was for the first time considered by Euler in his famous work of 1744 where the method of broken lines was used to derive the relation (2). As a matter of fact, this was the main subject of the work. Undoubtedly, this work already contained the Lagrange multiplier rule in its incipiency.

In conclusion, we shall briefly consider the problem with higher derivatives:

$$\mathcal{J}(x(\cdot)) = \int_{t_0}^{t_1} f(t, x, \dot{x}, \ddot{x}, \ldots, x^{(n)})\,dt \to \text{extr},$$

$$x^{(j)}(t_i)=x_i^j, \quad i=0, 1, j=0, \ldots, n-1. \tag{6}$$

For greater detail we refer the reader to [2] or [3]. We shall investigate this problem in the space $C^n([t_0, t_1])$ of functions continuous together with their derivatives up to the order n inclusive on the closed interval $[t_0, t_1]$. We shall assume that the function f and its derivatives with respect to x, $\dot{x},\ldots,x^{(n)}$ are continuous jointly with respect to their arguments. Let $\hat{x}(\cdot)$ be a function suspected to yield an extremum. Let us calculate the first Lagrange variation of the functional:

$$\delta\mathcal{J}(\hat{x}(\cdot), x(\cdot))=\int_{t_0}^{t_1}\left(\sum_{j=0}^{n} p_j(t)\, x^{(j)}(t)\right)dt, \tag{7}$$

where $p_j(t)=\frac{\partial f}{\partial x^{(j)}}(t, \hat{x}(t), \ldots, \hat{x}^{(n)}(t))$.

The conditions of local extremality imply that $\delta\mathcal{J}(\hat{x}(\cdot), x(\cdot))=0$, provided that $x^{(j)}(t_i) = 0$, $i = 0, 1$, $j = 0, 1,\ldots,n - 1$. Integrating by parts in (7) [let the reader state the necessary requirement concerning the smoothness of $p_j(\cdot)$ for the integration by parts to be legitimate], we bring the first variation to the form

$$\delta\mathcal{J}(\hat{x}(\cdot), x(\cdot))=\int_{t_0}^{t_1}\left(\sum_{j=0}^{n}(-1)^j p_j^{(j)}(t)\right)x(t)\,dt.$$

Further, applying the generalization of the fundamental lemma of the classsical calculus of variations to the case of functions belonging to $C^{(n)}([t_0, t_1])$ and satisfying the conditions $x^{(j)}(t_i) = 0$, $i = 0, 1$, $j = 0, 1,\ldots,n - 1$ (let the reader think over this generalization), we conclude that the desired necessary condition for the extremality of $x(\cdot)$ is the following equation:

$$\sum_{j=0}^{n}(-1)^j\left(\frac{d}{dt}\right)^j\frac{\partial f}{\partial x^{(j)}}(t, \hat{x}(t), \dot{\hat{x}}(t), \ldots, \hat{x}^{(n)}(t))=0.$$

It is called the *Euler–Poisson equation*.

1.5. Lagrange Problem and the Basic Optimal Control Problem

1.5.1. Statements of the Problems.

The publication in 1788 of the treatise by Joseph Louis Lagrange "Analytical Mechanics" was an important stage of the development of natural science. In particular, the treatise also played an exceptional role in the development of the calculus of variations. Namely, the following conditional extremum problem was stated there:

$$\mathcal{J}(x(\cdot))=\int_{t_0}^{t_1} f(t, x(t), \dot{x}(t))\,dt\longrightarrow \text{extr}, \tag{1}$$

$$\begin{gathered}\Phi(t, x(t), \dot{x}(t))=0 \Leftrightarrow \Phi_j(t, x(t), \dot{x}(t))=0,\\ j=1, \ldots, m,\\ x(t_0)=x_0,\ x(t_1)=x_1.\end{gathered} \tag{2}$$

Here $x(\cdot)$: $[t_0, t_1] \to \mathbf{R}^n$, f: $\mathbf{R} \times \mathbf{R}^n \times \mathbf{R}^n \to \mathbf{R}$, and Φ: $\mathbf{R} \times \mathbf{R}^n \times \mathbf{R}^n \to \mathbf{R}^m$. Later problem (1) and its various modifications connected with some additional constraints (some other boundary conditions, additional integral relations, etc.; also see Section 1.2.6) were called the *Lagrange problems*. To solve problem (1) Lagrange used the basic method described in Section 1.3.2, i.e., the multiplier rule. However, he did not prove it rigorously, and it took more than 100 years to give Lagrange's reasoning the form of a rigorously proved theorem.

Let us indicate two most important special types of constraints embraced by the general expression (2). In the calculus of variations a constraint of the form $\Phi(t, x) = 0$ in which the function Φ involved in (2) is independent of $\dot{x}$ is called a *phase constraint*. In mechanics the phase constraints are also called *holonomic constraints*.

The other case is the one in which the relation (2) can be resolved with respect to the derivatives $\dot{x}$. Then this constraint is written in the form of equations $\dot{x} = \varphi(t, x, u)$, $x(\cdot)\colon [t_0, t_1] \to \mathbf{R}^n$, $u(\cdot)\colon [t_0, t_1] \to \mathbf{R}^r$, $\varphi\colon \mathbf{R}\times\mathbf{R}^n\times\mathbf{R}^r \to \mathbf{R}^n$. Here the variables $x(\cdot)$ are called *phase coordinates* and the variables $u(\cdot)$ are called *controls*. Attention will be primarily paid to this most important case. More precisely, further we shall consider the problem

$$\mathcal{J}(x(\cdot), u(\cdot)) = \int_{t_0}^{t_1} f(t, x(t), u(t))\,dt + \psi_0(x(t_0), x(t_1)) \to \text{extr}, \tag{1'}$$

$$\dot{x} - \varphi(t, x, u) = 0,\ \psi(x(t_0), x(t_1)) = 0\ (\Leftrightarrow \psi_j(x(t_0), x(t_1)) = 0,\ \ j = 1, \ldots, s). \tag{2'}$$

Formulas (1') and (2') involve functions $f\colon V \to \mathbf{R}$, $\varphi\colon V \to \mathbf{R}^n$, $\psi\colon W \to \mathbf{R}^s$, where V and W are open sets in the spaces $\mathbf{R} \times \mathbf{R}^n \times \mathbf{R}^r$ and $\mathbf{R}^n \times \mathbf{R}^n$, respectively. Here we shall assume the instants of time t_0 and t_1 to be fixed.

The constraint $\dot{x} - \varphi(t, x, u) = 0$ is called a *differential constraint*, and the constraints $\psi(x(t_0), x(t_1)) = 0$ are called *boundary conditions*. Problem (1'), (2') will be called the *Lagrange problem in Pontryagin's form*.

We shall assume that all the functions f, φ, and ψ are continuously differentiable.

Later in Chapter 4 a still more general case will be considered in which t_0 and t_1 may also vary and isoperimetric equality and inequality constraints, etc. are admitted.

Problem (1'), (2') will be considered in the Banach space $Z = C^1([t_0, t_1], \mathbf{R}^n) \times C([t_0, t_1], \mathbf{R}^r)$. In other words, we shall consider the pairs $(x(\cdot), u(\cdot))$ where $x(\cdot)$ is a continuously differentiable n-dimensional vector function and $u(\cdot)$ is a continuous r-dimensional vector function. For a pair $(x(\cdot), u(\cdot))$ we shall sometimes use the abbreviated notation z.

An element $z = (x(\cdot), u(\cdot)) \in Z$ will be called a *controlled process* in the problem (1'), (2') if $\dot{x}(t) = \varphi(t, x(t), u(t))$, $\forall t \in [t_0, t_1]$, and will be called an *admissible controlled process* if, in addition, the boundary conditions are satisfied. An admissible element $\hat{z} = (\hat{x}(\cdot), \hat{u}(\cdot))$ will be called an *optimal process in the weak sense* or will be said to yield a *weak minimum* in the problem (1'), (2') if it yields a local minimum in the problem, i.e., if there is $\varepsilon > 0$ such that $\|x - \hat{x}\|_1 < \varepsilon$ and $\|u - \hat{u}\|_0 < \varepsilon$ imply $\mathcal{J}(z) \geqslant \mathcal{J}(\hat{z})$.

1.5.2. Necessary Conditions in the Lagrange Problem. Let us try to apply the general method of Lagrange mentioned in Section 1.3.2 to the problem (1'), (2') of the foregoing section. By analogy with the finite-dimensional case, the Lagrange function should be written thus:

$$\mathcal{L}(x(\cdot), u(\cdot);\ p(\cdot), \mu, \lambda_0) = \int_{t_0}^{t_1} L\,dt + l, \tag{1}$$

where

$$L(t, x, \dot{x}, u) = p(t)(\dot{x} - \varphi(t, x, u)) + \lambda_0 f(t, x, u), \tag{2}$$

$$l(x_0, x_1) = \sum_{j=0}^{s} \mu_j \psi_j(x_0, x_1),\ \lambda_0 = \mu_0. \tag{3}$$

There is no doubt that the "endpoint part" l of the Lagrange function is of the form (3): here there is a complete similarity between the case under consideration and the finite-dimensional case. As to the constraint $\dot{x} = \varphi(t, x, u)$, it must be fulfilled for all $t \in [t_0, t_1]$, and, by analogy, the corresponding "Lagrange multiplier" $p(\cdot)$ must be a function of t, and its contribution to the Lagrange function has the form of an integral instead of a sum.

Thus, we have formed the Lagrange function. Now, following Lagrange's recipe we must seek extremum conditions for the resultant expression "as if the variables were independent." In other words, we should consider the problem

$$\mathscr{L}(x(\cdot), u(\cdot); \hat{p}(\cdot), \hat{\mu}, \hat{\lambda}_0) \to \text{extr}, \tag{4}$$

where the Lagrange multipliers are assumed to be fixed. The problem (4) is a Bolza problem considered in Section 1.4.2. The application of the extremum conditions written in that section leads to the correct equations called the *Euler–Lagrange equations*. More precisely, the following theorem holds:

The Euler–Lagrange Theorem. If $\hat{z} = (\hat{x}(\cdot), \hat{u}(\cdot))$ is an optimal process in the weak sense in the problem (1'), (2') of Section 1.5.1, then there are Lagrange multipliers $\hat{\lambda}_0 = \hat{\mu}_0 \geqslant 0$ (in a minimum problem) or $\hat{\lambda}_0 = \mu_0 \leqslant 0$ (in a maximum problem), $\hat{p}(\cdot) \in C^1([t_0, t_1], \mathbf{R}^n)$, and $\hat{\mu} = (\hat{\mu}_1, \ldots, \hat{\mu}_s)$, not all zero,* such that there hold the Euler equations

$$-\frac{d}{dt} L_{\dot{x}}(t, \hat{x}(t), \dot{\hat{x}}(t), \hat{u}(t)) + L_x(t, \hat{x}(t), \dot{\hat{x}}(t), \hat{u}(t)) = 0, \tag{5}$$

$$L_u(t, \hat{x}(t), \dot{\hat{x}}(t), \hat{u}(t)) = 0, \tag{6}$$

and the transversality conditions

$$L_{\dot{x}}(t_k, \hat{x}(t_k), \dot{\hat{x}}(t_k), \hat{u}(t_k)) = (-1)^k \frac{\partial l}{\partial x_k}(\hat{x}(t_0), \hat{x}(t_1)), \quad k = 0,1. \tag{7}$$

The Euler equation in u has the degenerate form (6) because L is independent of $\dot{u}$ and l is independent of u, and there is no transversality condition "in u" at all.

This theorem (even in a more general form) will be proved in Section 4.1 as a direct consequence of a general theorem concerning the Lagrange multiplier rule for smooth infinite-dimensional problems.

It should be noted that the Euler–Lagrange theorem implies a necessary condition for an extremum in the isoperimetric problem (with an arbitrary number of isoperimetric conditions) and the Euler–Poisson equation for problems with higher derivatives. The isoperimetric constraints

$$\mathscr{I}_i(x(\cdot)) = \int_{t_0}^{t_1} f_i(t, x, \dot{x})\, dt = \alpha_i, \quad i = 1, \ldots, m; \quad x \in \mathbf{R}^n,$$

can be taken into account by introducing new phase coordinates connected with the old ones by the differential constraint $\dot{x}_{n+i}(t) = f_i(t, x_1(t), \ldots, x_n(t), \dot{x}_1(t), \ldots, \dot{x}_n(t))$ and satisfying the boundary conditions

$$x_{n+i}(t_1) - x_{n+i}(t_0) = \alpha_i, \quad i = 1, \ldots, m.$$

*An advocate of mathematical purism may notice a glaring inaccuracy in this statement: $\hat{\lambda}_0$ and $\hat{\mu}_i$ are numbers while $\hat{p}(\cdot)$ is an element of a function space, and therefore they cannot be simultaneously equal to one and the same "zero." Each of them may or may not be equal to the zero element of its own space. However, we are used to identifying zeros of all the space, and this statement does not irritate us.

Now the application of the Euler–Lagrange theorem leads to the desired necessary conditions in the isoperimetric problem. When the problem with higher derivatives is investigated, it can be brought to the form (1'), (2') by putting $x = x_1$, $\dot{x}_1 = x_2, \ldots, \dot{x}_n = u$ and applying the Euler–Lagrange theorem.

1.5.3. The Pontryagin Maximum Principle. In the Fifties the demands of applied sciences (technology, economics, etc.) stimulated the statement and the analysis of a new class of extremal problems which were called *optimal control problems*. A necessary extremum condition for this class of problems, known as the "maximum principle," was stated by L. S. Pontryagin in 1953 and was proved and developed later by him and his pupils and collaborators (see [12]). It is important to note that this condition is of an essentially different form in comparison with the classical equations of Euler and Lagrange: the solution of an optimal control problem includes as a necessary stage the solution of an auxiliary maximum problem (this accounts for the name the "maximum principle").

Here we shall consider a special case of the general statement of an optimal control problem in which the Lagrange problem in Pontryagin's form [see (1') and (2') in Section 1.5.1] involves one more additional condition on the control: $u \in \mathfrak{U}$. More precisely, we shall consider the following problem:

$$\mathcal{J}(x(\cdot), u(\cdot)) = \int_{t_0}^{t_1} f(t, x(t), u(t))\,dt + \psi_0(x(t_0), x(t_1)) \to \inf, \tag{1}$$

$$\dot{x}(t) - \varphi(t, x(t), u(t)) = 0, \quad \psi_j(x(t_0), x(t_1)) = 0, \quad j = 1, \ldots, s, \tag{2}$$

$$u \in \mathfrak{U}. \tag{3}$$

Here the functions f, φ, and ψ_j are the same as in problem (1'), (2') of Section 1.5.1 and $\mathfrak{U}$ is a fixed set in $\mathbf{R}^r$. A more general problem will be considered in Chapter 4.

The Lagrange problem was considered in a Banach space. However, here we want to use the simplest means, and the admissible elements will be described in another way similar to the one mentioned in Section 1.4.3, where the simplest problem of the classical calculus of variations was extended to a problem in the space $KC^1([t_0, t_1])$ of piecewise-differentiable functions.

A pair $(x(\cdot), u(\cdot))$ will be called a *controlled process* in problem (1)-(3) if the control $u(\cdot): [t_0, t_1] \to \mathfrak{U}$ is a piecewise-continuous function, the phase trajectory $x(\cdot)$ is piecewise-continuously differentiable, and, moreover, the function $x(\cdot)$ satisfies the differential equation $\dot{x}(t) = \varphi(t, x(t), u(t))$ everywhere except at the points of discontinuity of the control $u(\cdot)$. A controlled process is said to be *admissible* if, in addition, the boundary conditions are satisfied.

An admissible process $(\hat{x}(\cdot), \hat{u}(\cdot))$ is said to be *optimal* if there is $\varepsilon > 0$ such that the inequality $\mathcal{J}(x(\cdot), u(\cdot)) \geqslant \mathcal{J}(\hat{x}(\cdot), \hat{u}(\cdot))$ holds for every admissible controlled process $(x(\cdot), u(\cdot))$ such that $|x(t) - \hat{x}(t)| < \varepsilon$, $\forall t \in [t_0, t_1]$.

Let us again try to apply the Lagrange general method discussed in Section 1.3.2 to problem (1)-(3). The Lagrange function of problem (1)-(3) is of the same form as in the Lagrange problem: the constraints of the form of inclusions ($u \in \mathfrak{U}$) are not involved in the Lagrange function. Thus

$$\mathcal{L}(x(\cdot), u(\cdot); p(\cdot), \mu, \lambda_0) = \int_{t_0}^{t_1} L\,dt + l, \tag{4}$$

where

$$L(t, x, \dot{x}, u) = p(t)(\dot{x} - \varphi(t, x, u)) + \lambda_0 f(t, x, u), \tag{5}$$

$$l(x_0, x_1) = \sum_{j=0}^{s} \mu_j \psi_j(x_0, x_1), \quad \mu_0 = \lambda_0. \tag{6}$$

Further, as usual, we have to consider the problem

$$\mathscr{L}(x(\cdot), u(\cdot); \hat{p}(\cdot), \hat{\mu}, \hat{\lambda}_0) \to \inf \tag{7}$$

(where the Lagrange multipliers are assumed to be fixed) "as if the variables were independent." Naturally, problem (7) splits into two partial problems:

$$\mathscr{L}(x(\cdot), \hat{u}(\cdot); \hat{p}(\cdot), \hat{\mu}, \hat{\lambda}_0) \to \inf \text{ (with respect to } x(\cdot)), \tag{8}$$

$$\mathscr{L}(\hat{x}(\cdot), u(\cdot); \hat{p}(\cdot), \hat{\mu}, \hat{\lambda}_0) \to \inf \text{ (with respect to } u(\cdot) \in \mathcal{U}), \tag{9}$$

where $\mathcal{U}$ designates the set of piecewise-smooth functions with values bebelonging to $\mathfrak{U}$. Problem (8) is again a Bolza problem; as can easily be understood, problem (9) has the following simpler form:

$$\int^{t_1} \chi(t, u(t))\,dt \to \inf, \; u(\cdot) \in \mathcal{U}, \tag{10}$$

where $\chi(t, u) = \hat{\lambda}_0 f(t, \hat{x}(t), u) - \hat{p}(t)\varphi(t, \hat{x}(t), u)$.

It is quite evident that a necessary (and sufficient) condition in the problem (10) can be stated thus: *$\hat{u}(\cdot) \in \mathcal{U}$ yields a minimum in the problem (10) if and only if the relation*

$$\min_{u \in \mathfrak{U}} \chi(t, u) = \chi(t, \hat{u}(t)) \Leftrightarrow \min_{u \in \mathfrak{U}} L(t, \hat{x}(t), \dot{\hat{x}}(t), u) = L(t, \hat{x}(t), \dot{\hat{x}}(t), \hat{u}(t)) \Leftrightarrow$$

$$\Leftrightarrow \max_{u \in \mathfrak{U}} [\hat{p}(t)\varphi(t, \hat{x}(t), u) - \hat{\lambda}_0 f(t, \hat{x}(t), u)] = \hat{p}(t)\varphi(t, \hat{x}(t), \hat{u}(t)) - \hat{\lambda}_0 f(t, \hat{x}(t), \hat{u}(t)) \tag{11}$$

holds everywhere except at the points of discontinuity of $\hat{u}(\cdot)$.

The combination of the necessary conditions for extremum in problem (8) and relation (11) leads to the necessary conditions for extremum in problem (1)-(3) which were called the *Pontryagin maximum principle* [because of the specific form of condition (11)]. More precisely, the following theorem holds:

THEOREM (Pontryagin Maximum Principle). If $(\hat{x}(\cdot), \hat{u}(\cdot))$ is an optimal process in problem (1)-(3), then there are Lagrange multipliers $\hat{\lambda}_0 = \hat{\mu}_0 \geqslant 0$, $\hat{p}(\cdot) \in KC^1([t_0, t_1], \mathbf{R}^{n*})$, $\hat{\mu} = (\hat{\mu}_1, \ldots, \hat{\mu}_s)$, not all zero, such that there hold the Euler equation

$$\frac{d}{dt} L_{\dot{x}}(t, \hat{x}(t), \dot{\hat{x}}(t), \hat{u}(t)) = L_x(t, \hat{x}(t), \dot{\hat{x}}(t), \hat{u}(t)), \tag{12}$$

the maximum principle (11), and the transversality conditions

$$L_{\dot{x}}(t, \hat{x}(t), \dot{\hat{x}}(t), \hat{u}(t))|_{t=t_k} = (-1)^k \frac{\partial l}{\partial x_k}(\hat{x}(t_0), \hat{x}(t_1)), \quad k = 0, 1. \tag{13}$$

1.5.4. Proof of the Maximum Principle for the Problem with a Free End Point. In this section the Pontryagin maximum principle will be proved for the simplest situation when there is no endpoint part in the functional, one of the end points is fixed, and the other is free, i.e., when $\psi_0 = 0$ in relation (1) of Section 1.5.3 and $\psi_j(x_0, x_1) = x_{0j} - \bar{x}_{0j}$, $j = 1, \ldots, n$, in relation (2) of that section.

Thus, we deal with the problem

$$\mathcal{J}(x(\cdot), u(\cdot)) = \int_{t_0}^{t_1} f(t, x(t), u(t))\,dt \to \inf, \tag{1}$$

$$\dot{x}(t)-\varphi(t, x(t), u(t))=0, \quad x(t_0)=\bar{x}_0, \tag{2}$$

$$u \in \mathfrak{U}. \tag{3}$$

As in the foregoing section, the functions f, f_x, φ, and φ_x are assumed to be continuous jointly with respect to their arguments. Let us see what the Pontryagin maximum principle is like in this case. Since the function $l(x_0, x_1) = \sum_{i=1}^{m} \hat{\mu}_i (x_{0i} - \bar{x}_{0i})$ does not depend on x_1, the transversality conditions (13) of Section 1.5.3 result in

$$L_{\dot{x}}(t_1, \hat{x}(t_1), \hat{\dot{x}}(t_1), \hat{u}(t_1)) = \hat{p}(t_1) = 0. \tag{4}$$

Further, Eq. (12) of Section 1.5.3 takes the form

$$-\frac{d}{dt} L_{\dot{x}} + L_x = 0 \Leftrightarrow -\dot{\hat{p}}(t) = \hat{p}(t)\hat{\varphi}_x(t) - \hat{\lambda}_0 \hat{f}_x(t), \tag{5}$$

where we have introduced the following abbreviated notation: $\hat{f}_x(t) = f_x(t, \hat{x}(t), \hat{u}(t))$, $\hat{\varphi}_x(t) = \varphi_x(t, \hat{x}(t), \hat{u}(t))$. If we suppose that $\hat{\lambda}_0 = 0$, then, by the uniqueness of the solution of Cauchy's problem for the homogeneous equation (5), there must be $\hat{p}(\cdot) \equiv 0$, and hence, by virtue of the transversality condition at the left end point [see the relation (13) in Section 1.5.3] there must also be $\hat{\mu} = 0$. But this contradicts the condition of the theorem according to which the Lagrange multipliers do not vanish simultaneously. Therefore, $\hat{\lambda}_0 \neq 0$, and we can assume that $\hat{\lambda}_0 = 1$. However, then $\hat{p}(\cdot)$ is uniquely determined by Eq. (5) and the boundary condition (4) (due to the uniqueness of the solution of Cauchy's problem for a nonhomogeneous linear system of differential equations). Summing up, we can state the maximum principle thus:

THEOREM (Pontryagin Maximum Principle for the Problem with a Free End Point. If a process $(\hat{x}(\cdot), \hat{u}(\cdot))$ is optimal in problem (1)-(3), then for the solution $\hat{p}(\cdot)$ of the system

$$-\dot{p}(t) = p(t)\hat{\varphi}_x(t) - \hat{f}_x(t) \tag{6}$$

with the boundary condition

$$p(t_1) = 0 \tag{6a}$$

there holds the maximum principle at the points of continuity of the control $\hat{u}(\cdot)$:

$$\max_{u \in \mathfrak{U}} (\hat{p}(t)\varphi(t, \hat{x}(t), u) - f(t, \hat{x}(t), u)) = \hat{p}(t)\varphi(t, \hat{x}(t), \hat{u}(t)) - f(t, \hat{x}(t), \hat{u}(t)). \tag{7}$$

As in the preceding cases, to prove the theorem we shall apply the method of variations. To begin with, we shall state the definition of an elementary (or Weierstrass' or needlelike) variation analogous to the one used in Section 1.4.4. Let us denote by T_0 the set of those points belonging to (t_0, t_1) at which $\hat{u}(\cdot)$ is continuous. Let us fix a point $\tau \in T_0$, an element $v \in \mathfrak{U}$, and a number $\lambda > 0$ so small that $[\tau-\lambda, \tau] \subset [t_0, t_1]$.

The control

$$u_\lambda(t) = u_\lambda(t; \tau, v) = \begin{cases} \hat{u}(t) & \text{for} \quad t \notin [\tau-\lambda, \tau), \\ v & \text{for} \quad t \in [\tau-\lambda, \tau), \end{cases} \tag{8}$$

will be called an *elementary needlelike variation of the control* $\hat{u}(\cdot)$. Let $x_\lambda(t) = x_\lambda(t; \tau, v)$ be the solution of the equation $\dot{x} = \varphi(t, x, u_\lambda(t))$ with the initial condition $x_\lambda(t_0) = \hat{x}(t_0) = x_0$. The function $x_\lambda(t)$ will be called an *elementary needlelike variation of the trajectory*, and the pair $(x_\lambda(t), \hat{u}(\cdot))$ will be called an *elementary variation of the process* $(\hat{x}(\cdot), \hat{u}(\cdot))$. Finally, the pair (τ, v) specifying this variation will be called an *elementary needle*.

<u>Proof.</u> As usual, we split the proof of the theorem into stages. The first two stages entirely belong to the theory of ordinary differential equations.

<u>A) The Lemma on the Properties of an Elementary Variation.</u> Let an elementary needle (τ, v) be fixed. Then there exists $\lambda_0 > 0$ such that for $0 \leqslant \lambda \leqslant \lambda_0$:

1) the trajectory $x_\lambda(\cdot)$ is defined throughout the closed interval $[t_0, t_1]$, and $x_\lambda(t) \to \hat{x}(t)$ for $\lambda \downarrow 0$ uniformly on $[t_0, t_1]$;

2) for $t > \tau$, $0 \leqslant \lambda \leqslant \lambda_0$, the derivative

$$\frac{d}{d\lambda} x_\lambda(t;\ \tau,\ v) = z_\lambda(t;\ \tau,\ v)$$

exists and is continuous with respect to λ (for $\lambda = 0$ it is understood as the right-hand derivative);

3) the function $t \to y(t) = z_0(t;\ \tau,\ v)$ satisfies the differential equation

$$\dot{y} = \hat{\varphi}_x(t)\, y \tag{9}$$

on the closed interval $[\tau, t_1]$ and the initial condition

$$y(\tau) = \varphi(\tau,\ \hat{x}(\tau),\ v) - \varphi(\tau,\ \hat{x}(\tau),\ \hat{u}(\tau)). \tag{10}$$

The *proof* of the lemma is based on two well-known facts of the theory of ordinary differential equations: the local existence and uniqueness theorem and the theorem on the continuous differentiability of the solution with respect to the initial data. These theorems are included in standard textbooks on ordinary differential equations in the form suitable for our aims (see [10, 11, 15]). We also refer the reader to the material of Section 2.5.

First we shall prove the lemma for the case when the function $\hat{u}(\cdot)$ is continuous.

Let us consider the differential equations

$$\dot{x} = \varphi(t,\ x,\ u_\lambda(t)), \tag{11}$$

$$\dot{x} = \varphi(t,\ x,\ \hat{u}(t)). \tag{12}$$

According to (8), the right-hand sides of these equations coincide for $t < \tau - \lambda$, and since $x_\lambda(t_0) = \bar{x}_0 = x(t_0)$, the uniqueness theorem for the solution of Cauchy's problem implies that $x_\lambda(t) \equiv \hat{x}(t)$ for $t < \tau - \lambda$, and, by the continuity,

$$\xi(\lambda) = x_\lambda(\tau - \lambda) = \hat{x}(\tau - \lambda). \tag{13}$$

In particular, $\xi(\lambda)$ is differentiable with respect to λ and

$$\xi(0) = \hat{x}(\tau),\ \xi'(0) = -\dot{\hat{x}}(\tau) = -\varphi(\tau, \hat{x}(\tau),\ u\ (\tau)). \tag{14}$$

Let us denote by $\Xi(t, s, \xi)$ the solution of Cauchy's problem for the differential equation with a fixed control v:

$$\dot{x} = \varphi(t,\ x,\ v),\quad x(s) = \xi. \tag{15}$$

According to the local existence and uniqueness theorem, there are $\varepsilon_1 > 0$ and $\delta_1 > 0$ such that $\Xi(t, s, \xi)$ is defined for

$$|t - \tau| < \delta_1,\ |s - \tau| < \delta_1,\ |\xi - \hat{x}(\tau)| < \varepsilon_1,$$

and, by virtue of the theorem on the continuous differentiability of the solution with respect to the initial data, Ξ is a continuously differentiable function.

According to (8) and (13), to define $x_\lambda(t)$ on the closed interval $[\tau - \lambda, \tau]$ we must put $\xi = \xi(\lambda) = \hat{x}(\tau - \lambda)$ in (15), and if $\lambda_1 < \delta_1$ is chosen so that $|\xi(\lambda) - \hat{x}(\tau)| < \varepsilon_1$ for $0 \leqslant \lambda \leqslant \lambda_1$, then

$$x_\lambda(t) = \Xi(t, \tau-\lambda, \xi(\lambda)), \quad \tau-\lambda \leqslant t \leqslant \tau.$$

In particular, since the function

$$\eta(\lambda) = x_\lambda(\tau) = \Xi(\tau, \tau-\lambda, \xi(\lambda)) \tag{16}$$

is a composition of continuously differentiable functions, it is also continuously differentiable with respect to λ and, by virtue of (14),

$$\eta(0) = \hat{x}(\tau), \quad \eta'(0) = -\Xi_s(\tau, \tau, \xi(0)) + \Xi_\xi(\tau, \tau, \xi(0))\,\xi'(0) =$$
$$= -\Xi_s(\tau, \tau, \xi(0)) - \Xi_\xi(\tau, \tau, \xi(0))\,\varphi(\tau, \hat{x}(\tau), \hat{u}(\tau)). \tag{17}$$

The solution $\Xi(t, s, \xi)$ of Cauchy problem (15) satisfies the equivalent integral equation

$$\Xi(t, s, \xi) = \xi + \int_s^t \varphi(\sigma, \Xi(\sigma, s, \xi), v)\, d\sigma. \tag{18}$$

Differentiating (18) with respect to s, we obtain

$$\Xi_s(t, s, \xi) = -\varphi(s, \Xi(s, s, \xi), v) + \int_s^t \varphi_x(\sigma, \Xi(\sigma, s, \xi), v)\, \Xi_s(\sigma, s, \xi)\, d\sigma$$

and putting $t = s = \tau$, $\xi = \hat{x}(\tau)$ we derive [taking into account the obvious identity $\Xi(t, t, \xi) \equiv \xi$] the relation

$$\Xi_s(\tau, \tau, \hat{x}(\tau)) = -\varphi(\tau, \hat{x}(\tau), v). \tag{19}$$

Similarly, differentiating (18) with respect to ξ and substituting the same values of the arguments, we obtain

$$\Xi_\xi(\tau, \tau, \hat{x}(\tau)) = I. \tag{20}$$

Here $I = (\partial\xi_i/\partial\xi_k) = (\delta_{ik})$ is the unit matrix (which is sometimes denoted by E). The substitution of (19) and (20) into (17) results in

$$\eta(0) = \hat{x}(\tau), \quad \eta'(0) = \varphi(\tau, \hat{x}(\tau), v) - \varphi(\tau, \hat{x}(\tau), \hat{u}(\tau)). \tag{21}$$

Further, the function Ξ is continuous at the point $(\tau, \tau, \hat{x}(\tau))$ and, moreover, $\Xi(\tau, \tau, \hat{x}(\tau)) = \hat{x}(\tau)$, and $\hat{x}(\cdot)$ is continuous at the point τ. Therefore, for any $\varepsilon > 0$ there exists $\delta > 0$ such that for

$$|t-\tau| < \delta, \quad |s-\tau| < \delta, \quad |\xi - \hat{x}(\tau)| < \delta$$

the inequalities

$$|\Xi(t, s, \xi) - \hat{x}(\tau)| < \varepsilon/2, \quad |\hat{x}(t) - \hat{x}(\tau)| < \varepsilon/2 \tag{22}$$

hold.

Let us take a positive number $\lambda_1 < \delta$ so small that the inequality $|\xi(\lambda) - \hat{x}(\tau)| < \delta$ holds for $0 \leqslant \lambda \leqslant \lambda_1$. Then for $\tau - \lambda \leqslant t \leqslant \tau$, $s = \tau - \lambda$, and $\xi = \xi(\lambda)$ inequalities (22) hold, whence

$$|x_\lambda(t) - \hat{x}(t)| = |\Xi(t, \tau-\lambda, \xi(\lambda)) - \hat{x}(t)| \leqslant |\Xi(t, \tau-\lambda, \xi(\lambda)) - \hat{x}(\tau)| + |\hat{x}(\tau) - \hat{x}(t)| < \varepsilon/2 + \varepsilon/2 = \varepsilon.$$

Since $x_\lambda(t) \equiv \hat{x}(t)$ for $t_0 \leqslant t \leqslant \tau - \lambda$, we have

$$0 \leqslant \lambda \leqslant \lambda_1 \Rightarrow |x_\lambda(t) - \hat{x}(t)| < \varepsilon, \quad t_0 \leqslant t \leqslant \tau,$$

which proves the first assertion of the lemma.

Now let us denote by $X(\cdot, \eta)$ the solution of Cauchy's problem for Eq. (12) with the initial condition $x(\tau) = \eta$. By the theorem on the continuous differentiability with respect to the initial data, there exists ε_2 such that $X(t, \eta)$ is defined for $|\eta - \hat{x}(\tau)| < \varepsilon_2$, $\tau \leqslant t \leqslant t_1$ and is a continuousl differentiable function. According to (8) and (16), and also by virtue of the uniqueness theorem, $x_\lambda(t) \equiv X(t, \eta(\lambda))$. Again, since the function $x_\lambda(t)$ is a composition of continuously differentiable functions, it is continuously differentiable with respect to (t, λ) for $\tau \leqslant t \leqslant t_1$ and $0 \leqslant \lambda \leqslant \lambda_2$, where λ_2 is chosen so that the inequality $|\eta(\lambda) - \hat{x}(\tau)| < \varepsilon_2$ holds for $0 \leqslant \lambda \leqslant \lambda_2$. Putting $\lambda_0 = \min(\lambda_1, \lambda_2)$ we conclude that the second assertion of the lemma is true.

Passing from Eq. (12) to the equivalent integral equation and taking into account (16), we obtain the equation

$$x_\lambda(t) = \eta(\lambda) + \int_\tau^t \varphi(\sigma, x_\lambda(\sigma), \hat{u}(\sigma))\, d\sigma.$$

Differentiating this equation with respect to λ and then putting $\lambda = 0$ and denoting, just as in the condition of the lemma, $y(t) = \frac{d}{d\lambda} x_\lambda(t)\Big|_{\lambda=0}$, we find

$$y(t) = \eta'(0) + \int_\tau^t \varphi_x(\sigma, \hat{x}(\sigma), \hat{u}(\sigma))\, y(\sigma)\, d\sigma.$$

The last integral equation is equivalent to Eq. (9) with the initial condition $y(\tau) = \eta'(0)$ coinciding with (10) by virtue of (21).

If the control $\hat{u}(\cdot)$ is a piecewise-continuous function, we resort to the following reasoning. For the sake of simplicity, let there be two points of discontinuity, say α_1 and α_2, and let τ be a point [at which $\hat{u}(\cdot)$ must be continuous] lying between them: $t_0 < \alpha_1 < \tau < \alpha_2 < t_1$. Systems (11) and (12) coincide in the strip $t_0 \leqslant t \leqslant \alpha_1$ [for $t = \alpha_1$ the control in these systems must be understood as its limiting value $\hat{u}(\alpha_1 - 0) = \lim_{t \to \alpha_1 - 0} \hat{u}(t)$], and, by the uniqueness theorem, $x_\lambda(t) \equiv \hat{x}(t)$.

Now we pass to the strip $\alpha_1 < t < \alpha_2$ [the control is again assumed to be equal to the limiting values $\hat{u}(\alpha_1 + 0)$ and $\hat{u}(\alpha_2 - 0)$ at the boundaries $t = \alpha_1$ and $t = \alpha_2$ of the strip]. Here Eqs. (11) and (12) are solved with the initial condition $x = \hat{x}(\alpha_1)$. The foregoing assertions are applicable to the present case, and thus we see that $x_\lambda(t)$ is continuously differentiable with respect to λ.

Finally, in the strip $\alpha_2 \leqslant t \leqslant t_1$ (under the same agreement on the value of the control at $t = \alpha_2$) the differential equations are solved with the initial data $x_\lambda(\alpha_2)$ and $\hat{x}(\alpha_2)$. Once again referring to the theorem on the continuous differentiability of the solution with respect to the initial data, we prove the continuous differentiability of $x_\lambda(t)$ with respect to λ for $\alpha_2 \leqslant t \leqslant t_1$ and compute $y(t) = \frac{d}{d\lambda} x_\lambda(t)\Big|_{\lambda=0}$ in the same way as before. ■

B) The Lemma on the Increment of the Functional. Let us put $\chi(\lambda) = \mathcal{J}(x_\lambda(\cdot), u_\lambda(\cdot))$ and prove that this function is differentiable on the right at the point $\lambda = 0$.

Let $\hat{p}(\cdot)$ be the solution of system (6) with the boundary condition (6a). Then

$$\chi'(+0) = \frac{d}{d\lambda} \mathcal{J}(x_\lambda(\cdot), u_\lambda(\cdot))\Big|_{\lambda=+0} = a(\tau, v),$$

where

$$a(\tau, v) = f(\tau, \ddot{x}(\tau), v) - f(\tau, \ddot{x}(\tau), \ddot{u}(\tau)) - \ddot{p}(\tau)[\varphi(\tau, \ddot{x}(\tau), v) - \varphi(\tau, \ddot{x}(\tau), \ddot{u}(\tau))]. \quad (23)$$

Proof. Since

$$\chi(\lambda) - \chi(0) = \int_{t_0}^{t_1} f(t, x_\lambda(t), u_\lambda(t))\, dt - \int_{t_0}^{t_1} f(t, \hat{x}(t), \hat{u}(t))\, dt =$$

$$=\int_{\tau}^{t_1}[f(t, x_\lambda(t), \hat{u}(t))-f(t, \hat{x}(t), \hat{u}(t))]\,dt+\int_{\tau-\lambda}^{\tau}[f(t, x_\lambda(t), v)-f(t, \hat{x}(t), \hat{u}(t)]\,dt,$$

we have

$$\chi'(0)=\lim_{\lambda\downarrow 0}\frac{\chi(\lambda)-\chi(0)}{\lambda}=\frac{d}{d\lambda}\int_{\tau}^{t_1}f(t, x_\lambda(t), \hat{u}(t))\,dt\Big|_{\lambda=+0}+\lim_{\lambda\downarrow 0}\frac{1}{\lambda}\int_{\tau-\lambda}^{\tau}[f(t, x_\lambda(t), v)-f(t, \hat{x}(t), \hat{u}(t)]\,dt.$$

By the lemma of Section A), $x_\lambda(t)$ is continuously differentiable with respect to λ, and therefore the ordinary differentiation rule under the integral sign is applicable to the first summand; to the second summand we apply the mean value theorem; this results in

$$\chi'(+0)=\int_{\tau}^{t_1}f_x(t, \hat{x}(t), \hat{u}(t))\frac{d}{d\lambda}x_\lambda(t)\Big|_{\lambda=0}dt+\lim_{\lambda\downarrow 0}[f(c, x_\lambda(c), v)-f(c, \hat{x}(c), \hat{u}(c)]=$$

$$=\int_{\tau}^{t_1}f_x(t, \hat{x}(t), \hat{u}(t))\,y(t)\,dt+f(\tau, \hat{x}(\tau), v)-f(\tau, \hat{x}(\tau), \hat{u}(\tau)) \tag{24}$$

[here we have used the fact that $\tau-\lambda \leqslant c \leqslant \tau$, which implies $c \to \tau$; $x_\lambda(c) \to \hat{x}(\tau)$ by virtue of the first assertion of the lemma in Section A); $y(t)$ denotes the same as in that lemma].

Further, $\hat{p}(\cdot)$ satisfies system (6), and $y(\cdot)$ satisfies system (9). Therefore

$$\frac{d}{dt}\hat{p}(t)\,y(t)=\dot{\hat{p}}(t)\,y(t)+\hat{p}(t)\,\dot{y}(t)=-\hat{p}(t)\,\hat{\varphi}_x(t)\,y(t)+\hat{f}_x(t)\,y(t)+\hat{p}(t)\,\hat{\varphi}_x(t)\,y(t)=\hat{f}_x(t)\,y(t).$$

Integrating this equality from τ to t_1 and taking into account the boundary condition (6a) for $\hat{p}(\cdot)$ and the condition (10) for $y(\cdot)$, we obtain

$$\int_{\tau}^{t_1}\hat{f}_x(t)\,y(t)\,dt=\int_{\tau}^{t_1}\frac{d}{dt}[\hat{p}(t)\,y(t)]\,dt=\hat{p}(t)\,y(t)\Big|_{\tau}^{t_1}=-\hat{p}(\tau)\,[\varphi(\tau, \hat{x}(\tau), v)-\varphi(\tau, \hat{x}(\tau), \hat{u}(\tau))]. \tag{25}$$

Comparing (24), (25), and (23) we conclude that $\chi'(+0) = a(\tau, v)$. ■

C) Completion of the Proof. If the right end point is free, then, by virtue of the lemma in Section A), every elementary variation is admissible (for sufficiently small λ). Hence, if $(\hat{x}(\cdot), \hat{u}(\cdot))$ is an optimal process, then for small λ we have

$$\mathcal{J}(x_\lambda(\cdot), u_\lambda(\cdot))\geqslant\mathcal{J}(\hat{x}(\cdot), \hat{u}(\cdot))\Leftrightarrow\chi(\lambda)\geqslant\chi(0)\Rightarrow\chi'(+0)\geqslant 0.$$

Applying the lemma of Section B) we conclude that the condition $a(\tau, v) \geqslant 0$ is necessary for the optimality of $(\hat{x}(\cdot), \hat{u}(\cdot))$. However, $\tau\in T_0$ and $v\in\mathfrak{U}$ were chosen arbitrarily, and consequently we have proved that for any t belonging to the set of the points of discontinuity of the control $\hat{u}(\cdot)$ and for any $u\in\mathfrak{U}$ there holds the inequality

$$\hat{p}(t)\,\varphi(t, \hat{x}(t), u)-f(t, \hat{x}(t), u)\leqslant\hat{p}(t)\,\varphi(t, \hat{x}(t), \hat{u}(t))-f(t, \hat{x}(t), \hat{u}(t)),$$

equivalent to the inequality $a(t, u) \geqslant 0$ and (7). The proof of the theorem is completed.

1.6. Solutions of the Problems

The problems mentioned at the beginning of this chapter were posed for different purposes and at different times; here we shall present a unified approach to these problems following a standard scheme implied by the Lagrange principle.

This scheme consists of six stages:

1. Writing the formalized problem and discussing the question of the existence and the uniqueness of the solution.

2. Forming the Lagrange function.

3. Applying the Lagrange principle.

4. Analyzing the possibility of the equality $\hat{\lambda}_0 = 0$.

5. Finding the stationary points, i.e. solving the equations implied by the Lagrange principle.

6. Investigating stationary points, choosing the solution, and writing the answer.

For all the problems that will be discussed the Lagrange principle is a rigorously justified theorem which either has already been proved or will be proved in the following chapters. The applicability of this unified scheme to the problems so different in their subject stresses the universality of the Lagrange principle. Of course, when considering some other problems researchers may encounter situations which are not embraced by any of the known concrete schemes (the classical calculus of variations, optimal control theory, linear programming, etc.). Nevertheless, even in these cases the Lagrange principle understood in one way or another may prove valid or can at least serve as a useful guide. The understanding of the general ideas and the situations to which the principle is applicable (they will be discussed in Chapters 3 and 4) may help to find necessary conditions for extremum in a new situation as well. However, it is also important to realize that the Lagrange principle is not always valid and therefore it must not be applied automatically.

1.6.1. Geometrical Extremal Problems. In this section all the problems posed in Section 1.1.2 and formalized in Section 1.2.2 are solved. The first stage of the above scheme includes the discussion of the question of the existence of the solution. The geometrical problems of the present section are finite-dimensional, and the existence of their solutions is guaranteed by the following theorem:

Weierstrass' Theorem. Let a function $f\colon \mathbf{R}^N \to \mathbf{R}$ be continuous and let the level set $\mathscr{L}_\alpha f = \{x \mid f(x) \leqslant \alpha\}$ be nonempty and bounded for some α. Then there exists a solution of the problem $f(x) \to \inf$.

The *proof* of this theorem is an evident consequence of the classical Weierstrass theorem on the existence of a minimum of a continuous function on a bounded and closed subset of $\mathbf{R}^N$ (see [14, Vol. 1, pp. 176 and 370] and [9, Vol. 1]) because the set $\mathscr{L}_\alpha f$ is obviously closed.

Now we proceed to the solution of the problem.

The *problem of Euclid* on the inscribed parallelogram. The problem was formalized thus [see formula (1) of Section 1.2.2]:

1. $f_0(x) = x(x - b) \to \inf,\ 0 \leqslant x \leqslant b.$

We have dropped the inessential factor H/b and reduced the problem to a minimization problem. The function f_0 is continuous, and the interval $[0, b]$ is bounded and closed. By Weierstrass' theorem, there exists a solution of the problem. Let this solution be $\hat{x}$. It is clear that $\hat{x} \neq 0$ and $\hat{x} \neq b$ because $f_0(0) = f_0(b) = 0$ and the function assumes negative values as well. Consequently, $\hat{x} \in (0, b)$. The function f_0 is smooth. Therefore, one should seek the stationary points of the problem $f_0(x) \to \inf$.

2-5. $f_0'(\hat{x}) = 0 \Rightarrow x = b/2.$

6. By the uniqueness of the stationary point, $\hat{x}=b/2\in[0, b]$ is the single solution of the problem; i.e., the sought-for parallelogram $\hat{A}\hat{D}\hat{E}\hat{F}$ is characterized by the property that $|A\hat{F}| = |AC|/2$, i.e., $\hat{F}$ *is the midpoint of the line segment* [AC]. This is the fact that was established in Section 1.1.2 purely geometrically.

Remark. The above problem turned out to be an elementary smooth problem. Therefore, we did not write the Lagrange function, and Sections 2-5 "merged." We used the Fermat theorem (Section 1.3.1).

The *problem of Archimedes* on the spherical segments with a given surface area (Section 1.1.2). Here the solution is quite similar to that of the problem of Euclid, and we present it without comments.

1. $f_0(h)=ha/2-\pi h^3/3\rightarrow \sup;\ 0\leqslant h\leqslant\sqrt{a/\pi}$.

2-5. $f_0'(\hat{h})=0\Rightarrow \hat{h}=\sqrt{a/2\pi}$.

6. The value of f_0 at zero is equal to zero, and its value at the point $\sqrt{a/\pi}$ is smaller than the value at the stationary point $\sqrt{a/2\pi}$. Consequently $\hat{h} = \sqrt{a/2\pi}$ is a solution of the problem. Recalling that $a = 2\pi R\hat{h}$ we obtain $\hat{h} = R$; i.e., *the sought-for spherical segment is a hemisphere* (its altitude is equal to the radius).

The *problem of Apollonius* on the shortest distance from a point to an ellipse. It was formalized thus [see the formula (6) in Section 1.2.2]:

1. $f_0(x_1, x_2) = (x_1-\xi_1)^2 + (x_2-\xi_2)^2 \rightarrow \inf;\ f_1(x_1, x_2) = (x_1/a_1)^2 + (x_2/a_2)^2 = 1$. The ellipse is a bounded and closed set, the function f_0 is continuous, and therefore, by Weierstrass' theorem, there exists a solution. The functions f_0 and f_1 are smooth.

2. $\mathscr{L}=\lambda_0((x_1-\xi_1)^2+(x_2-\xi_2)^2)+\lambda_1((x_1/a_1)^2+(x_2/a_2)^2-1)$.

3. $\mathscr{L}_{x_1}=0\Rightarrow\hat{\lambda}_0(\hat{x}_1-\xi_1)+\hat{\lambda}_1\hat{x}_1/a_1^2=0,\ \mathscr{L}_{x_2}=0\Rightarrow\hat{\lambda}_0(\hat{x}_2-\xi_2)+\hat{\lambda}_1\hat{x}_2/a_2^2=0$.

4. Let $\hat{\lambda}_0 = 0$. Then $\hat{\lambda}_1 \neq 0$ (the Lagrange multipliers are not equal to zero simultaneously). Hence, the third stage implies that $\hat{x}_1 = \hat{x}_2 = 0 \Rightarrow f_1(0, 0) = f_1(\hat{x}_1, \hat{x}_2) = 0 \neq 1$. Thus, $\hat{\lambda}_0 \neq 0$, and we can put $\lambda_0 = 1$. Let us denote $\hat{\lambda}_1 = \lambda$.

5. (For simplicity, we confine ourselves to the case when $\xi_1\xi_2 \neq 0$.)

$$(\hat{x}_i-\xi_i)+\lambda\hat{x}_i/a_i^2=0,\quad i=1,2\Rightarrow\hat{x}_i=\frac{\xi_i a_i^2}{(a_i^2+\lambda)},\quad i=1,2.$$

Substituting this into the equation of the ellipse, we obtain

$$\varphi(\lambda)=\frac{\xi_1^2a_1^2}{(a_1^2+\lambda)^2}+\frac{\xi_2^2a_2^2}{(a_2^2+\lambda)^2}=1. \tag{1}$$

6. The number of stationary points of the problem [i.e., the points corresponding to those λ which satisfy Eq. (1)] does not exceed four [see Fig. 26; the inequality $\varphi(0)=\xi_1^2/a_1^2+\xi_2^2/a_2^2>1$ shows that the graph of $\varphi(\lambda)$ is depicted for a point (ξ_1, ξ_2) lying outside the ellipse]. To complete the solution of the problem one must solve Eq. (1), find λ_i, find the corresponding points $x(\lambda_i)$, substitute these points into f_0, and choose the smallest of the resultant numbers.

Remarks. 1. Problem 1 is a smooth problem with equality constraints. In Sections 2-5 we used the Lagrange multiplier rule (Section 1.3.2).

2. The relations $(\hat{x}_i - \xi_i) + \lambda\hat{x}_i/a_i^2 = 0$ have an obvious geometrical meaning: the vector $\xi - \hat{x}$ joining the point ξ and the point of minimum $\hat{x}$ is proportional to the gradient vector of the function f_0 at the point $\hat{x}$, i.e., the vector $\xi - \hat{x}$ lies on the normal to the ellipse. Apollonius was the first to establish this fact.

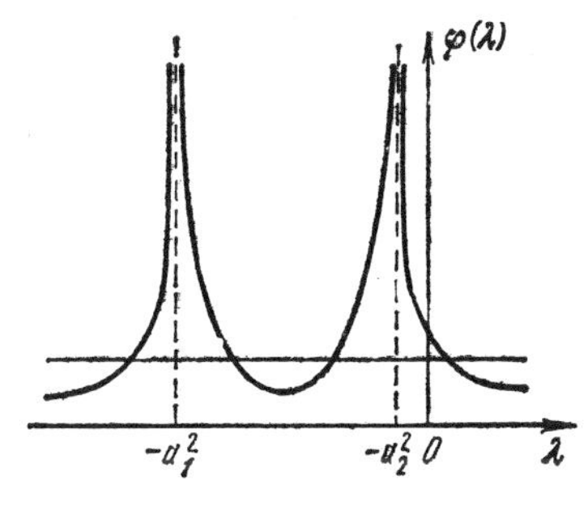

Fig. 26

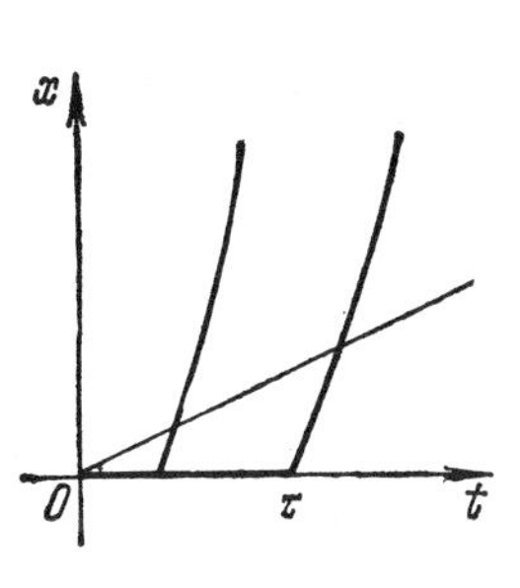

Fig. 27

3. Let us use the relations obtained to derive the equation of the curve "separating" those points ξ toward which two normals can be drawn and the points toward which four normals can be drawn. The separation obviously takes place for λ satisfying relation (1) for which

$$\varphi'(\lambda) = -\frac{\xi_1^2 a_1^2}{(a_1^2+\lambda)^3} - \frac{\xi_2^2 a_2^2}{(a_2^2+\lambda)^3} = 0, \quad \lambda \in (-a_1^2, -a_2^2). \tag{2}$$

From (2) we obtain

$$a_1^2+\lambda = A(\xi_1 a_1)^{2/3}, \quad a_2^2+\lambda = -A(\xi_2 a_2)^{2/3},$$

where

$$A = (a_1^2 - a_2^2)/((\xi_1 a_1)^{2/3} + (\xi_2 a_2)^{2/3}).$$

Substituting this into (1), we obtain the equation of the separating curve:

$$(\xi_1 a_1)^{2/3} + (\xi_2 a_2)^{2/3} = (a_2^2 - a_1^2)^{2/3}.$$

This is the equation of an *astroid* (see Fig. 7 in Section 1.1). To every point there correspond two normals when the point lies outside the astroid, four normals when it lies inside the astroid, and three normals on the astroid itself (with the exception of the cusps at which there are two normals). This result was also obtained by Apollonius in his "Conics."

Kepler's problem on an inscribed cylinder [see the formula (3) in Section 1.2.2]. The solution of this problem is similar to that of the problem of Euclid and we make no comments on it.

1. $f_0(x) = x(x^2-1) \to \inf, \quad 0 \leqslant x \leqslant 1.$

2-5. $f_0'(\hat{x}) = 0 \Rightarrow 3\hat{x}^2 = 1 \Rightarrow \hat{x} = \sqrt{3}/3.$

6. By virtue of the uniqueness of the stationary point in (0, 1), the point $\hat{x} = \sqrt{3}/3$ is the single solution in the problem: *the sought-for cylinder is characterized by the property that the ratio of its altitude* $2\hat{x}$ *to the radius* $\sqrt{1-\hat{x}^2}$ *is equal to* $\sqrt{2}$.

The Light Refraction Problem. This problem was posed and solved with the aid of Huygens' method in Section 1.1.3. Here we present its standard solution, which goes back to Leibniz [see the formula (4) in Section 1.2.2].

1. $f_0(x) = \sqrt{\alpha^2+x^2}/v_1 + \sqrt{\beta^2+(\xi-x)^2}/v_2 \to \inf.$

All the level sets $\mathscr{L}_a f_0$ of the continuous function f_0 are compact, and hence, by Weierstrass' theorem, there exists a solution of the problem.

2-5. $f'(\hat{x}) = 0 \Rightarrow \dfrac{\hat{x}}{v_1\sqrt{\alpha^2+\hat{x}^2}} = \dfrac{\xi-\hat{x}}{v_2\sqrt{\beta^2+(\xi-\hat{x})^2}} \Leftrightarrow \dfrac{\sin\varphi_1}{v_1} = \dfrac{\sin\varphi_2}{v_2}$

(see Fig. 19 in Section 1.2.2).

6. The point $\hat{x}$ satisfying the last equation is unique (check this!), and hence it is the single solution of the problem. Thus the point of refraction of a ray of light at the interface between two media is characterized by the property that *the ratio of the sine of the angle of incidence to the sine of the angle of refraction is equal to the ratio of the speeds of light in the corresponding media.* This is known as *Snell's law.*

Steiner's Problems. It was formalized thus (see the formula (5) in Section 1.2.2):

1. $f_0(x)=|x-\xi_1|+|x-\xi_2|+|x-\xi_3|\to\inf,\quad x\in\mathbf{R}^2,$
$\xi_i\in\mathbf{R}^2,\quad i=1, 2, 3,\quad |x|=\sqrt{x_1^2+x_2^2}.$

By Weierstrass' theorem, there exists a solution of this problem (check this!). There are two possible cases here: the solution coincides either with one of the points ξ_i, i = 1, 2, 3, or with none of them. We shall solve the problem for the latter case. Then the function is smooth in a neighborhood of the point $\hat{x}$ (check this!).

$$2\text{-}5.\quad f_0'(\hat{x})=0\Rightarrow\frac{\xi_1-\hat{x}}{|\hat{x}-\xi_1|}+\frac{\xi_2-\hat{x}}{|\hat{x}-\xi_2|}+\frac{\xi_3-\hat{x}}{|\hat{x}-\xi_3|}=0. \tag{3}$$

Equation (3) means that the sum of the three unit vectors drawn from $\hat{x}$ toward ξ_1, ξ_2, and ξ_3 is equal to zero. Hence, these vectors are parallel to the sides of an equilateral triangle, and consequently at the point $\hat{x}$ the angles of vision $\widehat{\xi_1\hat{x}\xi_2}$, $\widehat{\xi_2\hat{x}\xi_3}$, and $\widehat{\xi_3\hat{x}\xi_1}$ of the sides of the given triangle are equal to 120°. Thus, the point $\hat{x}$ is the *Torricelli point.* It can be found as the point of intersection of the two circular arcs with the chords $[\xi_1, \xi_2]$ and $[\xi_1, \xi_3]$ subtending angles of 120°. This construction is possible when none of the angles of the given triangle exceeds 120°. If otherwise, the arcs do not intersect, and hence it is impossible that the point $\hat{x}$ coincides with none of the points ξ_i, i = 1, 2, 3. Then it must obviously coincide with the vertex of the obtuse angle because the side of the triangle opposite to the obtuse angle is the greatest.

Answer: The sought-for point is the Torricelli point if all the angles of the triangle are less than 120° and it coincides with the vertex of the obtuse angle in all the other cases.

We have thus solved all the geometrical problems mentioned in Section 1.1.2 and also the light refraction problem (Section 1.1.3). In addition, let us solve *Tartaglia's problem*:

1. $f_0(x)=x(8-x)(8-2x)\to\sup,\quad 0\leqslant x\leqslant 4.$

2-5. $f_0'(\hat{x})=0\Rightarrow 3\hat{x}^2-24\hat{x}+32=0\Rightarrow\hat{x}=4-4/\sqrt{3}.$

6. Answer: One of the numbers is equal to $4-4/\sqrt{3}$ and the other is equal to $4+4/\sqrt{3}$.

1.6.2. Newton's Aerodynamic Problem. This problem was posed in Section 1.1.5 and formalized in Section 1.2.3:

$$1.\quad \int_0^T\frac{t\,dt}{1+u^2}\to\inf,\quad \dot{x}=u,\quad x(0)=0,\quad x(T)=\xi,\quad u\in\mathbf{R}_+.$$

It is by far not a simple task to prove directly the existence theorem for problems of this kind. The main obstacle is the nonconvexity of the integrand $t/(1+u^2)$ with respect to u for $u\geqslant 0$. However, we shall present an exhaustive solution of this problem. The plan of the solution is the following. First we shall suppose that there exists a solution of the problem and apply the Lagrange principle to the hypothetical solution. In this way we shall find that there is a single admissible stationary curve in the problem (i.e., an admissible curve for which all the conditions implied by the Lagrange principle hold). Then we shall show by means of a direct calculation that this very curve yields an absolute minimum. In conclusion we shall recall Newton's words and verify that the resultant solution coincides exactly with the one described by Newton in 1687.

2. $\mathscr{L} = \int_0^T L\,dt + \mu_0 x(0) + \mu_1 (x(T) - \xi),$

$L = \frac{\lambda_0 t}{1+u^2} + p(\dot{x} - u).$

3. *The Euler equation*:

$$-\frac{d}{dt} L_{\dot{x}} + L_x = 0 \Rightarrow \hat{p}(t) = \text{const} = p_0. \quad (1)$$

The transversality condition:

$$p_0 = \hat{\mu}_0 = -\hat{\mu}_1. \quad (2)$$

The minimality condition in u:

$$\frac{\hat{\lambda}_0 t}{1+u^2} - p_0 u \geqslant \frac{\hat{\lambda}_0 t}{1+\hat{u}^2(t)} - p_0 \hat{u}(t), \quad \forall u \geqslant 0. \quad (3)$$

4. If we assume that $\hat{\lambda}_0 = 0$, then it is necessary that $p_0 \neq 0$ [because, if otherwise, (2) would imply the equalities $\hat{\lambda}_0 = p_0 = \hat{\mu}_0 = \hat{\mu}_1 = 0$ while the Lagrange multipliers are not all zero]. In case $\hat{\lambda}_0 = 0$ and $p_0 \neq 0$, the relation (3) implies that $\hat{u}(t) \equiv 0$, and hence $\hat{x}(t) = \int_0^t \hat{u}(t)\,d\tau \equiv 0$. But then the sought-for body has no "length"; i.e., it is a plane. If $\xi > 0$, then $\hat{\lambda}_0 \neq 0$, and we can assume that $\hat{\lambda}_0 = 1$.

It should also be noted that the inequality $p_0 \geqslant 0$ is also impossible because in this case the function $t/(1 + u^2) - p_0 u$ is monotone decreasing and (3) does not hold for $u > u(t)$.

5. From (3) (with $\hat{\lambda}_0 = 1$) it follows that until an instant of time the optimal control is equal to zero [verify that for $p_0 < 0$ and small t the function $u \to (t/(1 + u^2)) - p_0 u$ attains its minimum at $u = 0$]. After that the optimal control $\hat{u}(\cdot)$ must be found from the equation

$$-p_0 = \frac{2ut}{(1+u^2)^2}, \quad (4)$$

which is obtained from the equation $L_u = 0$. The instant of time τ corresponding to the corner point of the control is determined by the condition that the function $u \to (\tau/(1 + u^2)) - p_0 u$ has two equal minima: at zero and at the point found from (4) for $t = \tau$. In other words, at time τ the relations

$$-p_0 = \frac{2\hat{u}(\tau)\tau}{(1+\hat{u}^2(\tau))^2}, \quad \frac{\tau}{1+\hat{u}^2(\tau)} - p_0 \hat{u}(\tau) = \tau \quad (5)$$

must be satisfied [here and henceforth $\hat{u}(\tau)$ designates $\hat{u}(\tau + 0) \neq 0$]. From the second equation we obtain $-\hat{u}^2(\tau)\tau/(1 + \hat{u}^2(\tau)) = p_0\hat{u}(\tau)$, whence $p_0 = -\tau\hat{u}(\tau)/(1 + \hat{u}^2(\tau))$. Substituting this relation into the first equation (5), we find that $\hat{u}^2(\tau) = 1 \Rightarrow \hat{u}(\tau) = 1$ (because $\hat{u} \geqslant 0$), and then the first equation (5) yields the equality $\tau = -2p_0$.

For the time after the corner point the optimal solution satisfies the relation (4) which implies that

$$t = -\frac{p_0(1+u^2)^2}{2u} = -\frac{p_0}{2}\left(\frac{1}{u} + 2u + u^3\right). \quad (6)$$

Now we write

$$\frac{dx}{dt} = u \Rightarrow \frac{dx}{du} = \frac{dx}{dt}\frac{dt}{du} = u\frac{dt}{du} = -\frac{p_0}{2}\left(-\frac{1}{u} + 2u + 3u^3\right).$$

Integrating this relation and taking into account the equalities $\hat{x}(\tau) = 0$ and $\hat{u}(\tau) = 1$, we obtain parametric equations of the sought-for optimal curve:

$$\hat{x}(t, p_0) = -\frac{p_0}{2}\left(\lg\frac{1}{u} + u^2 + \frac{3}{4}u^4\right) + \frac{7}{8}p_0,$$
$$t = -\frac{p_0}{2}\left(\frac{1}{u} + 2u + u^3\right), \quad p_0 < 0. \tag{7}$$

6. The curve (7) is called *Newton's curve*. Moreover, $u \in [1, \infty)$ in (7). It can easily be understood that there is a single point of intersection of the straight line $x = \alpha t$ and Newton's curve corresponding to the value $p_0 = -1$ of the parameter. Indeed, the function $x(\cdot, -1)$ is continuous and convex because

$$\frac{d^2x}{dt^2} = \frac{du}{dt} = \frac{1}{dt/du} \geqslant 0.$$

On the other hand, it is seen from the formulas (7) that Newton's curve $\hat{x}(\cdot, p_0)$ is obtained from the curve $\hat{x}(\cdot, -1)$ by means of the homothetic transformation with center at (0, 0) and the coefficient $|p_0|$ (Fig. 27). Hence, to draw a curve belonging to the family (7) through the given point (ξ, T) we must find the point of intersection of the straight line $x = \xi t/T$ and the curve $\hat{x}(\cdot, -1)$ and then perform the corresponding homothetic transformation of the curve $\hat{x}(\cdot, -1)$. In this way we obtain an admissible curve $\hat{x}(\cdot)$. Let us show that it yields an absolute minimum in the problem. To this end we come back to the relation (3) with $\hat{\lambda}_0 = 1$. Let $x(\cdot)$ be an arbitrary admissible curve [i.e., $x(\cdot) \in KC^1([0, T])$, $x(0) = 0$, and $x(T) = \xi$]. Then, by virtue of (3),

$$\frac{t}{1+\dot{x}^2(t)} - p_0\dot{x}(t) \geqslant \frac{t}{1+\hat{u}(t)^2} - p_0\hat{u}(t).$$

Integrating this relation and taking into account that $\hat{u}(t) = \dot{\hat{x}}(t)$ and $\int_0^T \dot{x}(t)\,dt = \int_0^T \dot{\hat{x}}(t)\,dt = \xi$, we obtain

$$\int_0^T \frac{t\,dt}{1+\dot{x}^2} \geqslant \int_0^T \frac{t\,dt}{1+\dot{\hat{x}}^2}.$$

The problem has been solved completely.

Remarks. 1. In Sections 2-5 we applied the Lagrange principle for optimal control problems which is reducible to the Pontryagin maximum principle (Section 1.5.3).

2. Let us compare the solution obtained with the solution described by Newton himself. Let us recall Newton's words quoted in Section 1.1.5 and look once again at his diagram shown in Fig. 18. In the figure, besides the letters set by Newton himself, some additional symbols are written. We have $|MN| = t$, $|BM| = x$, $|BG| = \tau$, and $BGP = \varphi$; now Newton's construction implies

$$\tan\varphi = \dot{x}(t), \quad |BP|/|BG| = \tan\varphi \Rightarrow |BP| = \tau\dot{x},$$
$$|GP|^2 = |BG|^2 + |BP|^2 = (\dot{x}^2 + 1)\tau^2.$$

Substituting our symbols into Newton's proportion

$$|MN| : |GP| = |GP|^3 : 4(|BP| \times |GB|^2),$$

we obtain

$$\frac{t}{(\dot{x}^2+1)^{1/2}\tau} = \frac{\tau^3(\dot{x}^2+1)^{3/2}}{4\tau\dot{x}\tau^2} \Leftrightarrow \frac{\dot{x}t}{(\dot{x}^2+1)^2} = \frac{\tau}{4}. \tag{8}$$

We see that this is nothing other than the relation (4) into which the value $p_0 = -\tau/2$ is substituted. With the aid of the same reasoning as before, we find from (8), by means of integration, the expression (7) for Newton's curve. It should also be noted that the "bluntness" of the curve and the condition on the jump at the point $G \Leftrightarrow \tau$ (where the angle is equal to 135°) were in fact indicated by Newton in his "Scholium" on a frustum of a cone.

We see that Newton solved his problem completely, but the meaning of the solution was beyond the understanding of his contemporaries and many of his followers (up to the present time).

1.6.3. The Simplest Time-Optimal Problem. This problem was posed in Section 1.1.7 and was formalized in Section 1.2.4 in the following way:

$$T \to \inf, \quad m\ddot{x} = u, \quad u \in [u_1, u_2], \qquad (1)$$
$$x(0) = x_0, \quad \dot{x}(0) = v_0, \quad x(T) = \dot{x}(T) = 0$$

[see the formulas (5) in Section 1.2.4]. The case $u_1 = u_2$ is of no interest because in this situation for some pairs (x_0, v_0) there exist no functions $x(\cdot)$ satisfying all the constraints (1), and if such a function exists and is not identically equal to zero, the value of T is determined uniquely, and so there is no minimization problem.

For $u_1 < u_2$ we can reduce the number of the parameters in the problem by means of the substitution $x(t) = A\xi(t) + B(t - T)^2$. In terms of the function $\xi(\cdot)$ the general form of the problem (1) does not change, but the parameters x_0, v_0, u_1, and u_2 assume some other values. In particular, if we put

$$A = (u_1 - u_2)/2m, \quad B = (u_1 + u_2)/4m,$$

then $\ddot{\xi} \in [-1, 1]$. Taking into consideration what has been said, we shall assume that $m = 1$, $u_1 = -1$, and $u_2 = +1$ in (1). Moreover, we denote $x = x_1$ and $\dot{x} = x_2$.

Now we proceed to the realization of our standard scheme:

$$1. \quad T = \int_0^T 1 \cdot dt \to \inf, \quad \dot{x}_1 = x_2, \quad \dot{x}_2 = u, \quad u \in [-1, 1], \qquad (2)$$
$$x_1(0) = x_0, \quad x_2(0) = v_0, \quad x_1(T) = x_2(T) = 0.$$

The existence of the solution will be treated in exactly the same way as in the foregoing section: on finding with the aid of the Lagrange principle a function $x(\cdot)$ "suspected" to be optimal, we verify directly that it yields the solution in the problem.

$$2. \quad \mathscr{L} = \int_0^T L\,dt + \mu_1(x_1(0) - x_0) + \mu_2(x_2(0) - v_0) + \nu_1 x_1(T) + \nu_2 x_2(T),$$

where $L = \lambda_0 + p_1(\dot{x}_1 - x_2) + p_2(\dot{x}_2 - u)$.

3. *The Euler–Lagrange equations*:

$$-\frac{d}{dt}L_{\dot{x}_i} + L_{x_i} = 0, \quad i = 1, 2 \Rightarrow \frac{d\hat{p}_1}{dt} = 0, \quad \frac{d\hat{p}_2}{dt} = -\hat{p}_1. \qquad (4)$$

The transversality conditions:

$$\hat{p}_1(0) = \hat{\mu}_1, \quad \hat{p}_2(0) = \hat{\mu}_2, \quad \hat{p}_1(T) = -\hat{\nu}_1, \quad \hat{p}_2(T) = -\hat{\nu}_2. \qquad (5)$$

The Maximum Principle: dropping the summands independent of u, we can write this condition in the form

$$\hat{p}_2(t)\hat{u}(t) = \max_{-1 \le u \le 1} \{\hat{p}_2(t) u\} = |\hat{p}_2(t)|,$$

or

$$\hat{u}(t)=\begin{cases}\operatorname{sgn}\hat{p}_2(t), & \text{if } \hat{p}_2(t)\neq 0,\\ \text{an arbitrary value belonging to } [-1,1], & \text{if } \hat{p}_2(t)=0.\end{cases} \tag{6}$$

Moreover, the terminal instant of time T is also variable in the problem under consideration (this is the quantity that must be minimized), and so, formally, this problem lies out of the framework of Section 1.5.3. However, it can be shown (this will be done in Chapter 4) that in such a situation the condition $\mathscr{L}_T=0$ should be added (the stationarity condition of the Lagrange function with respect to T), which is in complete agreement with the general idea of the Lagrange principle. Differentiating (3) with respect to T and taking into account the equalities $x_2(T) = u(T)$ and $\dot{x}_1(T) = x_2(T) = 0$, we obtain

$$\mathscr{L}_T=\hat{\lambda}_0+\hat{v}_2\hat{u}(T)=0. \tag{7}$$

4. It is unimportant in the problem under consideration whether or not $\hat{\lambda}_0$ vanishes because $\hat{\lambda}_0$ is not involved in (4)-(6).

5. From Eqs. (4) we conclude that $\hat{p}_1(t) \equiv$ const and $\hat{p}_2(\cdot)$ is an arbitrary linear function. Moreover, $p_2(t) \not\equiv 0$ because

$$\hat{p}_2(t)\equiv 0\overset{(4)}{\Rightarrow}\hat{p}_1(t)\equiv 0\overset{(5)}{\Rightarrow}\hat{\mu}_1=\hat{\mu}_2=\hat{v}_1=\hat{v}_2=0\overset{(7)}{\Rightarrow}\hat{\lambda}_0=0$$

and all the Lagrange multipliers turn out to be zero.

A linear function which does not vanish identically has not more than one zero on the closed interval [0, T]. Therefore, from (6) we obtain the following possible cases for the optimal control:

a) $$\hat{u}(t)\equiv 1;$$

b) $$\hat{u}(t)\equiv -1$$

($\hat{p}_2(\cdot)$ does not vanish on [0, T]);

c) $$\hat{u}(t)=\begin{cases}1, & 0\leqslant t<\tau,\\ -1, & \tau<t\leqslant T;\end{cases}$$

d) $$\hat{u}(t)=\begin{cases}-1, & 0\leqslant t<\tau,\\ 1, & \tau<t\leqslant T\end{cases}$$

[$\hat{p}_2(\cdot)$ vanishes at the point $t = \tau$, the value of the control at the point τ is of no importance because its change at one point does not affect the function $\hat{x}(\cdot)$ (why?); for the same reason, the cases $\tau = T$ and $\tau = 0$ can be dropped].

It is convenient to carry out the further analysis using the phase plane (x_1, x_2) (Fig. 28). Solving Cauchy's problem

$$\dot{x}_1=x_2,\quad \dot{x}_2=\hat{u}(t),\quad x_1(T)=x_2(T)=0 \tag{8}$$

for one of the controls a)-d) we obtain a single solution together with a single initial point $(x_0, v_0) = (x_1(0), x_2(0))$ corresponding to this solution. It can easily be verified that for all the possible values of τ and T these initial points cover the whole plane in a one-to-one manner. First of all, we have

$$\dot{x}_1=x_2,\quad \dot{x}_2=1\Rightarrow x_1=x_2^2/2+C_1, \tag{9}$$

$$\dot{x}_1=x_2,\quad \dot{x}_2=-1\Rightarrow x_1=-x_2^2/2+C_2, \tag{10}$$

and hence for the intervals of constancy of the control the phase trajectories lie on the parabolas of one of the families (9) and (10).

The initial points corresponding to the control a) lie on the arc OFA (Fig. 28): $x_0 = v_0^2/2$, $v_0 < 0$; the points corresponding to the control b) lie on the arc ODB: $x_0 = -v_0^2/2$, $v_0 > 0$.

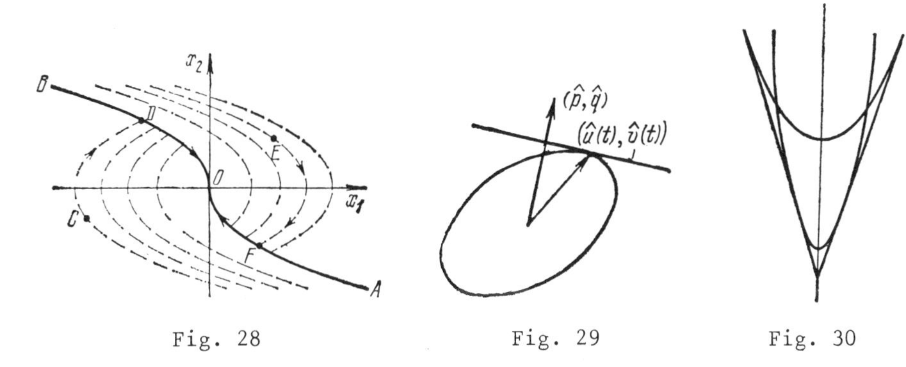

Fig. 28 Fig. 29 Fig. 30

The initial points C lying to the left of the separating curve BDOFA ($x_0 = -v_0|v_0|/2$) correspond to the controls c: $\hat{u}(t) \equiv 1$ on the arc CD of the family (9); to time $t = \tau$ there corresponds the point D at which the switching of the control takes place; after that the moving point on the phase trajectory traces the arc DO with $\hat{u}(t) \equiv -1$. Similarly, to the initial points E lying to the right of the separating curve there corresponds the controls d).

6. It remains to show that the single extremal $\hat{x}(t) = \hat{x}_1(t)$ we have found [corresponding to the given initial point (x_0, v_0)] does in fact yield a solution in the problem (2).

Let us suppose that a function $x(\cdot)$ is defined on a closed interval $[0, \tilde{T}]$ where $\tilde{T} \leqslant T$, possesses the piecewise-continuous second derivative, and satisfies the conditions $x(0) = x_0$, $\dot{x}(0) = v_0$, $x(\tilde{T}) = \dot{x}(\tilde{T}) = 0$. Let us extend $x(\cdot)$ to the interval $[\tilde{T}, T]$ (for $\tilde{T} \leqslant T$) by putting $x(t) \equiv 0$, $t \in [\tilde{T}, T]$. After that the two functions $x(\cdot)$ and $\hat{x}(\cdot)$ will be defined on one and the same closed interval $[0, T]$ and will have the same boundary values:

$$x(0)=\hat{x}(0)=x_0, \quad \dot{x}(0)=\dot{\hat{x}}(0)=v_0, \\ x(T)=\hat{x}(T)=\dot{x}(T)=\dot{\hat{x}}(T)=0. \tag{11}$$

Let us show that if $|\ddot{x}| \leqslant 1$, then $x(\cdot) = \hat{x}(\cdot)$, and, in particular, the inequality $\tilde{T} < T$ is impossible. This will prove the optimality of $x(\cdot)$. By virtue of the symmetry of the problem, we may confine ourselves to the control c) [or to its limiting case a)]. If $|\ddot{x}| \leqslant 1$, then integrating twice the inequality $\ddot{x}(t) \leqslant 1$ and taking into account (11), we obtain

$$\hat{x}(\tau)-x(\tau)=\int_0^\tau\int_0^t(1-\ddot{x}(s))\,ds\,dt\geqslant 0, \tag{12}$$

where the equality can only take place if $\ddot{x}(s) \equiv 1$ at all the points of continuity, and then $x(t)\equiv\hat{x}(t)$, $t\in[0,\tau]$.

Similarly, integrating twice the inequality $\ddot{x}(t) \geqslant -1$, we obtain

$$\hat{x}(\tau)-x(\tau)=\int_\tau^T\int_t^T(-1-\ddot{x}(s))\,ds\,dt\leqslant 0, \tag{13}$$

and here too the equality can take place only if $\ddot{x}(s) \equiv -1$ and $x(t) \equiv \hat{x}(t)$, $t\in[\tau, T]$.

However, comparing (12) and (13), we conclude that $\hat{x}(\tau) = x(\tau)$, and then, as was already mentioned, $\hat{x}(t)\equiv x(t)$, $t\in[0, T]$.

Remark. Relations (5)-(7) imply the equality $\hat{\lambda}_0 = |p_2(T)|$, and hence the equality $\lambda_0 = 0$ turns out to be possible in this problem when $p_2(\cdot)$ vanishes for $t = T$. Then $p_2(\cdot) \not\equiv 0$ (because, if otherwise, all the Lagrange multipliers would be equal to zero), and, by virtue of the equality $p_2(T) = 0$, the function $p_2(\cdot)$ does not change sign, and thus there are no switchings of the control at all. Consequently, the case $\hat{\lambda}_0 = 0$ corresponds to the the motion along the lines of switching AFO and BDO.

1.6.4. The Classical Isoperimetric Problem and the Chaplygin Problem. The most ancient extremal problem (the first problem mentioned in the title of this section) was posed in Section 1.1.1 and formalized in various ways in Section 1.2.4. In particular, the formalization (2) in Section 1.2.4 reduced the problem to a more general problem embracing the Chaplygin problem as well. Proceeding from this formalization, we shall present the solution of the two problems using our standard scheme. Thus:

1.
$$S=\frac{1}{2}\int_0^T (xv-yu)\,dt \to \sup;$$
$$\dot{x}=u,\quad \dot{y}=v,\quad (u,\, v)\in A,\quad x(0)=x(T),\quad y(0)=y(T). \tag{1}$$

We shall assume that the set A of the admissible velocities is a closed, convex, and bounded set in $\mathbf{R}^2$.

Exercise. Show that for the solvability of the problem stated it is necessary that $0\in A$. Hint: one can make use of the result of Exercise 2 in Section 2.2.3.

2.
$$\mathcal{L}=\int_0^T L\,dt+\mu\,(x(0)-x(T))+\nu\,(y(0)-y(T)), \tag{2}$$

where

$$L=-\frac{\lambda_0}{2}(xv-yu)+p\,(\dot{x}-u)+q\,(\dot{y}-v).$$

3. *The Euler–Lagrange equations*:

$$-\frac{d}{dt}L_{\dot{x}}+L_x=0\Rightarrow-\dot{\hat{p}}-\frac{\hat{\lambda}_0}{2}\hat{v}=0,$$
$$-\frac{d}{dt}L_{\dot{y}}+L_y=0\Rightarrow-\dot{\hat{q}}+\frac{\hat{\lambda}_0}{2}\hat{u}=0. \tag{3}$$

The transversality conditions:

$$\hat{p}(0)=\hat{p}(T)=\hat{\mu},\quad \hat{q}(0)=\hat{q}(T)=\hat{\nu}.$$

The maximum principle:

$$\left(\hat{p}(t)-\frac{\hat{\lambda}_0}{2}\hat{y}(t)\right)\hat{u}(t)+\left(\hat{q}(t)+\frac{\hat{\lambda}_0}{2}\hat{x}(t)\right)\hat{v}(t)=$$
$$=\max_{(u,\,v)\in A}\left(\left(\hat{p}(t)-\frac{\hat{\lambda}_0}{2}\hat{y}(t)\right)u+\left(\hat{q}(t)+\frac{\hat{\lambda}_0}{2}\hat{x}(t)\right)v\right). \tag{4}$$

4. If we assume that $\hat{\lambda}_0 = 0$, then the Euler–Lagrange equations imply that $\hat{p}(t) \equiv \text{const}$, $\hat{q}(t) \equiv \text{const}$, and, moreover, $\hat{p}^2 + \hat{q}^2 > 0$ because, if otherwise, the transversality conditions would imply that all the Lagrange multipliers are equal to zero. Now the maximum principle takes the form

$$\hat{p}\hat{u}(t)+\hat{q}\hat{v}(t)=\max_{(u,\,v)\in A}(\hat{p}u+\hat{q}v),$$

whence it is seen that the point $(\hat{u}(t),\, \hat{v}(t))$ belongs to one and the same straight line all the time, namely, to one of the two straight lines supporting to the set A which are perpendicular to the vector $(\hat{p},\, \hat{q})$ (Fig. 29). Therefore,

$$\dot{\hat{x}}(t)=\hat{u}(t)=\hat{u}(0)-\alpha(t)\,\hat{q},$$
$$\dot{\hat{y}}(t)=\hat{v}(t)=\hat{v}(0)+\alpha(t)\,\hat{p}.$$

Integrating these equalities, we obtain

$$\hat{x}(t)=\hat{x}(0)+\hat{u}(0)t-\int_0^t \alpha(t)\,dt\cdot\hat{q},$$
$$\hat{y}(t)=\hat{y}(0)+\hat{v}(0)t+\int_0^t \alpha(t)\,dt\cdot\hat{p}. \tag{5}$$

Putting $t = T$ in (5) and using the boundary conditions $x(0) = x(T)$ and $y(0) = y(T)$, we conclude that the vector $(\hat{q}, -\hat{p})$ is proportional to $(\hat{u}(0), \hat{v}(0))$, after which it is seen from (5) that the point $(\hat{x}(t), \hat{y}(t))$ all the time lies on the straight line parallel to the vector $(\hat{q}, -\hat{p})$ and passing through the point $(\hat{x}(0), \hat{y}(0))$. Consequently, the closed curve $\{(\hat{x}(t), \hat{y}(t)), 0 \leqslant t \leqslant T\}$ degenerates (lies on a straight line), and the area bounded by it is equal to zero (prove this analytically proceeding from the indicated expression for S!). Consequently, this curve cannot be optimal.

5. The 4th stage implies that we can put $\hat{\lambda}_0 = 1$. Then from (3) it follows that

$$\dot{\hat{p}}+\frac{\hat{v}}{2}=0\Rightarrow\hat{p}(t)+\frac{\hat{y}(t)}{2}=b=\text{const},$$
$$\dot{\hat{q}}-\frac{\hat{u}}{2}=0\Rightarrow\hat{q}(t)-\frac{\hat{x}(t)}{2}=-a=\text{const}.$$

Substituting these expressions into (4), we derive

$$(\hat{x}(t)-a)\hat{v}(t)-(\hat{y}(t)-b)\hat{u}(t)=\max_{(u,\,v)\in A}\{(\hat{x}(t)-a)v-(\hat{y}(t)-b)u\}. \tag{6}$$

Now we lay aside the general case and confine ourselves to two special versions.

a) <u>The Classical Isoperimetric Problem.</u> Here

$$A=\{(u,\,v)\,|\,u^2+v^2\leqslant 1\}$$

and from (6) we find

$$\dot{\hat{x}}(t)=\hat{u}(t)=-\frac{\hat{y}(t)-b}{\mathcal{H}},\quad \dot{\hat{y}}(t)=\hat{v}(t)=\frac{\hat{x}(t)-a}{\mathcal{H}}, \tag{7}$$

where

$$\mathcal{H}=((\hat{x}(t)-a)^2+(\hat{y}(t)-b)^2)^{1/2}$$

[the scalar product of the vectors $(\hat{x} - a, \hat{y} - b)$ and $(v, -u)$ assumes its maximum value when the vectors are in one direction and the second of them has the maximum possible length, i.e., is of unit length]. From (7) we derive

$$\frac{d}{dt}(\mathcal{H}^2)=0,$$

i.e.,

$$\mathcal{H}(\hat{x}(t),\hat{y}(t))=((\hat{x}(t)-a)^2+(\hat{y}(t)-b)^2)^{1/2}=$$
$$=\max_{(u,\,v)\in A}((\hat{x}(t)-a)v-(\hat{y}(t)-b)u)=R=\text{const} \tag{8}$$

on the optimal trajectory. It follows that *the optimal trajectory is a circle of radius* R *with center at the point* (*a*, b). Since the angular velocity of the motion along this circle is equal to

$$\frac{d\varphi}{dt}=\frac{1}{R^2}((x-a)\dot{y}-(y-b)\dot{x})\equiv\frac{1}{R},$$

and since the moving point must return to the initial position (possibly after it has made several revolutions) in time T, we have $2\pi n = T/R$ whence $R = T/2\pi n$. The area bounded by this curve is equal to

$$S=\frac{1}{2}\int_0^T (xv-yu)dt=\frac{1}{2}\int_0^T (x\dot{y}-y\dot{x})dt=\frac{1}{2}\int_0^T ((x-a)\dot{y}-(y-b)\dot{x})dt=\frac{R^2}{2}\frac{T}{R}=\frac{T^2}{4\pi n}.$$

Consequently, $n = 1$ (the circle is traced only once), and $S = T^2/4\pi$. It should be noted that here T is the length of the curve.

b) The Chaplygin Problem. Here

$$A=\{(u, v)\,|\,(u-u_0)^2+v^2\leqslant V^2\},$$

$(u_0, 0)$ is the wind velocity, and $V \geqslant |u_0|$ is the maximum velocity of the airplane. To solve the auxiliary extremal problem

$$(\hat{x}(t)-a)v-(\hat{y}(t)-b)u\rightarrow \sup,$$
$$(u-u_0)^2+v^2\leqslant V^2,$$

we can make the substitution $u = u_0 + \xi$, $v = \eta$, and use the same geometrical consideration as in the foregoing case. This results in the equations

$$\dot{\hat{x}}(t)=\hat{u}(t)=u_0-V\frac{\hat{y}(t)-b}{R(t)},\quad \dot{\hat{y}}(t)=\hat{v}(t)=V\frac{\hat{x}(t)-a}{R(t)},$$

where the quantity

$$R(t)=((\hat{x}(t)-a)^2+(\hat{y}(t)-b)^2)^{1/2}$$

is no longer constant. However,

$$\frac{dR(t)}{dt}=\frac{1}{R(t)}((\hat{x}(t)-a)\dot{\hat{x}}(t)+(\hat{y}(t)-b)\dot{\hat{y}}(t))=\frac{\hat{x}(t)-a}{R(t)}u_0=\frac{u_0}{V}\frac{d\hat{y}(t)}{dt},$$

and so we obtain

$$\mathscr{H}(\hat{x}(t), \hat{y}(t))=R(t)-\frac{u_0}{V}\hat{y}(t)=((\hat{x}(t)-a)^2+(\hat{y}(t)+b)^2)^{1/2}-\frac{u_0}{V}\hat{y}(t)=\text{const.} \tag{9}$$

Consequently, *the optimal curve is an ellipse determined by the equation*

$$((x-a)^2+(y-b)^2)^{1/2}-\frac{u_0}{V}y=\text{const.} \tag{10}$$

As in the foregoing case, it can be shown that the moving point $(\hat{x}(t), \hat{y}(t))$ traces the ellipse only once. It should be noted that if the velocity of the airplane were less than the wind velocity ($V < |u_0|$), the airplane could not return to the initial point (prove this; cf. the exercise at the beginning of the present section), and the curve (10) would be a hyperbola, i.e., nonclosed.

Remark. Contrary to our scheme, we have disregarded the question of the existence of the solution. The matter is that a simple reference to Weierstrass' theorem will not do here. Let us outline briefly a possible course of reasoning.

First of all, without loss of generality, we can assume that $x(0) = y(0) = 0$.

Using Arzela's theorem (see [KF, pp. 110 and 111]), we can show that the set of the pairs of functions $\{(x(\cdot), y(\cdot)\}$ satisfying the generalized Lipschitz condition

$$\left(\frac{x(t_1)-x(t_2)}{t_1-t_2}, \frac{y(t_1)-y(t_2)}{t_1-t_2}\right)\in A,\ t_1\neq t_2,\ 0\leqslant t_1,\ t_2\leqslant T \tag{11}$$

on the closed interval $[0, T]$ and the boundary conditions

$$x(0)=x(T)=0, \quad y(0)=y(T)=0$$

is compact in the space $C([0, T]) \times C([0, T])$.

Further, the "area" functional is defined on this set and is a lower semicontinuous function. Applying the corresponding generalization of Weierstrass' theorem, we prove the existence of the solution $(\hat{x}(\cdot), \hat{y}(\cdot))$.

Finally, the functions satisfying the condition (11) are differentiable almost everywhere, and moreover $(\dot{\hat{x}}(t), \dot{\hat{y}}(t)) \in A$. The maximum principle can be extended to this situation, and consequently the above reasoning is applicable to the solution $(\hat{x}(\cdot), \hat{y}(\cdot))$.

1.6.5. The Brachistochrone Problem and Some Geometrical Problems. Let us consider the following series of the simplest problems of the classical calculus of variations:

1. $$\mathcal{J}(y(\cdot))=\int_{x_0}^{x_1} y^\alpha \sqrt{1+y'^2}\,dx \to \inf; \quad y(x_0)=y_0, \quad y(x_1)=y_1. \tag{1}$$

It embraces a number of interesting problems: for $\alpha = 0$ we obtain the problem on the shortest lines in a plane, for $\alpha = -1/2$ we obtain the brachistochrone problem (see Section 1.2.4), for $\alpha = 1$ we obtain the minimum-surface-of-revolution problem, and for $\alpha = -1$ this is the problem on the shortest lines (the geodesics) on the Lobachevskian plane in Poincaré's interpretation (for the Poincaré half-plane see [13, pp. 131 and 132]).

The integrand $f_\alpha = y^\alpha\sqrt{1 + y'^2}$ in the problems (1) does not depend on x. Therefore, the Euler equation (Section 1.4.1) possesses the energy integral from which we find all the extremals lying in the upper half-plane.

2-4. $$y' f_{\alpha y'} - f = \text{const} \Rightarrow y^{-2\alpha}(1+y'^2)=D^2. \tag{2}$$

5. First we shall consider the case of negative α: $\alpha = -\beta$, $\beta > 0$. Then

$$\frac{y^\beta\,dy}{\sqrt{D^2-y^{2\beta}}}=dx \Rightarrow x-C_1=\int \frac{y^\beta\,dy}{\sqrt{D^2-y^{2\beta}}}. \tag{3}$$

Now we change the variable of integration:

$$y^{2\beta}=D^2\sin^2 t \Rightarrow y=D^{1/\beta}\sin^{1/\beta} t \Rightarrow dy=\frac{D^{1/\beta}}{\beta}\sin^{1/\beta-1} t\cos t\,dt.$$

Substituting the resultant expressions into (3), we arrive at the following relations (in which $D^{1/\beta} = C$):

$$y=C\sin^{1/\beta} t, \quad x=C_1+\frac{C}{\beta}\int_0^t \sin^{1/\beta} s\,ds.$$

In particular, if $\beta = 1/2$ (the brachistochrone problem), we obtain a family of cycloids represented parametrically ($\tau = 2t$):

$$\hat{y}(\tau, C, C_1)=\frac{C}{2}(1-\cos\tau), \quad \hat{x}(\tau, C, C_1)=C_1+\frac{C}{2}(\tau-\sin\tau).$$

If $\beta = 1$ (the case of the Poincaré half-plane), we obtain a family of semicircles:

$$\hat{y}(t, C, C_1)=C\sin t, \quad \hat{x}(t, C, C_1)=C_1-C\cos t \Rightarrow (\hat{x}-C_1)^2+\hat{y}^2=C^2.$$

Among the problems with $\alpha > 0$, we shall only consider the minimum-surface-of-revolution problem. From (2) we derive

$$\frac{dy}{\sqrt{D^2y^2-1}}=dx.$$

Using the substitution

$$Dy = \cosh t \Rightarrow D\,dy = \sinh t\,dt, \quad \sqrt{D^2y^2 - 1} = \sinh t,$$

we find the two-parameter family of catenaries:

$$y = \frac{1}{D} \cosh (Dx + D_1).$$

For all the cases we have obtained two-parameter families of extremals (solutions of the Euler equation), and now we must choose the constants of integration in such a way that the boundary conditions $y(x_i) = y_i$ are satisfied and then pass to the 6th stage of our scheme, i.e., analyzing the solutions we have found and writing the answer.

However, here we shall not do this. If, in addition, the reader recalls that we have again evaded the question of the existence of the solution, he will have even more ground than in the foregoing section to declare: "Something is rotten in the state of Extremum!"* Indeed:

For $\alpha > 0$ the problem (1) may have no ordinary solution. For example, since the boundary conditions in the minimum-surface-of-revolution problem ($\alpha = 1$) are symmetric ($y(-x_0) = y(x_0)$), one should seek a symmetric extremal, i.e., $y = \cosh Dx/D$. All the extremals of this form are obtained from the extremal $y = \cosh x$ by means of a homothetic transformation, and their collection does not cover the whole plane and only fills an angular region (Fig. 30). Therefore, a problem with such boundary conditions, say $y(-x_0) = y(x_0) = 1$, is unsolvable for sufficiently large x_0.

In the case $\alpha < 0$ the integrand in the problem (1) tends to infinity for $y \to 0$. Therefore, for instance, the brachistochrone problem ($\alpha = -1/2$) with ordinary boundary conditions $y(x_0) = 0$, $y(x_1) > 0$ (Section 1.1.4) does in fact lie outside the framework of the standard scheme.

The complete solution of the problems of the series (1) requires some additional work.

*"Something is rotten in the state of Denmark" (*Shakespeare*, Hamlet, Act I, Sc. IV).

Chapter 2

MATHEMATICAL METHODS OF THE THEORY OF EXTREMAL PROBLEMS

Now the reader is familiar with various statements of extremal problems, with basic notions and general ideas, and with some methods of solving these problems. This chapter is devoted to a systematic and rather formal presentation of the corresponding mathematical theory.

The generality of the theory and its comparative simplicity are due to the free use of facts and notions of the related divisions of mathematics and, particularly, of functional analysis. Here they are presented in the form suitable for our aims. A considerable part of the material of this chapter comprises what could be called "elements of differential calculus."

2.1. Background Material of Functional Analysis

In this section we gather the facts of functional analysis, which we need for the further presentation of the theory of extremal problems. As a rule, proofs that can be found in the textbook by A. N. Kolmogorov and S. V. Fomin [KF] are omitted.

2.1.1. Normed Linear Spaces and Banach Spaces.

We remind the reader that a linear space X is said to be *normed* if a functional $\|\cdot\|: X \to \mathbf{R}$ (called a *norm*) is defined on X which satisfies the following three axioms:

a) $$\|x\| \geqslant 0, \quad \forall x \in X \text{ and } \|x\| = 0 \Leftrightarrow x = 0,$$

b) $$\|\alpha x\| = |\alpha| \, \|x\|, \quad \forall x \in X, \quad \forall \alpha \in \mathbf{R}, \tag{1}$$

c) $$\|x_1 + x_2\| \leqslant \|x_1\| + \|x_2\|, \quad \forall x_1, \quad x_2 \in X.$$

Sometimes in order to stress that the norm is defined on X we shall write $\|\cdot\|_X$.

Every normed linear space becomes a *metric space* if it is equipped with the distance (metric)

$$\rho(x_1, x_2) = \|x_1 - x_2\|.$$

A normed linear space complete with respect to the distance it is equipped with is called a *Banach space*.

Exercises. Let $X = \mathbf{R}^2$. Which of the functions enumerated below (and for what values of the parameters) specify a norm in X?

1. $N(x)=(|x_1|^p+|x_2|^p)^{1/p},\quad 0<p<\infty;$

2. $N(x)=|a_{11}x_1+a_{12}x_2|+|a_{21}x_1+a_{22}x_2|;$

3. $N(x)=\max\{|a_{11}x_1+a_{12}x_2|,\quad |a_{21}x_1+a_{22}x_2|\}.$

4. Describe all the norms in $\mathbf{R}^2$.

5. Prove that all the norms in $\mathbf{R}^2$ are equivalent. (Two norms N_1 and N_2 are said to be *equivalent* if there are $c > 0$ and $C > 0$ such that $cN_1(x) \leqslant N_2(x) \leqslant CN_1(x),\ \forall x \in X$.)

The set X^* of all continuous linear functionals on X (it is called the *space conjugate to* X) is a Banach space if the norm in X^* is defined as

$$\|x^*\|_{X^*} = \sup_{\|x\|_X \leqslant 1} \langle x^*, x\rangle,$$

where $\langle x^*, x\rangle$ is the value of the functional x^* assumed on x.

For greater detail, see [KF, Chapter 4, Section 2].

Exercise 6. Find the norms in the spaces conjugate to those described in exercises 1-4.

For our further aims, the following Banach spaces will be of great importance.

Example 1. The space $C(K, \mathbf{R}^n)$ of all continuous vector functions $x(\cdot):K \to \mathbf{R}^n$ defined on a compact set K with the norm

$$\|x(\cdot)\|_0 = \max_{t\in K} |x(t)|.$$

The space $C(K, \mathbf{R})$ will be denoted by $C(K)$.

Example 2. The space $C^r([t_0, t_1], \mathbf{R}^n)$ of r times continuously differentiable vector functions $x(\cdot):[t_0, t_1] \to \mathbf{R}^n$ defined on a finite closed interval $[t_0, t_1] \subset \mathbf{R}$, and equipped with the norm

$$\|x(\cdot)\|_r = \max(\|x(\cdot)\|_0, \ldots, \|x^{(r)}(\cdot)\|_0).$$

The space $C^r([t_0, t_1], \mathbf{R})$ will be denoted by $C^r([t_0, t_1])$.

Exercises. 7. Prove that every finite-dimensional normed space is a Banach space.

8. Prove that the unit ball in a finite-dimensional normed space is a convex, closed, bounded, and centrally symmetric set for which the origin is an interior point, and, conversely, for every convex, closed, bounded, and centrally symmetric set containing the origin as an interior point there is a norm in which this set is the unit ball.

9. Give an example of a normed space which is not a Banach space.

2.1.2. Product Space. Factor Space. Let X and Y be linear spaces. Their (Cartesian) product $X \times Y$, i.e., the set of all the pairs (x, y), $x \in X$, $y \in Y$, becomes a linear space if the operations of addition and multiplication by a number are defined coordinatewise:

$$(x_1, y_1) + (x_2, y_2) = (x_1 + x_2, y_1 + y_2), \quad \alpha(x, y) = (\alpha x, \alpha y).$$

If X and Y are normed spaces, then the product space $X \times Y$ can be equipped with a norm, for instance, by putting

$$\|(x, y)\|_{X\times Y} = \max\{\|x\|_X, \|y\|_Y\}. \tag{1}$$

Exercise. Verify that $\|x\|_X + \|y\|_Y$ and $\sqrt{\|x\|_X^2 + \|y\|_Y^2}$ are also norms in $X \times Y$ and are equivalent to norm (1).

The two lemmas below are quite obvious.

Lemma 1. If X and Y are Banach spaces, then $X \times Y$ is also a Banach space.

The proof is left to the reader.

LEMMA 2. Every linear functional $\Lambda \in (X \times Y)^*$ can be uniquely represented in the form

$$\langle \Lambda, (x, y)\rangle = \langle x^*, x\rangle + \langle y^*, y\rangle, \tag{2}$$

where $x^* \in X^*$ and $y^* \in Y^*$.

Proof. Let us put $\langle x^*, x\rangle = \langle \Lambda, (x, 0)\rangle$ and $\langle y^*, y\rangle = \langle \Lambda, (0, y)\rangle$. These functionals are obviously linear, and their continuity ($\Leftrightarrow$ boundedness) follows from the inequalities

$$|\langle x^*, x\rangle| \leqslant \|\Lambda\| \|(x, 0)\| = \|\Lambda\| \|x\|$$

and

$$|\langle y^*, y\rangle| \leqslant \|\Lambda\| \|(0, y)\| = \|\Lambda\| \|y\|.$$

The uniqueness of the representation (2) is also quite evident. ■

Since formula (2) specifies a linear functional $\Lambda \in (X \times Y)^*$ for any $x^* \in X^*$ and $y^* \in Y^*$, we have obtained an exhaustive description of the space $(X \times Y)^*$. It can be briefly expressed by the formula $(X \times Y)^* = X^* \oplus Y^*$.

Now let X be a linear space, and let L be its subspace. We shall put $x \sim x'$ if $x - x' \in L$. The relation thus introduced is an equivalence relation (since we obviously have $x \sim x$, $x_1 \sim x_2 \Rightarrow x_2 \sim x_1$ and $x_1 \sim x_2$, $x_2 \sim x_3 \Rightarrow x_1 \sim x_3$), and therefore [KF, Chapter 1, Section 2], a decomposition (partition) of X into equivalence classes is defined. A class of equivalent elements with respect to this equivalence relation is called a *residue class* generated by the subspace L. The operations of addition and multiplication by a number are introduced in the set of the residue classes in a natural way [KF, Chapter 3, Section 1, Subsection 4], and the axioms of linear space hold. Thus, the set of the residue classes becomes a linear space; it is called the *factor space* (or the *quotient space*) of X relative to L and is denoted by X/L.

The residue class $\pi(x)$ to which an element $x \in X$ belongs is the class $x + L$. The mapping $\pi: X \to X/L$ is linear (prove this!). It is called the *canonical mapping* of X on X/L. (The mapping π is of course an epimorphism.†) It is evident that $L = \operatorname{Ker} \pi$.

Now let X be a normed space, and let L be its subspace. Let us put

$$\|\xi\|_{X/L} = \inf_{x \in \xi} \|x\|_X \tag{3}$$

or

$$\|\xi\|_{X/L} = \inf_{\pi x = \xi} \|x\|_X = \inf_{\substack{\pi x_0 = \xi \\ x' \in L}} \|x_0 + x'\|. \tag{3'}$$

From definitions (3) and (3') it readily follows that the operator π is continuous ($\|\pi x\| \leqslant \|x\|$), and for any $\xi \in X/L$ there is $x \in X$, such that

$$\pi x = \xi, \quad \|x\| \leqslant 2\|\xi\|_{X/L}. \tag{4}$$

Theorem on the Factor Space. Let X be a normed space, and let L be its closed subspace. Then the function $\|\cdot\|_{X/L}$ defined by relation (3) is a norm on X/L. If X is a Banach space, then the factor space X/L with the norm $\|\cdot\|_{X/L}$ is also a Banach space.

Proof. We have to prove that, in the first place, the functional $\|\cdot\|_{X/L}$ satisfies the axioms of the norm and that, in the second place, the completeness of X implies the completeness of X/L. Let us prove the first

†That is, π maps X onto the whole factor space X/L.

assertion of the theorem, i.e., verify that Axioms (1) of Section 2.1.1 are fulfilled.

Axiom a): $\|\xi\|_{X/L}\geqslant 0$ $(\forall\xi)$ since the norm in X is nonnegative. If $\xi = 0$, then $x = 0$ can be taken as $x\in\xi$, and therefore $\|0\|_{X/L} = 0$. Let $\|\xi\|_{X/L} = 0$. Then (3) implies that there is a sequence $\{x_n\}$, $x_n\in\xi$, such that $\|x_n\| \to 0$, i.e. $x_n \to 0$. Since ξ is closed, we have $0\in\xi$; i.e., is the zero element of X/L.

Axiom b): Let $\alpha\in\mathbf{R}$. Then if $\pi x = \xi$, we have

$$\|\alpha\xi\|_{X/L}=\inf_{\pi y=\alpha\xi}\|y\|=\inf_{\pi x=\xi}\|\alpha x\|=|\alpha|\inf_{\pi x=\xi}\|x\|=|\alpha|\|\xi\|.$$

Axiom c):

$$\|\xi_1+\xi_2\|=\inf_{\substack{\pi x_i=\xi_i\\ x'\in L}}\|x_1+x_2+x'\|=\inf_{x_i'\in L}\|x_1+x_1'+x_2+x_2'\|\leqslant$$

$$\leqslant\inf_{x_1'\in L}\|x_1+x_1'\|+\inf_{x_2'\in L}\|x_2+x_2'\|=\|\xi_1\|_{X/L}+\|\xi_2\|_{X/L}.$$

Let us prove the second assertion of the theorem, i.e., the completeness of X/L. Let $\{\xi_n\}$ be a fundamental (Cauchy) sequence in X/L, i.e.,

$$\forall\varepsilon>0\ \exists N(\varepsilon):\ n>N(\varepsilon)\Rightarrow\|\xi_{n+m}-\xi_n\|_{X/L}<\varepsilon,\quad\forall m\geqslant 1.$$

Now we choose $\varepsilon_k = 2^{-k}$ and indices n_k such that $n_{k+1} > n_k$, $\|\xi_{n_k+m} - \xi_{n_k}\| \leqslant 2^{-k}$, $k \geqslant 1$. Then $\|\xi_{n_1} - \xi_{n_2}\| \leqslant 1/2$, and from (4) it follows that there exist representatives $x_i\in\xi_{n_i}$ such that $\|x_2 - x_1\| \leqslant 1$. In a similar way we construct elements $\{x_n\}$, $n \geqslant 3$, such that $\|x_k-x_{k-1}\|\leqslant 2^{-(k-1)}$, $x_k\in\xi_{n_k}$, $k = 3, 4,\ldots$. The sequence $\{x_k\}_{k=1}^{\infty}$ is fundamental in X (check this!), and, by the hypothesis, X is complete. Hence, the limit $x_0=\lim_{k\to\infty}x_k$ exists. Let us consider the class $\xi_0 = \pi x_0$. We have

$$\|\xi_{n_k}-\xi_0\|_{X/L}=\inf_{x'\in L}\|x_k-x_0-x'\|_X\leqslant\|x_k-x_0\|_X\underset{k\to\infty}{\longrightarrow}0.$$

Thus, $\xi_{n_k} \to \xi_0$, and therefore we also have $\xi_n \to \xi_0$ since $\{\xi_n\}$ is fundamental. Consequently, X/L is complete. ■

Exercises. 1. Let $X = C([0, 1])$, $0 \leqslant \tau_1 < \tau_2 < \ldots < \tau_m \leqslant 1$, and let the subspace L consist of the functions $x(\cdot) \in C([0, 1])$ such that $x(\tau_i) = 0$, $i = 1,\ldots,m$. Find the factor space X/L and determine $\|\cdot\|_{X/L}$.

2. Generalize the result of the exercise 1 and prove that if F is a closed subspace of the closed interval [0, 1] and

$$L_F=\{x(\cdot)\in C([0,1])\mid x(t)\equiv 0,\ t\in F\},$$

then the space $C([0, 1])/L_F$ is isometrically isomorphic to the space $C(F)$.

2.1.3. Hahn–Banach Theorem and Its Corollaries. In the theory of extremal problems an important role is played by separation theorems and some other facts of convex analysis. Most of them are consequences of the Hahn–Banach theorem, which is often called the first principle of linear analysis. Since this theorem is included in all standard courses in functional analysis, we shall limit ourselves to its statement and to some of its most important corollaries.

In this section X is a linear space and R is the extended real line, i.e. $\overline{\mathbf{R}}=\mathbf{R}\cup\{-\infty,+\infty\}$.

Definition 1. A function $p:X\to\overline{\mathbf{R}}$ is said to be *convex and homogeneous* if

$$p(x+y)\leqslant p(x)+p(y)\quad\text{for any}\quad x, y\in X,\tag{1}$$

and

$$p(\alpha x)=\alpha p(x) \qquad \text{for any} \qquad x\in X \text{ and } \alpha>0. \tag{2}$$

Examples. 1) X is a normed space and $p(x) = \|x\|$.

2) X is a linear space, L is a linear subspace of X, and

$$p(x)=\begin{cases} 0, & x\in L,\\ +\infty, & x\notin L.\end{cases}$$

3) X is a linear space, $l_1,\ldots,l_m$ is a set of linear functionals on X, and $p(x)=\max(\langle l_1, x\rangle,\ldots, \langle l_m, x\rangle)$.

The following definition implies an important example.

Definition 2. Let X be a linear space, and let A be its convex subset containing 0. The *Minkowski function* $\mu A(\cdot)$ of A is defined by the equality

$$\mu A(x)=\inf\{t>0 \mid x/t\in A\} \tag{3}$$

[if there are no $t > 0$ such that $x/t\in A$, then $\mu A(x) = +\infty$].

Proposition 1. The Minkowski function is nonnegative, convex, and homogeneous;

$$\{x \mid \mu A(x)<1\}\subseteq A\subseteq\{x\mid \mu A(x)\leqslant 1\}. \tag{4}$$

If X is a topological linear space,* then $\mu A(\cdot)$ is continuous at the point 0 if and only if $0\in \operatorname{int} A$.

Proof. If $\mu A(x)$ or $\mu A(y)$ is equal to $+\infty$ then (1) holds. Therefore, let $\mu A(x) < +\infty$ and $\mu A(y) < +\infty$. By definition, for any $\varepsilon > 0$ there are t and s such that

$$0<t<\mu A(x)+\varepsilon/2 \text{ and } x/t\in A,$$
$$0<s<\mu A(y)+\varepsilon/2 \text{ and } y/s\in A.$$

It follows that

$$\frac{x+y}{t+s}=\frac{x}{t}\frac{t}{t+s}+\frac{y}{s}\frac{s}{t+s}\in A,$$

since A is convex, and consequently

$$\mu A(x+y)\leqslant t+s<\mu A(x)+\mu A(y)+\varepsilon.$$

By the arbitrariness of ε, condition (1) holds. Further, for $\alpha > 0$ we have

$$\mu A(\alpha x)=\inf\{t>0 \mid \alpha x/t\in A\}=\inf\{\alpha s>0 \mid x/s\in A\}=\alpha\inf\{s>0 \mid x/s\in A\}=\alpha\mu A(x),$$

and hence (2) is true.

The nonnegativity of $\mu A(x)$ and the second inclusion in (4) follow directly from the definition. If $\mu A(x) < 1$, then there is $t\in(0,1)$ for which $x/t\in A$, and since $0\in A$ and A is convex, the inclusion

$$x=0\cdot(1-t)+\frac{x}{t}\cdot t\in A$$

holds, i.e., the first inclusion in (4) is true.

Now let X be a topological linear space. Then the following chain of equivalent assertions takes place: $\mu A(\cdot)$ is continuous at $0 \Leftrightarrow$

$$\forall\varepsilon>0,\ \exists U_\varepsilon\ni 0,\ \forall x\in U_\varepsilon,\quad \mu A(x)<\varepsilon \Leftrightarrow$$
$$\exists U_1\ni 0,\ \forall x\in U_1,\quad \mu A(x)<1 \Leftrightarrow$$
$$\exists U_1\ni 0,\quad U_1\subset A \Leftrightarrow 0\in \operatorname{int} A$$

*On topological linear spaces see [KF, Chapter 3, Section 5].

[here U_ε and U_1 are neighborhoods of 0; the second equivalence relation holds because of the homogeneity of $\mu A(\cdot)$ and the possibility of putting $U_\varepsilon = \varepsilon U_1$]. ■

Proposition 2. For a linear functional x^* on a topological linear space X to be continuous it is necessary and sufficient that there exist a convex homogeneous function $p(\cdot)$ continuous at the point 0 such that the inequality

$$\langle x^*, x\rangle \leqslant p(x) \tag{5}$$

holds for all $x \in X$.

Proof. The necessity is established at once by putting $p(x) = |\langle x^*, x\rangle|$.

Now let us suppose that (5) holds and $p(x)$ is continuous at the point 0. Then for any $\varepsilon > 0$ there is a neighborhood $U \ni 0$ such that $p(x) < \varepsilon$ for all $x \in U$. Since $0 \in U$ and $0 = (-0) \in (-U)$, there is a neighborhood W such that $0 \in W \subset U \cap (-U)$. If $x \in W$, then $x \quad -x \in U$, and, by virtue of (5),

$$\langle x^*, x\rangle \leqslant p(x) < \varepsilon,$$
$$\langle x^* - x\rangle \leqslant p(-x) < \varepsilon.$$

Consequently, $|\langle x^*, x\rangle| < \varepsilon$ for all $x \in W$, i.e., x^* is continuous at 0. It remains to take into consideration that a linear functional continuous at one point is continuous on the whole space X [KF, p. 174].

Hahn–Banach Theorem [KF, pp. 134-137]. Let $p: X \to \overline{\mathbf{R}}$ be a convex homogeneous function on a linear space X, and let $l: L \to \mathbf{R}$ be a linear functional on a subspace L of the space X such that

$$\langle l, x\rangle \leqslant p(x) \quad \text{for all } x \in L. \tag{6}$$

Then there exists a linear functional Λ defined on the whole space X which is an extension of l, i.e.,

$$\langle \Lambda, x\rangle \equiv \langle l, x\rangle, \quad x \in L, \tag{7}$$

and satisfies the inequality

$$\langle \Lambda, x\rangle \leqslant p(x) \quad \text{for all } x \in X. \tag{8}$$

COROLLARY 1. Let X be a normed space and $x_0 \in X$, $x_0 \neq 0$. Then there is an element $\Lambda \in X^*$ such that $\|\Lambda\| = 1$ and $\langle \Lambda, x_0\rangle = \|x_0\|$.

Proof. Let us define a linear functional l on the subspace

$$L = \{x \mid x = \alpha x_0, \alpha \in \mathbf{R}\}$$

by putting

$$\langle l, \alpha x_0\rangle = \alpha \|x_0\|. \tag{9}$$

The function $p(x) = \|x\|$ is convex and homogeneous, and $\langle l, \alpha x_0\rangle = \alpha\|x_0\| \leqslant |\alpha|\|x_0\| = \|\alpha x_0\| = p(\alpha x_0)$; therefore, inequality (6) holds. By the Hahn–Banach theorem, l can be extended to a linear functional Λ on the whole space X, and relations (7) and (8) hold. Recalling the definition of the norm of a functional (Section 2.1.1), we obtain from (8) the inequality

$$\|\Lambda\| = \sup_{\|x\|=1} \langle \Lambda, x\rangle \leqslant \sup_{\|x\|=1} p(x) = 1,$$

whence, in particular, $\Lambda \in X^*$.

On the other hand, (7) and (9) imply

$$\langle \Lambda, x_0\rangle = \langle l, x_0\rangle = \|x_0\|,$$

and thus the assertion of the corollary holds and, moreover,

$$\|\Lambda\| = \sup_{\|x\|=1} \langle\Lambda, x\rangle \geqslant \left\langle\Lambda, \frac{x_0}{\|x_0\|}\right\rangle = 1.$$

Therefore, $\|\Lambda\| = 1$. ■

Corollary 1 immediately implies:

COROLLARY 2. If a normed linear space X is nontrivial (i.e., $X \neq \{0\}$) then the space X^* conjugate to X is also nontrivial.

2.1.4. Separation Theorems. In this section X is a topological linear space and X^* is its conjugate space consisting of all the continuous linear functionals on X.

Definition 1. We say that a linear functional $x^* \in X^*$ *separates sets* $A \subset X$ and $B \subset X$ if

$$\sup_{x \in A} \langle x^*, x\rangle \leqslant \inf_{x \in B} \langle x^*, x\rangle, \tag{1}$$

and that it *strongly separates* A *and* B if

$$\sup_{x \in A} \langle x^*, x\rangle < \inf_{x \in B} \langle x^*, x\rangle. \tag{2}$$

The geometrical meaning of inequality (1) is that the hyperplane

$$H(x^*, c) = \{x \mid \langle x^*, x\rangle = c\},$$

where $\sup_{x \in A} \langle x^*, x\rangle \leqslant c \leqslant \inf_{x \in B} \langle x^*, x\rangle$ separates the sets A and B in the sense that A lies in one half-space ($H_+(x^*, c) = \{x \mid \langle x^*, x\rangle \leqslant c\}$) generated by $H(x^*, c)$ and B in the other half-space ($H_-(x^*, c) = \{x \mid \langle x^*, x\rangle \geqslant c\}$) (Fig. 31); the inequality (2) means that c can be chosen so that A and B lie inside the corresponding half-spaces and have no common points with $H(x^*, c)$ (Fig. 32).

First Separation Theorem. If sets $A \subset X$ and $B \subset X$ are convex and nonempty and do not intersect, and A is open, then there exists a nonzero continuous linear functional separating A and B.

Proof. A) Since A and B are nonempty, there exist points $a_0 \in A$ and $b_0 \in B$. The set

$$C = (A - a_0) - (B - b_0) = \{x \mid x = a - a_0 - b + b_0,\ a \in A,\ b \in B\},$$

is obviously convex (also see Proposition 1 and Exercise 1 in Section 2.6.1), contains 0, and is open. Indeed, if $\hat{x} = \hat{a} - a_0 - \hat{b} + b_0$ and $\hat{a} \in A$, there is a neighborhood U, $\hat{a} \in U \subset A$, and then $\hat{x} \in U - a_0 - \hat{b} + b_0 \subset C$. Moreover, we have $c = b_0 - a_0 \notin C$, because, if otherwise, then $b_0 - a_0 = \hat{a} - a_0 - \hat{b} + b_0$ for some $\hat{a} \in A$ and $\hat{b} \in B$, whence $\hat{a} = \hat{b} \in A \cap B$, i.e., these sets intersect, which contradicts the hypothesis.

B) Let us denote by $p(x)$ the Minkowski function of the set C. According to Proposition 1 of Section 2.1.3, the function $p(x)$ is nonnegative, convex, homogeneous, and continuous at the point 0. Moreover, $p(x) \leqslant 1$ for all $x \in C$.

C) Let us define a linear functional l on the subspace

$$L = \{x \mid x = \alpha c = \alpha(b_0 - a_0),\ \alpha \in \mathbf{R}\}$$

by putting $\langle l, \alpha c\rangle = \alpha p(c)$. Then for $\alpha > 0$ we have $\langle l, \alpha c\rangle = \alpha p(c) = p(\alpha c)$ and for $\alpha \leqslant 0$ we have $\langle l, \alpha c\rangle = \alpha p(c) \leqslant p(\alpha c)$ since $p(\cdot)$ is nonnegative. Consequently, the inequality $\langle l, x\rangle \leqslant p(x)$ holds for all $x \in L$, and, by the Hahn–Banach theorem, l can be extended to a linear functional Λ such that

$$\langle\Lambda, \alpha c\rangle = \langle l, \alpha c\rangle = \alpha p(c), \quad \alpha \in \mathbf{R}, \tag{3}$$

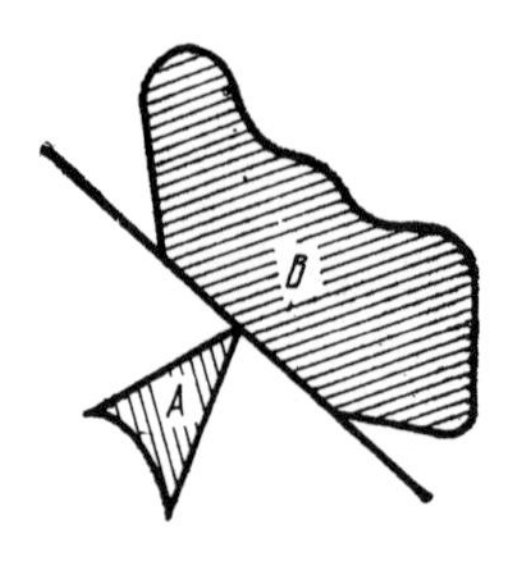

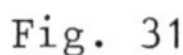
Fig. 31

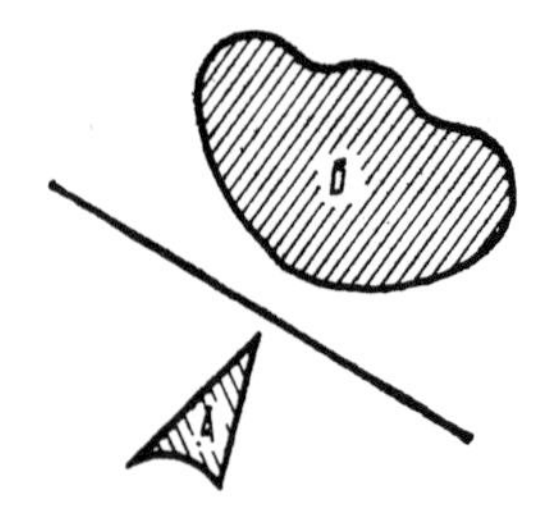

Fig. 32

and

$$\langle \Lambda, x \rangle \leqslant p(x), \quad x \in X. \tag{4}$$

Since p(·) is continuous at 0, inequality (4) implies the continuity of the functional Λ (Proposition 2 of Section 2.1.3).

D) For any $a \in A$ and $b \in B$ we have

$$\langle \Lambda, a-b \rangle = \langle \Lambda, a-a_0-b+b_0 \rangle + \langle \Lambda, a_0-b_0 \rangle \leqslant$$
$$\leqslant p(a-a_0-b+b_0) + \langle l, (-1)(b_0-a_0) \rangle \leqslant 1 - p(b_0-a_0),$$

since $a-a_0-b+b_0 \in C$, and $p(x) \leqslant 1$ on C; besides, we have used (4). However, for $0 < t \leqslant 1$ the point $(b_0 - a_0)/t = c/t$ cannot belong to the set C because C is convex and does not contain 0 while $[0, c/t]$ contains the point $c = b_0 - a_0 \notin C$. Therefore,

$$p(b_0-a_0) = \inf\left\{t > 0 \,\Big|\, \frac{b_0-a_0}{t} \in C\right\} \geqslant 1. \tag{5}$$

Consequently, $\langle \Lambda, a-b \rangle \leqslant 1 - p(b_0-a_0) \leqslant 0$ for any $a \in A$ and $b \in B$. The elements $a \in A$ and $b \in B$ in the inequality $\langle \Lambda, a \rangle \leqslant \langle \Lambda, b \rangle$ can be chosen independently and therefore

$$\sup_{a \in A} \langle \Lambda, a \rangle \leqslant \inf_{b \in B} (\Lambda, b).$$

Moreover, according to (3) and (5), we have $\langle \Lambda, b_0-a_0 \rangle = p(b_0-a_0) \geqslant 1$, and therefore $\Lambda \neq 0$. Consequently, Λ separates A and B. ■

<u>Remark.</u> In Section 1.3.3 we already proved the separation theorem for the case of a finite-dimensional space X. Comparing its conditions with those of the theorem we have just proved we see that in the finite-dimensional case the condition that A is open can be omitted (the fact that in Section 1.3.3 one of the sets was a point was insignificant). We can weaken the conditions of the first separation theorem and require that $\operatorname{int} A \neq \emptyset$ and $(\operatorname{int} A) \cap B = \emptyset$. However, it is impossible to do without the existence of interior points of at least one of the sets.

<u>Exercises.</u> 1. Let A be a convex set of a topological linear space X, and let $\operatorname{int} A \neq \emptyset$ and $x^* \in X^*$. Prove that

$$\sup_{x \in A} \langle x^*, x \rangle = \sup_{x \in \operatorname{int} A} \langle x^*, x \rangle. \tag{6}$$

2. Derive from (6) the separation theorem for the case of convex A and B such that $\operatorname{int} A \neq \emptyset$, $B \neq \emptyset$, $B \cap \operatorname{int} A = \emptyset$.

3. Prove that there is no continuous linear functional on the space $X = l_2$ separating the compact ellipsoid

$$A = \left\{ x = [x_k] \in l_2 \,\Big|\, \sum_{k=1}^{\infty} k^2 x_k^2 \leqslant 1 \right\}$$

and the ray

$$B = \{x = [x_k] \mid x_k = t/k, \quad t > 0\}.$$

Second Separation Theorem. Let X be a locally convex topological linear space [KF, p. 169], let A be a nonempty closed convex subset of X, and let $\hat{x} \in X$ be a point not belonging to A. Then there is a nonzero continuous linear functional strongly separating A and $\hat{x}$.

Proof. Since $\hat{x} \notin A$ and A is closed, there is a neighborhood $V \ni \hat{x}$ such that $A \cap V = \varnothing$. Since the space X is locally convex, there exists a convex neighborhood $B \subset V$ of the point $\hat{x}$. It is evident that $B \cap A = \varnothing$, and, by the first separation theorem, there exists a nonzero linear functional x* separating A and B:

$$\sup_{x \in A} \langle x^*, x \rangle \leqslant \inf_{x \in B} \langle x^*, x \rangle.$$

It remains to use the fact that

$$\inf_{x \in B} \langle x^*, x \rangle < \langle x^*, \hat{x} \rangle$$

because the infimum of the nonzero linear functional x* cannot be attained at the interior point $\hat{x}$ of the set B. ■

Definition 2. By the *annihilator* $A^\perp$ of a subset A of a linear space X is meant the set of those linear functionals l on X for which $\langle l, x \rangle = 0$ for all $x \in A$.

It should be noted that $A^\perp$ is always closed and contains $0 \in X^*$.

Lemma on the Nontriviality of the Annihilator. Let L be a closed subspace of a locally convex topological linear space X, and let $L \neq X$. Then the annihilator $L^\perp$ contains a nonzero element $x^* \in X^*$.

Proof. Let us take an arbitrary point $\hat{x} \notin L$. By the second separation theorem, there exists a nonzero functional $x^* \in X^*$, strongly separating $\hat{x}$ and L (L is a subspace of a linear space and consequently it is convex):

$$\sup_{x \in L} \langle x^*, x \rangle < \langle x^*, \hat{x} \rangle. \tag{7}$$

If there existed $x_0 \in L$, for which $\langle x^*, x_0 \rangle \neq 0$ then, since $\alpha x_0 \in L$ for any $\alpha \in \mathbf{R}$, we would have

$$\sup_{x \in L} \langle x^*, x \rangle \geqslant \sup_{\alpha \in \mathbf{R}} \langle x^*, \alpha x_0 \rangle = +\infty,$$

which contradicts (7). Consequently, $\langle x^*, x \rangle \equiv 0$ on L, and therefore $x^* \in L^\perp$. ■

2.1.5. Inverse Operator Theorem of Banach and the Lemma on the Right Inverse Mapping. The theorem below is the second fundamental principle of linear analysis.

Inverse Operator Theorem of Banach. Let X and Y be Banach spaces, and let $\Lambda: X \to Y$ be a continuous linear operator. If Λ is a monomorphism, i.e.,

$$\operatorname{Ker} \Lambda = \{x \mid \Lambda x = 0\} = \{0\},$$

and an epimorphism, i.e.,

$$\operatorname{Im} \Lambda = \{y \mid y = \Lambda x, \quad x \in X\} = Y,$$

then Λ is an isomorphism between X and Y, i.e., there exists a continuous linear operator $M = \Lambda^{-1}: Y \to X$ such that

$$M\Lambda = I_X, \quad \Lambda M = I_Y.$$

For the *proof*, see [KF, p. 225]. In the present book the following consequence of the Banach theorem will be frequently used:

Lemma on the Right Inverse Mapping. Let X and Y be Banach spaces, and let Λ be a continuous linear operator from X into Y which is an epimorphism. Then there exists a mapping $M: Y \to X$ (in the general case this mapping is discontinuous and nonlinear) satisfying the conditions

$$\Lambda \circ M = I_Y,$$

and

$$\|M(y)\| \leqslant C\|y\| \quad \text{for some} \quad C > 0.$$

It should be noted that, by virtue of the second condition, the mapping M is continuous *at zero*.

Proof. The operator Λ is continuous, and consequently its kernel $\operatorname{Ker}\Lambda$, which is the inverse image of the point zero, is a closed subspace. By the theorem on the factor space (Section 2.1.2), $X/\operatorname{Ker}\Lambda$ is a Banach space. Let us define an operator $\tilde{\Lambda}: X/\operatorname{Ker}\Lambda \to Y$ by putting $\tilde{\Lambda}\xi = \Lambda x$ on the residue class $\xi = \pi x$. This definition is correct: $x_1, x_2 \in \xi \Rightarrow \pi x_1 = \pi x_2 = \xi \Rightarrow x_1 - x_2 \in \operatorname{Ker}\Lambda \Rightarrow \Lambda x_1 = \Lambda x_2$. The operator $\tilde{\Lambda}$ is linear: if $\pi x_i = \xi_i$, $i = 1, 2$, then

$$\tilde{\Lambda}(\alpha_1\xi_1 + \alpha_2\xi_2) = \Lambda(\alpha_1 x_1 + \alpha_2 x_2) = \alpha_1\Lambda x_1 + \alpha_2\Lambda x_2 = \alpha_1\tilde{\Lambda}\xi_1 + \alpha_2\tilde{\Lambda}\xi_2.$$

The operator $\tilde{\Lambda}$ is continuous: for any $x \in \xi$ we have

$$\|\tilde{\Lambda}\xi\| = \|\Lambda x\| \leqslant \|\Lambda\|\|x\|,$$

whence

$$\|\tilde{\Lambda}\xi\| \leqslant \|\Lambda\| \inf_{x\in\xi}\|x\| = \|\Lambda\|\|\xi\|_{X/\operatorname{Ker}\Lambda}.$$

Finally, the operator $\tilde{\Lambda}$ is bijective (one-to-one). Indeed, Λ is an epimorphism, and hence so is $\tilde{\Lambda}$: $\forall y \in Y\ \exists x \in X,\ \Lambda x = y \Rightarrow\ \tilde{\Lambda}\xi = y$ for $\xi = \pi(x)$. On the other hand, $\operatorname{Ker}\tilde{\Lambda} = 0$: $\tilde{\Lambda}\xi = 0,\ \pi x = \xi \Rightarrow \Lambda x = 0 \Rightarrow x \in \operatorname{Ker}\Lambda \Rightarrow \xi = 0$.

We have thus proved that $\tilde{\Lambda}$ is a continuous bijective operator from $X/\operatorname{Ker}\Lambda$ into Y. By the inverse operator theorem of Banach, there exists a continuous operator $\tilde{M} = \tilde{\Lambda}^{-1}$, $\tilde{M}: Y \to X/\operatorname{Ker}\Lambda$, inverse to $\tilde{\Lambda}$. According to relation (4) of Section 2.1.2, for the element $\xi = \tilde{M}y$ there is an element $x = x(\xi)$ such that $x \in \xi$ and $\|x\| \leqslant 2\|\xi\|_{X/\operatorname{Ker}\Lambda}$. Let us put $M(y) = x(\xi)$. This results in

$$(\Lambda \circ M)(y) = \Lambda x(\xi) = \tilde{\Lambda}\xi = \tilde{\Lambda}\tilde{M}y = y,$$

$$\|M(y)\| = \|x(\xi)\| \leqslant 2\|\xi\|_{X/\operatorname{Ker}\Lambda} = 2\|\tilde{M}y\|_{X/\operatorname{Ker}\Lambda} \leqslant 2\|\tilde{M}\|\|y\|. \quad \blacksquare$$

2.1.6. Lemma on the Closed Image. Let X, Y, and Z be Banach spaces, and let $A: X \to Y$ and $B: X \to Z$ be continuous linear operators. The equality

$$Cx = (Ax, Bx)$$

specifies a continuous linear operator $C: X \to Y \times Z$.

Exercise. Prove that if the norm in $Y \times Z$ is defined by equality (1) of Section 2.1.2, then $\|C\| = \max\{\|A\|, \|B\|\}$.

Lemma on the Closed Image. *If the subspace* $\operatorname{Im} A$ *is closed in* Y *and the subspace* $B\operatorname{Ker} A$* *is closed in* Z, *then the subspace* $\operatorname{Im} C$ is closed in $Y \times Z$.

Proof. The closed subspace $\tilde{Y} = \operatorname{Im} A$ of the Banach space Y is itself a Banach space, and $A: X \to \tilde{Y}$ is an epimorphism. By the lemma of Section 2.1.5, there exists the right inverse mapping $M: \tilde{Y} \to X$ to A. Let (y, z) belong to the closure of the image of the operator C. This means that there is a sequence $\{x_n | n \geqslant 1\}$ such that $y = \lim_{n\to\infty} Ax_n$, $z = \lim_{n\to\infty} Bx_n$. Let us set

$$\xi_n = M(Ax_n - y)$$

---

*$B\operatorname{Ker} A$ is the image of the kernel of the operator A under the mapping B.

($y \in \tilde{Y}$ because y is the limit of the elements $Ax_n \in \tilde{Y}$). Taking into account the properties of M, we obtain

$$A(x_n - \xi_n) = Ax_n - (A \circ M)(Ax_n - y) = y,$$
$$\|\xi_n\| = \|M(Ax_n - y)\| \leqslant C_M \|Ax_n - y\| \to 0.$$

Therefore, $B\xi_n \to 0$ and $\lim_{n\to\infty} B(x_n - \xi_n) = \lim_{n\to\infty} Bx_n = z$, i.e., z belongs to the closure of the set

$$\Sigma = \{\zeta = Bx \mid Ax = y\}.$$

This set is a translate of the subspace B Ker A:

$$\zeta_1, \zeta_2 \in \Sigma \Rightarrow \zeta_i = Bx_i, \quad Ax_i = y, \quad i = 1, 2 \Rightarrow$$
$$\Rightarrow \zeta_2 - \zeta_1 = B(x_2 - x_1), \quad A(x_2 - x_1) = 0 \Rightarrow \zeta_2 = \zeta_1 + (\zeta_2 - \zeta_1),$$
$$\zeta_1 \in \Sigma, \quad \zeta_2 - \zeta_1 \in B \operatorname{Ker} A.$$

However, by the hypothesis, B Ker A is closed, and hence the set Σ is also closed. Consequently, the element z belonging to the closure of Σ belongs to Σ itself, which implies the existence of an element $\hat{x}$ for which $A\hat{x} = y$, $B\hat{x} = z$. Thus $(y, z) \in \operatorname{Im} C$. ■

2.1.7. Lemma on the Annihilator of the Kernel of a Regular Operator. Let X and Y be Banach spaces, and let $A: X \to Y$ be a continuous linear epimorphism. Then $(\operatorname{Ker} A)^{\perp} = \operatorname{Im} A^*$.

Proof. A) Let $x^* \in \operatorname{Im} A^* \Leftrightarrow x^* = A^* y^*$. Then $x \in \operatorname{Ker} A$ implies $\langle x^*, x\rangle = \langle A^*y^*, x\rangle = \langle y^*, Ax\rangle = 0$, i.e., $\operatorname{Im} A^* \subset (\operatorname{Ker} A)^{\perp}$.

B) Let $x^* \in (\operatorname{Ker} A)^{\perp}$, i.e., $Ax = 0 \Rightarrow \langle x^*, x\rangle = 0$. Let us consider the mapping $C: X \to \mathbf{R} \times Y$, $Cx = (\langle x^*, x\rangle, Ax)$. The image of the kernel of A under the mapping x^* is zero, i.e., a closed set, and, by the lemma on the closed image (Section 2.1.6), Im C is a closed subspace of $\mathbf{R} \times Y$. It does not coincide with $\mathbf{R} \times Y$ since, for instance, $(1, 0) \notin \operatorname{Im} C$ [indeed, if $(\alpha, 0) = (\langle x^*, x\rangle, Ax)$ then $Ax = 0 \Rightarrow \alpha = \langle x^*, x\rangle = 0$].

By the lemma on the nontriviality of the annihilator of a closed proper subspace (Section 2.1.4), there exists a nonzero continuous linear functional $\Lambda \in (\operatorname{Im} C)^{\perp} \subset (\mathbf{R} \times Y)^*$, and, according to the representation lemma for the general linear functional on a product space (Section 2.1.2), there exist a number $\hat{\lambda}_0 \in \mathbf{R}^* = \mathbf{R}$ and an element $\hat{y}^* \in Y^*$ such that

$$\langle \hat{\lambda}_0 x^* + A^* \hat{y}^*, x\rangle = \hat{\lambda}_0 \langle x^*, x\rangle + \langle \hat{y}^*, Ax\rangle = \langle \Lambda, Cx\rangle \equiv 0, \ \forall x.$$

The case $\hat{\lambda}_0 = 0$ is impossible because Im A = Y, and hence

$$\langle \hat{y}^*, Ax\rangle \equiv 0 \Rightarrow \hat{y}^* = 0 \Rightarrow \Lambda = 0.$$

Consequently, $x^* = A^*(-\hat{y}^*/\hat{\lambda}_0)$, i.e., $x^* \in \operatorname{Im} A^*$ and $\operatorname{Ker} A)^{\perp} \subset \operatorname{Im} A^*$. ■

2.1.8. Absolutely Continuous Functions. In the theory of optimal control, we permanently deal with differential equations of the form

$$\dot{x} = \varphi(t, x, u(t)), \tag{1}$$

where $u(\cdot)$ is a given function called a *control*. Although the function $\varphi(t, x, u)$ is usually assumed to be continuous (and even smooth), the control $u(\cdot)$ does not possess these properties in the general case. It may be piecewise-continuous, and it is sometimes convenient to assume that $u(\cdot)$ is only a measurable function. Hence, the right-hand side of Eq. (1) may be discontinuous, and therefore it is necessary to state more precisely in what sense its solution should be understood.

Remark. In the present section such notions as "measurability," "integrability," "almost everywhere," etc. are understood in the ordinary Lebesgue sense [KF, Chapter 5, Sections 4 and 5]. The same rule is retained in the remaining part of the book with some rare exceptions, which are carefully stipulated: for instance, the integral in the formulas of the next section is understood in the Lebesgue or the Lebesgue–Stieltjes sense [KF, Chapter 6, Section 6]. When these notions are applied to vector-valued or matrix functions, then each of their components must possess the corresponding properties. For instance, a function $x(\cdot) = (x_1(\cdot), \ldots, x_n(\cdot)) : [\alpha, \beta] \to \mathbf{R}^n$ is measurable and integrable if each of the scalar functions $x_i(\cdot) : [\alpha, \beta] \to \mathbf{R}$ possesses the same properties, and, by definition,

$$\int_\alpha^\beta x(t)\,dt = \left(\int_\alpha^\beta x_1(t)\,dt, \; \ldots, \; \int_\alpha^\beta x_n(t)\,dt \right). \tag{2}$$

It is most natural to assume (and this is what we are doing) that the differential equation (1) together with the initial condition $x(t_0) = x_0$ is equivalent to the integral equation

$$x(t) = x_0 + \int_{t_0}^t \varphi(s, x(s), u(s))\,ds. \tag{3}$$

For this assumption to hold the function $x(\cdot)$ must be equal to the indefinite integral of its derivative, i.e., the Newton–Leibniz formula must hold:

$$x(t) - x(\tau) = \int_\tau^t \dot{x}(s)\,ds. \tag{4}$$

In the Lebesgue theory of integration [KF, Chapter 6, Section 4] the functions $x(\cdot)$ for which (4) holds are called absolutely continuous. It is important that the following equivalent and more effectively verifiable definition can be stated for these functions:

Definition. Let X be a normed linear space, and let $\Delta = [\alpha, \beta]$ be a closed interval on the real line. A function $x(\cdot) : \Delta \to X$ is said to be *absolutely continuous* if for any $\varepsilon > 0$ there is $\delta > 0$ such that for any finite system of pairwise disjoint open intervals $(\alpha_k, \beta_k) \subset [\alpha, \beta]$, $k = 1, 2, \ldots, N$, the sum of whose lengths does not exceed δ:

$$\sum_{k=1}^N (\beta_k - \alpha_k) < \delta, \tag{5}$$

the inequality

$$\sum_{k=1}^N \| x(\beta_k) - x(\alpha_k) \| < \varepsilon \tag{6}$$

holds.

Example. A function $x(\cdot) : \Delta \to X$ satisfying the Lipschitz condition

$$\| x(t_1) - x(t_2) \| \leqslant K | t_1 - t_2 |, \quad t_1, t_2 \in \Delta,$$

is absolutely continuous. Indeed, here $\delta = \varepsilon / K$. From the definition it is readily seen that an absolutely continuous function is uniformly continuous on Δ (we take $N = 1$).

Proposition 1. Let X and Y be normed linear spaces, and let Δ be a closed interval of the real line.

If functions $x_i(\cdot) : \Delta \to X$, $i = 1, 2$, are absolutely continuous and $c_i \in \mathbf{R}$, then the function $c_1 x_1(\cdot) + c_2 x_2(\cdot)$ is also absolutely continuous.

If functions $x(\cdot) : \Delta \to X$ and $A(\cdot) : \Delta \to \mathscr{L}(X, Y)$ are absolutely continuous, then the function $A(\cdot) x(\cdot) : \Delta \to Y$ is also absolutely continuous.

If a function $\Phi(\cdot) : G \to Y$ satisfies the Lipschitz condition on a set $\mathscr{K} \subset G$, and a function $x(\cdot) : \Delta \to G$ is absolutely continuous and $\operatorname{im} x(\cdot) = \{ y \mid y = x(t), t \in \Delta \} \subset \mathscr{K}$, then the function $\Phi(x(\cdot)) : \Delta \to Y$ is absolutely continuous.

Proof. The first assertion follows at once from the definition (also see [KF, p. 343]). To prove the second assertion we note that since the functions $x(\cdot)$ and $A(\cdot)$ are continuous, they are bounded, and therefore

$$\|x(t)\| \leqslant M_x, \quad \|A(t)\| \leqslant M_A, \quad t \in \Delta.$$

Now for any system of disjoint open intervals $(\alpha_k, \beta_k) \subset \Delta$, $k = 1, \ldots, N$, we have

$$\sum_{k=1}^{N} \|A(\beta_k)x(\beta_k) - A(\alpha_k)x(\alpha_k)\| \leqslant \sum_{k=1}^{N} \|A(\beta_k)x(\beta_k) - A(\beta_k)x(\alpha_k)\| +$$
$$+ \sum_{k=1}^{N} \|A(\beta_k)x(\alpha_k) - A(\alpha_k)x(\alpha_k)\| \leqslant$$
$$\leqslant \sum_{k=1}^{N} \|A(\beta_k)\| \|x(\beta_k) - x(\alpha_k)\| + \sum_{k=1}^{N} \|A(\beta_k) - A(\alpha_k)\| \|x(\alpha_k)\| \leqslant$$
$$\leqslant M_A \sum_{k=1}^{N} \|x(\beta_k) - x(\alpha_k)\| + M_x \sum_{k=1}^{N} \|A(\beta_k) - A(\alpha_k)\|.$$

Let us choose $\delta > 0$ so that (5) implies the inequalities

$$\sum_{k=1}^{N} \|x(\beta_k) - x(\alpha_k)\| < \frac{\varepsilon}{2M_A}, \quad \sum_{k=1}^{N} \|A(\beta_k) - A(\alpha_k)\| < \frac{\varepsilon}{2M_x}.$$

Then

$$\sum_{k=1}^{N} \|A(\beta_k)x(\beta_k) - A(\alpha_k)x(\alpha_k)\| < \frac{\varepsilon}{2M_A} M_A + \frac{\varepsilon}{2M_x} M_x = \varepsilon,$$

and hence $A(\cdot)x(\cdot)$ is absolutely continuous.

Similarly, if

$$\|\Phi(x_1) - \Phi(x_2)\| \leqslant K \|x_1 - x_2\|, \quad \forall x_1, x_2 \in \mathcal{K},$$

then

$$\sum_{k=1}^{N} \|\Phi(x(\beta_k)) - \Phi(x(\alpha_k))\| \leqslant \sum_{k=1}^{N} \|x(\beta_k) - x(\alpha_k)\| < \varepsilon$$

if $\delta > 0$ is chosen so that (5) implies (6), where ε is replaced by ε/K. ■

We shall state the basic theorem of this section only for a finite-dimensional space X to avoid the definition of the operation of integration in an arbitrary normed space [for $X = \mathbf{R}^n$ it is sufficient to take into consideration (2)]. In what follows we shall need only this case.

Lebesgue's Theorem [KF, p. 345]. If a function $x(\cdot): \Delta \to \mathbf{R}^n$ is absolutely continuous, it is differentiable almost everywhere and its derivative $\dot{x}(\cdot)$ is integrable on Δ, and equality (4) holds for all $t, \tau \in \Delta$.

If a function $\xi(\cdot): \Delta \to \mathbf{R}^n$ is integrable on Δ and $\tau \in \Delta$, then the function $x(t) = \int_\tau^t \xi(s)\,ds$ is absolutely continuous and $\dot{x}(t) = \xi(t)$ almost everywhere.

In accordance with this theorem, in what follows we shall say that a function $x(\cdot)$ is a solution of the differential equation (1) if it is absolutely continuous and satisfies (1) almost everywhere. Indeed, by Lebesgue's theorem, to this end it is sufficient that (1) imply (3). Conversely, if $x(\cdot)$ satisfies (3) [i.e., in particular, $t \to \varphi(t, x(t), u(t))$ is an integrable function] then, by Lebesgue's theorem, $x(\cdot)$ is absolutely continuous and (1) holds almost everywhere.

To conclude this section we shall state one more assertion whose scalar analogue is well known in the Lebesgue theory of integration and whose proof is left to the reader as an exercise.

Proposition 2. A measurable function $x(\cdot):\Delta \to \mathbf{R}^n$ is integrable if and only if $|x(\cdot)|$ is integrable, and then

$$\left|\int_\Delta x(t)\,dt\right| \leqslant \int_\Delta |x(t)|\,dt. \tag{7}$$

2.1.9. Riesz' Representation Theorem for the General Linear Functional on the Space C. Dirichlet's Formula. The reader is supposed to be familiar with the definitions of a function of bounded variation and Stieltjes' integral [KF, Chapter 6, Sections 2 and 6]. A function of bounded variation $v(\cdot):[\alpha, \beta] \to \mathbf{R}$ will be called *canonical* if it is continuous from the right at all the points of the open interval (α, β) and $v(\alpha) = 0$.

Riesz' Representation Theorem [KF, p. 369]. To every continuous linear functional $x^* \in C([\alpha, \beta])^*$ there corresponds a canonical function of bounded variation $v(\cdot):[\alpha, \beta] \to \mathbf{R}$ such that

$$\langle x^*, x(\cdot)\rangle = \int_\alpha^\beta x(t)\,dv(t) \tag{1}$$

for all $x(\cdot) \in C([\alpha, \beta])$, and the representation (1) is unique: if

$$\int_\alpha^\beta x(t)\,dv(t) = 0$$

for all $x(\cdot) \in C([\alpha, \beta])$ and for a canonical $v(\cdot)$, then $v(t) \equiv 0$.

This theorem can readily be generalized to the vector case. Let $e_1 = (1, 0,\dots, 0)$, $e_2 = (0, 1,\dots,0),\dots$ be the unit vectors of the standard basis in $\mathbf{R}^n$. An arbitrary element $x(\cdot) = (x_1(\cdot), \dots, x_n(\cdot)) \in C([\alpha, \beta], \mathbf{R}^n)$ is represented in the form

$$x(\cdot) = \sum_{k=1}^n x_k(\cdot)\,e_k.$$

Now if $x^* \in C([\alpha, \beta], \mathbf{R}^n)^*$, then

$$\langle x^*, x(\cdot)\rangle = \sum_{k=1}^n \langle x^*, x_k(\cdot)\,e_k\rangle = \sum_{k=1}^n \langle x_k^*, x_k(\cdot)\rangle,$$

where the functionals $x_k^* \in C([\alpha, \beta])^*$ are determined by the equalities $\langle x_k^*, \xi(\cdot)\rangle = \langle x^*, \xi(\cdot)e_k\rangle$. Applying Riesz' theorem to x_k^*, we obtain the representation

$$\langle x^*, x(\cdot)\rangle = \sum_{k=1}^n \int_\alpha^\beta x_k(t)\,dv_k(t). \tag{2}$$

The collection of functions of bounded variation $v(\cdot) = (v_1(\cdot),\dots,v_n(\cdot))$ can naturally be regarded as a vector-valued function of bounded variation $v(\cdot):[\alpha, \beta] \to \mathbf{R}^{n*}$ (whose values are row vectors). Then formula (2) can be rewritten in the form

$$\langle x^*, x(\cdot)\rangle = \int_\alpha^\beta dv(t)\,x(t) \tag{3}$$

coinciding with (1) to within the order of the factors. As in Riesz' theorem, the correspondence between x^* and $v(\cdot)$ is one-to-one if the condition that the function $v(\cdot)$ is "canonical" holds.

Every function of bounded variation $v(\cdot):[\alpha, \beta] \to \mathbf{R}$ specifies a (generalized) signed measure (or "charge") [KF, Chapter 6, Section 5]. Stieltjes' integral (1) is nothing other than the integral with respect to such a measure. Similarly, a vector-valued function of bounded variation $v(\cdot)$: $[\alpha, \beta] \to \mathbf{R}^{n*}$ specifies a measure on $[\alpha, \beta]$ with vector values, (3) being the integral with respect to this measure. The measure corresponding to a function of bounded variation v(t) is usually denoted by dv(t).

If $v_1(\cdot)$ and $v_2(\cdot)$ are two functions of bounded variation on a closed interval $[\alpha, \beta]$, then the product measure $dv_1 \times dv_2$ is defined on the square $[\alpha, \beta] \times [\alpha, \beta]$, and Fubini's theorem holds [KF, p. 317]. In particular, there holds the well-known formula for interchanging the limits of integration ("Dirichlet's formula"):

$$\int_\alpha^\beta \left\{ \int_\alpha^t f(t, s)\, dv_1(s) \right\} dv_2(t) = \int_\alpha^\beta \left\{ \int_s^\beta f(t, s)\, dv_2(t) \right\} dv_1(s). \tag{4}$$

In the generalization of the formula (4) to vector-valued functions (which we shall need in what follows) attention should be paid to the order of the factors. For instance, if

$$f(\cdot, \cdot)\colon [\alpha, \beta] \times [\alpha, \beta] \to \mathscr{L}(\mathbf{R}^n, \mathbf{R}^n),$$
$$v_1(\cdot)\colon [\alpha, \beta] \to \mathbf{R}^n,\ v_2(\cdot)\colon [\alpha, \beta] \to \mathbf{R}^{n*},$$

then (4) should be written thus:

$$\int_\alpha^\beta dv_2(t) \left\{ \int_\alpha^t f(t, s)\, dv_1(s) \right\} = \int_\alpha^\beta \left\{ \int_s^\beta dv_2(t) f(t, s) \right\} dv_1(s). \tag{5}$$

2.2. Fundamentals of Differential Calculus in Normed Linear Spaces

Differential calculus is a branch of mathematical analysis in which the local properties of smooth mappings are studied. At an intuitive level, the smoothness of a mapping means the possibility of approximating this mapping locally by a properly chosen affine mapping, i.e., the differential, or, more generally, by a polynomial mapping, which gives some information about the properties of the mapping under consideration. Differential calculus as well as convex analysis is a basic mathematical tool of the theory of extremal problems.

The foundation of differential calculus of scalar functions on the real line was laid by Newton and Leibniz. The problems of the calculus of variations required the generalization of the notions of the derivative and the differential to mappings of many-dimensional and infinite-dimensional spaces, which was done by Lagrange in the 19th century; in our time this theory was developed more completely by Fréchet, Gateaux, Lévy, and other mathematicians.

2.2.1. Directional Derivative, First Variation, the Fréchet and the Gateaux Derivatives, Strict Differentiability. As is known, for real functions of one real variable the two definitions based on the existence of a finite limit

$$\lim_{h \to 0} \frac{F(\hat{x}+h) - F(\hat{x})}{h} \tag{1}$$

and on the possibility of the asymptotic expansion

$$F(\hat{x}+h) = F(\hat{x}) + F'(\hat{x})\, h + o(h) \tag{2}$$

for $h \to 0$ lead to one and the same notion of differentiability. For functions of several variables and all the more so for functions with infinite-dimensional domains the situation is not so simple. The definition (1) and the definition of a partial derivative generalizing (1) lead to the notions of the directional derivative, the first variation, and the Gateaux derivative while the generalization of the definition (2) leads to the notions of the Fréchet derivative and the strict differentiability.

Let X and Y be normed linear spaces, let U be a neighborhood of a point $\hat{x}$ in X, and let F be a mapping from U into Y.

Definition 1. If the limit

$$\lim_{\lambda \downarrow 0} \frac{F(\hat{x}+\lambda h)-F(\hat{x})}{\lambda} \tag{3}$$

exists, it is called the *derivative of* F *at the point* $\hat{x}$ *in the direction* h and is denoted by $F'(\hat{x}; h)$ [and also by $(\nabla_h F)(\hat{x})$, $D_h F(\hat{x})$, and $\partial_h F(\hat{x})$]. For real functions (Y = **R**) we shall understand (1) in a broader sense by admitting $-\infty$ and $+\infty$ as possible values of the limit (3).

Definition 2. Let us suppose that for any $h \in X$ there exists the directional derivative $F'(\hat{x}; h)$. The mapping $\delta_+ F(\hat{x}; \cdot): X \to Y$ determined by the formula $\delta_+ F(x; h) = F'(x; h)$ is called the *first variation of* F *at the point* $\hat{x}$ (in this case the function F is said to have the first variation at the point $\hat{x}$).

If $\delta_+(\hat{x}, -h) = -\delta_+(\hat{x}, h)$ for all h or, in other words, if for any $h \in X$ there exists the limit

$$\delta F(\hat{x}, h) = \lim_{\lambda \to 0} \frac{F(\hat{x}+\lambda h)-F(\hat{x})}{\lambda},$$

then the mapping $h \to \delta F(\hat{x}; h)$ is called the *Lagrange first variation* of the mapping F at the point $\hat{x}$ (see Section 1.4.1).

Definition 3. Let F have the first variation at a point $\hat{x}$; let us suppose that there exists a continuous linear operator $\Lambda \in \mathscr{L}(X, Y)$ such that $\delta_+ F(x; h) \equiv \Lambda h$.

Then the operator Λ is called the *Gateaux derivative* of the mapping F at the point x and is denoted by $F'_\Gamma(x)$.

Thus, $F'_\Gamma(x)$ is an element of $\mathscr{L}(X, Y)$ such that, given any $h \in X$, the relation

$$F(x+\lambda h) = F(x) + \lambda F'_\Gamma h + o(\lambda) \tag{4}$$

holds for $\lambda \downarrow 0$.

Despite the resemblance between relations (2) and (4) there is a significant distinction between them. As will be seen below, even for $X = \mathbf{R}^2$ a Gateaux differentiable function may be discontinuous. The matter is that the expression $o(\lambda)$ in the expansion (4) is not necessarily uniformly small with respect to h.

Definition 4. Let it be possible to represent a mapping F in a neighborhood of a point $\hat{x}$ in the form

$$F(\hat{x}+h) = F(\hat{x}) + \Lambda h + \alpha(h)\|h\|, \tag{5}$$

where $\Lambda \in \mathscr{L}(X, Y)$ and

$$\lim_{\|h\| \to 0} \|\alpha(h)\| = \|\alpha(0)\| = 0. \tag{6}$$

Then the mapping $F(\cdot)$ is said to be *Fréchet differentiable at the point* $\hat{x}$, and we write $F \in D^1(\hat{x})$. The operator Λ is called the *Fréchet derivative* (or simply the derivative) of the mapping F at the point $\hat{x}$ and is denoted by $F'(\hat{x})$.

Relations (5) and (6) can also be written thus:

$$F(\hat{x}+h) = F(\hat{x}) + F'(\hat{x})h + o(\|h\|). \tag{7}$$

It readily follows that the Fréchet derivative is determined uniquely (for the Gateaux derivative this is obvious since the directional derivatives are determined uniquely) because the equality $\|\Lambda_1 h - \Lambda_2 h\| = o(\|h\|)$ for operators $\Lambda_i \in \mathscr{L}(X, Y)$, i = 1, 2, is only possible when $\Lambda_1 = \Lambda_2$.

Finally, in terms of the ε, δ formalism the relations (5) and (6) are stated thus: given an arbitrary $\varepsilon > 0$, there is $\delta > 0$ for which the inequality

$$\|F(x+h)-F(x)-\Lambda h\| \leqslant \varepsilon \|h\| \tag{8}$$

holds for all h such that $\|h\| < \delta$. This naturally leads to a further strengthening:

Definition 5. A mapping F is said to be *strictly differentiable* at a point $\hat{x}$ [and we write $F \in SD^1(\hat{x})$] if there is an operator $\Lambda \in \mathscr{L}(X, Y)$ such that for any $\varepsilon > 0$ there is $\delta > 0$ such that for all x_1 and x_2 satisfying the inequalities $\|x_1 - \hat{x}\| < \delta$ and $\|x_2 - \hat{x}\| < \delta$ the inequality

$$\|F(x_1)-F(x_2)-\Lambda(x_1-x_2)\| \leqslant \varepsilon \|x_1-x_2\| \tag{9}$$

holds.

Putting $x_2 = \hat{x}$ and $x_1 = \hat{x} + h$ in (9), we obtain (8), and hence a strictly differentiable function is Fréchet differentiable and $\Lambda = F'(x)$.

By definition, the (Gateaux, Fréchet, or strict) derivative $F'(x)$ is a linear mapping from X into Y. The value of this mapping assumed on a vector $h \in X$ will often be denoted by $F'(\hat{x})[h]$. For scalar functions of one variable (when $X = Y = \mathbf{R}$) the space $\mathscr{L}(X, Y) = \mathscr{L}(\mathbf{R}, \mathbf{R})$ of continuous linear mappings from **R** into **R** is naturally identified with **R** (with a linear function $y = kx$ its slope k is associated). In this very sense the derivative at a point in elementary mathematical analysis is a number (equal to the slope of the tangent) and the corresponding linear mapping from **R** into **R** is the differential $dy = F'(\hat{x})dx$ (here dx and dy are elements of the one-dimensional space **R**).

As is also known, for functions of one variable Definition 3 (or the definition of the first Lagrange variation which coincides with Definition 3 in this case) and Definition 4 are equivalent (a function possesses the derivative at a point if and only if it possesses the differential at that point) and lead to one and the same notion of the derivative which was in fact introduced by Newton and Leibniz. In this case Definition 2 is applicable to functions possessing the two one-sided derivatives at a point $\hat{x}$, which do not necessarily coincide. In elementary mathematical analysis for functions of several variables ($1 < \dim X < \infty$), we use Definition 1 [if h is a unit vector, then (3) is the ordinary definition of the directional derivative; if h is a base vector along an axis OX_k, then replacing in (3) the condition $\lambda \downarrow 0$ by $\lambda \to 0$ we arrive at the definition of the partial derivative $\partial F/\partial x_k$] and Definition 4 (the existence of the total differential).

The generalization of the notion of the derivative to infinite-dimensional spaces was stimulated by the problems of the calculus of variations. The definition of the first variation $\delta F(\hat{x}; h)$ and its notation and name were introduced by Lagrange. In terms of functional analysis the same definition was stated by Gateaux and therefore the first Lagrange variation is often called the Gateaux differential. The requirement of linearity in the definition of the Gateaux derivative has been generally accepted after the works of P. Lévy. The definition most commonly used in both finite-dimensional and infinite-dimensional analysis is that of the Fréchet derivative. However, for some purposes (in particular, for our further aims) the existence of the Fréchet derivative at a point is insufficient and some strengthening is needed and this leads to the notion of the strict differentiability.

Proposition 1. For Definitions 1-5 and the continuity of functions the following implications hold:

$$(1) \Leftarrow (2) \Leftarrow (3) \Leftarrow (4) \quad \Leftarrow \quad (5)$$

$(4) \Downarrow$ continuity at the point $\hat{x}$

$(5) \Downarrow$ continuity in a neighborhood of the point $\hat{x}$

and none of them can be converted.

Proof. The positive part of the proof (i.e., the existence of the implications) follows at once from the definitions, and the negative part (the impossibility of converting the implications) is confirmed by the set of the counterexamples below whose thorough study is left to the reader as an exercise.

1) 2 does not imply 3: f_1: $\mathbf{R} \to \mathbf{R}$, $f_1(x) = |x|$.

The variation is nonlinear at the point x = 0. The same example shows that the continuity at a point does not imply the Fréchet and the Gateaux differentiability.

2) 3 does not imply 4: f_2: $\mathbf{R}^2 \to \mathbf{R}$,

$$f_2(x_1, x_2) = \begin{cases} 1, & x_1 = x_2^2, \ x_2 > 0, \\ 0 & \text{at all the other points.} \end{cases}$$

This example shows that a Gateaux differentiable function [at the point (0, 0)] is not necessarily continuous.

3) 4 does not imply 5: f_3: $\mathbf{R} \to \mathbf{R}$,

$$f_3(x) = \begin{cases} x^2 & \text{if } x \text{ is rational,} \\ 0 & \text{if } x \text{ is irrational.} \end{cases}$$

This function is (Fréchet) differentiable at the point x = 0 but is not strictly differentiable. ■

Exercises.

1. Show that the function f_4: $\mathbf{R}^2 \to \mathbf{R}$ defined on polar coordinates on $\mathbf{R}^2$ by the equality

$$f_4(x) = r \cos 3\varphi \qquad (x = (x_1, x_2) = (r \cos \varphi, r \sin \varphi))$$

possesses the first Lagrange variation at the point (0, 0) but is not Gateaux differentiable.

2. If $F \in SD^1(\hat{x})$, then the function F satisfies the Lipschitz condition in a neighborhood of the point $\hat{x}$.

3. If $F \in SD^1(\hat{x})$ is a scalar function (X = Y = $\mathbf{R}$), then $F'(\hat{x})$ exists at almost all the points of a neighborhood of the point $\hat{x}$.

Proposition 2. If mappings $F_i: U \to Y$, i = 1, 2, and A: $U \to \mathscr{L}(Y, Z)$, where Z is a normed linear space, are differentiable in the sense of one of the Definitions 1-5 (one and the same definition for all the three mappings), then for any $\alpha_i \in \mathbf{R}$, i = 1, 2 the mapping $F = \alpha_1 F_1 + \alpha_2 F_2$ is differentiable at the point $\hat{x}$ in the same sense, and, moreover,

$$F'(\hat{x}) = \alpha_1 F_1'(\hat{x}) + \alpha_2 F_2'(\hat{x})$$

or

$$F'(\hat{x}; h) = \alpha_1 F_1'(\hat{x}; h) + \alpha_2 F_2'(\hat{x}; h).$$

The mapping

$$\Phi = AF_i$$

is differentiable at the point $\hat{x}$ in the same sense, and

$$\Phi'(\hat{x};\ h)=A'(\hat{x};\ h)F_i'(\hat{x})+A(\hat{x})F_i'(\hat{x};\ h).$$

Proof. The proof follows at once from the corresponding definitions. ■

In the special case when X = **R** and $A\colon U\to\mathscr{L}(Y,\mathbf{R})=Y^*$ the expression $\Phi(x) = \langle A(x), F_i(x)\rangle$ is a scalar function and the last formula can be rewritten thus:

$$\langle A(x),\ F_i(x)\rangle'\big|_{x=\hat{x}}=\langle A'(\hat{x}),\ F_i(\hat{x})\rangle+\langle A(\hat{x}),\ F_i'(\hat{x})\rangle,$$

which corresponds to the ordinary differentiation formula.

Below we give two simple examples of computing derivatives.

1) Affine Mapping. A mapping $A:X \to Y$ of one linear space into another is said to be *affine* if there exist a linear mapping $\Lambda:X \to Y$ and a constant $a\in Y$ such that

$$A(x)=\Lambda x+a.$$

If X and Y are normed spaces and $\Lambda\in\mathscr{L}(X,Y)$, then the mapping A is strictly differentiable at any point x, and, moreover,

$$A'(x)=\Lambda.$$

This assertion is verified directly. In particular, if A is linear (a = 0), then $A'(x)[h] = A(h)$, and the derivative of a constant mapping ($\Lambda = 0$) is equal to zero.

Exercise 1. Prove the converse: if the derivative (it is sufficient to understand it in Gateaux' sense) of a mapping $A:X \to Y$ exists at each point $x\in X$ and is one and the same for all x, then A is an affine mapping.

2) Multilinear Mapping. Let X and Y be topological linear spaces, and let $\mathscr{L}^n(X,Y)$ be the linear space of the continuous multilinear mappings of the Cartesian product $X^n=\underbrace{X\times\ldots\times X}_{\text{n times}}$ into Y. We remind the reader that a mapping $\Pi(x_1,\ldots,x_n)$ is said to be *multilinear* if the mapping

$$x_i\mapsto\Pi(\hat{x}_1,\ \ldots,\ \hat{x}_{i-1},\ x_i,\ \hat{x}_{i+1},\ \ldots,\ \hat{x}_n)$$

is linear for every i. If X and Y are normed spaces, the continuity of Π equivalent to its boundedness, i.e., the finiteness of the number

$$\|\Pi\|=\sup_{\|x_1\|\leqslant 1,\ldots,\|x_n\|\leqslant 1}\|\Pi(x_1,\ \ldots,\ x_n)\|. \tag{10}$$

In this case $\mathscr{L}^n(X,\ Y)$ is a normed linear space with the norm (10), and the inequality

$$\|\Pi(x_1,\ \ldots,\ x_n)\|\leqslant\|\Pi\|\|x_1\|\ldots\|x_n\| \tag{11}$$

holds.

The function $Q_n(x) = \Pi(x,\ldots,x)$ is called a *form of degree* n (a quadratic form for n = 2 and a ternary form for n = 3) corresponding to the multilinear mapping Π. It follows from the definitions that

$$Q_n(\hat{x}+h)=Q_n(\hat{x})+\sum_{i=1}^{n}\Pi(\hat{x},\ \ldots,\ h,\ \ldots,\ \hat{x})+o(\|h\|^2), \tag{12}$$

and therefore $Q_n(\cdot)$ is Fréchet differentiable at each point $\hat{x}$ and

$$Q_n'(\hat{x})[h]=\sum_{i=1}^{n}\Pi(\hat{x},\ \ldots,\ h,\ \ldots,\ \hat{x}) \tag{13}$$

[in the formulas (12) and (13) the argument h occupies the i-th place in the i-th term of the sum, the other arguments being equal to $\hat{x}$]. If the mapping Π is symmetric, i.e., $\Pi(x_1,\ldots,x_n)$ does not change under an arbitrary permutation of the arguments x_i, then (13) turns into

$$Q_n'(\hat{x})[h] = n\Pi(h, \hat{x}, \ldots, \hat{x}).$$

In particular, if X is a real Hilbert space with a scalar product $(\cdot|\cdot)$ and if $A \in \mathscr{L}(X, X)$ and $\Pi(x_1, x_2) = (Ax_1|x_2)$, then $Q_2(x) = (Ax|x)$ (the quadratic form of the operator A) is differentiable and

$$Q_2'(\hat{x})[h] = (A\hat{x}|h) + (Ah|\hat{x}) = (A\hat{x} + A^*\hat{x}|h),$$

and therefore $Q_2'(\hat{x})$ can naturally be identified with the vector $A\hat{x} + A^*\hat{x}$. For instance, for $Q_2(x) = {}^1/_2(x|x) = {}^1/_2\|x\|^2$ the derivative is $Q'(x) = x$.

Exercise 2. Prove that $Q_n(x) = \Pi(x,\ldots,x)$, where $\Pi \in \mathscr{L}^n(X, Y)$ is strictly differentiable for all x.

The proposition below implies a more complicated example.

Proposition 3. Let $U \subset \mathscr{L}(X, Y)$ be the set of those continuous linear operators $A: X \to Y$ for which the inverse operators $A^{-1} \in \mathscr{L}(Y, X)$ exist. If at least one of the spaces X and Y is a Banach space, then U is open and the function $\Phi(A) = A^{-1}$ is Fréchet differentiable at any point $A \in U$, and, moreover, $\Phi'(\hat{A})[H] = -\hat{A}^{-1}H\hat{A}^{-1}$.

Proof. If X is a Banach space, then the series $\sum_{k=0}^{\infty}(-\hat{A}^{-1}H)^k$ is convergent in $\mathscr{L}(X, X)$ for $\|H\| < \|\hat{A}^{-1}\|^{-1}$ and

$$(\hat{A}+H)\left[\sum_{k=0}^{\infty}(-\hat{A}^{-1}H)^k\hat{A}^{-1}\right] = I_Y,$$

$$\left[\sum_{k=0}^{\infty}(-\hat{A}^{-1}H)^k\hat{A}^{-1}\right](\hat{A}+H) = I_X,$$

which can readily be verified. Consequently,

$$(\hat{A}+H)^{-1} = \sum_{k=0}^{\infty}(-\hat{A}^{-1}H)^k\hat{A}^{-1} \tag{14}$$

for $\|H\| < \|\hat{A}^{-1}\|^{-1}$, and therefore $\mathring{B}(\hat{A}, \|\hat{A}^{-1}\|^{-1}) \subset U$ and U is open.

If Y is a Banach space, then (14) should be replaced by

$$(\hat{A}+H)^{-1} = \hat{A}^{-1}\sum_{k=0}^{\infty}(-H\hat{A}^{-1})^k.$$

Further, from (14) we obtain

$$\|(\hat{A}+H)^{-1} - \hat{A}^{-1} + \hat{A}^{-1}H\hat{A}^{-1}\| = \left\|\sum_{k=2}^{\infty}(-\hat{A}^{-1}H)^k\hat{A}^{-1}\right\| \leqslant$$

$$\leqslant \sum_{k=2}^{\infty}(\|\hat{A}^{-1}\|\|H\|)^k\|\hat{A}^{-1}\| = \frac{\|\hat{A}^{-1}\|^3\|H\|^2}{1-\|\hat{A}^{-1}\|\|H\|} = o(\|H\|)$$

for $H \to 0$, and consequently the function $\Phi(\hat{A}) = \hat{A}^{-1}$ is differentiable and its Fréchet derivative is $\Phi'(\hat{A})[\cdot] = -\hat{A}^{-1}(\cdot)\hat{A}^{-1}$. ■

Exercise 3. Prove that $\Phi(A)$ is strictly differentiable at any point $\hat{A} \in U$.

2.2.2. Theorem on the Composition of Differentiable Mappings. The Composition Theorem. Let X, Y, and Z be normed spaces; let U be a neighborhood of a point $\hat{x}$ in X, let V be a neighborhood of a point $\hat{y}$ in Y, let $\varphi: U \to V$, $\varphi(\hat{x}) = \hat{y}$, and $\psi: V \to Z$, and let $f = \psi \circ \varphi: U \to Z$ be the composition of the mappings φ and ψ.

If ψ is Fréchet differentiable at the point $\hat{y}$ and φ is Fréchet differentiable (or is Gateaux differentiable or possesses the first variation or possesses the derivative in a direction h) at the point $\hat{x}$, then f possesses the same property as φ at the point $\hat{x}$, and, moreover,

$$f'(\hat{x}) = \psi'(\hat{y}) \circ \varphi'(\hat{x}) \tag{1}$$

or

$$f'(\hat{x};\ h) = \psi'(\hat{y})[\varphi'(\hat{x};\ h)]. \tag{2}$$

If ψ is strictly differentiable at $\hat{y}$ and φ is strictly differentiable at $\hat{x}$, then f is strictly differentiable at $\hat{x}$.

Proof. We shall consider in detail the two extreme cases: the directional derivative and the strict differentiability.

A) By the definition of the Fréchet derivative,

$$\psi(y) = \psi(\hat{y}) + \psi'(\hat{y})[y - \hat{y}] + \alpha(y)\|y - \hat{y}\|,$$

where

$$\lim_{y \to \hat{y}} \alpha(y) = \alpha(\hat{y}) = 0.$$

If the limit

$$\varphi'(\hat{x};\ h) = \lim_{\lambda \downarrow 0} \frac{\varphi(\hat{x}+\lambda h) - \varphi(\hat{x})}{\lambda} = \lim_{\lambda \downarrow 0} \frac{\varphi(\hat{x}+\lambda h) - \hat{y}}{\lambda}$$

exists, then

$$\lim_{\lambda \downarrow 0} \frac{f(\hat{x}+\lambda h) - f(\hat{x})}{\lambda} = \lim_{\lambda \downarrow 0} \frac{\psi(\varphi(\hat{x}+\lambda h)) - \psi(\hat{y})}{\lambda} =$$
$$= \lim_{\lambda \downarrow 0} \frac{\psi'(\hat{y})[\varphi(\hat{x}+\lambda h) - \hat{y}] + \alpha(\varphi(\hat{x}+\lambda h))\|\varphi(\hat{x}+\lambda h) - \hat{y}\|}{\lambda} =$$
$$= \psi'(\hat{y})\left[\lim_{\lambda \downarrow 0} \frac{\varphi(\hat{x}+\lambda h) - \hat{y}}{\lambda}\right] + \lim_{\lambda \downarrow 0} \alpha(\varphi(\hat{x}+\lambda h)) \lim_{\lambda \downarrow 0} \left\| \frac{\varphi(\hat{x}+\lambda h) - \hat{y}}{\lambda} \right\| =$$
$$= \psi'(\hat{y})[\varphi'(\hat{x};\ h)] + \alpha(\hat{y})\|\varphi'(\hat{x};\ h)\| = \psi'(\hat{y})[\varphi'(\hat{x};\ h)],$$

which proves (2).

B) For brevity, we shall denote $A_1 = \varphi'(\hat{x})$, $A_2 = \psi'(\hat{y})$. By the definition of the strict differentiability, for any $\varepsilon_1 > 0$, $\varepsilon_2 > 0$ there are $\delta_1 > 0$, $\delta_2 > 0$ such that

$$\|x_1 - \hat{x}\| < \delta_1, \|x_2 - \hat{x}\| < \delta_1 \Rightarrow \|\varphi(x_1) - \varphi(x_2) - A_1(x_1 - x_2)\| \leqslant \varepsilon_1 \|x_1 - x_2\| \tag{3}$$

and

$$\|y_1 - \hat{y}\| < \delta_2,\ \|y_2 - \hat{y}\| < \delta_2 \Rightarrow \|\psi(y_1) - \psi(y_2) - A_2(y_1 - y_2)\| \leqslant \varepsilon_2 \|y_1 - y_2\|. \tag{4}$$

Now for any $\varepsilon > 0$ we choose $\varepsilon_1 > 0$ and $\varepsilon_2 > 0$ so that the inequality

$$\varepsilon_1 \|A_2\| + \varepsilon_2 \|A_1\| + \varepsilon_1 \varepsilon_2 < \varepsilon$$

holds, and for these ε_1, ε_2 we find $\delta_1 > 0$ and $\delta_2 > 0$ such that the relations (3) and (4) hold; finally we put

$$\delta = \min\left(\delta_1, \frac{\delta_2}{\varepsilon_1 + \|A_1\|}\right).$$

If $\|x_1 - \hat{x}\| < \delta$ and $\|x_2 - \hat{x}\| < \delta$, then, by virtue of (3),

$$\|\varphi(x_1) - \varphi(x_2)\| \leqslant \|\varphi(x_1) - \varphi(x_2) - A_1(x_1 - x_2)\| + \|A_1(x_1 - x_2)\| \leqslant$$
$$\leqslant \varepsilon_1 \|x_1 - x_2\| + \|A_1\| \|x_1 - x_2\| = (\|A_1\| + \varepsilon_1)\|x_1 - x_2\|. \tag{5}$$

Putting $x_1 = \hat{x}$ and $x_2 = \hat{x}$ in this inequality, we obtain

$$\|\varphi(x_i) - \varphi(\hat{x})\| = \|\varphi(x_i) - \hat{y}\| < (\|A_1\| + \varepsilon_1)\delta \leqslant \delta_2,\ i = 1,\ 2,$$

and thus (4) holds for $y_i = \varphi(x_i)$. Now using (4), (3), and (5) we obtain

$$\|f(x_1) - f(x_2) - A_2 A_1(x_1 - x_2)\| \leqslant \|\psi(\varphi(x_1)) - \psi(\varphi(x_2)) - A_2(\varphi(x_1) - \varphi(x_2))\| +$$
$$+ \|A_2(\varphi(x_1) - \varphi(x_2)) - A_2 A_1(x_1 - x_2)\| \leqslant$$

$$\leqslant \varepsilon_2 \|\varphi(x_1)-\varphi(x_2)\| + \|A_2\| \|\varphi(x_1)-\varphi(x_2)-A_1(x_1-x_2)\| \leqslant$$
$$\leqslant \varepsilon_2 (\|A_1\|+\varepsilon_1)\|x_1-x_2\| + \|A_2\| \varepsilon_1 \|x_1-x_2\| =$$
$$= (\varepsilon_2 \|A_1\| + \varepsilon_1\varepsilon_2 + \|A_2\| \varepsilon_1)\|x_1-x_2\| \leqslant \varepsilon \|x_1-x_2\|,$$

which implies the strict differentiability of f at the point $\hat{x}$.

Putting $x_2 = \hat{x}$ and $x_1 = \hat{x} + h$ in this argument, we arrive at the proof of the theorem for the case of the Fréchet differentiability of φ. The remaining assertions are proved by analyzing equality (2), which we have already proved. ■

The counterexample below shows that in the general case the composition theorem does not hold when ψ is only Gateaux differentiable.

Example. Let $X = Y = \mathbf{R}^2$, $Z = \mathbf{R}$;

$$\varphi(x) = (\varphi_1(x_1, x_2), \varphi_2(x_1, x_2)) = (x_1^2, x_2);$$
$$\psi(y) = \psi(y_1, y_2) = \begin{cases} 1 & \text{for } y_1 = y_2^2,\ y_2 > 0, \\ 0 & \text{in all the other cases} \end{cases}$$

(cf. the counterexample in the proof of Proposition 1 of Section 2.1.1). Here φ is Fréchet differentiable at the point (0, 0) and even strictly differentiable (check this!) and ψ is Gateaux differentiable at (0, 0); however, the function

$$f(x) = (\psi \circ \varphi)(x) = \psi(x_1^2, x_2) = \begin{cases} 1 & \text{for } x_2 = |x_1| > 0, \\ 0 & \text{in all the other cases} \end{cases}$$

is not Gateaux differentiable at (0, 0) [and even does not possess the derivatives in the directions h = (1, 1) and h = (−1, 1) at that point].

COROLLARY. Let X and Y be normed spaces, and let U be a neighborhood of a point $\hat{y}$ in Y. Let $f:U \to \mathbf{R}$ be Fréchet differentiable at the point $\hat{y}$, and let $\Lambda \in \mathscr{L}(X, Y)$. Then $f \circ \Lambda: \Lambda^{-1}(U) \to \mathbf{R}$ is Fréchet differentiable at the point $\hat{x} = \Lambda^{-1}(\hat{y})$ and

$$(f \circ \Lambda)'(\hat{x}) = \Lambda^*(f' \circ \Lambda)(\hat{x}). \tag{6}$$

Proof. According to the definition,

$$(f' \circ \Lambda)(\hat{x}) = f'(\Lambda\hat{x}) = f'(\hat{y}) \in \mathscr{L}(Y, \mathbf{R}) = Y^*$$

and

$$(f \circ \Lambda)'(\hat{x}) = f'(\Lambda\hat{x}) \circ \Lambda = f'(\hat{y}) \circ \Lambda \in \mathscr{L}(X, \mathbf{R}) = X^*.$$

Now for any $h \in X$ we have

$$\langle \Lambda^*(f' \circ \Lambda)(\hat{x}), h \rangle = \langle (f' \circ \Lambda)(\hat{x}), \Lambda h \rangle = f'(\hat{y})[\Lambda h] =$$
$$= (f'(\hat{y}) \circ \Lambda) h = \langle f'(\hat{y}) \circ \Lambda, h \rangle = \langle (f \circ \Lambda)'(\hat{x}), h \rangle,$$

which proves (6).

2.2.3. Mean Value Theorem and Its Corollaries. As is known, for scalar functions of one variable Lagrange's theorem holds (it is also called the *mean value theorem* or the *formula of finite increments*): if a function $f:[a, b] \to \mathbf{R}$ is continuous on a closed interval $[a, b]$ and differentiable in the open interval (a, b), then there exists a point $c \in (a, b)$ such that

$$f(b) - f(a) = f'(c)(b - a). \tag{1}$$

It can also be easily shown that the formula (1) remains valid for scalar functions f(x) whose argument belongs to an arbitrary topological linear space.

In this case

$$[a, b] = \{x \mid x = a + t(b - a),\ 0 \leqslant t \leqslant 1\},$$

and the open interval (a, b) is defined analogously; the differentiability may be understood in Gateaux' sense. Putting

$$\Phi(t)=f(a+t(b-a)),$$

we reduce the proof to the case of one real variable.

For vector-valued functions the situation is quite different.

Example. Let the mapping $f:\mathbf{R}\to\mathbf{R}^2$ be determined by the equality $f(t) = (\sin t, -\cos t)$. Then for every t there exists the (strict) Fréchet derivative $f'(t)$ (prove this!):

$$f'(t)[\alpha]=(\cos t,\ \sin t)\alpha=(\alpha\cos t,\ \alpha\sin t).$$

At the same time, for any c we have

$$f(2\pi)-f(0)=0\neq f'(c)[2\pi-0]=(2\pi\cos c,\ 2\pi\sin c),$$

and hence the formula (1) does not hold.

However, it should be noted that formula (1) itself is used in mathematical analysis much rarer than the inequality $|f(b) - f(a)| \leqslant M|b - a|$ [where $M = \sup|f'(x)|$] following from it. We shall show that in this weakened form the proposition can be extended to the case of arbitrary normed spaces. By tradition, the name the "mean value theorem" is retained, although it would be, of course, more correct to call it a "theorem on the estimation of a finite increment."

The Mean Value Theorem. Let X and Y be normed linear spaces, and let an open set $U\subseteq X$ contain a closed interval $[a, b]$.

If $f:U\to Y$ is a Gateaux differentiable function at each point $x\in[a, b]$, then

$$\|f(b)-f(a)\|\leqslant\sup_{c\in[a,b]}\|f'(c)\|\|b-a\|. \tag{2}$$

Proof. Let us take an arbitrary $y^*\in Y^*$ and consider the function

$$\Phi(t)=\langle y^*,\ f(a+t(b-a))\rangle.$$

This function possesses the left-hand and the right-hand derivatives at each point of the closed interval $[0, 1]$:

$$\Phi'_-(t)=\lim_{\alpha\downarrow 0}\frac{\Phi(t-\alpha)-\Phi(t)}{-\alpha}=\lim_{\alpha\downarrow 0}\left\langle y^*,\ \frac{f(a+(t-\alpha)(b-a))-f(a+t(b-a))}{-\alpha}\right\rangle=$$

$$=-\left\langle y^*,\ \lim_{\alpha\downarrow 0}\frac{f(a+t(b-a)-\alpha(b-a))-f(a+t(b-a))}{\alpha}\right\rangle=$$

$$=-\langle y^*,\ f'(a+t(b-a))[-(b-a)]\rangle=\langle y^*,\ f'(a+t(b-a))[b-a]\rangle,$$

and similarly

$$\Phi'_+(t)=\lim_{\alpha\downarrow 0}\frac{\Phi(t+\alpha)-\Phi(t)}{\alpha}=\langle y^*,\ f'(a+t(b-a))[b-a]\rangle.$$

Since these derivatives coincide, the function $\Phi(t)$ is differentiable (in the ordinary sense) on the closed interval $[0, 1]$ and therefore it is continuous on the interval. By Lagrange's formula, there exists $\theta\in(0, 1)$ such that

$$\langle y^*,\ f(b)-f(a)\rangle=\Phi(1)-\Phi(0)=\Phi'(\theta)=\langle y^*,\ f'(a+\theta(b-a))[b-a]\rangle.$$

Now we shall make use of Corollary 1 of the Hahn–Banach theorem (Section 2.1.3), according to which for any element $y\in Y$ there is a linear functional $y^*\in Y^*$ such that $\|y^*\| = 1$ and $\langle y^*, y\rangle = \|y\|$. Taking this very functional y^* for the element $y = f(b) - f(a)$, we obtain

$$\|f(b)-f(a)\|=\langle y^*,\ f(b)-f(a)\rangle=\langle y^*,\ f'(a+\theta(b-a))[b-a]\rangle\leqslant$$

$$\leqslant\|y^*\|\|f'(a+\theta(b-a))\|\|b-a\|\leqslant\sup_{c\in[a,b]}\|f'(c)\|\|b-a\|,$$

which is what we had to prove. ■

We shall present several corollaries of the mean value theorem.

COROLLARY 1. Let all the conditions of the mean value theorem be fulfilled, and let $\Lambda \in \mathscr{L}(X, Y)$. Then

$$\|f(b)-f(a)-\Lambda(b-a)\| \leqslant \sup_{c \in [a, b]} \|f'(c)-\Lambda\| \|b-a\|. \tag{3}$$

Proof. The proof reduces to the application of the mean value theorem to the mapping $g(x) = f(x) - \Lambda x$. ■

COROLLARY 2. Let X and Y be normed spaces, let U be a neighborhood of a point $\hat{x}$ in X, and let a mapping $f: U \to Y$ be Gateaux differentiable at each point $x \in U$. If the mapping $x \to f'_\Gamma(x)$ is continuous [with respect to the uniform topology of operators in the space $\mathscr{L}(X, Y)$] at the point $\hat{x}$, then the mapping f is strictly differentiable at $\hat{x}$ (and consequently it is Fréchet differentiable at that point).

Proof. Given $\varepsilon > 0$, there is $\delta > 0$ such that the relation

$$\|x-\hat{x}\| < \delta \Rightarrow \|f'_\Gamma(x) - f'_\Gamma(\hat{x})\| < \varepsilon \tag{4}$$

holds. If $\|x_1 - \hat{x}\| < \delta$ and $\|x_2 - \hat{x}\| < \delta$, then for any $x = x_1 + t(x_2 - x_1) \in [x_1, x_2], 0 \leqslant t \leqslant 1$, we have

$$\|x-\hat{x}\| = \|x_1 + t(x_2-x_1) - \hat{x}\| = \|t(x_2-\hat{x}) + (1-t)(x_1-\hat{x})\| \leqslant$$
$$\leqslant t\|x_2-\hat{x}\| + (1-t)\|x_1-x\| < t\delta + (1-t)\delta = \delta,$$

and therefore, by virtue of (4), we have $\|f'_\Gamma(x) - f'(\hat{x})\| < \varepsilon$.

Applying Corollary 1 to $\Lambda = f'_\Gamma(\hat{x})$, we obtain

$$\|f(x_1)-f(x_2)-f'_\Gamma(\hat{x})(x_1-x_2)\| \leqslant \sup_{x \in [x_1, x_2]} \|f'_\Gamma(x) - f'_\Gamma(\hat{x})\| \|x_1 - x_2\| \leqslant \varepsilon \|x_1 - x_2\|,$$

which implies the strict differentiability of f at $\hat{x}$. ■

Definition. Let X and Y be normed spaces. A mapping $F: U \to Y$ defined on an open subset $U \subset X$ is said to belong to the class $C^1(U)$ if it possesses the derivative at each point $x \in U$ and the mapping $x \to F'(x)$ is continuous (in the uniform topology of operators).

Corollary 2 indicates that here it is unnecessary to stipulate which derivative is meant. This fact is always used when the differentiability of concrete functionals is verified: the existence of the Gateaux derivative is proved and its continuity is verified, which guarantees the strict differentiability (and hence the existence of the Fréchet derivative).

Exercises. Let the conditions of the mean value theorem hold, let $Y = \mathbf{R}^n$, and let the mapping $x \to f'_\Gamma(x)$ be continuous in $U \subset X$. Then

1)
$$f(b)-f(a) = \int_0^1 f'_\Gamma(a+t(b-a))\,dt\,[b-a]. \tag{5}$$

2) There exists an operator $\Lambda \in \mathscr{L}(X, Y)$ belonging to the convex closure (see Section 2.6.2) of the set $\{f'_\Gamma(x), x \in [a, b]\}$ such that

$$f(b)-f(a) = \Lambda(b-a).$$

3) Generalize 1) and 2) to the case of an arbitrary Banach space Y by defining appropriately the integral in (5).

2.2.4. Differentiation in a Product Space. Partial Derivatives. The Theorem on the Total Differential. In this section X, Y, and Z are normed spaces. We shall begin with the case of a mapping whose values belong to the product space Y × Z, i.e., $F: U \to Y \times Z$, $U \subset X$. Since a point of Y × Z is a pair (y, z), the mapping F also consists of two components:

$F(x) = (G(x); H(x))$, where $G:U \to Y$ and $H:U \to Z$. The corresponding definitions immediately imply:

Proposition 1. Let X, Y, and Z be normed spaces, let U be a neighborhood of a point $\hat{x}$ in X, and let $G:U \to Y$ and $H:U \to Z$.

For the mapping $F = (G, H):U \to Y \times Z$ to be differentiable at the point $\hat{x}$ in the sense of one of the Definitions 1-5 of Section 2.2.1 it is necessary and sufficient that G and H possess the same property. Moreover, in this case

$$F'(\hat{x}) = (G'(\hat{x}), H'(\hat{x}))$$

or

$$F'(\hat{x}, h) = (G'(\hat{x}, h), H'(\hat{x}, h)).$$

Now we pass to the case when the domain of the mapping $F:U \to Z$ belongs to the product space: $U \subset X \times Y$.

Definition. Let X, Y, and Z be normed spaces, let U be a neighborhood of a point $(\hat{x}, \hat{y})$ in $X \times Y$, and let $F:U \to Z$.

If the mapping $x \to F(x, \hat{y})$ is (Fréchet, Gateaux, or strictly) differentiable at the point $\hat{x}$, its derivative is called the *partial derivative of the mapping* F *with respect to* x *at the point* $(\hat{x}, \hat{y})$ and is denoted $F_x(\hat{x}, \hat{y})$ or $\frac{\partial F}{\partial x}(\hat{x}, \hat{y})$.

The partial derivative

$$F_y(\hat{x}, \hat{y}) = \frac{\partial F}{\partial y}(\hat{x}, \hat{y})$$

with respect to y is defined in an analogous manner.

Theorem on the Total Differential. Let X, Y, and Z be normed spaces, let U be a neighborhood in $X \times Y$, and let $F:U \to Z$ be a mapping possessing the partial derivatives $F_x(x, y)$ and $F_y(x, y)$ in Gateaux' sense at each point $(x, y) \in W$.

If the mappings $(x, y) \to F_x(x, y)$ and $(x, y) \to F_y(x, y)$ are continuous at a point $(\hat{x}, \hat{y}) \in U$ in the uniform topology of operators, then F is strictly differentiable at the point, and moreover,

$$F'(\hat{x}, \hat{y})[\xi, \eta] = F_x(\hat{x}, \hat{y})[\xi] + F_y(\hat{x}, \hat{y})[\eta].$$

Proof. Given an arbitrary $\varepsilon > 0$, let us choose $\delta > 0$ so small that the "rectangular" neighborhood

$$V = \mathring{B}(\hat{x}, \delta) \times \mathring{B}(\hat{y}, \delta) = \{(x, y) \mid \|x - \hat{x}\| < \delta, \|y - \hat{y}\| < \delta\}$$

of the point $(\hat{x}, \hat{y})$ is contained in U and the inequalities

$$\|F_x(x, y) - F_x(\hat{x}, \hat{y})\| < \varepsilon, \quad \|F_y(x, y) - F_y(\hat{x}, \hat{y})\| < \varepsilon \tag{1}$$

hold in that neighborhood. Now we have

$$\begin{aligned}\Delta = F(x_1, y_1) - F(x_2, y_2) - F_x(\hat{x}, \hat{y})[x_1 - x_2] - F_y(\hat{x}, \hat{y})[y_1 - y_2] = \\ = [F(x_1, y_1) - F(x_2, y_1) - F_x(\hat{x}, \hat{y})[x_1 - x_2]] + \\ + [F(x_2, y_1) - F(x_2, y_2) - F_y(\hat{x}, \hat{y})[y_1 - y_2]].\end{aligned}$$

It can readily be seen that if the points (x_1, y_1), (x_2, y_2) belong to V, then $(x_2, y_1) \in V$, and moreover the line segments $[(x_1, y_1), (x_2, y_1)]$ and $[(x_2, y_1), (x_2, y_2)]$ are contained in $V \subset U$. Therefore, the functions $x \to F(x, y_1)$ and $y \to F(x_2, y)$ are Gateaux differentiable: the first of them possesses the derivative F_x on $[x_1, x_2]$ and the other possesses the derivative F_y on $[y_1, y_2]$. Applying the mean value theorem to these functions [in the form of inequality (3) of Section 2.2.3 with the corresponding Λ] we obtain, by virtue of (1), the relations

$$(x_1, y_1) \in V, (x_2, y_2) \in V \Rightarrow \|\Delta\| \leqslant \sup_{\xi \in [x_1, x_2]} \{\|F_x(\xi, y_1) - F_x(\hat{x}, \hat{y})\| \|x_1 - x_2\|\} +$$
$$+ \sup_{\eta \in [y_1, y_2]} \{\|F_y(x_2, \eta) - F_y(\hat{x}, \hat{y})\| \|y_1 - y_2\|\} \leqslant \varepsilon \|x_1 - x_2\| + \varepsilon \|y_1 - y_2\|. \blacksquare$$

COROLLARY. For F to be of class $C^1(U)$ it is necessary and sufficient that the partial derivatives F_x and F_y be continuous in U.

Both Proposition 1 and the theorem on the total differential can easily be generalized to the case of a product of an arbitrary finite number of spaces.

Let us dwell on the finite-dimensional case. Let $F:U \to \mathbf{R}^m$ be defined on an open set $U \subset \mathbf{R}^n$. Since we naturally have

$$\mathbf{R}^m = \underbrace{\mathbf{R} \times \ldots \times \mathbf{R}}_{\text{m times}}, \quad \mathbf{R}^n = \underbrace{\mathbf{R} \times \ldots \times \mathbf{R}}_{\text{n times}},$$

it is possible to use both Proposition 1 and the theorem on the total differential. If

$$F(x) = \begin{pmatrix} F_1(x) \\ F_2(x) \\ \cdot\ \cdot \\ \cdot \\ \cdot \\ F_m(x) \end{pmatrix} = \begin{pmatrix} F_1(x_1, \ldots, x_n) \\ F_2(x_1, \ldots, x_n) \\ \cdot\ \cdot\ \cdot\ \cdot\ \cdot\ \cdot\ \cdot \\ F_m(x_1, \ldots, x_n) \end{pmatrix} \text{ and } h = \begin{pmatrix} h_1 \\ h_2 \\ \cdot \\ \cdot \\ \cdot \\ h_n \end{pmatrix},$$

then

$$F'(\hat{x})[h] = \begin{pmatrix} F_1'(\hat{x})[h] \\ F_2'(\hat{x})[h] \\ \cdot\ \cdot\ \cdot\ \cdot \\ F_m'(\hat{x})[h] \end{pmatrix} = \begin{bmatrix} \sum_{j=1}^n \frac{\partial F_1}{\partial x_j}(\hat{x}) h_j \\ \sum_{j=1}^n \frac{\partial F_2}{\partial x_j}(\hat{x}) h_j \\ \cdot\ \cdot\ \cdot\ \cdot\ \cdot\ \cdot\ \cdot \\ \sum_{j=1}^n \frac{\partial F_m}{\partial x_j}(\hat{x}) h_j \end{bmatrix} = \begin{bmatrix} \frac{\partial F_1}{\partial x_1}(\hat{x}) & \ldots & \frac{\partial F_1}{\partial x_n}(\hat{x}) \\ \cdot & & \\ \cdot & & \\ \cdot & & \\ \frac{\partial F_m}{\partial x_1}(\hat{x}) & \ldots & \frac{\partial F_m}{\partial x_n}(\hat{x}) \end{bmatrix} \begin{pmatrix} h_1 \\ \cdot \\ \cdot \\ \cdot \\ h_n \end{pmatrix} = \frac{\partial F}{\partial x}(\hat{x}) \cdot h. \quad (2)$$

The (m × n) matrix

$$\frac{\partial F}{\partial x}(\hat{x}) = \left(\frac{\partial F_i}{\partial x_j}(\hat{x})\right)$$

is called *Jacobi's matrix* of the mapping F at the point x. It is readily seen that this is nothing other than the matrix of the linear operator $F'(\hat{x}):\mathbf{R}^n \to \mathbf{R}^m$ corresponding to the standard bases in $\mathbf{R}^n$ and $\mathbf{R}^m$.

In classical mathematical analysis the notation

$$h = dx = \begin{pmatrix} dx_1 \\ \cdot \\ \cdot \\ \cdot \\ dx_n \end{pmatrix}$$

is used and formula (2) has the form

$$dF(\hat{x}) = \sum_{j=1}^n \frac{\partial F}{\partial x_j}(\hat{x})\, dx_j.$$

The assertion proved in this section is a generalization of the well-known theorem stating that the existence of continuous partial derivatives is a sufficient condition for the differentiability of a function of several variables.

2.2.5. Higher-Order Derivatives. Taylor's Formula. In this section X and Y are normed spaces and U is an open set in X. The differentiability is understood in Fréchet's sense.

If a mapping $f:U \to Y$ is differentiable at each point $x \in U$, then the mapping $f'(x): U \longrightarrow \mathscr{L}(X, Y)$ is defined. Since $\mathscr{L}(X, Y)$ is also a normed space, the question of the existence of the second derivative

$$f''(x) = (f')'(x) \in \mathscr{L}(X, \mathscr{L}(X, Y))$$

can be posed. The higher-order derivatives are defined by induction: if $f^{(n-1)}(x)$ has already been defined in U, then

$$f^{(n)}(x) = (f^{(n-1)})'(x) \in \underbrace{\mathscr{L}(X, \ldots, \mathscr{L}(X, Y)}_{\text{n times}} \ldots).$$

Definition 1. Let $f:U \to Y$. We shall say that $f^{(n)}$ exists at a point $\hat{x} \in U$ if $f'(x), f''(x), \ldots, f^{(n-1)}(x)$ exist in a neighborhood of that point and $f^{(n)}(\hat{x})$ exists.

If $f^{(n)}(x)$ exists at each point $x \in U$ and the mapping $x \to f^{(n)}(x)$ is continuous in the uniform topology of the space $\mathscr{L}(X, \ldots, \mathscr{L}(X, Y)\ldots)$ (generated by the norm), then f is said to be a mapping of class $C^{(n)}(U)$.

In what follows we shall need some properties of continuous multilinear mappings [see Example 2) of Section 2.2.1].

Proposition 1. The normed spaces $\mathscr{L}^n(X, \mathscr{L}^m(X, Y))$ and $\mathscr{L}^{n+m}(X, Y)$ are isometrically isomorphic.

Proof. If $\pi \in \mathscr{L}^n(X, \mathscr{L}^m(X, Y))$, then $\pi(x_1, \ldots, x_n) \in \mathscr{L}^m(X, Y)$ and the equality

$$\Pi(x_1, \ldots, x_n, x_{n+1}, \ldots, x_{n+m}) = \pi(x_1, \ldots, x_n)[x_{n+1}, \ldots, x_{n+m}] \quad (1)$$

specifies a multilinear mapping Π of the space $X^{n+m} = \underbrace{X \times \ldots \times X}_{n+m \text{ times}}$ into Y. Conversely, every such mapping Π specifies [with the aid of equality (1)] a multilinear mapping π of the space $X^n = \underbrace{X \times \ldots \times X}_{n \text{ times}}$ into the space of multilinear mappings of X^m into Y. It only remains to note that

$$\|\Pi\| = \sup_{\substack{\|x_1\| \leqslant 1 \\ \cdots \\ \|x_{n+m}\| \leqslant 1}} \|\Pi(x_1, \ldots, x_{n+m})\| =$$

$$= \sup_{\substack{\|x_1\| \leqslant 1 \\ \cdots \\ \|x_n\| \leqslant 1}} \sup_{\substack{\|x_{n+1}\| \leqslant 1 \\ \cdots \\ \|x_{n+m}\| \leqslant 1}} \pi(x_1, \ldots, x_n)[x_{n+1}, \ldots, x_{n+m}] =$$

$$= \sup_{\substack{\|x_1\| \leqslant 1 \\ \cdots \\ \|x_n\| \leqslant 1}} \|\pi(x_1, \ldots, x_n)\| = \|\pi\|. \blacksquare$$

COROLLARY 1. $\underbrace{\mathscr{L}(X, \mathscr{L}(X, \ldots, \mathscr{L}}_{\text{n times}}(X, Y) \ldots))$ is isometrically isomorphic to $\mathscr{L}^n(X, Y)$.

Thus, we can assume that $f^{(n)}(x) \in \mathscr{L}^n(X, Y)$. The values of this multilinear mapping on the vectors $(\xi_1, \ldots, \xi_n)$ will be denoted by $f^{(n)}(x)[\xi_1, \ldots, \xi_n]$. In accordance with the inductive definition,

$$f^{(n)}(x_0)[\xi_1, \ldots, \xi_n] = \lim_{\alpha \downarrow 0} \frac{f^{(n-1)}(x_0 + \alpha\xi_1)[\xi_2, \ldots, \xi_n] - f^{(n-1)}(x_0)[\xi_2, \ldots, \xi_n]}{\alpha}.$$

For an arbitrary $\Pi \in \mathscr{L}^n(X, Y)$ and an arbitrary collection of k indices $(i_1, \ldots, i_k)$ each of which assumes one of the values 1, 2, ..., n we denote

$$\Pi_{i_1\ldots i_k}(x;h_1,\ldots,h_k)=\Pi(x_1,\ldots,x_n)\Big|_{\substack{x_{i_l}=h_l,\ l=1,2,\ldots,k,\\ x_i=x,\ i\neq i_1,\ldots,i_k.}}$$

Proposition 2. If $\Pi\in\mathscr{L}^n(X,Y)$ and $Q(x) = \Pi(x,\ldots,x)$ then

$$\begin{aligned} &Q'(x)[h_1]=\sum_{i=1}^{n}\Pi_i(x;h_1),\\ &Q''(x)[h_1,h_2]=\sum_{i=1}^{n}\sum_{\substack{j=1\\ j\neq i}}^{n}\Pi_{ij}(x;h_1,h_2),\\ &\ldots\ldots\ldots\ldots\ldots\ldots\ldots\ldots\\ &Q^{(n)}(x)[h_1,\ldots,h_n]=\sum_{(i_1,\ldots,i_n)}\Pi(h_{i_1},\ldots,h_{i_n}),\\ &Q^{(k)}(x)\equiv 0\ \text{ for }\ k>n \end{aligned}\tag{2}$$

(in the formula for $Q^{(n)}$ the summation extends over all the permutations of the indices).

These formulas are proved by means of the direct calculation (also cf. the example of Section 2.2.1). It should be noted that, as (2) shows, each of the derivatives $Q^{(l)}$ is a symmetric function of $(h_1,\ldots,h_l)$. As will be seen below, this is not accidental.

Proposition 3. If a mapping $\Pi\in\mathscr{L}^n(X,Y)$ is symmetric and $Q(x) = \Pi(x,\ldots,x)\equiv 0$, then $\Pi\equiv 0$.

Proof. Let us put

$$\begin{aligned}\varphi(t_1,\ldots,t_n)&=Q(t_1x_1+\ldots+t_nx_n)=\sum_{i_1,\ldots,i_n}t_{i_1}\ldots t_{i_n}\Pi(x_{i_1}\ldots x_{i_n})=\\ &=\sum_{\substack{0\leqslant\alpha_i\leqslant n\\ \Sigma\alpha_i=n}}t_1^{\alpha_1}\ldots t_n^{\alpha_n}\Sigma'_\alpha\Pi(x_{i_1},\ldots,x_{i_n})=\\ &=\sum_{\substack{0\leqslant\alpha_i\leqslant n\\ \Sigma\alpha_i=n}}t_1^{\alpha_1}\ldots t_n^{\alpha_n}N_{\alpha_1\ldots\alpha_n}\Pi(\underbrace{x_1,\ldots,x_1}_{\alpha_1\text{ times}},\ldots,\underbrace{x_n,\ldots,x_n}_{\alpha_n\text{ times}}),\end{aligned}$$

where in the sum Σ'_α the terms are collected among whose indices $i_1,\ldots,i_n$ there are α_1 indices equal to unity,...,n $-$ α_n indices equal to n, and $N_{\alpha_1\ldots\alpha_n}$ is the number of the terms in this sum (in the passage to the last equality we have used the symmetry of Π).

By the hypothesis, we have $\varphi(t_1,\ldots,t_n)\equiv 0$, and therefore all the coefficients of the polynomial φ are equal to zero. In particular, the coefficient $n!\Pi(x_1,\ldots,x_n)$ in $t_1t_2\ldots t_n$ is also equal to zero. ■

Theorem on the Mixed Derivatives. If a function $f:U\to Y$ possesses the second derivative $f''(\hat{x})$, then

$$f''(\hat{x})[\xi,\eta]=f''(\hat{x})[\eta,\xi]\tag{3}$$

for all $\xi,\eta\in X$.

Proof. According to the definition of the second derivative,

$$f'(x)-f'(\hat{x})=(f')'(\hat{x})[x-\hat{x}]+\alpha(x)\|x-\hat{x}\|,$$

where $\alpha(x)\in\mathscr{L}(X,Y)$ and

$$\lim_{x\to\hat{x}}\alpha(x)=\alpha(x)=0.\tag{4}$$

For sufficiently small η and x close to $\hat{x}$, then function

$$\varphi(x)=f(x+\eta)-f(x)$$

is defined and we have

$$\begin{aligned}\varphi'(x)&=f'(x+\eta)-f'(x)=f'(x+\eta)-f'(\hat{x})-(f'(x)-f'(\hat{x}))=\\ &=(f')'(\hat{x})[x+\eta-\hat{x}]+\alpha(x+\eta)\|x+\eta-\hat{x}\|-\end{aligned}$$

$$-(f')'(\hat{x})[x-\hat{x}]-\alpha(x)\|x-\hat{x}\|=(f')'(\hat{x})[\eta]+\alpha(x+\eta)\|x-\hat{x}+\eta\|-\alpha(x)\|x-\hat{x}\|.$$

In particular,

$$\varphi'(\hat{x})=(f')'(\hat{x})[\eta]+\alpha(\hat{x}+\eta)\|\eta\|, \tag{5}$$

and therefore

$$\varphi'(x)-\varphi'(\hat{x})=\alpha(x+\eta)\|x-\hat{x}+\eta\|-\alpha(x)\|x-\hat{x}\|-\alpha(\hat{x}+\eta)\|\eta\|. \tag{6}$$

By virtue of (4), for any $\varepsilon > 0$ there is $\delta > 0$ such that

$$\|x-\hat{x}\|<\delta\Rightarrow\|\alpha(x)\|<\varepsilon, \tag{7}$$

whence, by virtue of (6), it follows that

$$\|x-\hat{x}\|<\delta/2,$$

$$\|\eta\|<\delta/2\Rightarrow\|\varphi'(x)-\varphi'(\hat{x})\|<2\varepsilon(\|x-\hat{x}\|+\|\eta\|). \tag{8}$$

For sufficiently small ξ and η, the second difference

$$\Delta(\eta;\ \xi)=f(\hat{x}+\xi+\eta)-f(\hat{x}+\xi)-f(\hat{x}+\eta)+f(\hat{x}) \tag{9}$$

makes sense, and using (5), (7), and (8) we obtain for $\|\xi\| < \delta/2$, $\|\eta\| < \delta/2$ the inequalities

$$\begin{aligned}\|\Delta(\eta,\ \xi)-f''(\hat{x})[\eta,\ \xi]\|&=\|\varphi(\hat{x}+\xi)-\varphi(\hat{x})-((f')'(\hat{x})[\eta])[\xi]\|=\\ &=\|\varphi(\hat{x}+\xi)-\varphi(\hat{x})-\varphi'(\hat{x})[\xi]+\alpha(\hat{x}+\eta)[\xi]\|\eta\|\|\leqslant\\ &\leqslant\sup_{x\in[\hat{x},\ \hat{x}+\xi]}\|\varphi'(x)-\varphi'(\hat{x})\|\|\xi\|+\|\alpha(\hat{x}+\eta)\|\|\xi\|\|\eta\|\leqslant\\ &\leqslant 2\varepsilon(\|\xi\|+\|\eta\|)\|\xi\|+\varepsilon\|\xi\|\|\eta\|\leqslant 3\varepsilon(\|\xi\|+\|\eta\|)\|\xi\|.\end{aligned}$$

Replacing ξ and η by $t\xi$ and $t\eta$, where now ξ and η can be arbitrary fixed values and $t\in\mathbf{R}$ must be small, we obtain the inequality

$$\|\Delta(t\eta,\ t\xi)-t^2f''(\hat{x})[\eta,\ \xi]\|\leqslant 3\varepsilon t^2(\|\xi\|+\|\eta\|)\|\xi\|. \tag{10}$$

Interchanging ξ and η in the above argument, we obtain, in addition to (10), the inequality

$$\|\Delta(t\xi,\ t\eta)-t^2f''(\hat{x})[\xi,\ \eta]\|\leqslant 3\varepsilon t^2(\|\xi\|+\|\eta\|)\|\eta\|. \tag{11}$$

Now since the second difference (9) is symmetric with respect to ξ and η, we have $\Delta(t\eta,\ t\xi) = \Delta(t\xi,\ t\eta)$ and on canceling by t^2 we derive from (10) and (11) the inequality

$$\|f''(\hat{x})[\eta,\ \xi]-f''(\hat{x})[\xi,\ \eta]\|\leqslant 3\varepsilon(\|\xi\|+\|\eta\|)^2.$$

By the arbitrariness of ε, this implies (3). ■

The name given to this theorem is accounted for by the fact that in the classical mathematical analysis to this theorem there corresponds the theorem on the independence of the mixed derivative of the order of differentiation: $\partial^2 f/\partial x\partial y = \partial^2 f/\partial y\partial x$ (why?).

COROLLARY 2. The derivative $f^{(n)}(\hat{x})$ (provided that it exists) is a symmetric multilinear function.

Proof. For $n = 2$ this assertion was proved in the above theorem. The further proof is carried out by induction using the equalities

$$\begin{aligned}((f^{(n-1)})(\hat{x})[\xi])[\eta_1,\ \ldots,\ \eta_{n-1}]&=f^{(n)}(\hat{x})[\xi_1,\ \eta_1,\ \ldots,\ \eta_{n-1}]=\\ &=((f^{(n-2)})''(\hat{x})[\xi,\ \eta_1])[\eta_2,\ \ldots,\ \eta_{n-1}].\end{aligned}$$

By the induction hypothesis, the left-hand equality implies the invariance of $f^{(n)}(\hat{x})[\xi,\ \eta_1,\ldots,\eta_{n-1}]$ with respect to the permutation of the arguments η_i, and from Theorem 1 and the right-hand equality follows the invariance with respect to the transposition of ξ and η_1. This is sufficient for proving the symmetry of $f^{(n)}(\hat{x})$. ■

Passing from a symmetric multilinear function to the form corresponding to it (see the example of Section 2.2.1), we obtain the differential from the derivative:

Definition 2.

$$d^n f(x_0;\ h) = f^n(x_0)[h,\ \ldots,\ h].$$

Example. If $Q(x) = \Pi(x,\ldots,x)$, where $\Pi \in \mathscr{L}^n(X,Y)$, then, according to Proposition 2,

$$d^k Q(0;\ h) = 0 \quad \text{for} \quad k \neq n$$

and

$$d^n Q(0;\ h) = k!Q(h).$$

LEMMA. If $f^{(n)}(\hat{x})$ exists and $f(\hat{x}) = 0$, $f'(\hat{x}) = 0,\ldots,f^{(n)}(\hat{x}) = 0$, then

$$\lim_{h\to 0} \frac{\|f(\hat{x}+h)\|}{\|h\|^n} = 0.$$

Proof. For $n = 1$ the assertion is true by the definition of the derivative. The further proof is carried out by induction. Let $n > 1$. Let us put $g(x) = f'(x)$. Then $g(\hat{x}) = 0,\ldots,g^{(n-1)}(\hat{x}) = 0$, and, by the induction hypothesis, for any $\varepsilon > 0$ there is $\delta > 0$ such that

$$\|g(\hat{x}+h_1)\| < \varepsilon\|h_1\|^{n-1} \quad \text{for} \quad \|h_1\| < \delta.$$

Now, by the mean value theorem, for $\|h\| < \delta$ we obtain

$$\|f(\hat{x}+h)\| = \|f(\hat{x}+h) - f(\hat{x}) - f'(\hat{x})h\| \leqslant \sup_{h_1\in[0,h]} \|f'(\hat{x}+h_1) - f'(\hat{x})\|\|h\| =$$

$$= \sup_{h_1\in[0,h]} \|g(\hat{x}+h_1)\|\|h\| \leqslant \sup_{h_1\in[0,h]} \varepsilon\|h_1\|^{n-1}\|h\| = \varepsilon\|h\|^n. \blacksquare$$

Theorem on Taylor's Formula. If $f^{(n)}(\hat{x})$ exists, then

$$f(\hat{x}+h) = f(\hat{x}) + df(\hat{x};\ h) + \ldots + \frac{1}{n!}d^n f(\hat{x};\ h) + \alpha_n(h)\|h\|^n,$$

where

$$\lim_{h\to 0} \alpha(h) = \alpha_n(0) = 0.$$

Proof. Let us consider the function

$$g(\xi) = f(\hat{x}+\xi) - f(\hat{x}) - df(\hat{x};\ \xi) - \ldots - \frac{1}{n!}d^n f(\hat{x};\ \xi).$$

Using the above example and the equality $d^k f(\hat{x};\ \xi) = f^{(k)}(\hat{x})[\xi,\ \ldots,\ \xi]$, $f^{(k)}(\hat{x}) \in \mathscr{L}^k(X,\ Y)$, we obtain $d^k g(0;\ h) = d^k f(\hat{x};\ h) - \frac{1}{k!}(k!d^k f(\hat{x};\ h)) = 0$.

Applying Corollary 1, according to which $g^{(k)}(0)$ is a symmetric function belonging to $\mathscr{L}^k(X,Y)$, and Proposition 3, according to which a symmetric function vanishes when the corresponding form vanishes, we conclude that $g(0) = 0,\ldots,g^{(n)}(0) = 0$. The lemma implies that $\|g(\xi)\| = o(\|\xi\|^n)$, and the assertion of the theorem is thus proved, and, moreover, $\alpha_n(h) = g(h)/\|h\|^n$ for $h \neq 0$. ■

Exercise. If $f^{(n)}(x)$ exists in a starlike neighborhood of a point $\hat{x}$, then

$$1)\ \left\|f(\hat{x}+h) - f(\hat{x}) - df(\hat{x};\ h) - \ldots - \frac{1}{(n-1)!}d^{n-1}f(\hat{x};\ h)\right\| \leqslant \sup_{c\in[\hat{x},\ \hat{x}+h]} \|f^{(n)}(c)\|\frac{\|h\|^n}{n!}$$

and

$$2)\ \left\|f(\hat{x}+h) - f(\hat{x}) - df(\hat{x};\ h) - \ldots - \frac{1}{n!}d^n f(\hat{x},\ h)\right\| \leqslant \sup_{c\in[\hat{x},\ \hat{x}+h]} \|f^{(n)}(c) - f^{(n)}(\hat{x})\|\frac{\|h\|^n}{n!}.$$

Hint. One of the possible ways of solving the problem is the following: for $n > 1$ the derivative $f'(x)$ is continuous in a neighborhood of the point $\hat{x}$, and we can make use of the equality

$$f(\hat{x}+h)-f(\hat{x})=\int_0^1 f'(x_0+th)\,dt\,[h]$$

(see the exercise in Section 2.2.3).

The inequalities of this problem are generalizations of Taylor's classical formula with the remainder in Lagrange's form, whereas the above theorem implies the same formula with the remainder in Peano's form.

The proof of the existence of Fréchet higher-order derivatives for concrete mappings is often connected with cumbersome calculations. At the same time, the calculation of the derivatives, or, more precisely, the differentials, i.e., the corresponding homogeneous forms, is usually carried out comparatively easily. Most often this is performed with the aid of the notion of the Lagrange variation.

Definition 3. A function $f:U \to Y$ is said to have the n-th *Lagrange variation* at a point $\hat{x} \in U$ if for any $h \in X$ the function $\varphi(\alpha)=f(\hat{x}+\alpha h)$ is differentiable at the point $\alpha = 0$ up to the order n inclusive. By the n-th variation of f at the point $\hat{x}$ we mean the mapping $h \to \delta^n f(\hat{x}; h)$ determined by the equality

$$\delta^n f(\hat{x};\ h)=\varphi^{(n)}(0)=\frac{d^n}{d\alpha^n} f(\hat{x}+\alpha h)\Big|_{\alpha=0}. \tag{12}$$

COROLLARY 3. If $f^{(n)}(\hat{x})$ exists, then f possesses the n-th Lagrange variation at the point $\hat{x}$, and, moreover

$$\delta^n f(\hat{x};\ h)=d^n f(\hat{x};\ h)=f^{(n)}(\hat{x})[h,\ \ldots,\ h]. \tag{13}$$

Proof. The proof is readily carried out by induction with the aid of Taylor's formula because this formula implies

$$f(\hat{x}+\alpha h)=f(\hat{x})+\frac{\alpha}{1!}df(\hat{x};\ h)+\ldots+\frac{\alpha^n}{n!}d^n f(\hat{x};\ h)+o(\alpha^n).$$

Thus, if it is known in advance that the derivative $f^{(n)}(\hat{x})$ exists, then $d^{(n)}f(\hat{x};\ h)$ can be computed with the aid of formulas (12) and (13) on the basis of the differential calculus of functions of a scalar argument, which is of course much simpler. According to Proposition 3, the derivative $f^{(n)}(\hat{x})$ (i.e., the corresponding multilinear function) is uniquely determined by the differential $d^n f(\hat{x};\ h)$.

The converse is not, of course, true: a function possessing the n-th variation does not necessarily have the Fréchet derivative $f^{(n)}$.

Let the reader find what additional properties the n-th variation must possess for the analogue of the sufficient condition for differentiability (Corollary 2 of Section 2.2.3) to hold.

2.3. Implicit Function Theorem

The generalization of the classical implicit function theorem, which is proved in this section, can be fruitfully used in various branches of mathematical analysis. In what follows we shall have a chance to see this.

2.3.1. Statement of the Existence Theorem for an Implicit Function. Let X be a topological space, let Y and Z be Banach spaces, let W be a neighborhood of a point $(x_0,\ y_0)$ in $X \times Y$, let Ψ be a mapping from W into Z, and let $\Psi(x_0,\ y_0) = z_0$.

If

1) the mapping $x \to \Psi(x, y_0)$ is continuous at the point x_0;

2) there exists a continuous linear operator $\Lambda: Y \to Z$ such that, given any $\varepsilon > 0$, there is a number $\delta > 0$ and a neighborhood Ξ of the point x_0 possessing the property that the condition $x \in \Xi$ and the inequalities $\|y' - y_0\| < \delta$ and $\|y'' - y_0\| < \delta$ imply the inequality

$$\|\Psi(x, y') - \Psi(x, y'') - \Lambda(y' - y'')\| < \varepsilon \|y' - y''\|;$$

3) $\Lambda Y = Z$;

then there is a number $K > 0$, a neighborhood $\mathcal{U}$ of the point (x_0, z_0) in $X \times Z$ and a mapping $\varphi: \mathcal{U} \to Y$ such that:

a) $\Psi(x, \varphi(x, z)) = z$;

and

b) $\|\varphi(x, z) - y_0\| \leqslant K \|\Psi(x, y_0) - z\|$.

2.3.2. Modified Contraction Mapping Principle.

LEMMA. Let T be a topological space, let Y be a Banach space, let V be a neighborhood of a point (t_0, y_0) in $T \times Y$, and let Φ be a mapping from V into Y. Then if there exist a neighborhood U of the point t_0 in T, a number $\beta > 0$, and a number θ, $0 < \theta < 1$, such that $t \in U$, and $\|y - y_0\| < \beta$ imply

a) $$(t, \Phi(t, y)) \in V,$$

b) $$\|\Phi(t, \Phi(t, y)) - \Phi(t, y)\| \leqslant \theta \|\Phi(t, y) - y\|,$$

c) $$\|\Phi(t, y_0) - y_0\| < \beta(1 - \theta),$$

then the sequence $\{y_n(t) \mid n \geqslant 0\}$ defined by the recurrence relations

$$y_0(t) = y_0, \quad y_n(t) = \Phi(t, y_{n-1}(t))$$

is contained in the ball $\mathring{B}(y_0, \beta) = \{y \mid \|y - y_0\| < \beta\}$ for any $t \in U$ and is convergent to a mapping $t \to \varphi(t)$, uniformly with respect to $t \in U$, and, moreover,

$$\|\varphi(t) - y_0\| \leqslant \frac{\|\Phi(t, y_0) - y_0\|}{1 - \theta}.$$

Proof. The lemma is proved by induction. Let t be an element belonging to U. Let us denote by Γ_n the assertion "the element $y_k(t)$ is defined for $0 \leqslant k \leqslant n$ and belongs to $\mathring{B}(y_0, \beta)$." Since $y_0(t) = y_0 \in \mathring{B}(y_0, \beta)$, the assertion Γ_0 is true. Let Γ_n be true. Then for $1 \leqslant k \leqslant n$ we have

$$\begin{aligned}\|y_{k+1}(t) - y_k(t)\| &\overset{\text{def}}{=} \|\Phi(t, y_k(t)) - \Phi(t, y_{k-1}(t))\| \overset{\text{def}}{=} \\ &= \|\Phi(t, \Phi(t, y_{k-1}(t)) - \Phi(t, y_{k-1}(t))\| \leqslant \\ &\leqslant \theta \|\Phi(t, y_{k-1}(t)) - y_{k-1}(t)\| \overset{\text{def}}{=} \theta \|y_k(t) - y_{k-1}(t)\|.\end{aligned}$$

Whence, continuing the above arguments, we obtain

$$\begin{aligned}&\|y_{k+1}(t) - y_k(t)\| \leqslant \theta \|y_k(t) - y_{k-1}(t)\| \leqslant \\ &\leqslant \theta^2 \|y_{k-1}(t) - y_{k-2}(t)\| \leqslant \ldots \leqslant \theta^k \|y_1(t) - y_0\| = \theta^k \|\Phi(t, y_0) - y_0\|.\end{aligned}$$

Now let $k \geqslant 1$, $k + l \leqslant n + 1$; then, by the triangle inequality,

$$\begin{aligned}\|y_{k+l}(t) - y_k(t)\| &= \|y_{k+l}(t) - y_{k+l-1}(t) + y_{k+l-1}(t) - \ldots + y_{k+1}(t) - y_k(t)\| \leqslant \\ &\leqslant (\theta^{k+l-1} + \ldots + \theta^k) \|\Phi(t, y_0) - y_0\| < \\ &< \left(\sum_{s=k}^{\infty} \theta^s\right) \|\Phi(t, y_0) - y_0\| = \frac{\theta^k}{1 - \theta} \|\Phi(t, y_0) - y_0\| < \theta^k \beta.\end{aligned} \tag{1}$$

In particular, putting $k + l = n + 1$, $k = 1$, in (1) and using the condition c), we obtain

$$\|y_{n+1}(t)-y_0\|=\|y_{n+1}(t)-y_1(t)+y_1(t)-y_0\|\leqslant$$
$$\leqslant\|y_{n+1}(t)-y_1(t)\|+\|y_1(t)-y_0\|<\theta\beta+\beta(1-\theta)=\beta.$$

Hence Γ_{n+1} is true, whence, by induction, the assertion Γ_n is true for all n. Therefore, we obtain from (1) the relation

$$\|y_{k+l}(t)-y_k(t)\|\leqslant\frac{\theta^k}{1-\theta}\|\Phi(t,\ y_0)-y_0\|,\quad \forall l,\quad k\geqslant 1. \tag{2}$$

From (2) it follows that the sequence $\{y_k(t)\,|\,k \geqslant 0\}$ is fundamental and, by virtue of the completeness of Y, it is convergent to an element which we denote by $\varphi(t)$. Passing to the limit in (2) for $l \to \infty$, we conclude that

$$\|\varphi(t)-y_k(t)\|\leqslant\frac{\theta^k}{1-\theta}\|\Phi(t,\ y_0)-y_0\|<\theta^k\beta. \tag{3}$$

From (3) it follows that the mapping $t\to y_k(t)$, $t\in U$, is uniformly convergent to the mapping $t\to\varphi(t)$. Finally, putting $k = 1$ in (3), we obtain

$$\|\varphi(t)-y_0\|=\|\varphi(t)-y_1(t)+y_1(t)-y_0\|\leqslant$$
$$\leqslant\frac{\theta}{1-\theta}\|\Phi(t,\ y_0)-y_0\|+\|\Phi(t,\ y_0)-y_0\|=\frac{\|\Phi(t,\ y_0)-y_0\|}{1-\theta}.$$

2.3.3. Proof of the Theorem. We break the proof of the theorem stated in Section 2.3.1 into several sections.

A) According to the condition of the theorem, Λ is an epimorphism from Y into Z. Consequently, by the lemma on the right inverse mapping, there exists a mapping $M:Z \to Y$ such that

$$(\Lambda\circ M)(z)=z \tag{1}$$

and

$$\|M(z)\|\leqslant C\|z\| \tag{2}$$

for some $C > 0$. Now we put $T = X \times Z$ and verify the applicability of the lemma of Section 2.3.2 to the mapping

$$\Phi(t,\ y)=\Phi(x,\ z,\ y)=y+M(z-\Psi(x,\ y)).$$

B) On choosing θ, $0 < \theta < 1$, and $\varepsilon = \theta/C$, where C is the constant in (2), we find a neighborhood Ξ_1 of the point x_0 and a number $\beta_0 > 0$ such that the inequalities $\|y' - y_0\| < \beta_0$ and $\|y'' - y_0\| < \beta_0$ imply [in accordance with the condition 2) of the theorem] the inequality

$$\|\Psi(x,\ y')-\Psi(x,\ y'')-\Lambda(y'-y'')\|<\varepsilon\|y'-y''\| \tag{3}$$

for $x\in\Xi_1$.

Let us denote $V=\Xi_1\times Z\times\mathring{B}(y_0,\ \beta_0)$.

C) Let $\beta = \beta_0/(C\|\Lambda\| + 2)$. Using the continuity of the function $(x, z) \to \Psi(x, y_0) - z$ at the point (x_0, y_0) [see the condition 1) of the theorem] we choose a neighborhood $\Xi_2\subset\Xi_1$ of the point x_0 and $\gamma > 0$ such that $(x, z)\in\mathcal{U}=\Xi_2\times\mathring{B}(z_0,\gamma)$ implies the inequality

$$\|\Psi(x,\ y_0)-z\|<\beta(1-\theta)/C. \tag{4}$$

D) If $(x,\ z)\in\mathcal{U}$ and $y\in\mathring{B}(y_0,\ \beta)$, then

$$\|\Phi(x,\ z,\ y)-y_0\|\overset{\text{def}}{=}\|y-y_0+M(z-\Psi(x,\ y))\|\overset{(2)}{\leqslant}$$
$$\leqslant\|y-y_0\|+C\|z-\Psi(x,\ y)\|\leqslant\|y-y_0\|+$$
$$+C\{\|z-\Psi(x,\ y_0)-\Psi(x,\ y)+\Psi(x,\ y_0)+\Lambda(y-y_0)-$$
$$-\Lambda(y-y_0)\|\}\leqslant\|y-y_0\|+C\{\|z-\Psi(x,\ y_0)\|+$$

$$+\|\Psi(x, y)-\Psi(x, y_0)-\Lambda(y-y_0)\|+\|\Lambda(y-y_0)\|\}\overset{(3)}{\leqslant}$$
$$\leqslant\|y-y_0\|+C\{\|z-\Psi(x, y_0)\|+\varepsilon\|y-y_0\|+\|\Lambda\|\|y-y_0\|\}\leqslant$$
$$\leqslant(1+\theta+C\|\Lambda\|)\|y-y_0\|+C\|z-\Psi(x, y_0)\| \tag{5}$$

(because $\varepsilon C = \theta$).

By virtue of (4), we conclude from (5) that, in the first place,

$$\|\Phi(x, z, y)-y_0\|<(1+\theta+C\|\Lambda\|)\beta+C\frac{\beta(1-\theta)}{C}=(2+C\|\Lambda\|)\beta=\beta_0. \tag{5'}$$

Hence the condition a) of the lemma of Section 2.3.2 is fulfilled: $(x, z, \Phi(x, z, y))\in V$. In the second place, for $y = y_0$ we obtain

$$\|\Phi(x, z_0, y_0)-y_0\|\leqslant C\|z-\Psi(x, y_0)\|\overset{(4)}{<}C\frac{\beta(1-\theta)}{C}=\beta(1-\theta). \tag{5''}$$

Therefore, the condition c) of the lemma of Section 2.3.2 is fulfilled.

E) If again $(x, z)\in\mathcal{U}$, $y\in\mathring{B}(y_0, \beta)$, then

$$\Phi(x, z, \Phi(x, z, y))-\Phi(x, z, y)\overset{\text{def}}{=}$$
$$=\Phi(x, z, y)+M(z-\Psi(x, \Phi(x, z, y)))-\Phi(x, z, y)=M(z-\Psi(x, \Phi(x, z, y))). \tag{6}$$

Here (1) implies

$$\Lambda\Phi(x, z, y)=\Lambda y+\Lambda M(z-\Psi(x, y))=\Lambda y+z-\Psi(x, y),$$

whence

$$z=\Psi(x, y)+\Lambda(\Phi(x, z, y)-y). \tag{7}$$

Substituting (7) into (6) and taking into account that, by virtue of (5'), the relation $\Phi(x, z, y)\in\mathring{B}(y_0, \beta_0)$ holds, we obtain

$$\|\Phi(x, z, \Phi(x, z, y))-\Phi(x, z, y)\|\overset{(6)}{=}$$
$$=\|M(z-\Psi(x, \Phi(x, z, y)))\|\overset{(2)}{\leqslant}C\|z-\Psi(x, \Phi(x, z, y))\|\overset{(7)}{=}$$
$$=C\|\Psi(x, \Phi(x, z, y))-\Psi(x, y)-\Lambda(\Phi(x, z, y)-y)\|\overset{(3)}{\leqslant}$$
$$\leqslant C\varepsilon\|y-\Phi(x, z, y)\|=\theta\|y-\Phi(x, z, y)\|. \tag{8}$$

Thus, the condition b) of the lemma of Section 2.3.2 is fulfilled.

F) Let us again compare our construction with the lemma of Section 2.3.2. To the topological space T of the lemma there corresponds the product space $X \times Z$, to the point t_0 there corresponds the point (x_0, z_0), to the neighborhood U of the point t_0 there corresponds the neighborhood $\mathcal{U}$ of the point (x_0, z_0), to the neighborhood V in the lemma there corresponds the neighborhood V constructed in B), to the mapping $(t, y) \to \Phi(t, y)$ there corresponds the mapping $(x, z, y) \to \Phi(x, z, y)$, and to the number β in the lemma there corresponds the number β constructed in C).

Therefore, relation (5') means that condition a) of the lemma is fulfilled, relation (5") means that condition c) of the lemma is fulfilled, and finally relation (8) means that condition b) is fulfilled. Hence, the lemma is applicable.

G) By the lemma, the sequence $\{y_n(x, z)|n\geqslant 0\}$, $(x, z)\in\mathcal{U}$, specified by the recurrence relations

$$y_0(x, z)=y_0, \quad y_{n+1}(x, z)=\Phi(x, z, y_n(x, z))$$

is contained in the ball $\mathring{B}(y_0, \beta)$ for all $n \geqslant 0$ and is uniformly convergent to a function $\varphi(x, z)$, and, moreover,

$$\|\varphi(x, z)-y_0\|\leqslant\|\Phi(x, z, y_0)-y_0\|/(1-\theta)\overset{\text{def}}{=}$$

$$=\|M(z-\Psi(x, y_0))\|/(1-\theta)\overset{(2)}{\leqslant}\frac{C}{1-\theta}\|z-\Psi(x, y_0)\|.$$

Therefore, the assertion b) of the theorem with $K = C/(1 - \theta)$ holds.

H)

$$\|\Psi(x, \varphi(x, z))-z\|\leqslant\|\Psi(x, \varphi(x, z))-\Psi(x, y_n(x, z))\|+\|\Psi(x, y_n(x, z))-z\|\overset{(1)}{\leqslant}$$
$$\leqslant\|\Psi(x, \varphi(x, z))-\Psi(x, y_n(x, z))-\Lambda(\varphi(x, z)-y_n(x, z))\|+$$
$$+\|\Lambda(\varphi(x, z)-y_n(x, z))\|+\|\Lambda M(z-\Psi(x, y_n(x, z)))\|\overset{(3),(2)}{<}$$
$$<\varepsilon\|\varphi(x, z)-y_n(x, z)\|+\|\Lambda\|\|\varphi(x, z)-y_n(x, z)\|+\|\Lambda\|\|y_{n+1}(x, z)-y_n(x, z)\|\to 0$$

for $n \to \infty$. Consequently, $\Psi(x, \varphi(x, z))=z$. ■

2.3.4. Classical Implicit Function and Inverse Mapping Theorems. From the general theorem proved above, two classical theorems can easily be derived which are constantly used in various applications of mathematical analysis, for example, in differential geometry. In comparison with Section 2.3.1, here we shall strengthen the requirements imposed on the function Ψ in two aspects. Firstly, we shall assume that Ψ is continuously differentiable and, secondly, and most significantly, we shall suppose that the operator Λ involved in the statement of the theorem of Section 2.3.1 is invertible. All this will guarantee the smoothness and the uniqueness of the implicit function.

A) Classical Implicit Function Theorem. Let X, Y, and Z be Banach spaces, let W be a neighborhood in $X \times Y$, and let $\Psi: W \to Z$ be a mapping of class $C^1(W)$. If:

1) $\Psi(x_0, y_0) = 0$; (1)

2) there exists the inverse operator $[\Psi_y(x_0, y_0)]^{-1}\in\mathscr{L}(Z, Y)$, then there exist $\varepsilon > 0$, $\delta > 0$ and a mapping $\varphi: \mathring{B}(x_0, \delta)\to Y$ of class $C^1(\mathring{B}(x_0, \delta))$ such that:

a) $\varphi(x_0)=y_0$; (2)

b) $\|x-x_0\|<\delta\Rightarrow\|\varphi(x)-y_0\|<\varepsilon$ and $\Psi(x, \varphi(x))\equiv 0$;

c) the equality $\Psi(x, y) = 0$ is possible in the "rectangle" $\mathring{B}(x_0, \delta) \times \mathring{B}(y_0, \varepsilon)$ only for $y=\varphi(x)$;

d) $\varphi'(x)=-[\Psi_y(x, \varphi(x))]^{-1}\Psi_x(x, \varphi(x))$. (3)

Proof. The continuous differentiability of Ψ in W implies that there hold the conditions 1) and 2) of the existence theorem for an implicit function of Section 2.3.1 [to verify 2) we must put $\Lambda = \Psi_y(x_0, y_0)$ and make use of the mean value theorem]. Condition 3) is fulfilled because $\Psi_y^{-1}(x_0, y_0)$ exists. Hence, according to the implicit function theorem, there is a number $K > 0$, a neighborhood $U\ni x_0$, and a mapping $\varphi: U\to Y$ such that

$$\Psi(x, \varphi(x))\equiv 0 \tag{4}$$

and

$$\|\varphi(x)-y_0\|\leqslant K\|\Psi(x, y_0)\|. \tag{5}$$

Putting $x = x_0$ in (5) and using (1), we obtain (2), and (4) implies the second half of assertion b).

B) Since $\Psi\in C^1(W)$, the mapping Ψ is strictly differentiable at the point (x_0, y_0) (Corollary 2 of Section 2.2.3). Consequently, given any $\varkappa>0$, there is $\varepsilon(\varkappa)>0$, such that

$$\|x_i-x_0\|<\varepsilon(\varkappa),\ \|y_i-y_0\|<\varepsilon(\varkappa),\quad i=1, 2, \Rightarrow$$
$$\Rightarrow\|\Psi(x_1, y_1)-\Psi(x_2, y_2)-\Psi_x(x_0, y_0)(x_1-x_2)-$$
$$-\Psi_y(x_0, y_0)(y_1-y_2)\|\leqslant\varkappa\max\{\|x_1-x_2\|, \|y_1-y_2\|\}. \tag{6}$$

First we shall put $\varkappa_0 = 1/2\|\Psi_y^{-1}(x_0, y_0)\|^{-1}$ and find the corresponding $\varepsilon = \varepsilon(\varkappa_0)$. Since $\Psi(x, y)$ is continuous and $\Psi(x_0, y_0) = 0$, there is δ, $0 < \delta < \varepsilon$, such that

$$\mathring{B}(x_0, \delta) = \{x \mid \|x - x_0\| < \delta\} \subset U$$

and

$$\|x - x_0\| < \delta \Rightarrow \|\Psi(x, y_0)\| < \frac{\varepsilon}{K} \overset{(5)}{\Rightarrow} \|\varphi(x) - y_0\| < \varepsilon, \tag{7}$$

and consequently the first half of the assertion b) holds.

Now let us suppose that $\Psi(\hat{x}, \hat{y}) = 0$ at a point $(\hat{x}, \hat{y}) \in \mathring{B}(x_0, \delta) \times \mathring{B}(y_0, \varepsilon)$. Applying (6) with $\varkappa_0$ indicated above $[x_1 = x_2 = \hat{x},\ y_1 = \hat{y},\ y_2 = \varphi(\hat{x})]$, we obtain

$$\begin{aligned}\|\hat{y} - \varphi(\hat{x})\| &= \|\Psi_y^{-1}(x_0, y_0)\Psi_y(x_0, y_0)(\hat{y} - \varphi(\hat{x}))\| \leqslant \\ &\leqslant \|\Psi_y^{-1}(x_0, y_0)\| \|\Psi(\hat{x}, \hat{y}) - \Psi(\hat{x}, \varphi(\hat{x})) - \Psi_y(x_0, y_0)(\hat{y} - \varphi(\hat{x}))\| \overset{(6)}{\leqslant} \\ &\leqslant \|\Psi_y(x_0, y_0)^{-1}\|\varkappa_0\|\hat{y} - \varphi(\hat{x})\| = \frac{1}{2}\|\hat{y} - \varphi(\hat{x})\|,\end{aligned}$$

which is only possible for $\hat{y} = \varphi(\hat{x})$, i.e., assertion c) is true.

C) Let us put $x_1 = x$, $x_2 = x_0$, $y_1 = y_2 = y_0$ in (6) for the same $\varkappa_0$, ε, and δ as above. Then

$$\begin{aligned}\|x - x_0\| < \delta &\overset{(6)}{\Rightarrow} \|\Psi(x, y_0) - \Psi(x_0, y_0) - \Psi_x(x_0, y_0)(x - x_0)\| \leqslant \\ &\leqslant \varkappa_0\|x - x_0\| \Rightarrow \|\Psi(x, y_0)\| \leqslant (\|\Psi_x(x_0, y_0)\| + \varkappa_0)\|x - x_0\| \overset{(5)}{\Rightarrow} \\ &\Rightarrow \|\varphi(x) - y_0\| \leqslant K(\|\Psi_x(x_0, y_0)\| + \varkappa_0)\|x - x_0\|.\end{aligned} \tag{8}$$

Now we take an arbitrary $\varkappa_1 > 0$ and apply (6) to

$$x_1 = x, \quad x_2 = x_0, \quad y_1 = \varphi(x), \quad y_2 = y_0.$$

For

$$\delta_1 = \min\left\{\varepsilon(\varkappa_1), \quad \frac{\varepsilon(\varkappa_1)}{K(\|\Psi_x(x_0, y_0)\| + \varkappa_0)}\right\}$$

we have

$$\begin{aligned}\|x - x_0\| < \delta_1 &\Rightarrow \|\Psi_y^{-1}(x_0, y_0)\Psi_x(x_0, y_0)(x - x_0) + \varphi(x) - y_0\| = \\ &= \|\Psi_y^{-1}(x_0, y_0)\{\Psi_x(x_0, y_0)(x - x_0) + \Psi_y(x_0, y_0)(\varphi(x) - y_0)\}\| \leqslant \\ &\leqslant \|\Psi_y^{-1}(x_0, y_0)\| \|\Psi(x, \varphi(x)) - \Psi(x_0, y_0) - \\ &- \Psi_x(x_0, y_0)(x - x_0) - \Psi_y(x_0, y_0)(\varphi(x) - y_0)\| \overset{(6)}{\leqslant} \\ &\leqslant \|\Psi_y^{-1}(x_0, y_0)\|\varkappa_1 \max\{\|x - x_0\|, \|\varphi(x) - y_0\|\} \overset{(8)}{\leqslant} C\varkappa_1\|x - x_0\|,\end{aligned}$$

where

$$C = \|\Psi_y^{-1}(x_0, y_0)\| \max\{1, K(\|\Psi_x(x_0, y_0)\| + \varkappa_0)\}.$$

Since $\varkappa_1$ can be chosen arbitrarily, the inequality we have proved implies

$$\Psi_y^{-1}(x_0, y_0)\Psi_x(x_0, y_0)(x - x_0) + \varphi(x) - y_0 = o(x - x_0)$$

for $x \to x_0$, i.e.,

$$\varphi(x) = y_0 + [-\Psi_y^{-1}(x_0, y_0)\Psi_x(x_0, y_0)](x - x_0) + o(x - x_0),$$

and consequently the Fréchet derivative $\varphi'(x_0)$ exists and equality (3) holds for $x = x_0$.

To prove the differentiability of φ for the other values of x, we take into account that the set of the operators Λ for which the inverse operators exist is open in $\mathscr{L}(X, Y)$ (Proposition 3 of Section 2.2.1).

Therefore, diminishing δ, if necessary, we can assume that $\Psi_y^{-1}(x, \varphi(x))$ exists for $\|x - x_0\| < \delta$.

Now we can repeat the whole proof replacing the point (x_0, y_0) by the point $(\hat{x}, \varphi(\hat{x}))$, where $\|\hat{x} - x_0\| < \delta$. By virtue of the uniqueness proved above, we obtain the same function $\varphi(x)$ (at least if $\|x - \hat{x}\|$ is sufficiently small), and formula (3) holds for this function at $x = \hat{x}$. Since $\hat{x}$ is arbitrary, the assertion d) holds for all $x \in \mathring{B}(x_0, \delta)$.

Finally, the derivative $\varphi'(x)$ is continuous because, according to (3), it can be represented as the composition of the continuous mappings $x \mapsto (x, \varphi(x))$, $(x, y) \mapsto \Psi_x(x, y)$, $(x, y) \mapsto \Psi_y(x, y)$, and $A \to A^{-1}$ (the last of them is continuous since it is Fréchet differentiable according to Proposition 3 of Section 2.2.1). ■

Remark. If, in addition to the conditions of the theorem,

$$\Psi \in C^r(W), \quad r \geqslant 2, \text{ then } \varphi \in C^r(\mathring{B}(x_0, \delta)).$$

Indeed, from formula (3) and the composition theorem it follows that the derivative

$$\varphi''(x) = -\{\Psi_y(x, \varphi(x))^{-1} \circ \Psi_x(x, \varphi(x))\}'$$

exists and is continuous. The further part of the proof is carried out by induction.

Theorem on the Inverse Mapping. Let Y and Z be Banach spaces, let W be a neighborhood of a point y_0 in Y, let $\Psi: W \to Z$ be a mapping of class $C^1(W)$, and let $z_0 = \Psi(y_0)$.

If there exists the inverse operator $\Psi'(y_0)^{-1} \in \mathscr{L}(Z, Y)$ then:

a) there exists $\varepsilon > 0$ such that $\mathring{B}(y_0, \varepsilon) \subset W$ and $V = \{z \mid z = \Psi(y), \ \|y - y_0\| < \varepsilon\} = \Psi(\mathring{B}(y_0, \varepsilon))$ is a neighborhood of z_0 in Z;

b) there exists a mapping $\Phi: V \to Y$ of class $C^1(V)$ inverse to $\Psi | \mathring{B}(y_0, \varepsilon)$, i.e., such that

$$y = \Phi(z) \Leftrightarrow z = \Psi(y), \quad (y, z) \in \mathring{B}(y_0, \varepsilon) \times V; \tag{9}$$

c)

$$\Phi'(z) = [\Psi'(\Phi(z))]^{-1}. \tag{10}$$

Proof. A) The mapping $\Psi \in C^1(W)$ is strictly differentiable at each point of W and, in particular, at y_0. Therefore, for every $\varkappa > 0$ there is $\varepsilon(\varkappa) > 0$ such that

$$\|y_i - y_0\| < \varepsilon(\varkappa), \quad i = 1, 2, \Rightarrow \|\Psi(y_1) - \Psi(y_2) - \Psi'(y_0)(y_1 - y_2)\| \leqslant \varkappa \|y_1 - y_2\|. \tag{11}$$

Let us choose $\varepsilon > 0$ such that for all $y \in \mathring{B}(y_0, \varepsilon)$ there exists the inverse operator $\Psi(y)^{-1}$ (see Proposition 3 of Section 2.2.1) and $\varepsilon < \varepsilon(\varkappa_0)$, where $\varkappa_0 = \frac{1}{2}\|\Psi'(y_0)^{-1}\|^{-1}$.

The last fact guarantees the injectivity of the mapping Ψ on $\mathring{B}(y_0, \varepsilon)$:

$$y_1, y_2 \in \mathring{B}(y_0, \varepsilon), \ \Psi(y_1) = \Psi(y_2) \Rightarrow \|y_1 - y_2\| = \|\Psi'(y_0)^{-1}\Psi'(y_0)(y_1 - y_2)\| \leqslant$$

$$\leqslant \|\Psi'(y_0)^{-1}\| \|\Psi(y_1) - \Psi(y_2) - \Psi'(y_0)(y_1 - y_2)\| \overset{(11)}{\leqslant}$$

$$\leqslant \|\Psi'(y_0)^{-1}\| \frac{1}{2\|\Psi'(y_0)^{-1}\|} \|y_1 - y_2\| = \frac{1}{2}\|y_1 - y_2\| \Rightarrow y_1 = y_2.$$

Further we shall prove that the ball $\mathring{B}(y_0, \varepsilon)$ is the one mentioned in Section a) of the theorem; at present we put $V = \Psi(\mathring{B}(y_0, \varepsilon))$. Since the mapping $\Psi | \mathring{B}(y_0, \varepsilon)$ is injective, it is invertible, and hence there exists $\Phi: V \to \mathring{B}(y_0, \varepsilon)$ possessing property (9).

B) For an arbitrary point $\hat{y} \in \mathring{B}(y_0, \varepsilon)$ let us apply the theorem of Section 2.3.1 to the mapping $\Psi(y)$ (this mapping is independent of x and therefore the space X can be arbitrary) in a neighborhood of the point $\hat{y}$. The condition 1) of this theorem is trivial in this case; the condition 2) turns into the condition for the strict differentiability at the point $\hat{y}$ and, as was indicated above, it is fulfilled, and we have $\Lambda = \Psi'(\hat{y})$; condition 3) is fulfilled since we have $\hat{y} \in \mathring{B}(y_0, \varepsilon)$, and, according to the choice of ε, there exists $\Psi'(\hat{y})^{-1}$.

By the theorem of Section 2.3.1, there is a neighborhood U of the point $\hat{z} = \Psi(\hat{y})$ and a mapping $\varphi: U \to Y$ such that

$$\Psi(\varphi(z)) \equiv z \tag{12}$$

and

$$\|\varphi(z) - \hat{y}\| \leqslant K\|z - \hat{z}\| \tag{13}$$

for some $K > 0$. Let us take a sufficiently small $\delta > 0$ such that $\mathring{B}(\hat{z}, \delta) \subset U$ and $\delta < (\varepsilon - \|y_0 - \hat{y}\|)/K$. Then

$$\|z - \hat{z}\| < \delta \overset{(13)}{\Rightarrow} \|\varphi(z) - \hat{y}\| < K\left(\frac{\varepsilon - \|y_0 - \hat{y}\|}{K}\right) \Rightarrow$$
$$\Rightarrow \|\varphi(z) - y_0\| < \varepsilon \Rightarrow z = \Psi(\varphi(z)) \in \Psi(\mathring{B}(y_0, \varepsilon)) = V \Rightarrow \mathring{B}(\hat{z}, \delta) \subset V.$$

Thus, every point $\hat{z} \in V$ possesses a neighborhood $\mathring{B}(\hat{z}, \delta) \subset V$, and hence V is open. Moreover, for $z \in \mathring{B}(z_0, \delta)$ we have

$$\Psi(\varphi(z)) \overset{(12)}{=} z = \Psi(\Phi(z)) \Rightarrow \varphi(z) = \Phi(z) \tag{14}$$

since Ψ is bijective.

C) Now we shall prove the differentiability of the mapping Φ and the formula (10). By virtue of the strict differentiability of Ψ at the point $\hat{y} = \Phi(\hat{z})$, an analogue of (11) holds: for any $\varkappa > 0$ there is $\varepsilon(\varkappa) > 0$ such that

$$\|y_i - \hat{y}\| < \varepsilon(\varkappa), \quad i = 1, 2, \Rightarrow \|\Psi(y_1) - \Psi(y_2) - \Psi'(\hat{y})(y_1 - y_2)\| \leqslant \varkappa\|y_1 - y_2\|. \tag{15}$$

Putting $y_1 = \Phi(z)$ and $y_2 = \hat{y} = \Phi(\hat{z})$ in (15), we obtain

$$\|z - \hat{z}\| < \frac{\varepsilon(\varkappa)}{K} \overset{(13)(14)}{\Longrightarrow} \|\Phi(z) - \hat{y}\| < \varepsilon(\varkappa) \overset{(15)}{\Rightarrow}$$
$$\Rightarrow \|\Psi(\Phi(z)) - \Psi(\Phi(\hat{z})) - \Psi'(\hat{y})(\Phi(z) - \Phi(\hat{z}))\| \leqslant \varkappa\|\Phi(z) - \hat{y}\| \overset{(13)}{\leqslant} \varkappa K\|z - \hat{z}\| \overset{(12)}{\Rightarrow}$$
$$\Rightarrow \|z - \hat{z} - \Psi'(\hat{y})(\Phi(z) - \Phi(\hat{z}))\| \leqslant \varkappa K\|z - \hat{z}\| \Rightarrow$$
$$\Rightarrow \|\Phi(z) - \Phi(\hat{z}) - \Psi'(\hat{y})^{-1}(z - \hat{z})\| =$$
$$= \|\Psi'(\hat{y})^{-1}(z - \hat{z} - \Psi'(\hat{y})(\Phi(z) - \Phi(\hat{z})))\| \leqslant \|\Psi'(\hat{y})^{-1}\| \varkappa K\|z - \hat{z}\|,$$

and therefore

$$\Phi(z) - \Phi(\hat{z}) - \Psi'(\hat{y})^{-1}(z - \hat{z}) = o(z - \hat{z})$$

for $z \to \hat{z}$ since $\varkappa$ is arbitrary. Therefore, the Fréchet derivative $\Phi'(\hat{z}) = \Psi'(\hat{y})^{-1}$ exists, and thus (10) is proved. The derivative $\Phi'(z)$ is continuous on V because, according to (10), it can be represented in the form of the composition of the continuous mappings $z \to \Phi(z)$, $y \to \Psi'(y)$, and $A \to A^{-1}$ (the last of them is continuous by virtue of Proposition 3 of Section 2.2.1).

2.3.5. Tangent Space and Lyusternik's Theorem. In this section X is a normed space and M is its subset.

Definition. An element $h \in X$ is called a *tangent vector to* M *at a point* $x_0 \in M$, if there exist $\varepsilon > 0$ and a mapping $r: [-\varepsilon, \varepsilon] \to X$ such that:

a) $x_0 + \alpha h + r(\alpha) \in M$,

b) $r(\alpha) = o(\alpha)$ for $\alpha \to 0$.

An element $h \in X$ is said to be a *one-sided tangent vector to* M *at a point* $x_0 \in M$ if there exist $\varepsilon > 0$ and a mapping $r: (0, \varepsilon) \to X$ such that a)

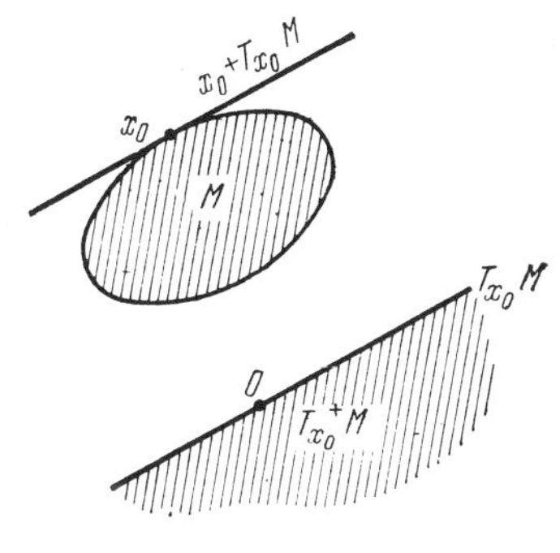

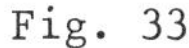
Fig. 33

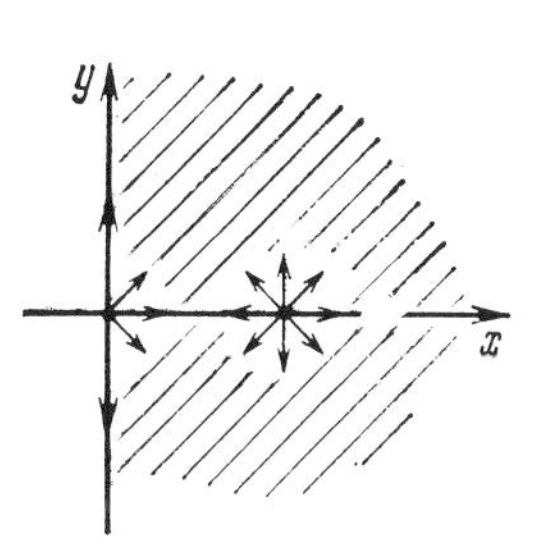

Fig. 34

is fulfilled and b) holds for $\alpha \downarrow 0$.

The set of all the tangent vectors to M at a point x_0 is denoted by $T_{x_0}M$, and the set of the one-sided tangent vectors is denoted by $T^+_{x_0}M$. It is obvious that $T_{x_0}M \subset T^+_{x_0}M$ and $T_{x_0}M = T^+_{x_0}M \cap (-T^+_{x_0}M)$. If the set $T_{x_0}M$ is a subspace of X, then it is called a *tangent space* to M at the point x_0.

Remark. In geometry, by a tangent line, a tangent plane, etc. is usually meant the affine manifold $x_0 + T_{x_0}M$ (Fig. 33) and not $T_{x_0}M$.

Examples (the detailed proofs are left to the reader).

1) $X = \mathbf{R}^2$, $M = \{(x, y) \mid x \geqslant 0\}$ (Fig. 34); $T^+_{(0,0)}M = M$,
$T_{(0,0)}M = \{(0, b) \mid b \in \mathbf{R}\}$, $T_{(1,0)}M = T^+_{(1,0)}M = \mathbf{R}^2$.
2) $X = \mathbf{R}^2$, $M = \{(x, y) \mid x \geqslant 0,\ y \geqslant 0\} = \mathbf{R}^2_+$ (Fig. 35);
$T_{(1,0)}M = \{(a, 0) \mid a \in \mathbf{R}\}$, $T_{(0,0)}M = \{0\}$, $T_{(0,1)}M = \{(0, b) \mid b \in \mathbf{R}\}$,
$T^+_{(1,0)}M = \{(a, b) \mid a \in \mathbf{R},\ b \geqslant 0\}$, $T^+_{(0,0)}M = M$, $T^+_{(0,1)}M = \{(a, b) \mid a \geqslant 0,\ b \in \mathbf{R}\}$.
3) $X = Y \times \mathbf{R}$, where Y is a normed space and $M = \{(y, z) \mid z = \|y\|\}$ is a cone (Fig. 36).

$$T_{(0,0)}M = \{0\}, \quad T^+_{(0,0)}M = M.$$

Exercises.

1. Assuming that Y is a Hilbert space (for instance, $Y = \mathbf{R}^2$) find $T_{(y,\|y\|)}M$ and $T^+_{(y,\|y\|)}M$ in the example 3) for $y \neq 0$ (Fig. 36).

2. Find T_aM and T^+_aM, where:

a) $M = \{2^{-n} \mid n = 1, 2, \ldots\} \cup \{0\} \subset \mathbf{R}$, $a = 0$;

b) $M = \left\{\frac{1}{n} \,\middle|\, n = 1, 2, \ldots\right\} \cup \{0\} \subset \mathbf{R}$, $a = 0$;

c) $M = \{(x, y) \mid x^2 = y^3\} \subset \mathbf{R}^2$, $a = (0, 0)$ and $a = (1, 1)$;

d) $M = \{(x, y) \mid x^2 \leqslant y^3\} \subset \mathbf{R}^2$, $a = (0, 0)$ and $a = (1, 1)$;

e) $M = \{(x, y) \mid x^2 \geqslant y^3\} \subset \mathbf{R}^2$, $a = (0, 0)$ and $a = (1, 1)$;

f) $M = \{(x, y, z) \mid z = 1$, if x and y are rational; $z = 0$ in the other cases$\}$, $a = (0, 0, 1)$.

In many cases including those important for the theory of extremal problems, the set of tangent vectors can be found with the aid of the following general theorem.

Lyusternik's Theorem. Let X and Z be Banach spaces, let U be a neighborhood of a point x_0 in X, and let $F: U \to Z$, $F(x_0) = 0$.

If F is strictly differentiable at the point x_0 and $F'(x_0)$ is an epimorphism, then the set $M = \{x \mid F(x) = 0\}$ has the tangent space

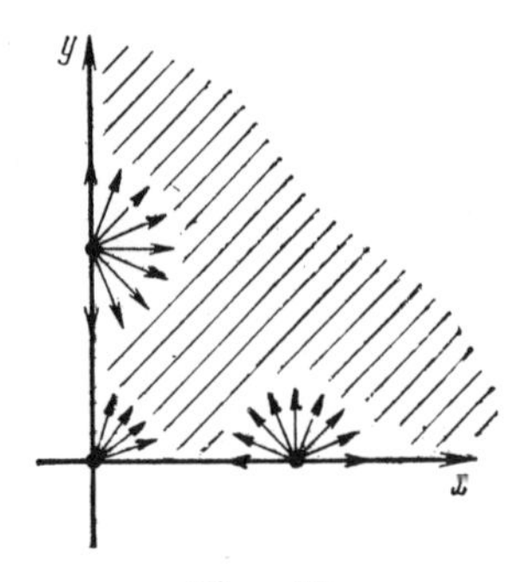

Fig. 35

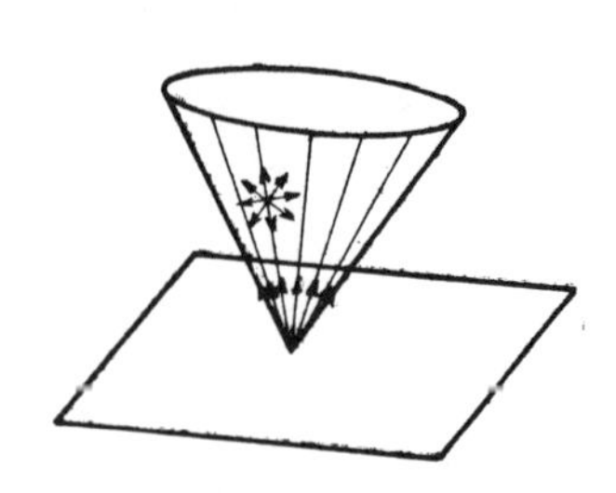

Fig. 36

$$T_{x_0}M = T^+_{x_0}M = \operatorname{Ker} F'(x_0).$$

at the point x_0.

Proof. A) Let $h \in T^+_{x_0}M$. Then if $r(\cdot)$ is the mapping mentioned in the definition of a one-sided tangent vector, it follows, by the definition of the strict differentiability, that

$$0 = F(x_0 + \alpha h + r(\alpha)) = F(x_0) + \alpha F'(x_0)h + o(\alpha) = \alpha F'(x_0)h + o(\alpha).$$

Consequently, $F'(x_0)h = 0$ and $T^+_{x_0}M \subset \operatorname{Ker} F'(x_0)$.

B) Let us apply the theorem of Section 2.3.1 to the mapping $\Psi(x, y) = F(x + y)$. The strict differentiability of F implies that the mapping $x \to \Psi(x, 0)$ is continuous at the point x_0 and

$$\|\Psi(x, y') - \Psi(x, y'') - F'(x_0)(y' - y'')\| \leqslant \varepsilon \|y' - y''\|$$

if $\|x - x_0\| < \delta$, $\|y'\| < \delta$, $\|y''\| < \delta$. The condition 3) of the theorem of Section 2.3.1 is also fulfilled because $F'(x_0)$ maps X into Z. According to this theorem, there is a mapping $\varphi\colon U \to X$ of the neighborhood $U \subset X$ of the point x_0 such that

$$\Psi(x, \varphi(x)) \equiv 0 \Leftrightarrow F(x + \varphi(x)) \equiv 0,$$
$$\|\varphi(x)\| \leqslant K\|\Psi(x, 0)\| = K\|F(x)\|.$$

It only remains to put $r(\alpha) = \varphi(x_0 + \alpha h)$. Then

$$F(x_0 + \alpha h + r(\alpha)) = F(x_0 + \alpha h + \varphi(x_0 + \alpha h)) = 0,$$
$$\|r(\alpha)\| = \|\varphi(x_0 + \alpha h)\| \leqslant K\|F(x_0 + \alpha h)\| =$$
$$= K\|F(x_0 + \alpha h) - F(x_0)\| = K\|F'(x_0)\alpha h + o(\|\alpha h\|)\| = K\|\alpha F'(x_0)h + o(\alpha)\|,$$

and if $h \in \operatorname{Ker} F'(x_0)$, then $\|r(\alpha)\| = o(\alpha)$. Consequently,

$$\operatorname{Ker} F'(x_0) \subset T_{x_0}M \subset T^+_{x_0}M. \blacksquare$$

Lyusternik's theorem is intuitively quite clear from the geometrical viewpoint. Neglecting the rigorousness, we can define the tangent space L to a set M at a point x_0 by saying that M is approximated by the affine manifold $x_0 + L$ with an "accuracy to within infinitesimals of higher order." In an "infinitesimal neighborhood" of the point x_0 the function F is approximated by its linear part:

$$F(x) \approx F(x_0) + F'(x_0)(x - x_0) = F'(x_0)(x - x_0) \tag{1}$$

[since $x_0 \in M$, we have $F(x_0) = 0$], and the set $M = \{x \mid F(x) = 0\}$ is approximated by the set

$$\tilde{M} = \{x \mid F'(x_0)(x - x_0) = 0\} = \operatorname{Ker} F'(x_0) + x_0.$$

Hence M coincides with the manifold $\tilde{M} = x_0 + \operatorname{Ker} F'(x_0)$ "to within infinitesimals of higher order," and consequently $\operatorname{Ker} F'(x_0)$ is the tangent space to M at x_0.

2.4. Differentiability of Certain Concrete Mappings

2.4.1. Nemytskii's Operator and a Differential Constraint Operator. Let U be an open set in $\mathbf{R} \times \mathbf{R}^n$, and let a function $f(t, x): U \to \mathbf{R}^m$ and its partial derivative $f_x(t, x)$ be continuous in U. Let us consider the equality

$$\mathcal{N}(x(\cdot))(t) \equiv f(t, x(t)), \quad t_0 \leqslant t \leqslant t_1, \tag{1}$$

specifying a mapping $\mathcal{N}: \mathcal{U} \to C([t_0, t_1], \mathbf{R}^m)$ defined on the set

$$\mathcal{U} = \{x(\cdot) \in C([t_0, t_1], \mathbf{R}^n) \mid (t, x(t)) \in U, \ t_0 \leqslant t \leqslant t_1\}. \tag{2}$$

This mapping will be called *Nemytskii's operator*.

Exercise 1. Check that set (2) is open in $C([t_0, t_1], \mathbf{R}^n)$.

Proposition 1. Nemytskii's operator specified by relation (1) is continuously differentiable on set (2), and, moreover,

$$\mathcal{N}'(\hat{x}(\cdot))[h(\cdot)](t) = \hat{f}_x(t)h(t), \tag{3}$$

where

$$\hat{f}_x(t) \Leftrightarrow \left(\frac{\partial f_i}{\partial x_j}(t, \hat{x}(t)) \quad \begin{matrix} i = 1, \ldots, m, \\ j = 1, \ldots, n \end{matrix}\right) \tag{4}$$

is Jacobi's matrix.

Proof. To begin with, we shall compute the first Lagrange variation of the mapping $\mathcal{N}$. According to the definition,

$$\delta\mathcal{N}(\hat{x}(\cdot); h(\cdot)) = \lim_{\alpha \to 0} \frac{\mathcal{N}(\hat{x}(\cdot) + \alpha h(\cdot)) - \mathcal{N}(\hat{x}(\cdot))}{\alpha}.$$

Here the left-hand and right-hand members are elements of the space $C([t_0, t_1], \mathbf{R}^m)$, i.e., some functions of $t \in [t_0, t_1]$. For a fixed t we have

$$\lim_{\alpha \to 0} \frac{\mathcal{N}(\hat{x}(\cdot) + \alpha h(\cdot))(t) - \mathcal{N}(\hat{x}(\cdot))(t)}{\alpha} =$$
$$= \lim_{\alpha \to 0} \frac{f(t, \hat{x}(t) + \alpha h(t)) - f(t, \hat{x}(t))}{\alpha} = f_x(t, \hat{x}(t))h(t) = \hat{f}_x(t)h(t). \tag{5}$$

In order to show that here the convergence takes place in the sense of the space $C([t_0, t_1], \mathbf{R}^m)$ as well, i.e., to prove that the convergence is uniform, we take into account that for some $a > 0$ we have

$$\mathcal{K} = \{(t, x) \mid |x - \hat{x}(t)| \leqslant a, \ t_0 \leqslant t \leqslant t_1\} \subset U,$$

and f_x is uniformly continuous on the compact set $\mathcal{K}$. This means that

$$\forall \varepsilon > 0, \exists \delta > 0, \quad |t' - t''| < \delta, \quad |x' - x''| < \delta \Rightarrow \|f_x(t', x') - f_x(t'', x'')\| < \varepsilon.$$

If $|\alpha| < \delta / \|h(\cdot)\|_C$, then, by the mean value theorem (Section 2.2.3),

$$\left\| \frac{\mathcal{N}(\hat{x}(\cdot) + \alpha h(\cdot)) - \mathcal{N}(\hat{x}(\cdot))}{\alpha} - \{f_x(t)h(t)\} \right\| =$$
$$= \max_{t \in [t_0, t_1]} \left\| \frac{f(t, \hat{x}(t) + \alpha h(t)) - f(t, \hat{x}(t)) - f_x(t, \hat{x}(t))\alpha h(t)}{\alpha} \right\| \leqslant$$
$$\leqslant \max_{t \in [t_0, t_1]} \max_{\theta \in [0, 1]} \|f_x(t, \hat{x}(t) + \theta\alpha h(t)) - f_x(t, \hat{x}(t))\| |h(t)| \leqslant \varepsilon \|h\|_C. \tag{6}$$

Therefore, the convergence in (5) is in fact uniform, and the existence of the Lagrange first variation

$$\delta\mathcal{N}(\hat{x}(\cdot), h(\cdot))(t) \equiv \hat{f}_x(t)h(t)$$

is proved. Since the first variation is specified by a linear operator

$$\mathcal{N}'_\Gamma(\hat{x}(\cdot)): C([t_0, t_1], \mathbf{R}^n) \to C([t_0, t_1], \mathbf{R}^m)$$

[namely, $\mathcal{N}'_\Gamma(\hat{x}(\cdot))$ is the operator of multiplication by the matrix function $\hat{f}_x(t)$] and this operator is bounded:

$$\|\mathcal{N}'_\Gamma(\hat{x}(\cdot))\| \leqslant \max_{t\in[t_0, t_1]} \|f_x(t, \hat{x}(t))\|, \tag{7}$$

we see that $\mathcal{N}(x(\cdot))$ is Gateaux differentiable. To prove the continuity of the Gateaux derivative, we shall estimate the norm of the difference

$$\|\mathcal{N}'_\Gamma(x(\cdot)) - \mathcal{N}'_\Gamma(\hat{x}(\cdot)) = \sup_{\|h(\cdot)\|_C \leqslant 1} \|\mathcal{N}'_\Gamma(x(\cdot))[h(\cdot)] - \mathcal{N}'_\Gamma(\hat{x}(\cdot))[h(\cdot)]\| =$$

$$= \sup_{\|h(\cdot)\|_C \leqslant 1} \max_{t\in[t_0, t_1]} \|f_x(t, x(t))h(t) - f_x(t, \hat{x}(t))h(t)\| \leqslant$$

$$\leqslant \max_{t\in[t_0, t_1]} \|f_x(t, x(t)) - f_x(t, \hat{x}(t))\|.$$

Using again the compact set $\mathcal{K}$ and the uniform continuity of f_x, we conclude that $\mathcal{N}'_\Gamma(x(\cdot))$ is continuous with respect to $x(\cdot)\in\mathcal{U}$. Now applying Corollary 2 of Section 2.2.3, we see that $\mathcal{N}(x(\cdot))$ possesses the Fréchet derivative at each point of the set $\mathcal{U}$ and is strictly differentiable. The equality $\mathcal{N}'_\Gamma = \mathcal{N}'$ implies that the Fréchet derivative is continuous in $\mathcal{U}$. ■

Exercise 2. Prove that in (7) the equality takes place.

Now let U be an open set in $\mathbf{R} \times \mathbf{R}^n \times \mathbf{R}^r$, and let a function $f(t, x, u): U \to \mathbf{R}^m$ and its partial derivatives f_x and f_y be continuous in U. The mapping $\mathcal{N}: \mathcal{U} \to C([t_0, t_1], \mathbf{R}^m)$ defined on the set

$$\mathcal{U} = \{(x(\cdot), u(\cdot)) \mid (t, x(t), u(t)) \in U, \ t_0 \leqslant t \leqslant t_1\} \subset C^1([t_0, t_1], \mathbf{R}^n) \times C([t_0, t_1], \mathbf{R}^r) \tag{8}$$

by the equality

$$\mathcal{N}(x(\cdot), u(\cdot))(t) \equiv f(t, x(t), u(t)) \tag{9}$$

will also be called *Nemytskii's operator*.

Proposition 2. Nemytskii's operator determined by the relation (8) is continuously differentiable on set (9), and, moreover,

$$\mathcal{N}'(\hat{x}(\cdot), \hat{u}(\cdot))[h(\cdot), v(\cdot)](t) = \hat{f}_x(t)h(t) + \hat{f}_u(t)v(t), \tag{10}$$

where

$$\hat{f}_x(t) \Leftrightarrow \left(\frac{\partial f_i(t, \hat{x}(t), \hat{u}(t))}{\partial x_j}, \ i = 1, \ldots, m, \ j = 1, \ldots, n\right) \tag{11}$$

and

$$\hat{f}_u(t) \Leftrightarrow \left(\frac{\partial f_i(t, \hat{x}(t), \hat{u}(t))}{\partial u_k}, \ i = 1, \ldots, m, \ k = 1, \ldots, r\right). \tag{12}$$

Proof. As in the foregoing case, the mapping $\mathcal{N}$ possesses the partial derivatives

$$\mathcal{N}_x(\hat{x}(\cdot), \hat{u}(\cdot))[h(\cdot)](t) = \hat{f}_x(t)h(t)$$

and

$$\mathcal{N}_u(\hat{x}(\cdot), \hat{u}(\cdot))[v(\cdot)](t) = \hat{f}_u(t)v(t),$$

and the mappings

$$\mathcal{N}_x: \mathcal{U} \to \mathcal{L}(C^1([t_0, t_1], \mathbf{R}^n), C([t_0, t_1], \mathbf{R}^m))$$

and

$$\mathcal{N}_u: \mathcal{U} \to \mathcal{L}(C^1([t_0, t_1], \mathbf{R}^r); C([t_0, t_1], \mathbf{R}^m))$$

are continuous in $\mathcal{U}$. It remains to refer to the theorem on the total differential of Section 2.2.4. ■

Further, let U and $\mathcal{U}$ be the same as before, and let a function $\varphi(t, x, u): U \to \mathbf{R}^n$ and its partial derivatives φ_x and φ_u be continuous in U.

The mapping $\Phi: \mathcal{U} \to C([t_0, t_1], \mathbf{R}^n)$ determined by the equality

$$\Phi(x(\cdot), u(\cdot))(t) = \dot{x}(t) - \varphi(t, x(t), u(t)) \tag{13}$$

will be called a *differential constraint operator*.

Proposition 3. The differential constraint operator determined by the relation (13) is continuously differentiable on the set (8), and, moreover,

$$\Phi'(\hat{x}(\cdot), \hat{u}(\cdot))[h(\cdot), v(\cdot)](t) = \dot{h}(t) - \hat{\varphi}_x(t) h(t) - \hat{\varphi}_u(t) v(t), \tag{14}$$

where

$$\hat{\varphi}_x(t) \Leftrightarrow \left(\frac{\partial \varphi_i(t, \hat{x}(t), \hat{u}(t))}{\partial x_j}, \ i, j = 1, \ldots, n\right) \tag{15}$$

and

$$\hat{\varphi}_u(t) \Leftrightarrow \left(\frac{\partial \varphi_i(t, \hat{x}(t), \hat{u}(t))}{\partial u_k}, \ i = 1, \ldots, n, \ k = 1, \ldots, r\right). \tag{16}$$

Proof. The mapping Φ is the difference between the continuous linear operator $(x(\cdot), u(\cdot)) \to \dot{x}(\cdot)$ and Nemytskii's operator $\mathcal{N}(x(\cdot), u(\cdot))(t) = \varphi(t, x(t), u(t))$. Therefore, the assertion we have to prove follows from the general properties of the derivatives (Section 2.2.1) and Proposition 2. ■

2.4.2. Integral Functional. Let U be an open set in $\mathbf{R} \times \mathbf{R} \times \mathbf{R}^n$, and let a function $f(t, x, \dot{x}): U \to \mathbf{R}^m$ and its partial derivatives f_x and $f_{\dot{x}}$ be continuous in U. Let us define a mapping $\mathcal{I}: \mathcal{W} \to \mathbf{R}^m$ on the set

$$\mathcal{W} = \{x(\cdot) \in C^1([t_0, t_1], \mathbf{R}^n) \mid (t, x(t), \dot{x}(t)) \in U, \ t_0 \leqslant t \leqslant t_1\} \tag{1}$$

with the aid of the equality

$$\mathcal{I}(\dot{x}(\cdot)) = \int_{t_0}^{t_1} f(t, x(t), \dot{x}(t))\, dt. \tag{2}$$

Proposition 1. The integral mapping (2) is continuously differentiable on set (1), and, moreover,

$$\mathcal{I}'(\hat{x}(\cdot))[h(\cdot)] = \int_{t_0}^{t_1} \{\hat{f}_x(t) h(t) + \hat{f}_{\dot{x}}(t) \dot{h}(t)\}\, dt, \tag{3}$$

where

$$\hat{f}_x(t) \Leftrightarrow \left(\frac{\partial f_i}{\partial x_j}(t, \hat{x}(t), \dot{\hat{x}}(t)), \ i = 1, \ldots, m, \ j = 1, \ldots, n\right) \tag{4}$$

and

$$\hat{f}_{\dot{x}}(t) \Leftrightarrow \left(\frac{\partial f_i}{\partial \dot{x}_j}(t, \hat{x}(t), \dot{\hat{x}}(t)), \ i = 1, \ldots, m, \ j = 1, \ldots, n\right). \tag{5}$$

Proof. Let us represent mapping (2) in the form of the composition

$$\mathcal{I} = I \circ \mathcal{N} \circ D,$$

where

$$D: C([t_0, t_1], \mathbf{R}^m) \to C^1([t_0, t_1], \mathbf{R}^n) \times C([t_0, t_1], \mathbf{R}^n)$$

is the continuous linear operator determined by the equality

$$D(x(\cdot)) = (x(\cdot), \dot{x}(\cdot));$$

$$\mathcal{N}: \mathcal{U} \to C([t_0, t_1], \mathbf{R}^m)$$

is Nemytskii's operator determined by equality (9) of Section 2.4.1 (for r = n) and

$$I: C([t_0, t_1], \mathbf{R}^m) \to \mathbf{R}^m$$

is the operator of integration:

$$I(x(\cdot)) = \int_{t_0}^{t_1} x(t)\,dt$$

which is also linear and continuous. All these operators are differentiable (see Section 2.2.1 and Proposition 2 of Section 2.4.1), the derivative of the linear operator coinciding with that operator itself, and the derivative of Nemytskii's operator being expressed by the formula (10) of Section 2.4.1. By the composition theorem,

$$\mathcal{J}'(\hat{x}(\cdot)) = I \circ \mathcal{N}'(D\hat{x}(\cdot)) \circ D, \tag{6}$$

that is,

$$\mathcal{J}'(\hat{x}(\cdot))[h(\cdot)] = I\{\mathcal{N}'(D\hat{x}(\cdot))[Dh(\cdot)]\} = I\{\mathcal{N}'(\hat{x}(\cdot), \dot{\hat{x}}(\cdot))[h(\cdot), \dot{h}(\cdot)] =$$
$$= I\{\hat{f}_x(t)h(t) + \hat{f}_{\dot{x}}(t)\dot{h}(t)\} = \int_{t_0}^{t_1} \{\hat{f}_x(t)h(t) + \hat{f}_{\dot{x}}(t)\dot{h}(t)\}\,dt.$$

This proves formula (3). The continuity of $\mathcal{J}'(x(\cdot))$ follows from the continuity of the derivative of Nemytskii's operator and equality (6). ■

Exercise. Find the norms of operators D and I.

In problems of the classical calculus of variations and optimal control problems we also deal with integral functionals of form (2) with variable limits of integration t_0 and t_1. To include them in the general scheme we shall use the following procedure.

Let $f(t, x, \dot{x})$ satisfy the same conditions as before, let $\Delta \subseteq \mathbf{R}$ be a closed interval, and let

$$\mathcal{U} = \{(x(\cdot), t_0, t_1) \mid x(\cdot) \in C^1(\Delta, \mathbf{R}^n), (t, x(t), \dot{x}(t)) \in U, \quad t \in \Delta, t_0, t_1 \in \operatorname{int}\Delta\}. \tag{7}$$

Let us define a mapping $\mathcal{J}: \mathcal{U} \to \mathbf{R}^m$ with the aid of the equality

$$\mathcal{J}(x(\cdot), t_0, t_1) = \int_{t_0}^{t_1} f(t, x(t), \dot{x}(t))\,dt. \tag{8}$$

Proposition 2. The integral mapping (8) is continuously differentiable on set (7), and, moreover,

$$\mathcal{J}'(\hat{x}(\cdot), \hat{t}_0, \hat{t}_1)[h(\cdot), \tau_0, \tau_1] = \int_{t_0}^{t_1} (\hat{f}_x(t)h(t) + \hat{f}_{\dot{x}}(t)\dot{h}(t))\,dt + \hat{f}(\hat{t}_i)\tau_i \Big|_{i=0}^{i=1}, \tag{9}$$

where $\hat{f}_x(t)$ and $\hat{f}_{\dot{x}}(t)$ are determined by formulas (4) and (5) and

$$\hat{f}(t) = f(t, \hat{x}(t), \dot{\hat{x}}(t)). \tag{10}$$

Proof. We shall make use of the theorem on the total differential of Section 2.2.4. The partial derivatives exist according to Proposition 1 and the classical theorem on the derivatives of an integral with respect to its upper and lower limits of integration:

$$\mathcal{J}_{x(\cdot)}(\hat{x}(\cdot), \hat{t}_0, \hat{t}_1)[h(\cdot)] = \int_{\hat{t}_0}^{\hat{t}_1} \{\hat{f}_x(t)h(t) + \hat{f}_{\dot{x}}(t)\dot{h}(t)\}\,dt,$$

$$\mathcal{J}_{t_0}(\hat{x}(\cdot), \hat{t}_0, \hat{t}_1)[\tau_0] = -\hat{f}(\hat{t}_0)\tau_0,$$
$$\mathcal{J}_{t_1}(\hat{x}(\cdot), \hat{t}_0, \hat{t}_1)[\tau_1] = \hat{f}(\hat{t}_1)\tau_1.$$

Now we shall verify the continuity of the partial derivatives.

A)
$$\|\mathcal{I}_{t_0}(x(\cdot), t_0, t_1)-\mathcal{I}_{t_0}(\hat{x}(\cdot), \hat{t}_0, \hat{t}_1)\|=$$
$$=\sup_{\|\tau_0\|\leqslant 1}\|f(t_0, x(t_0), \dot{x}(t_0))\tau_0-f(\hat{t}_0, \hat{x}(\hat{t}_0), \dot{\hat{x}}(\hat{t}_0))\tau_0\|\leqslant$$
$$\leqslant|f(t_0, x(t_0), \dot{x}(t_0))-f(\hat{t}_0, \hat{x}(\hat{t}_0), \dot{\hat{x}}(\hat{t}_0))|. \quad (11)$$

Moreover,

$$|x(t_0)-\hat{x}(\hat{t}_0)|\leqslant|x(t_0)-\hat{x}(t_0)|+|\hat{x}(t_0)-\hat{x}(\hat{t}_0)|\leqslant\|\hat{x}(\cdot)-x(\cdot)\|_{C^1}+|\hat{x}(t_0)-\hat{x}(\hat{t}_0)| \quad (12)$$

and

$$|\dot{x}(t_0)-\dot{\hat{x}}(\hat{t}_0)|\leqslant|\dot{x}(t_0)-\dot{\hat{x}}(t_0)|+|\dot{\hat{x}}(t_0)-\dot{\hat{x}}(\hat{t}_0)|\leqslant\|\hat{x}(\cdot)-x(\cdot)\|_{C^1}+|\dot{\hat{x}}(t_0)-\dot{\hat{x}}(\hat{t}_0)|. \quad (13)$$

Therefore, if $t_0 \to \hat{t}_0$ and $x(\cdot) \to \hat{x}(\cdot)$ in $C^1(\Delta, \mathbf{R}^n)$, then $x(t_0) \to \hat{x}(\hat{t}_0)$, $\dot{x}(t_0) \to \dot{\hat{x}}(\hat{t}_0)$, and hence the right-hand side of equality (11) tends to zero. Consequently, $\mathcal{I}_{t_0}(x(\cdot), t_0, t_1)$ depends continuously on $(x(\cdot), t_0, t_1)$ (this derivative does not depend on t_1 at all). The continuity of $\mathcal{I}_{t_1}(x(\cdot), t_0, t_1)$ is verified analogously.

B) Let us choose a number $a > 0$ such that

$$\mathcal{K}=\{(t, x, u)\,|\,|x-\hat{x}(t)|\leqslant a,\ |u-\dot{\hat{x}}(t)|\leqslant a,\ t\in\Delta\}\subset U.$$

The derivatives $f_x(t, x, \dot{x})$ and $f_{\dot{x}}(t, x, \dot{x})$ are uniformly continuous and bounded on the compact set $\mathcal{K}$. Now for $\|x(\cdot) - x(\cdot)\|_{C^1} \leqslant a$ we have

$$\|\mathcal{I}_{x(\cdot)}(x(\cdot), t_0, t_1)-\mathcal{I}_{x(\cdot)}(\hat{x}(\cdot), \hat{t}_0, \hat{t}_1)\|=$$
$$=\sup_{\|h(\cdot)\|_{C^1}\leqslant 1}\left|\int_{t_0}^{t_1}(f_xh+f_{\dot{x}}\dot{h})\,dt-\int_{\hat{t}_0}^{\hat{t}_1}(\hat{f}_xh+\hat{f}_{\dot{x}}\dot{h})\,dt\right|\leqslant$$
$$\leqslant\sup_{\|h(\cdot)\|_{C^1}\leqslant 1}\left\{\left|\int_{\hat{t}_1}^{t_1}\{f_xh+f_{\dot{x}}\dot{h}\}\,dt\right|+\left|\int_{t_0}^{\hat{t}_0}\{f_xh+f_{\dot{x}}\dot{h}\}\,dt\right|+\left|\int_{\hat{t}_1}^{\hat{t}_1}\{(f_x-\hat{f}_x)h+(f_{\dot{x}}-\hat{f}_{\dot{x}})\dot{h}\}\,dt\right|\right\}\leqslant$$
$$\leqslant\max_{\mathcal{K}}\{\|f_x(t, x, u)\|+\|f_{\dot{x}}(t, x, u)\|\}(|t_1-\hat{t}_1|+|t_0-\hat{t}_0|)+$$
$$+\max_{[\hat{t}_0, \hat{t}_1]}\{\|f_x(t, x(t), \dot{x}(t))-f_x(t, \hat{x}(t), \dot{\hat{x}}(t))\|+$$
$$+\|f_{\dot{x}}(t, x(t), x(t))-f_{\dot{x}}(t, \hat{x}(t), \dot{\hat{x}}(t))\|\}.$$

Here the first term tends to zero for $t_0 \to \hat{t}_0$ and $t_1 \to \hat{t}_1$. The second term is estimated using the uniform continuity as in the proof of Proposition 1 of Section 2.4.1, and hence this term tends to zero when $x(\cdot) \to \hat{x}(\cdot)$ in the space $C^1(\Delta, \mathbf{R}^n)$.

Applying the theorem on the total differential, we obtain (9). ■

2.4.3. Operator of Boundary Conditions. Let a function $\psi(t_0, x_0, t_1, x_1): W \to \mathbf{R}^s$ be continuously differentiable on an open set $W\subset\mathbf{R}\times\mathbf{R}^n\times\mathbf{R}\times\mathbf{R}^n$, and let

$$\mathcal{W}=\{(x(\cdot), t_0, t_1)\,|\,x(\cdot)\in C^1(\Delta, \mathbf{R}^n),\ t_0, t_1\in\operatorname{int}\Delta,\ (t_0, x(t_0), t_1, x(t_1))\in W\}. \quad (1)$$

The mapping $\Psi: \mathcal{W}\to\mathbf{R}^s$ determined by the equality

$$\Psi(x(\cdot), t_0, t_1)=\psi(t_0, x(t_0), t_1, x(t_1)) \quad (2)$$

will be called an *operator of boundary conditions*.

Proposition. The operator of boundary conditions (2) is continuously differentiable on set (1), and, moreover,

$$\Psi'(\hat{x}(\cdot), \hat{t}_0, \hat{t}_1)[h(\cdot), \tau_0, \tau_1]=\hat{\psi}_{t_0}\tau_0+\hat{\psi}_{x_0}[h(\hat{t}_0)+\dot{\hat{x}}(\hat{t}_0)\tau_0]+\hat{\psi}_{t_1}\tau_1+\hat{\psi}_{x_1}[h(\hat{t}_1)+\dot{\hat{x}}(\hat{t}_1)\tau_1], \quad (3)$$

where

$$\hat{\psi}_{t_i}=\psi_{t_i}(\hat{t}_0,\ \hat{x}(\hat{t}_0),\ \hat{t}_1,\ \hat{x}(\hat{t}_1)),\quad i=0,\ 1, \tag{4}$$

and

$$\hat{\psi}_{x_i}=\psi_{x_i}(\hat{t}_0,\ \hat{x}(\hat{t}_0),\ t_1,\ \hat{x}(\hat{t}_1)),\quad i=0,\ 1. \tag{5}$$

Proof. A) First we shall consider the simplest mapping $\mathrm{ev}: C^1(\Delta,\ \mathbf{R}^n)\times \operatorname{int}\Delta\to\mathbf{R}^n$ determined by the equality

$$\mathrm{ev}(x(\cdot),\ t_0)=x(t_0)$$

(the mapping of the values). The mapping is linear with respect to $x(\cdot)$, and therefore

$$\mathrm{ev}_{x(\cdot)}(\hat{x}(\cdot),\ \hat{t}_0)[h(\cdot)]=h(\hat{t}_0).$$

The partial derivative with respect to t_0 is an ordinary partial derivative:

$$\mathrm{ev}_{t_0}(\hat{x}(\cdot),\ \hat{t}_0)[\tau_0]=\dot{\hat{x}}(\hat{t}_0)\tau_0.$$

Let us verify the continuity:

$$\|\mathrm{ev}_{x(\cdot)}(x(\cdot),\ t_0)-\mathrm{ev}_{x(\cdot)}(\hat{x}(\cdot),\ \hat{t}_0)\|=\sup_{\|h(\cdot)\|_{C^1}\leqslant 1}\|h(t_0)-h(\hat{t}_0)\|\leqslant$$

$$\leqslant\sup_{\|h(\cdot)\|_{C^1}\leqslant 1}\ \sup_{\theta\in[t_0,\hat{t}_0]}\|\dot{h}(\theta)\|\,|t_0-\hat{t}_0|\leqslant|t_0-\hat{t}_0|\to 0$$

for $t_0\to\hat{t}_0$. Further,

$$\|\mathrm{ev}_{t_0}(x(\cdot),\ t_0)-\mathrm{ev}_{t_0}(\hat{x}(\cdot),\ t_0)\|=\sup_{|\tau_0|\leqslant 1}\|\dot{x}(t_0)\tau_0-\dot{\hat{x}}(\hat{t}_0)\tau_0\|=\|\dot{x}(t_0)-\dot{\hat{x}}(\hat{t}_0)\|\to 0$$

for $t_0\to\hat{t}_0$ and $x(\cdot)\to\hat{x}(\cdot)$ in the space $C^1(\Delta,\ \mathbf{R}^n)$ as was shown in the proof of Proposition 2 of Section 2.4.2 [inequalities (12) and (13)]. By virtue of the theorem on the total differential,

$$\mathrm{ev}(\hat{x}(\cdot),\ \hat{t}_0)'[h(\cdot),\ \tau_0]=h(\hat{t}_0)+\dot{\hat{x}}(\hat{t})\tau_0.$$

The continuous differentiability of mapping $(x(\cdot),\ t_1)\to x(t_1)$ is proved in a similar way.

B) Using the composition theorem, we prove the differentiability of mapping (2) and equality (3). ■

Exercise 1. Let $\mathrm{ev}: C^1([0,1])\times(0,1)\to\mathbf{R}$ be the mapping determined by the formula $\mathrm{ev}(x(\cdot),\ t_0)=x(t_0)$. Prove that:

a) for the second variation $\delta^2\mathrm{ev}(x(\cdot),\ t_0)$ to exist it is necessary and and sufficient that $\ddot{x}(t_0)$ exist and, moreover,

$$\delta^2\mathrm{ev}(x(\cdot),\ t_0)[h(\cdot),\ \tau]=2\dot{h}(t_0)\tau+\ddot{x}(t_0)\tau^2;$$

b) the mapping ev does not possess the second Fréchet derivative, although its first Fréchet derivative is Gateaux differentiable.

Hint. $\dot{h}(t_0+\tau)-\dot{h}(t_0)\neq 0((\|h\|^2+\tau^2)^{1/2})$.

2.5. Necessary Facts of the Theory of Ordinary Differential Equations

As was already mentioned in Section 2.1.8, differential equations $\dot{x}=\varphi(t,\ x,\ u(t))$ considered in optimal control problems possess certain specific properties. Since discontinuous controls $u(\cdot)$ are admitted, the right-hand sides of the equations do not necessarily satisfy the conditions of the standard theorems presented in courses of differential equations.

Moreover, the necessary facts concerning the solutions are not always presented in these courses in a form suitable for our aims. That is why for the completeness of the presentation of the material and in order to demonstrate the possibility of the application of the general theorems of Sections 2.3 and 2.4 we shall present in this section the proofs of the basic theorems on the existence, the uniqueness, and the differentiability of solutions and also some special assertions concerning linear systems. Some of the statements are given in a more general form than is usually done. Most often we deal with piecewise-continuous controls $u(\cdot)$, but even in this case it is convenient to speak about "measurability," "integrability," etc.

A function $x(\cdot)$ is said to be a *solution of the differential equation*

$$\dot{x}=F(t, x)$$

if it is absolutely continuous (see Section 2.1.8) and satisfies the equation almost everywhere.

In the equivalent form we can say that $x(\cdot)$ must be the solution of the integral equation

$$x(t)=x(t_0)+\int_{t_0}^{t} F(s, x(s))\, ds.$$

2.5.1. Basic Assumptions. Here and in Sections 2.5.2-2.5.7 we shall assume that G is a given open set in $\mathbf{R}\times\mathbf{R}^n$ and $F:G\to\mathbf{R}^n$ is a given function satisfying the following three conditions:

A) For any x the function $t\to F(t, x)$ defined on the section $G_x = \{t\,|\,(t, x)\in G\}$ is measurable and integrable on any finite closed interval contained in G_x.

B) For any t the function $x\to F(t, x)$ defined on the section $G_t = \{x\,|\,(t, x)\in G\}$ is differentiable (at least in Gateaux' sense).

C) For any compact set $\mathcal{K}\subset G$ there exists a locally integrable function $k(\cdot)$ such that

$$\|F_x(t, x)\|\leqslant k(t), \quad \forall (t, x)\in\mathcal{K}. \tag{1}$$

(A function is said to be *locally integrable* if it is integrable on any finite closed interval.)

A typical example of this kind is the function $F(t, x)=\varphi(t, x, u(t))$, where $\varphi(t, x, u)$ and $\varphi_x(t, x, u)$ are continuous on $G\times\mathcal{U}$, and $\mathbf{R}\to\mathcal{U}$ is piecewise-continuous. The conditions A) and B) are obviously fulfilled, and $k(t)$ can be defined as the maximum of $F_x(t, x)$ on $\mathcal{K}$. Another example is $F(t, x) = A(t)x$, where $A(\cdot)$ is a locally integrable measurable matrix function:

$$A:\ \mathbf{R}\to\mathscr{L}(\mathbf{R}^n, \mathbf{R}^n).$$

Remark. Conditions A)-C) are more restrictive than the well-known Carathéodory conditions [6] but they are more suitable, on the one hand, for our aims [the consideration of equations $\dot{x}=\varphi(t, x, u(t))$ with right-hand sides involving discontinuous controls $u(\cdot)$ for which, however, the derivatives φ_x exist] and, on the other hand, for our facilities (the apparatus of differential calculus presented in Section 2.2).

LEMMA 1. For any compact set $\mathcal{K}\subset G$ there exists a locally integrable function $\varkappa(\cdot)$ such that

$$|F(t, x)|\leqslant\varkappa(t), \quad \forall (t, x)\in\mathcal{K}. \tag{2}$$

Proof. Since the set $\mathcal{K}$ is compact, there exist $\delta > 0$ and $\varepsilon > 0$ such that for any point $(t_0, x_0)\in\mathcal{K}$ we have

$$C_{t_0x_0}=\{(t,\ x)\,|\,|t-t_0|\leqslant\delta,\ |x-x_0|\leqslant\varepsilon\}\subset G$$

for the "cylinder" $C_{t_0x_0}$. For this cylinder we find the corresponding locally integrable function $K_{t_0x_0}(t)$ indicated in C).

Now we again use the compactness and cover $\mathcal{K}$ by a finite number of cylinders: $\mathcal{K}\subset\bigcup\limits_{i=1}^{N}C_{t_ix_i}$.

The formula

$$\varkappa(t)=\sum_{i=1}^{N}[|F(t,\ x_i)|+k_{t_ix_i}(t)\cdot\varepsilon]\,\chi_{[t_i-\delta,\ t_i+\delta]}(t)$$

determines the sought-for function (here $\chi_{[\alpha,\beta]}(\cdot)$ is the characteristic function of the closed interval $[\alpha,\ \beta]$; $|F(t,\ x_i)|$ is integrable on $[t_i - \delta,\ t_i + \delta]$ by virtue of A)). Indeed,

$$(t,\ x)\in\mathcal{K}\Rightarrow\exists i,\ (t,\ x)\in C_{t_ix_i}\Rightarrow|F(t,\ x)|\leqslant|F(t,\ x_i)|+|F(t,\ x)-F(t,\ x_i)|\leqslant$$
$$\leqslant|F(t,\ x)|+\sup_{c\in[x,\ x_i]}\|F_x(t,\ c)\|\varepsilon\leqslant\varkappa(t)$$

(we have $[(t,\ x),\ (t,\ x_i)]\subset C_{t_ix_i}$, and the mean value theorem of Section 2.2.3 is applicable).

LEMMA 2. If a function $x:\Delta\to\mathbf{R}^n$ is continuous on a closed interval Δ and its graph $\{(t,\ x(t))\,|\,t\in\Delta\}$ is contained in an open set $G\subset\mathbf{R}\times\mathbf{R}^n$, and a function $F:G\to\mathbf{R}^n$ satisfies the conditions A)-C), then the function $t\to f(t) = F(t,\ x(t))$ is measurable and integrable on Δ.

Proof. For sufficiently small $\varepsilon > 0$ we have

$$\mathcal{K}=\{(t,\ x)\,|\,|x-x(t)|\leqslant\varepsilon,\ t\in\Delta\}\subset G.$$

Let $k(\cdot)$ be the function corresponding to the compact set $\mathcal{K}$ in accordance with the condition C). The continuous function $x(\cdot)$ is the uniform limit of the piecewise-constant functions $x_n(\cdot)$, and, without loss of generality, we can assume that the graphs of $x_n(\cdot)$ are contained in $\mathcal{K}$.

By condition A), the function $t\to f_n(t) = F(t,\ x_n(t))$ is measurable and integrable on every interval of constancy of $x_n(\cdot)$, and therefore $f_n(\cdot)$ is measurable and integrable on Δ. Using the mean value theorem (Section 2.2.3) and (1), we write

$$|f_n(t)-f(t)|=|F(t,\ x_n(t))-F(t,\ x(t))|\leqslant k(t)\,|x_n(t)-x(t)|,$$

whence $f_n(t)\to f(t)$ for all $t\in\Delta$, and consequently $f(\cdot)$ is measurable since it is a limit of measurable functions [KF, p. 284]. Finally,

$$|f(t)|\leqslant|f_n(t)|+k(t)\,|x_n(t)-x(t)|,$$

and consequently $f(t)$ is integrable.

2.5.2. Local Existence Theorem. Let the function $F:G\to\mathbf{R}^n$ satisfy the conditions A)-C) on the open set $G\subset\mathbf{R}\times\mathbf{R}^n$, and let $\mathcal{K}\subset G$ be a compact set.

Then there exist $\delta > 0$ and $\varepsilon > 0$ such that for any point $(\hat{t},\ \hat{x})\in\mathcal{K}$ and for $(t_0,\ x_0)$ satisfying the inequalities

$$|t_0-\hat{t}|<\delta,\quad |x_0-\hat{x}|<\varepsilon \tag{1}$$

the solution $X(t,\ t_0,\ x_0)$ of Cauchy's problem

$$\left.\begin{aligned}\dot{x}&=F(t,\ x),\\ x(t_0)&=x_0,\end{aligned}\right\} \tag{2}$$

is defined on the closed interval $[\hat{t} - \delta,\ \hat{t} + \delta]$ and is jointly continuous with respect to its arguments.

Proof. Let us apply the contraction mapping principle in the form stated in Section 2.3.2. Cauchy's problem (2) is equivalent to the integral equation

$$x(t)=x_0+\int_{t_0}^{t} F(s, x(s))\,ds, \tag{3}$$

and we shall apply the lemma of Section 2.3.2 to the mapping

$$(t_0, x_0, x(\cdot))\mapsto\Phi(t_0, x_0, x(\cdot))=x_0+\int_{t_0}^{(\cdot)} F(s, x(s))\,ds. \tag{4}$$

Let us choose γ and β such that

$$\mathcal{K}_1^\circ=\{(t, x)\,|\,|t-\hat{t}|\leqslant\gamma,\ |x-\hat{x}|\leqslant\beta,\ (\hat{t}, \hat{x})\in\mathcal{K}^\circ\}\subset G,$$

and let $k(\cdot)$ and $\varkappa(\cdot)$ be integrable functions corresponding to the compact set $\mathcal{K}_1^\circ$ by virtue of C) and Lemma 1 of Section 2.5.1. Let us check the conditions of the lemma of Section 2.3.2. Here the topological space T is the set (1), $U = T$ and $Y = C([\hat{t} - \delta, \hat{t} + \delta], \mathbf{R}^n)$, where δ, $0 < \delta \leqslant \gamma$, will be chosen later; $y_0(s) \equiv \hat{x}$, and

$$V=\{(t_0, x_0, x(\cdot))\,|\,|t_0-\hat{t}|<\delta,\ |x_0-\hat{x}|<\varepsilon,\ \|x(\cdot)-y_0(\cdot)\|_C<\beta\}. \tag{5}$$

The constants ε and θ are such that $0 < \theta < 1$ and $0 < \varepsilon < \beta(1 - \theta)$.

Condition c) of Section 2.3.2 is fulfilled if

$$\left|x_0+\int_{t_0}^{t} F(s, \hat{x})\,ds-\hat{x}\right|<\beta(1-\theta) \tag{6}$$

for $t\in[\hat{t}-\delta, \hat{t}+\delta]$. By Lemma 1 of Section 2.5.1, $|F(s, \hat{x})|\leqslant\varkappa(s)$, and taking a sufficiently small δ, we obtain

$$\int_{\hat{t}-\delta}^{\hat{t}+\delta} |F(s, \hat{x})|\,ds\leqslant \int_{\hat{t}-\delta}^{\hat{t}+\delta} \varkappa(s)\,ds<\beta(1-\theta)-\varepsilon, \tag{7}$$

whence follows (6) because

$$|x_0+\int_{t_0}^{t} F(s, \hat{x})\,ds-\hat{x}|\leqslant|x_0-\hat{x}|+\left|\int_{t_0}^{t} F(s, \hat{x})\,ds\right|\leqslant\varepsilon+\int_{\hat{t}-\delta}^{\hat{t}+\delta} |F(s, \hat{x})|\,ds<\beta(1-\theta).$$

Condition a) of Section 2.3.2 means that the mapping $(t_0, x_0, x(\cdot)) \to (t_0, x_0, \Phi(t_0, x_0, x(\cdot)))$ transforms V into itself. It holds if

$$\left|x_0+\int_{t_0}^{t} F(s, x(s))\,ds-\hat{x}\right|<\beta$$

for $(t_0, x_0, x(\cdot))\in V$ and $t\in[\hat{t}-\delta, \hat{t}+\delta]$. We shall estimate the left-hand member of the last inequality using (6), the mean value theorem and relation (1) of Section 2.5.1:

$$\left|x_0+\int_{t_0}^{t} F(s, x(s))\,ds-\hat{x}\right|\leqslant\left|x_0+\int_{t_0}^{t} F(s, \hat{x})\,ds-\hat{x}\right|+\left|\int_{t_0}^{t}\{F(s, x(s))-F(s, \hat{x})\}\,ds\right|<$$

$$<\beta(1-\theta)+\int_{\hat{t}-\delta}^{\hat{t}+\delta} k(s)|x(s)-\hat{x}|\,ds\leqslant\beta\left(1-\theta+\int_{\hat{t}-\delta}^{\hat{t}+\delta} k(s)\,ds\right)\leqslant\beta,$$

if the inequality

$$\int_{\hat{t}-\delta}^{\hat{t}+\delta} k(s)\,ds \leqslant \theta \tag{8}$$

is fulfilled, which can be achieved by taking, if necessary, a smaller δ.

Finally, since condition a) has already been verified, condition b) of Section 2.3.2 is fulfilled provided that

$$\|\Phi(t_0, x_0, x(\cdot)) - \Phi(t_0, x_0, y(\cdot))\| \leqslant \theta \|x(\cdot) - y(\cdot)\|$$

for any $(t_0, x_0, x(\cdot)) \in V$ and $(t_0, x_0, y(\cdot)) \in V$. By virtue of (8),

$$\|\Phi(t_0, x_0, x(\cdot)) - \Phi(t_0, x_0, y(\cdot))\| =$$

$$= \max_t \left| x_0 + \int_{t_0}^{t} F(s, x(s))\,ds - x_0 - \int_{t_0}^{t} F(s, y(s))\,ds \right| =$$

$$= \max_t \left| \int_{t_0}^{t} \{F(s, x(s)) - F(s, y(s))\}\,ds \right| \leqslant \int_{\hat{t}-\delta}^{\hat{t}+\delta} k(s)\,ds \|x(\cdot) - y(\cdot)\| \leqslant \theta \|x(\cdot) - y(\cdot)\|.$$

Hence the lemma of Section 2.3.2 is applicable, and therefore the sequence $X_n(\cdot, t_0, x_0)$ determined by the equalities $X_0(\cdot, t_0, x_0) \equiv x$ and

$$X_{n+1}(\cdot, t_0, x_0) = x_0 + \int_{t_0}^{(\cdot)} F(s, X_n(s, t_0, x_0))\,ds \tag{9}$$

is uniformly convergent in the space $C([\hat{t} - \delta, \hat{t} + \delta], \mathbf{R}^n)$ with respect to (t_0, x_0) satisfying (1), i.e., uniformly convergent with respect to $t \in [\hat{t} - \delta, \hat{t} + \delta]$. Passing to the limit in (9), we conclude that the function $X(t, t_0, x_0) = \lim_{n\to\infty} X_n(t, t_0, x_0)$ satisfies the integral equation (3) and, consequently, it is the solution of Cauchy's problem (2). It can easily be verified by induction that the functions $X_n(t, t_0, x_0)$ are continuous, and since the convergence is uniform, the function $X(t, t_0, x_0)$ is jointly continuous with respect to its arguments.

2.5.3. Uniqueness Theorem.

LEMMA (Gronwall's Inequality). Let nonnegative functions $\alpha(\cdot)$ and $\omega(\cdot)$ be measurable on a closed interval Δ, and let $\alpha(\cdot)\omega(\cdot)$ be integrable on Δ. If for some $b > 0$ and $\tau \in \Delta$ and for all $t \in \Delta$ the inequality

$$\omega(t) \leqslant \left| \int_{\tau}^{t} \alpha(s)\omega(s)\,ds \right| + b \tag{1}$$

is fulfilled, then

$$\omega(t) \leqslant b e^{\left| \int_{\tau}^{t} \alpha(s)\,ds \right|} \tag{2}$$

for all $t \in \Delta$.

Proof. We begin with $t \geqslant \tau$. By the hypothesis,

$$N = \int_{\Delta} \alpha(s)\omega(s)\,ds < \infty$$

and, according to (1),

$$\omega(t) \leqslant N + b.$$

By induction, we find that (1) implies the inequalities

$$\omega(t) \leqslant b + b\int_{\tau}^{t} \alpha(s)\,ds + \ldots + \frac{b}{(m-1)!}\left[\int_{\tau}^{t} \alpha(s)\,ds\right]^{m-1} + \frac{b+N}{m!}\left[\int_{\tau}^{t} \alpha(s)\,ds\right]^{m}. \tag{3}$$

Indeed, (3) holds for m = 0, and if (3) holds for some m, then substituting (3) into (1) and using the equality

$$\int_\tau \alpha(s)\left[\int_\tau^s \alpha(\sigma)\,d\sigma\right]^k ds = \left.\frac{\left[\int_\tau^s \alpha(\sigma)\,d\sigma\right]^{k+1}}{k+1}\right|_{s=\tau}^{s=t} = \frac{\left[\int_\tau^t \alpha(\sigma)\,d\sigma\right]^{k+1}}{k+1},$$

we conclude that (3) holds for m + 1. Passing to the limit in (3) for $m \to \infty$, we obtain (1).

For $t \leqslant \tau$ the argument is quite similar: it is only necessary to interchange the limits of integration everywhere. [It should be noted that the function $\alpha(\cdot)$ is not necessarily integrable, and therefore the right-hand side of (2) may be infinite.] ■

Uniqueness Theorem. Let function $F: G \to \mathbf{R}^n$ satisfy conditions A)-C) of Section 2.5.1 on the open set $G \subset \mathbf{R} \times \mathbf{R}^n$; let Δ_i, i = 1, 2, be intervals belonging to $\mathbf{R}$, and let $x_i(\cdot): \Delta_i \to \mathbf{R}^n$ be two solutions of Cauchy's problem (2) of Section 2.5.2 whose graphs are contained in G. Then $x_1(t) \equiv x_2(t)$ for all $t \in \Delta_1 \cap \Delta_2$.

Proof. The set $\Delta = \{t \mid x_1(t) = x_2(t)\}$ is obviously closed in $\Delta_1 \cap \Delta_2$ and nonempty because $t_0 \in \Delta$. It remains to verify that it is open, and the assertion of the theorem will follow from the connectedness of the interval $\Delta_1 \cap \Delta_2$.

Let $\hat{t} \in \Delta$; then $x_1(\hat{t}) = x_2(\hat{t}) = \hat{x}$. Let us choose $\gamma > 0$ and $\beta > 0$ such that $\mathscr{K} = \{(t, x) \mid |t - \hat{t}| \leqslant \gamma, |x - \hat{x}| \leqslant \beta\} \subset G$ and $(t, x_i(t)) \in \mathscr{K}$ for $|t - \hat{t}| \leqslant \gamma$; let $k(t)$ be the function corresponding to $\mathscr{K}$ by virtue of the condition C) of Section 2.5.1.

Applying the mean value theorem (Section 2.2.3) and inequality (1) of Section 2.5.1, we find that

$$|x_1(t) - x_2(t)| \leqslant \left|\int_{\hat{t}}^t [F(s, x_1(s)) - F(s, x_2(s))]\,ds\right| \leqslant \left|\int_{\hat{t}}^t k(s)\,|x_1(s) - x_2(s)|\,ds\right|.$$

Now if the lemma we have just proved $[\alpha(t) = k(t),\ b = 0]$ is applied to the function $\omega(t) = |x_1(t) - x_2(t)|$ on $[\hat{t} - \gamma, \hat{t} + \gamma]$, then, by virtue of (2), we obtain $x_1(t) \equiv x_2(t)$.

Consequently $(\hat{t} - \gamma, \hat{t} + \gamma) \subset \Delta$, and Δ is open. ■

2.5.4. Linear Differential Equations. In this section Δ is a closed interval on the real line and

$$A: \Delta \to \mathscr{L}(\mathbf{R}^n, \mathbf{R}^n),\ b: \Delta \to \mathbf{R}^n \text{ and } c: \Delta \to \mathbf{R}^{n*}$$

are an integrable matrix function and two integrable vector functions, $x = (x_1, \ldots, x_n) \in \mathbf{R}^n$ and $p = (p_1, \ldots, p_n) \in \mathbf{R}^{n*}$. We shall consider linear systems of differential equations

$$\dot{x} = A(t)x + b(t) \tag{1}$$

and

$$\dot{p} = -pA(t) + c(t) \tag{2}$$

[(2) is sometimes called the system *adjoint to the system* (1)]. As is known, for linear systems it is possible to prove a nonlocal existence theorem for the solutions (for the whole closed interval Δ in the case under consideration).

LEMMA. If the functions $A: \Delta \to \mathscr{L}(\mathbf{R}^n, \mathbf{R}^n)$, $b: \Delta \to \mathbf{R}^n$, and $c: \Delta \to \mathbf{R}^{n*}$ are measurable and integrable on the closed interval Δ, then any Cauchy's problem for Eq. (1) or Eq. (2) posed for an initial instant $t_0 \in \Delta$ has a unique solution which can be extended to the whole closed interval Δ.

Proof. We shall confine ourselves to Eq. (1) since for Eq. (2) the whole argument is quite analogous. For the sake of convenience, let us extend the functions $A(\cdot)$, $b(\cdot)$, and $c(\cdot)$ to the whole real line **R** assuming that they are equal to zero outside Δ. Then the function

$$F(t, x) = A(t)x + b(t)$$

will satisfy the conditions A)-C) of Section 2.5.1 in the domain $G = \mathbf{R} \times \mathbf{R}^n$, and therefore the uniqueness theorem of Section 2.5.3 and the local existence theorem of Section 2.5.2 will be applicable to the differential equation (1) at each point $(\hat{t}, \hat{x})$. Hence the solution of Eq. (1) with $x(\hat{t}) = \hat{x}$ is sure to be unique and defined on a closed interval $[\hat{t} - \delta, \hat{t} + \delta]$.

Now we note that if $|x| \leqslant M$, then

$$|F(t, \hat{x})| \leqslant \|A(t)\| M + |b(t)|$$

and

$$\|F_x(t, \hat{x})\| = \|A(t)\|.$$

Coming back to the proof of the theorem of Section 2.5.2, we can choose the constants γ, β, θ, and ε sufficiently arbitrarily by putting, for instance, $\theta = 1/2$, $\beta = 4$, $\varepsilon = \gamma = 1$, and then inequalities (7) and (8) of Section 2.5.2 specifying δ take the form

$$\begin{gathered} M \int_{\hat{t}-\delta}^{\hat{t}+\delta} \|A(s)\| ds + \int_{\hat{t}-\delta}^{\hat{t}+\delta} |b(s)| ds < 1, \\ \int_{\hat{t}-\delta}^{\hat{t}+\delta} \|A(s)\| ds \leqslant 1/2. \end{gathered} \tag{3}$$

Since $\|A(\cdot)\|$ and $|b(\cdot)|$ are integrable, we can choose a sufficiently small $\delta > 0$ (not exceeding $\gamma = 1$) such that inequalities (3) are fulfilled for all t [KF, p. 301]. Consequently, the solution of Eq. (1) with $x(\hat{t}) = \hat{x}$ is defined on a closed interval $[\hat{t} - \delta, \hat{t} + \delta]$, where δ is guaranteed to be one and the same for any $\hat{t}$ and any $\hat{x}$ such that $|\hat{x}| \leqslant M$.

Now let $t_0 \in \Delta$ and x_0 be given, and let a solution of Eq. (1) be sought which satisfies the condition

$$x(t_0) = x_0. \tag{4}$$

Let us put

$$M = \left(|x_0| + \int_\Delta |b(s)| ds\right) e^{\int_\Delta \|A(s)\| ds}. \tag{5}$$

Cauchy's problem (1), (4) is equivalent to the integral equation

$$x(t) = x_0 + \int_{t_0}^{t} A(s) x(s) ds + \int_{t_0}^{t} b(s) ds,$$

whence

$$|x(t)| \leqslant |x_0| + \left|\int_{t_0}^{t} A(s)x(s) ds\right| + \left|\int_{t_0}^{t} b(s) ds\right| \leqslant \left(|x_0| + \int_\Delta |b(s)| ds\right) + \left|\int_{t_0}^{t} \|A(s)\| |x(s)| ds\right|. \tag{6}$$

The solution $x(\cdot)$ of problem (1), (4) is sure to be defined on the closed interval $\Delta_1 = [t_0 - \delta, t_0 + \delta]$. Applying the lemma of Section 2.5.3 to the function $\omega(t) = |x(t)|$, $\left(\alpha(t) = \|A(t)\|, b = |x_0| + \int_\Delta |b(s)| ds\right)$ on that closed interval, we obtain the inequality

$$|x(t)| \leqslant \left(|x_0| + \int_{\Delta} |b(s)|\, ds\right) e^{\left|\int_{t_0}^{t} \|A(s)\| ds\right|} \tag{7}$$

and, in particular, according to (5), we have $|x(t_0 \pm \delta)| \leqslant M$. Therefore, the local existence theorem can again be applied for the points $(t_0 \pm \delta, x(t_0 \pm \delta))$, and the solution $x(\cdot)$ can be extended to the closed interval $\Delta_2 = [t_0 - 2\delta, t_0 + 2\delta]$.

Further this procedure is repeated. Inequality (6) holds for Δ_2, and hence (7) holds for Δ_2, and we have $|x(t_0 \pm 2\delta)| \leqslant M$, etc. After a finite number of steps we extend the solution of Cauchy's problem (1), (4) to the whole closed interval Δ. ■

The explicit formulas for the solutions of systems (1) and (2) are expressed in terms of the principle matrix solution of the homogeneous system

$$\dot{x} = A(t)x. \tag{8}$$

Definition. By the *principal matrix solution* $\Omega(t, \tau)$ of the system (8) is meant the matrix function $\Omega: \Delta \times \Delta \to \mathscr{L}(\mathbf{R}^n, \mathbf{R}^n)$, which is the solution of Cauchy's problem

$$\frac{\partial \Omega(t, \tau)}{\partial t} = A(t)\Omega(t, \tau), \tag{9}$$

$$\Omega(\tau, \tau) = I. \tag{10}$$

In other words, each of the columns of the matrix $\Omega(t, \tau)$ is a solution of system (8), and for $t = \tau$ these columns turn into the set of the unit vectors forming the standard basis in $\mathbf{R}^n$: $e_1 = (1, 0, \ldots, 0)$, $e_2 = (0, 1, \ldots, 0), \ldots, e_n = (0, 0, \ldots, 1)$.

THEOREM. If the matrix function $A(\cdot)$ is integrable on the closed interval Δ, then the principal matrix solution $\Omega(t, \tau)$ of system (8) exists and is continuous on the square $\Delta \times \Delta$, and, moreover:

1) $\Omega(t, s)\Omega(s, \tau) \equiv \Omega(t, \tau)$ for all $t, s, \tau \in \Delta$; (11)

2) $\Omega(t, \tau)$ is a solution of the differential equation

$$\frac{\partial \Omega(t, \tau)}{\partial \tau} = -\Omega(t, \tau)A(\tau) \tag{12}$$

for each t.

Moreover:

3) If the function $b: \Delta \to \mathbf{R}^n$ is integrable on the closed interval and $x(\cdot)$ is a solution of system (1), then

$$x(t) = \Omega(t, \tau)x(\tau) + \int_{\tau}^{t} \Omega(t, s)b(s)\, ds \tag{13}$$

for any $t, \tau \in \Delta$.

4) If function $c: \Delta \to \mathbf{R}^{n*}$ is integrable on the closed interval Δ and $p(\cdot)$ is a solution of system (1), then

$$p(t) = p(\tau)\Omega(\tau, t) - \int_{t}^{\tau} c(s)\Omega(s, t)\, ds \tag{14}$$

for any $t, \tau \in \Delta$.

Proof. Performing the direct substitution and using (9), we readily verify that the functions

$$x_1(t)=\Omega(t,s)\Omega(s,\tau)\xi \text{ and } x_2(t)=\Omega(t,\tau)\xi$$

are solutions of system (8) for any $\xi\in\mathbf{R}^n$. By virtue of (10),

$$x_1(s)=\Omega(s,s)\Omega(s,\tau)\xi=\Omega(s,\tau)\xi=x_2(s),$$

and therefore the uniqueness theorem implies

$$\Omega(t,s)\Omega(s,\tau)\xi=x_1(t)\equiv x_2(t)=\Omega(t,\tau)\xi.$$

Since ξ is arbitrary, (11) must hold. Putting $t = \tau$ in (11), we obtain $\Omega(t, s)\Omega(s, t) = I$, whence

$$\Omega(t,s)\equiv[\Omega(s,t)]^{-1}.$$

Fixing s in (11), we represent the function of two variables $\Omega(t, \tau)$ in the form of a product of two functions of one variable, and since these functions are both continuous, the function Ω is continuous with respect to (t, τ) on $\Delta \times \Delta$. In particular, it is bounded. According to Proposition 3 of Section 2.2.1, the matrix function $f(A) = A^{-1}$ is continuously differentiable on the set of invertible matrices and

$$f'(A)[H]=-A^{-1}HA^{-1}. \tag{15}$$

The set $\mathcal{K}=\{A\,|\,A=\Omega(s,t),\ s,t\in\Delta\}$ is the image of the compact set $\Delta \times \Delta$ under the continuous mapping $\Omega\colon \Delta\times\Delta\to\mathscr{L}(\mathbf{R}^n,\mathbf{R}^n)$, and therefore the set $\mathcal{K}$ is compact, and the function f is bounded on it ($\|f\| \leqslant M$) and satisfies the Lipschitz condition since

$$\|f(A)-f(B)\|=\|B^{-1}-A^{-1}\|=\|-B^{-1}(B-A)A^{-1}\|\leqslant M^2\|B-A\|.$$

According to Proposition 1 of Section 2.1.8, the function

$$\varphi(s)=\Omega(t,s)=\Omega(s,t)^{-1}=f(\Omega(s,t))$$

is absolutely continuous, and, by virtue of (15) and the composition theorem of Section 2.2.2, the relation

$$\frac{\partial\Omega(t,s)}{\partial s}=-[\Omega(s,t)]^{-1}\frac{\partial\Omega(s,t)}{\partial s}[\Omega(s,t)]^{-1}=$$
$$=-[\Omega(s,t)]^{-1}A(s)\Omega(s,t)[\Omega(s,t)]^{-1}=-\Omega(t,s)A(s)$$

holds almost everywhere, which proves (12).

Representing the right-hand side of (13) in the form

$$\Omega(t,\tau)\left[x(\tau)+\int_\tau^t\Omega(\tau,s)b(s)\,ds\right], \tag{16}$$

we see that it is absolutely continuous. Indeed, $\Omega(\tau, s)b(s)$ is integrable (as a function of s) on Δ, and the integral of an integrable function is absolutely continuous (Section 2.1.8). Consequently, the expression in the square brackets in (16) is absolutely continuous with respect to t. It remains to note that the product of two absolutely continuous functions is also absolutely continuous (Proposition 1 of Section 2.1.8). Taking into account (9) and (11) and differentiating (16), we see that, like $x(\cdot)$, the right-hand side of (13) is a solution of Eq. (1). Since these solutions of Eq. (1) coincide for $t = \tau$, the uniqueness theorem implies that they coincide for all $t\in\Delta$.

Equality (14) is proved in a similar way. ■

2.5.5. Global Existence and Continuity Theorem. Let us come back to Cauchy's problem

$$\dot{x}=F(t,x), \tag{1}$$
$$x(t_0)=x_0 \tag{2}$$

for differential equations whose right-hand sides satisfy the conditions A)-C) of Section 2.5.1. As it is done in courses of differential equations, we can use the local existence theorem and the uniqueness theorem to extend the solution $X(\cdot, t_0, x_0)$ of Cauchy's problem (1), (2) to the maximum interval (a, b) of its existence. From the theorem of Section 2.5.2 it follows that for $t \downarrow a$ and $t \uparrow b$ this solution must leave any compact set $\mathcal{K}\subset G$ because the possibility of extending the solution to an interval $(t - \delta, t + \delta)$ is guaranteed as long as $(t, x(t))\in\mathcal{K}$, whence $a + \delta < t < b + \delta$ since (a, b) is the maximum existence interval. The next theorem shows in what sense $X(\cdot, t_0, x_0)$ depends continuously on (t_0, x_0).

THEOREM. Let the function $F:G \to \mathbf{R}^n$ satisfy conditions A)-C) of Section 2.5.1 on the open set $G\subset\mathbf{R}\times\mathbf{R}^n$, and let $\hat{x}:[\hat{t}_0, \hat{t}_1] \to \mathbf{R}^n$ be a solution of Eq. (1) whose graph is contained in G: $\Gamma=\{(t, \hat{x}(t))|\hat{t}_0\leqslant t\leqslant\hat{t}_1\}\subset G$.

Then there exist $\delta > 0$ and a neighborhood $\hat{G}\supset\Gamma$ such that for any $(t_0, x_0)\in\hat{G}$ the solution $X(\cdot, t_0, x_0)$ of Cauchy's problem (1), (2) is defined on the closed interval $\Delta = [\hat{t}_0 - \hat{\delta}, \hat{t}_1 + \hat{\delta}]$, and the function $(t, t_0, x_0) \to X(t, t_0, x_0)$ is jointly continuous with respect to its arguments on $\Delta \times \hat{G}$. In particular, $X(t, t_0, x_0) \to \hat{x}(t)$ for $t_0 \to \tau$, $x_0 \to \hat{x}(\tau)$ uniformly with respect to $t\in\Delta$.

Proof. A) Applying the local existence theorem to the points $(\hat{t}_0, \hat{x}(\hat{t}_0))$ and $(\hat{t}_1, \hat{x}(\hat{t}_1))$ we extend the solution $\hat{x}(\cdot)$ to a closed interval $\Delta = [\hat{t}_0 - \hat{\delta}, \hat{t}_1 + \hat{\delta}]$ such that the graph of $\hat{x}(\cdot)$ remains inside G. After that we choose $\hat{\varepsilon}$ so that

$$\mathcal{K}=\{(t, x)|\,|x-\hat{x}(t)|\leqslant\hat{\varepsilon},\ t\in\Delta\}\subset G.$$

Applying the theorem of Section 2.5.2 to the compact set $\mathcal{K}$, we find $\delta > 0$ such that for $(t_0, x_0)\in\mathcal{K}$ the solution $X(\cdot, t_0, x_0)$ is defined on $[t_0 - \delta, t_0 + \delta]$. Finally, let $k(t)$ and $\varkappa(\cdot)$ be the locally integrable functions corresponding to $\mathcal{K}$ according to condition C) and Lemma 1 of Section 2.5.1.

Now we shall consider the domain

$$\hat{G}=\{(t, x)|\,t\in(\hat{t}_0-\hat{\delta}, \hat{t}_1+\hat{\delta}),\ |x-\hat{x}(t)|<\varepsilon\}$$

and show that for a sufficiently small $\varepsilon > 0$ this domain possesses the properties indicated in the statement of the theorem.

B) Let two solutions $x_i(\cdot)$, $i = 1, 2$, of Eq. (1) be defined on a closed interval $[\beta, \gamma]\subset\Delta$ such that $(t, x_i(t))\in\mathcal{K}$ for $\beta \leqslant t \leqslant \gamma$, and let $\tau \in [\beta, \gamma]$. Applying the mean value theorem (Section 2.2.3) and inequality (1) of Section 2.5.1, we obtain

$$\begin{aligned}|x_1(t)-x_2(t)|&=\left|x_1(\tau)+\int_\tau^t F(s, x_1(s))\,ds-x_2(\tau)-\int_\tau^t F(s, x_2(s))\,ds\right|\leqslant\\&\leqslant|x_1(\tau)-x_2(\tau)|+\left|\int_\tau^t [F(s, x_1(s))-F(s, x_2(s))]\,ds\right|\leqslant\\&\leqslant|x_1(\tau)-x_2(\tau)|+\left|\int_\tau^t k(s)|x_1(s)-x_2(s)|\,ds\right|.\end{aligned}\tag{3}$$

By the lemma of Section 2.5.3 $[\alpha(s)=k(s),\ \omega(s)=|x_1(s)-x_2(s)|;\ b=|x_1(\tau)-x_2(\tau)|]$, we have

$$|x_1(t)-x_2(t)|\leqslant|x_1(\tau)-x_2(\tau)|e^{\left|\int_\tau^t k(s)\,ds\right|},\quad \beta\leqslant t\leqslant\gamma.\tag{4}$$

C) Let us choose $\varepsilon > 0$ such that

$$\varepsilon e^{\int_\Delta k(s)\,ds} \leqslant \hat{\varepsilon}, \tag{5}$$

and let $(t_0, x_0) \in \hat{G}$. Since $\hat{G} \subset \mathcal{K}^\circ$, the solution $x(\cdot) = X(\cdot, t_0, x_0)$ is defined on $[\beta, \gamma] = [t_0 - \delta, t_0 + \delta] \cap \Delta$. Let us show that it remains in $\mathcal{K}$ on this closed interval. Indeed, let

$$T = \sup\{t \mid t_0 \leqslant t \leqslant \min\{t_0 + \delta, \hat{t}_1 + \delta\}; \quad (s, x(s)) \in \mathcal{K}^\circ, \ \forall s \in [t_0, t]\}. \tag{6}$$

The argument of Section B) of the proof is applicable to the solutions $x(\cdot)$ and $\hat{x}(\cdot)$ on the closed interval $[t_0, T]$, and, by virtue of (4), we have

$$|x(T) - \hat{x}(T)| \leqslant |x_0 - \hat{x}(t_0)|\, e^{\int_{t_0}^{T} k(s)\,ds} < \varepsilon e^{\int_\Delta k(s)\,ds} \leqslant \hat{\varepsilon}.$$

Therefore $(T, x(T)) \in \operatorname{int} \mathcal{K}^\circ$, and the point $(t, x(t))$ remains in $\mathcal{K}^\circ$ for $t > T$ close to T as well, which contradicts the definition (6). For $t \leqslant t_0$ the argument is quite similar.

Since $(\beta, x(\beta)) \in \mathcal{K}^\circ$ and $(\gamma, x(\gamma)) \in \mathcal{K}^\circ$, the solution $x(\cdot)$ is defined on the closed interval $[t_0 - 2\delta, t_0 + 2\delta] \cap \Delta$. Applying the same argument to this closed interval, we see that $(t, x(t))$ remains in $\mathcal{K}^\circ$. Continuing this procedure we conclude that $x(\cdot)$ is defined on Δ.

D) As has been proved, any solution $X(\cdot, t_0, x_0)$ remains in $\mathcal{K}^\circ$ if $(t_0, x_0) \in \hat{G}$. Using function $\varkappa(\cdot)$, we write the inequality

$$|X(t_1, t_0, x_0) - X(t, t_0, x_0)| = \left|\int_t^{t_1} F(s, X(s, t_0, x_0))\,ds\right| \leqslant \left|\int_t^{t_1} \varkappa(s)\,ds\right|. \tag{7}$$

Now let us take two solutions for which $(t_0, x_0) \in \hat{G}$ and $(\tilde{t}_0, \tilde{x}_0) \in \hat{G}$. Then, by virtue of (4) and (7),

$$\begin{aligned}
&|X(\tilde{t}, \tilde{t}_0, \tilde{x}_0) - X(t, t_0, x_0)| \leqslant \\
&\leqslant |X(\tilde{t}, \tilde{t}_0, \tilde{x}_0) - X(t, \tilde{t}_0, \tilde{x}_0)| + |X(t, \tilde{t}_0, \tilde{x}_0) - X(t, t_0, x_0)| \leqslant \\
&\leqslant \left|\int_t^{\tilde{t}} \varkappa(s)\,ds\right| + |\tilde{x}_0 - X(\tilde{t}_0, t_0, x_0)|\, e^{\left|\int_{\tilde{t}_0}^{t} k(s)\,ds\right|} \leqslant \\
&\leqslant \left|\int_t^{\tilde{t}} \varkappa(s)\,ds\right| + \left\{|\tilde{x}_0 - x_0| + \left|\int_{t_0}^{\tilde{t}_0} \varkappa(s)\,ds\right|\right\} e^{\int_\Delta k(s)\,ds}
\end{aligned} \tag{8}$$

It is evident that the right-hand member of (8) tends to zero for $\tilde{t} \to t$, $\tilde{t}_0 \to t_0$, $\tilde{x}_0 \to x_0$, which proves the continuity of $X(t, t_0, x_0)$. Putting $\tilde{t} = t$, $\tilde{t}_0 = \tau$, $\tilde{x}_0 = \hat{x}(\tau)$ in (8) and taking into consideration that, according to the uniqueness theorem, the identity $X(t, \tau, \hat{x}(\tau)) \equiv \hat{x}(t)$ holds, we obtain

$$|\hat{x}(t) - X(t, t_0, x_0)| \leqslant \left\{|\hat{x}(\tau) - x_0| + \int_{t_0}^{\tau} \varkappa(s)\,ds\right\} e^{\int_\Delta k(s)\,ds},$$

whence we see that $X(t, t_0, x_0) \to \hat{x}(t)$ uniformly for $x_0 \to \hat{x}(\tau)$, $t_0 \to \tau$. ■

Now we pass to the continuity of solutions with respect to parameters. Let us suppose that we are given a family of differential equations

$$\dot{x} = F_\alpha(t, x) \tag{9}$$

dependent on a parameter $\alpha \in \mathfrak{A}$, where $\mathfrak{A}$ is a topological space.

THEOREM 2. Let the functions of the family $\{F_\alpha\colon G\to\mathbf{R}^n \mid \alpha\in\mathfrak{A}\}$ satisfy conditions A) and B) of Section 2.5.1 on the open set $G\subset\mathbf{R}\times\mathbf{R}^n$ for each $\alpha\in\mathfrak{A}$: let $\hat{x}\colon[\hat{t}_0, \hat{t}_1]\to\mathbf{R}^n$ be a solution of Eq. (6) for $\alpha=\hat{\alpha}$, and let its graph be contained in G: $\Gamma=\{(t, \hat{x}(t)) \mid \hat{t}_0\leqslant t\leqslant\hat{t}_1\}\subset G$; also let the following conditions hold:

C') For any compact set $\mathcal{K}\subset G$ there exist locally integrable functions $\varkappa(\cdot)$ and $k(\cdot)$ such that for any $(t, x)\in\mathcal{K}$, $\alpha\in\mathfrak{A}$ the inequalities

$$|F_\alpha(t, x)|\leqslant\varkappa(t),\quad \|F_{\alpha x}(t, x)\|\leqslant k(t) \tag{10}$$

hold.

D)

$$\lim_{\alpha\to\hat{\alpha}}\int_{\hat{t}_0}^{\hat{t}_1}|F_\alpha(s, \hat{x}(s))-F_{\hat{\alpha}}(s, \hat{x}(s))|\,ds=0. \tag{11}$$

Then there exist $\hat{\delta}>0$, a neighborhood $\hat{G}\supset\Gamma$ in G and a neighborhood $U\ni\hat{\alpha}$ in $\mathfrak{A}$ such that for $(t_0, x_0)\in\hat{G}$, $\alpha\in U$ the solution $x_\alpha(t, t_0, x_0)$ of Cauchy's problem (9), (2) is defined on the closed interval $\Delta=[\hat{t}_0-\hat{\delta}, \hat{t}_1+\hat{\delta}]$, and

$$X_\alpha(t, t_0, x_0)\longrightarrow\hat{x}(t)$$

for $t_0\to\tau\in[\hat{t}_0, \hat{t}_1]$, $x_0\to\hat{x}(\tau)$, $\alpha\to\hat{\alpha}$ uniformly with respect to $t\in[\hat{t}_0, \hat{t}_1]$.

Proof. With a few changes the proof of the theorem is carried out according to the same scheme as the proof of Theorem 1. Namely, in Section A) we take into account that the number δ in the local existence theorem is specified by inequalities (7) and (8) of Section 2.5.2, and, by virtue of the condition C'), we can take one and the same δ for all $\alpha\in\mathfrak{A}$.

In Section B) we replace $x_1(\cdot)$ by the solution $x_\alpha(\cdot)$ of Eq. (9) and $x_2(\cdot)$ by the solution $x(\cdot)$. When estimating the difference between the solutions, we must take into account that the right-hand sides of the equations are different now and therefore

$$|x_\alpha(t)-\hat{x}(t)|\leqslant|x_\alpha(\tau)-\hat{x}(\tau)|+\left|\int_\tau^t\{F_\alpha(s, \hat{x}(s))-F_{\hat{\alpha}}(s, \hat{x}(s))\}\,ds\right|+$$
$$+\left|\int_\tau^t k(s)|x_\alpha(s)-\hat{x}(s)|\,ds\right|,$$

and (4) is replaced by the inequality

$$|x_\alpha(t)-\hat{x}(t)|\leqslant\left(|x_\alpha(\tau)-\hat{x}(\tau)|+\left|\int_\tau^t\{F_\alpha(s, \hat{x}(s))-F_{\hat{\alpha}}(s, \hat{x}(s))\}\,ds\right|\right)e^{\left|\int_\tau^t k(s)\,ds\right|}. \tag{4'}$$

In inequality (5) the number ε should be replaced by 2ε, and, according to the conditions C') and D), the number $\hat{\delta}$ and the neighborhood U should be chosen so that the inequalities

$$\left|\int_\tau^t\{F_\alpha(s, \hat{x}(s))-F_{\hat{\alpha}}(s, \hat{x}(s))\}\,ds\right|\leqslant\int_{\hat{t}_0-\hat{\delta}}^{\hat{t}_1+\hat{\delta}}|F_\alpha(s, \hat{x}(s)-F_{\hat{\alpha}}(s, \hat{x}(s))|\,ds\leqslant$$
$$\leqslant2\left(\int_{\hat{t}_0-\hat{\delta}}^{\hat{t}_0}+\int_{\hat{t}_1}^{\hat{t}_1+\hat{\delta}}\varkappa(s)\,ds\right)+\int_{\hat{t}_0}^{\hat{t}_1}|F_\alpha(s, \hat{x}(s))-F_{\hat{\alpha}}(s, \hat{x}(s))|\,ds<\varepsilon \tag{12}$$

hold. Then the argument of Section C) [with (4) replaced by (4')] remains valid for all $\alpha\in U$ simultaneously.

In inequality (8) of Section D) the correction corresponding to the replacement of (4) by (4') should be introduced, after which we obtain

$$|X_\alpha(t, t_0, x_0) - \hat{x}(t)| \leqslant \left\{ |x_0 - \hat{x}(\tau)| + \left| \int_{t_0}^{\tau} \varkappa(s)\, ds \right| + \right.$$

$$\left. + \int_{\hat{t}_0}^{\hat{t}_1} |F_\alpha(s, \hat{x}(s)) - F_{\hat{\alpha}}(s, \hat{x}(s))|\, ds \right\} e^{\int_{\hat{t}_0}^{\hat{t}_1} k(s)\, ds},$$

and the rest is obvious. ■

COROLLARY. Let all the conditions of Theorem 2 be fulfilled, the condition D) being taken in the strengthened form:

D') for any function $\tilde{x}:[\tilde{t}_0, \tilde{t}_1] \to \mathbf{R}^n$ whose graph is contained in G and any $\tilde{\alpha} \in \mathfrak{A}$ the relation

$$\lim_{\alpha \to \tilde{\alpha}} \int_{\tilde{t}_0}^{\tilde{t}_1} |F_\alpha(s, \tilde{x}(s)) - F_{\tilde{\alpha}}(s, \tilde{x}(s))|\, ds = 0$$

holds. Then the solution $x_\alpha(t, t_0, x_0)$ of Cauchy's problem (9), (2) is a continuous function of (t, t_0, x_0, α) on $\Delta \times \hat{G} \times U$.

Proof. The graph of the solution $\tilde{x}(\cdot) = X_{\tilde{\alpha}}(\cdot, \tilde{t}_0, \tilde{x}_0):\Delta \to \mathbf{R}^n$ is contained in G for any $(\tilde{t}_0, \tilde{x}_0) \in \hat{G}$ and $\tilde{\alpha} \in U$. Now we can apply Theorem 2 after replacing $\hat{x}(\cdot)$ by $\tilde{x}(\cdot)$ in the conditions of this theorem. ■

The following identity is an analogue of the basic property of the principal matrix solution [formula (11) of the foregoing section]:

$$X(t, \tau, X(\tau, t_0, x_0)) \equiv X(t, t_0, x_0). \tag{13}$$

This identity follows automatically from the uniqueness theorem because each of its members is a solution of Eq. (1), and for $t = \tau$ these solutions coincide. Let the reader analyze the connection between identity (13) and Huygens' principle stated in Section 1.1.3.

2.5.6. Theorem on the Differentiability of Solutions with Respect to the Initial Data. Up till now the presentation of the material has been based on the comparatively weak conditions A)-C) of Section 2.5.1. In order to carry out further analysis and to establish the differentiability of the solution $X(t, t_0, x_0)$ of Cauchy's problem with respect to the initial value x_0, these conditions should be strengthened to a certain extent. Then it becomes possible to apply the classical implicit function theorem. In the proof of the theorem of this section we shall not dwell on the existence of the solution $X(t, t_0, x_0)$ for $(t_0, x_0) \in \hat{G}$ and its extension to all $t \in \Delta$ since both the facts follow from Theorem 1 of Section 2.5.5. Nevertheless, we recommend the reader to show that these facts also follow from the implicit function theorem, i.e., it is possible to do without the reference to the foregoing section. (An analogous argument will be presented in full in Section 2.5.7, where the differentiability with respect to t_0 will be proved under still stronger assumptions and a somewhat different version of the implicit function theorem will be applied.)

THEOREM. Let function $F:G \to \mathbf{R}^n$ satisfy conditions A) and C) of Section 2.5.1 on the open set $G \subset \mathbf{R} \times \mathbf{R}^n$, and let the following condition hold:

B') for any t the function $x \to F(t, x)$ is continuously differentiable on the section $G_t = \{x \mid (t, x) \in G\}$.

Let $\hat{x}:[\hat{t}_0, \hat{t}_1] \to \mathbf{R}^n$ be a solution of Eq. (1) of Section 2.5.5 whose graph $\Gamma = \{(t, \hat{x}(t)) \mid \hat{t}_0 \leqslant t \leqslant \hat{t}_1\}$ is contained in G, and let δ, ε, and $\hat{G}$ be the same as in Theorem 1 of Section 2.5.5; let $\Delta = [\hat{t}_0 - \hat{\delta}, \hat{t}_1 + \delta]$.

Then for any fixed $t_0 \in \Delta$ the mapping $x_0 \to X(\cdot, t_0, x_0)$ from $\mathring{B}(\hat{x}(t_0), \varepsilon)$ into $C(\Delta, \mathbf{R}^n)$ is continuously differentiable in Fréchet's sense, and, moreover,

$$\frac{\partial X(t, t_0, x_0)}{\partial x_0} = \Omega(t, t_0), \tag{1}$$

where Ω is the principal matrix solution of the variational equation

$$\dot{z} = F_x(t, X(t, t_0, x_0))z. \tag{2}$$

Proof. Let the compact set $\mathcal{K}$ and the function $k(t)$ be the same as in Theorem 1 of Section 2.5.5. The set

$$\mathcal{G} = \{x(\cdot) \in C(\Delta, \mathbf{R}^n) | (t, x(t)) \in G, \ t \in \Delta\}$$

is obviously open in $C(\Delta, \mathbf{R}^n)$. Let us define a mapping $\mathcal{F}: \mathbf{R}^n \times \mathcal{G} \to C(\Delta, \mathbf{R}^n)$ with the aid of the equality

$$\mathcal{F}(x_0, x(\cdot))(t) = x_0 + \int_{t_0}^{t} F(s, x(s))\,ds - x(t). \tag{3}$$

It is evident that

$$\mathcal{F}(x_0, x(\cdot)) = 0 \Leftrightarrow x(t) \equiv X(t, t_0, x_0). \tag{4}$$

Let $\bar{x}_0 \in \mathring{B}(\hat{x}(t_0), \varepsilon)$ be fixed, and let $\bar{x}(t) \equiv X(t, t_0, \bar{x}_0)$. We shall verify that the classical implicit function theorem (Section 2.3.4) is applicable in the neighborhood of the point $(\bar{x}_0, \bar{x}(\cdot))$. Indeed, the derivative $\mathcal{F}_{x_0}(x_0, x(\cdot))[\xi_0] = \xi_0$ is obviously continuous. Now we compute the Gateaux derivative:

$$\mathcal{F}_{x(\cdot)}(x_0, x(\cdot))[\xi(\cdot)](t) = \lim_{\alpha \downarrow 0} \frac{\mathcal{F}(x_0, x(\cdot) + \alpha\xi(\cdot)) - \mathcal{F}(x_0, x(\cdot))}{\alpha}(t) =$$

$$= \lim_{\alpha \downarrow 0} \int_{t_0}^{t} \frac{F(s, x(s) + \alpha\xi(s)) - F(s, x(s))}{\alpha}\,ds - \xi(t) = \int_{t_0}^{t} F_x(s, x(s))\xi(s)\,ds - \xi(t). \tag{5}$$

To prove the existence of the uniform limit with respect to t we note that, according to the mean value theorem (Section 2.2.3),

$$\left| \int_{t_0}^{t} \frac{F(s, x(s)) + \alpha\xi(s)) - F(s, x(s))}{\alpha}\,ds - \int_{t_0}^{t} F_x(s, x(s))\xi(s)\,ds \right| \leqslant \int_{\Delta} r_\alpha(s)\,ds,$$

where

$$r_\alpha(s) = \max_{c \in [x(s), x(s) + \alpha\xi(s)]} \|F_x(s, c) - F_x(s, x(s))\| |\xi(s)|.$$

The condition B') implies that $r_\alpha(s) \to 0$ for any s and $\alpha \downarrow 0$, and, by virtue of the relation (1) of Section 2.5.1, we have $|r_\alpha(s)| \leqslant 2k(s)\|\xi\|$. By Lebesgue's bounded convergence theorem [KF, p. 302], limit (5) exists and is uniform.

The operator

$$\mathcal{F}_{x(\cdot)}(x_0, x(\cdot))[\xi(\cdot)](t) = \int_{t_0}^{t} F_x(s, x(s))\xi(s)\,ds$$

is bounded:

$$\|\mathcal{F}_{x(\cdot)}(x_0, x(\cdot))[\xi(\cdot)]\| \leqslant \max_t \left| \int_{t_0}^{t} \|F_x(s, x(s))\| |\xi(s)|\,ds \right| \leqslant \int_{\Delta} k(s)\,ds \|\xi\|.$$

It depends continuously on $(x_0, x(\cdot))$:

$$\|\mathcal{F}_{x(\cdot)}(\tilde{x}_0, \tilde{x}(\cdot)) - \mathcal{F}_{x(\cdot)}(x_0, x(\cdot))\| =$$

$$= \sup_{\|\xi\| \leqslant 1} \sup_t \left| \int_{t_0}^{t} F_x(s, \tilde{x}(s)) \xi(s) ds - \int_{t_0}^{t} F_x(s, x(s)) \xi(s) ds \right| \leqslant$$
$$\leqslant \int_{\Delta} \| F_x(s, \tilde{x}(s)) - F_x(s, x(s)) \| ds.$$

By virtue of B'), for $\|\tilde{x}(\cdot) - x(\cdot)\| \to 0$ we have

$$\| F_x(s, \tilde{x}(s)) - F_x(s, x(s)) \| \to 0, \quad \forall s \in \Delta.$$

Again using relation (1) of Section 2.5.1, we see that $\|F_x(s, \tilde{x}(s)) - F_x(s, x(s))\| \leqslant 2k(s)$, and therefore Lebesgue's theorem is applicable. Now Corollary 2 of Section 2.2.3 implies that the continuous Fréchet derivative $\mathcal{F}_{x(\cdot)}$ exists.

Let $\eta(\cdot) \in C(\Delta, \mathbf{R}^n)$ be arbitrary. According to (5),

$$\mathcal{F}_{x(\cdot)}(x_0, x(\cdot))[\xi(\cdot)] = \eta(\cdot) \Leftrightarrow \int_{t_0}^{t} F_x(s, x(s)) \xi(s) ds - \xi(t) = \eta(t), \tag{6}$$

and to apply the implicit function theorem we must show that the equation on the right-hand side of (6) is uniquely resolvable. The substitution $\xi(t) + \eta(t) = \zeta(t)$ transforms it into the integral equation

$$\zeta(t) = \int_{t_0}^{t} F_x(s, x(s)) [\zeta(s) - \eta(s)] ds,$$

which in its turn is equivalent to the linear differential equation

$$\dot{\zeta}(t) = F_x(t, x(t)) \zeta(t) - F_x(t, x(t)) \eta(t); \quad \zeta(t_0) = 0.$$

It remains to refer to the lemma of Section 2.5.4. By Banach's theorem (Section 2.1.5), $\mathcal{F}_{x(\cdot)}^{-1}$ is bounded.

By the implicit function theorem, there exists a mapping $\varphi: U \to C(\Delta, \mathbf{R}^n)$ continuously differentiable in a neighborhood $U \ni x_0$ such that $\mathcal{F}(x_0, \varphi(x_0)) \equiv 0$. According to (4), we have $\varphi(x_0)(t) \equiv X(t, t_0, x_0)$, and therefore the mapping $x_0 \to X(t, t_0, x_0)$ is in fact Fréchet differentiable at the point x_0. By virtue of the same implicit function theorem,

$$\frac{\partial X(t, t_0, x_0)}{\partial x_0} \xi_0 = \varphi'(x_0)[\xi_0](t) = -(\mathcal{F}_{x(\cdot)}^{-1} \circ \mathcal{F}_{x_0}[\xi_0])(t) = \xi(t),$$

and, as was already found, $\xi(t)$ is a solution of the integral equation (6), which takes the form

$$\int_{t_0}^{t} F_x(s, x(s)) \xi(s) ds - \xi(t) = -\xi_0.$$

Consequently,

$$\dot{\xi} = F_x(s, x(s)) \xi, \quad \xi(t_0) = \xi_0,$$

and the theorem of Section 2.5.4 implies that $\xi(t) = \Omega(t, t_0)\xi_0$, whence

$$\frac{\partial X}{\partial x_0}(t, t_0, x_0) = \Omega(t, t_0). \blacksquare$$

2.5.7. Classical Theorem on the Differentiability of Solutions with Respect to the Initial Data. In this section we again consider Cauchy's problem

$$\dot{x} = F(t, x), \tag{1}$$
$$x(t_0) = x_0, \tag{2}$$

but this time in the classical situation when the given function $F(t, x)$ and its partial derivative $F_x(t, x)$ are continuous in G. For the sake of convenience, the theorem stated below will be proved independently of the theorems of Sections 2.5.5 and 2.5.6. The references to the local existence theorem (Section 2.5.2) and the uniqueness theorem (Section 2.5.3) can be replaced by the reference to any textbook on differential equations.

THEOREM. Let function $F: G \to \mathbf{R}^n$ and its partial derivative F_x be continuous on the open set $G \subset \mathbf{R} \times \mathbf{R}^n$, and let $\hat{x}: [\hat{t}_0, \hat{t}_1] \to \mathbf{R}^n$ be a solution of Eq. (1) whose graph $\Gamma = \{(t, \hat{x}(t)) \mid \hat{t}_0 \leqslant t \leqslant \hat{t}_1\}$ is contained in G.

Then there exists $\hat{\delta} > 0$ and a neighborhood $\hat{G} \supset \Gamma$ such that for any $(t_0, x_0) \in \hat{G}$ the solution $X(t, t_0, x_0)$ of Cauchy's problem (1), (2) is defined on $\Delta = [\hat{t}_0 - \hat{\delta}, \hat{t}_1 + \hat{\delta}]$ and is a jointly continuously differentiable function with respect to its arguments in the domain $(\hat{t}_0 - \hat{\delta}, \hat{t}_1 + \hat{\delta}) \times \hat{G}$ and, moreover,

$$X(t, \tau, \hat{x}(\tau)) = \hat{x}(\tau), \tag{3}$$

$$\frac{\partial X}{\partial t}(t, \tau, \hat{x}(\tau)) = F(t, \hat{x}(t)), \tag{4}$$

$$\frac{\partial X}{\partial t_0}(t, t_0, \hat{x}(\tau)) \Big|_{t_0 = \tau} = -\Omega(t, \tau) F(\tau, \hat{x}(\tau)), \tag{5}$$

$$\frac{\partial X}{\partial x_0}(t, \tau, x_0) \Big|_{x_0 = \hat{x}(\tau)} = \Omega(t, \tau), \tag{6}$$

where $\Omega(t, \tau)$ is the principal matrix solution of the variational system of equations

$$\dot{z} = F_x(t, \hat{x}(t)) z.$$

Proof. A) Applying the local existence theorem to the points $(\hat{t}_0, \hat{x}(\hat{t}_0))$ and $(\hat{t}_1, \hat{x}(\hat{t}_1))$, we extend $\hat{x}(\cdot)$ to the closed interval $\Delta = [\hat{t}_0 - \hat{\delta}, \hat{t}_1 + \hat{\delta}]$ so that the graph $\{(t, \hat{x}(t)) \mid t \in \Delta\}$ remains inside G. After that we choose $\varepsilon > 0$ such that

$$\mathcal{K}^\circ = \{(t, x) \mid t \in \Delta, |x - \hat{x}(t)| \leqslant \varepsilon\} \subset G.$$

Functions $F(t, x)$ and $F_x(t, x)$ are uniformly continuous on the compact set $\mathcal{K}^\circ$.

B) Now we define a function $\mathcal{F}: \mathcal{G} \to C(\Delta, \mathbf{R}^n) \times \mathbf{R}^n$ in the domain

$$\mathcal{G} = \{(x(\cdot), t_0, x_0) \mid |x(t) - \hat{x}(t)| < \varepsilon, \forall t \in \Delta;\ t_0 \in \operatorname{int} \Delta,\ x_0 \in \mathbf{R}^n\} \subset C^1(\Delta, \mathbf{R}^n) \times \mathbf{R} \times \mathbf{R}^n$$

with the aid of the equality

$$\mathcal{F}(x(\cdot), t_0, x_0)(t) = \begin{pmatrix} \frac{dx(t)}{dt} - F(t, x(t)) \\ x(t_0) - x_0 \end{pmatrix}. \tag{7}$$

It is evident that the equality $\mathcal{F}(x(\cdot), t_0, x_0) = 0$ is equivalent to Cauchy's problem (1), (2). In particular, by the condition of the theorem, $\mathcal{F}(\hat{x}(\cdot), \tau, \hat{x}(\tau)) = 0$.

The function $\mathcal{F}$ is continuously differentiable in the domain $\mathcal{G}$.

Indeed, (7) involves the linear mappings $x(\cdot) \to dx(\cdot)/dt$ and $x_0 \to -x_0$, the operator of boundary conditions $(x(\cdot), t_0) \to x(t_0)$, and Nemytskii's operator $\{x(t)\} \to \{F(t, x(t))\}$, which are all continuously differentiable according to Sections 2.4.1 and 2.4.3, and, moreover,

$$\mathcal{F}_{x(\cdot)}(x(\cdot), t_0, x_0)[\xi(\cdot)] = \begin{pmatrix} \frac{d\xi}{dt} - F_x(t, x(t))\xi \\ \xi(t_0) \end{pmatrix}, \tag{8}$$

$$\mathcal{F}_{t_0}(x(\cdot), t_0, x_0)[\alpha] = \begin{pmatrix} 0 \\ \dot{x}(t_0)\alpha \end{pmatrix}, \tag{9}$$

$$\mathcal{F}_{x_0}(x(\cdot), t_0, x_0)[\xi_0] = \begin{pmatrix} 0 \\ -\xi_0 \end{pmatrix}. \tag{10}$$

C) Let $\hat{x} = \hat{x}(\hat{t})$; then $(\hat{t}, \hat{x}) \in \Gamma$. We shall verify that the operator $\mathcal{F}_{x(\cdot)}(\hat{x}(\cdot), \hat{t}, \hat{x})$ is invertible. The equality

$$\mathcal{F}_{x(\cdot)}(\hat{x}(\cdot), \hat{t}, \hat{x})[\xi(\cdot)] = \binom{\zeta(\cdot)}{\gamma} \in C(\Delta, \mathbf{R}^n) \times \mathbf{R}^n$$

is equivalent to Cauchy's problem

$$\dot{\xi} - F_x(t, \hat{x}(t))\xi = \zeta(t), \quad \xi(\hat{t}) = \gamma. \tag{11}$$

According to the theorem of Section 2.5.4, problem (11) possesses the solution

$$\xi(t) = \Omega(t, \hat{t})\gamma + \int_{\hat{t}}^{t} \Omega(t, s)\zeta(s)\,ds.$$

Thus, $\mathcal{F}_{x(\cdot)}$ maps the Banach space $C^1(\Delta, \mathbf{R}^n)$ onto the whole Banach space $C(\Delta, \mathbf{R}^n) \times \mathbf{R}^n$. Since problem (11) possesses only the zero solution for $\zeta(\cdot) = 0$ and $\gamma = 0$, the operator $\mathcal{F}_{x(\cdot)}$ is one-to-one. By Banach's theorem (Section 2.3.4), the operator $\mathcal{F}_{x(\cdot)}^{-1}$ exists and is bounded.

D) According to the classical implicit function theorem (Section 2.3.4), there exist $\delta > 0$ and a continuously differentiable mapping

$$\Phi\colon \{(t_0, x_0) \mid |t_0 - \hat{t}| < \delta, |x_0 - \hat{x}| < \delta\} \to C^1(\Delta, \mathbf{R}^n)$$

such that

$$\mathcal{F}(\Phi(t_0, x_0), t_0, x_0) \equiv 0.$$

Let us denote

$$X(t, t_0, x_0) = \Phi(t_0, x_0)(t).$$

Then $X(t, t_0, x_0)$ is the solution of Cauchy's problem (1), (2), and we have

$$\frac{\partial X}{\partial t} = F(t, X(t, t_0, x_0)), \tag{12}$$

$$\frac{\partial X}{\partial t_0} = \frac{\partial \Phi}{\partial t_0}(t_0, x_0)(t), \tag{13}$$

$$\frac{\partial X}{\partial x_0} = \frac{\partial \Phi}{\partial x_0}(t_0, x_0)(t). \tag{14}$$

If $(\tilde{t}_0, \tilde{x}_0) \to (t_0, x_0)$, then $\Phi(\tilde{t}_0, \tilde{x}_0) \to \Phi(t_0, x_0)$ in $C^1(\Delta, \mathbf{R}^n)$, i.e., $X(t, t_0, \tilde{x}_0) \to X(t, t_0, x_0)$ uniformly on Δ. It follows that $X(t, t_0, x_0)$ is jointly continuous with respect to its arguments. The same consideration and formulas (12)-(14) show that the derivatives of the function $X(t, t_0, x_0)$ are also continuous.

E) The foregoing argument was applicable to the initial data (t_0, x_0) belonging to the δ-neighborhood of the point $(\hat{t}, \hat{x}) \in \Gamma$. This point was chosen arbitrarily, and the whole graph Γ can be covered with such neighborhoods. If $X^{(1)}(t, t_0, x_0)$ and $X^{(2)}(t, t_0, x_0)$ are defined in two of these neighborhoods having a common point (t_0, x_0), then

$$X^{(1)}(t_0, t_0, x_0) = x_0 = X^{(2)}(t_0, t_0, x_0)$$

and the uniqueness theorem of Section 2.5.3 implies that

$$X^{(1)}(t, t_0, x_0) = X^{(2)}(t, t_0, x_0), \quad \forall t \in \Delta.$$

Consequently, the solution $X(t, t_0, x_0)$ is defined correctly for (t_0, x_0) belonging to a neighborhood of Γ and is obviously continuously differentiable in this neighborhood.

F) It remains to derive formulas (3)-(6). Equality (3) holds according to the uniqueness theorem, and then (4) is a consequence of (12). By the implicit function theorem,

$$\frac{\partial \Phi}{\partial t_0}(t_0, x_0) = -\mathscr{F}_{x(\cdot)}^{-1} \circ \mathscr{F}_{t_0},$$

and therefore (13) and (9) yield

$$\frac{\partial X}{\partial t_0}(t, \tau, \hat{x}(\tau))[\alpha] = -\mathscr{F}_{x(\cdot)}^{-1}(\hat{x}(\cdot), \tau, \hat{x}(\tau))\begin{pmatrix}0\\ \dot{\hat{x}}(\tau)\alpha\end{pmatrix}.$$

Therefore, $\frac{\partial X}{\partial t_0}(t, \tau, \hat{x}(\tau))$ (regarded as a function of t) is the solution of Cauchy's problem (11) with $\zeta = 0$ and $\gamma = -\dot{\hat{x}}(\tau) = -F(\tau, \hat{x}(\tau))$, and hence (5) takes place.

Similarly,

$$\frac{\partial \Phi}{\partial x_0}(t_0, x_0) = -\mathscr{F}_{x(\cdot)}^{-1} \circ \mathscr{F}_{x_0},$$

and therefore from (14) and (9), we obtain

$$\frac{\partial X}{\partial x_0}(t, \tau, \hat{x}(\tau))[\gamma] = -\mathscr{F}_{x(\cdot)}^{-1}(\hat{x}(\cdot), \tau, \hat{x}(\tau))\begin{pmatrix}0\\ -\gamma\end{pmatrix}.$$

Consequently, $\frac{\partial X}{\partial x_0}(t, \tau, \hat{x}(\tau))$ (regarded as a function of t) is the solution of Cauchy's problem (11) with $\zeta = 0$, whence

$$\frac{\partial X}{\partial x_0}(t, \tau, \hat{x}(\tau))[\gamma] = \Omega(t, \tau)\gamma,$$

which is equivalent to (6). ■

2.6*. Elements of Convex Analysis

Convex analysis is a branch of mathematics in which convex sets and functions are studied. The application of these concepts was already demonstrated in Section 1.3.3 (the Kuhn–Tucker theorem). Since the concept of convexity plays an important role in most of the works concerning extremal problems, we shall present here, besides elementary facts, fundamentals of the duality theory (the Fenchel–Moreau theorem). We shall also present the simplest properties of the subdifferential (a concept generalizing the concept of the differential to the case of convex functions). In Sections 3.3, 3.4, and 4.3 we shall demonstrate the application of these concepts. The presentation of the material in this section is primarily based on Sections 2.1.3 and 2.1.4.

2.6.1. Basic Definitions.

Let X be a real linear space.

Definition 1. a) A set $C \subset X$ is said to be *convex* if it contains, along with any two of its points x_1 and x_2, the whole (closed) line segment

$$[x_1, x_2] = \{x \mid x = \alpha x_1 + (1-\alpha)x_2,\ 0 \leqslant \alpha \leqslant 1\}.$$

b) A set $A \subset X$ is called an *affine manifold* if it contains, along with any two of its points x_1 and x_2, the whole straight line

$$\{x \mid x = \alpha x_1 + (1-\alpha)x_2,\ \alpha \in \mathbf{R}\}.$$

c) A set $K \subset X$ is called a *cone* (with vertex at the origin) if it contains, along with its any point x_0, the whole ray

$$\{\alpha x_0 \mid \alpha > 0\}.$$

By definition, an empty set is assumed to be a convex set, an affine manifold, and a cone. The family of all the convex sets belonging to X is denoted by $\mathfrak{B}(X)$.

Definition 1 immediately implies:

Proposition 1. a) The intersection of an arbitrary collection of convex sets (affine manifolds or cones) is itself a convex set (an affine manifold or a cone).

b) The image f(A) and the inverse image $f^{-1}(B)$ of convex sets (affine manifolds or cones) $A \subset X$ and $B \subset Y$ under a linear mapping $f: X \to Y$ is a convex set (an affine manifold or a cone).

c) A translate $C + \xi$ of a convex set (an affine manifold) C is a convex set (an affine manifold).

d) A cone K is a convex set if and only if

$$x_1, x_2 \in K \Rightarrow (x_1 + x_2) \in K.$$

e) An affine manifold A is a linear subspace of X if and only if $0 \in A$.

Exercise 1. Prove that an affine manifold is a translate of a linear subspace.

Definition 2. a) The intersection of all the convex sets C containing a given set M is called the *convex hull* of the set M and is denoted by conv M:

$$\operatorname{conv} M = \bigcap_{C \supset M} C, \quad C \in \mathfrak{V}(M). \tag{1}$$

b) If in (1), instead of the convex sets, all the convex cones $K \supset M$ (the affine manifolds $A \supset M$ or the linear subspaces $L \supset M$) are taken, then their intersection is called the *conic* (the *affine* or the *linear*) *hull* of M and is denoted by cone M (aff M or lin M).

Definition 3. Let $x_1, \ldots, x_n$ be elements of X. An element

$$x = \sum_{i=1}^{n} \lambda_i x_i \tag{2}$$

is called a *linear* or an *affine* or a *conic* or a *convex combination* of $x_1, \ldots, x_n$ if in (2):

λ_i are arbitrary (for a linear combination)

or

$\sum_{i=1}^{n} \lambda_i = 1$ (for an affine combination)

or

$\lambda_i \geqslant 0$ (for a conic combination)

or

$\sum_{i=1}^{n} \lambda_i = 1, \ \lambda_i \geqslant 0$ (for a convex combination).

It is proved by induction that if $x_1, \ldots, x_n$ belong to a convex set C (a convex cone K or an affine manifold A or a linear subspace L), then their convex (conic or affine or linear) combination belongs to C (K or A or L).

Proposition 2. a) The convex (the conic, the affine, or the linear) hull of a set M consists of all the convex (the conic, the affine, or the linear) combinations of the elements of M.

b) M is a convex set (a convex cone, an affine manifold, or a linear subspace) if and only if it coincides with its convex (conic, affine, or linear) hull.

Proof. We shall consider the case of a convex set M since in the other cases the proof is similar. Let us denote by $\tilde{M}$ the set of all the convex combinations of the points of M. We have $\tilde{M} \supset M$, and the set $\tilde{M}$ is convex because if

$$x_1 = \sum_{i=1}^{m_1} \alpha_{i1} x_{i1}, \quad x_2 = \sum_{i=1}^{m_2} \alpha_{i2} x_{i2}, \quad \alpha_{ij} \geqslant 0,$$
$$\sum_{i=1}^{m_j} \alpha_{ij} = 1, \quad j = 1, 2,$$

then for $\alpha \in [0, 1]$ we have

$$\alpha x_1 + (1-\alpha) x_2 = \sum_{i=1}^{m_1} \alpha \alpha_{i1} x_{i1} + \sum_{i=1}^{m_2} (1-\alpha) \alpha_{i2} x_{i2},$$

and the last sum is a convex combination of the points $x_{11}, \ldots, x_{m_1 1}$, $x_{12}, \ldots, x_{m_2 2}$. Consequently, $M \subset \tilde{M}$ (because conv M is the intersection of all the convex sets containing M). On the other hand, as was mentioned above, each point of $\tilde{M}$ is contained in every convex set containing M, and therefore $\tilde{M} \subset \operatorname{conv} M$.

If M is convex, then we can take C = M in (1), and consequently we obviously have conv M = M. Conversely, if conv M = M, then M is convex according to the definition of the convex hull and Proposition 1 a). ■

Exercise 2. Prove that the arithmetical sum

$$M_1 + \ldots + M_n = \left\{ x \mid x = \sum_{i=1}^{n} x_i, \; x_i \in M, \; i = 1, \ldots, n \right\}$$

of a finite number of convex sets (cones or affine manifolds or linear spaces) possesses the same property.

Examples. 1) The nonempty convex sets on a straight line are all the single-point sets and the intervals of all kinds (the closed intervals, the open intervals, the half-open intervals, the closed and the open rays, and the whole straight line itself).

2) Every linear subspace or affine manifold is a convex set.

Exercise 3. Prove these assertions.

3) The convex hull of finitely many points is called a *convex polytope*. An important special case of a convex polytope is an n-*dimensional simplex*, i.e., the convex hull of n + 1 affinely independent points (none of which is an affine combination of the others).

Exercise 4. Prove that the convex hull of three points in $\mathbf{R}^2$ not lying in one straight line (a two-dimensional simplex) is the triangle with vertices at these points. What happens when the points are in one straight line?

For a finite-dimensional space Proposition 2 can be strengthened. The corresponding assertion will be called Carathéodory's theorem although, strictly speaking, this name is only related to the part of the theorem concerning the convex hull.

Carathéodory's Theorem. Let n = dim (lin A) < ∞. Every point belonging to the convex (the conic, the affine, or the linear) hull of the set A is a convex (a conic, an affine, or a linear) combination of not more than s points $x_1, \ldots, x_s$ of A.

For the convex and the affine hulls, s = n + 1, and the points $x_1, \ldots,$ x_s can be regarded as affinely independent.

For the conic and the linear hulls, s = n, and the points $x_1, \ldots, x_s$ can be regarded as linearly independent.

Proof. A) We shall limit ourselves to the most complicated case of a convex hull; after the proof we shall indicate what changes must be made for the other three cases.

According to Proposition 2 a), every point $x \in \operatorname{conv} A$ can be represented in the form

$$x=\lambda_1 x_1+\ldots+\lambda_k x_k, \quad \lambda_k>0,$$
$$\sum_{i=1}^{k} \lambda_i=1, \quad x_i \in A \tag{3}$$

(in the definition of a convex hull, we have $\lambda_i \geqslant 0$, but it is clear that $\lambda_i = 0$ can be discarded).

We shall show that if $x_1,\ldots,x_k$ are affinely dependent, then the number of the points can be diminished while the representation (3) is retained. Indeed, if the points $x_1,\ldots,x_k$ are affinely dependent, then one of them is an affine combination of the others without loss of generality, we can assume that $x_k=\mu_1 x_1+\ldots+\mu_{k-1} x_{k-1}$, $\sum_{i=1}^{k-1} \mu_i=1$. Putting $\mu_k = -1$, we obtain $\sum_{i=1}^{k} \mu_i x_i=0$, $\sum_{i=1}^{k} \mu_i=0$. Now let

$$\alpha=\min\left\{-\frac{\lambda_i}{\mu_i}\,\middle|\,\mu_i<0\right\}.$$

Then

$$x=\sum_{i=1}^{k} \lambda_i x_i+\alpha \sum_{i=1}^{k} \mu_i x_i=\sum_{i=1}^{k}(\lambda_i+\alpha \mu_i) x_i=\sum_{i=1}^{k} \lambda_i' x_i,$$

and, by virtue of the choice of α, we have $\lambda_i' \geqslant 0$, $i = 1, 2,\ldots,k$, and at least one number λ_i' is equal to zero. Deleting from the set $\{x_1,\ldots,x_k\}$ those x_i for which $\lambda_i' = 0$, we arrive at a representation of form (3) with a smaller number of points $\left(\sum_{i=1}^{k} \lambda_i'=\sum_{i=1}^{k} \lambda_i+\alpha \sum_{i=1}^{k} \mu_i=1\right)$.

Let us denote by $s(x)$ the smaller power of the set $\{x_1,\ldots,x_k\}$ for which representation (3) is possible (any nonempty set of natural numbers contains the smallest number). Then the above argument shows that the points of the set $\{x_1,\ldots,x_{s(x)}\}$ must be affinely independent. It remains to take into account that in an n-dimensional space any $n + 2$ points $\{x_1,\ldots,x_{n+2}\}$ are affinely dependent. Indeed, the points $x_i - x_{n+2}$, $i = 1,\ldots,n + 1$, must be linearly dependent, and if, for instance, $x_{n+1}-x_{n+2}=\sum_{i=1}^{n} \lambda_i(x_i-x_{n+2})$, then $x_{n+1}=\left(1-\sum_{i=1}^{n} \lambda_i\right) x_{n+2}+\sum_{i=1}^{n} \lambda_i x_i$. Since $1-\sum_{i=1}^{n} \lambda_i+\sum_{i=1}^{n} \lambda_i=1$, the point x_{n+1} is an affine combination of the other points.

B) In the case of an affine hull, the part of the above argument connected with the inequalities $\lambda_i > 0$, $\lambda_i' \geqslant 0$ is dropped, and we can put $\alpha = -\lambda_i/\mu_i$ for any i with $\mu_i \neq 0$.

In the two remaining cases affine combinations are replaced by linear ones, and the number of linear independent points cannot exceed n. For a conic hull the argument with the inequalities is retained; for a linear hull it can be dropped. ■

Now we proceed to the description of convex functions. Here and henceforth in the present section by a function we shall mean a mapping

$f: X \to \overline{\mathbf{R}}$, where $\overline{\mathbf{R}} = \mathbf{R} \cup \{-\infty, +\infty\}$ is the extended real line.* With every function f the following two sets can be associated:

$$\operatorname{dom} f = \{x \in X \mid f(x) < +\infty\}$$

and

$$\operatorname{epi} f = \{(x, \alpha) \in X \times \mathbf{R} \mid \alpha \geqslant f(x),\ x \in \operatorname{dom} f\}.$$

They are called the *effective domain* and the *epigraph of the function* f, respectively.

Exercise 5. Prove that a set $C \subset X \times \mathbf{R}$ is the epigraph of a function $f: X \to \overline{\mathbf{R}}$ if and only if $\forall x$, $\{\alpha \mid (\alpha, x) \in C\}$ is $\emptyset$, $\mathbf{R}$ or $[a, +\infty)$, $\exists a$.

Definition 4. A function f for which $\operatorname{dom} f \neq \emptyset$ and $f(x) > -\infty$ everywhere is said to be *proper*; all the other functions are called *improper*. A function $f: X \to \overline{\mathbf{R}}$ on a linear space X is said to be *convex* if epi f is a convex set in $X \times \mathbf{R}$. The set of all the convex functions on X is denoted by $\mathscr{V}(X)$.

This definition immediately implies:

Proposition 3. a) For a proper function f to be convex it is necessary and sufficient that Jensen's inequality

$$f\left(\sum_{i=1}^{n} \alpha_i x_i\right) \leqslant \sum_{i=1}^{n} \alpha_i f_i(x_i) \tag{4}$$

hold for any points $x_i \in \operatorname{dom} f$ and any $\alpha_i \geqslant 0$, $i = 1, \ldots, n$, such that $\sum_{i=1}^{n} \alpha_i = 1$.

b) A sum $\sum_{i=1}^{n} f_i(x)$ of a finite number of convex functions and the upper bound $\vee_\alpha f_\alpha(x) = \sup_\alpha \{f_\alpha(x)\}$ of any family of convex functions are convex.

Examples. 1) Convex Functions on the Straight Line. Let $f_0(x): (\alpha, \beta) \to \mathbf{R}$, and let the derivative $f_0'(x)$ exist on the open interval $(\alpha, \beta) \subseteq \mathbf{R}$ and be nondecreasing. Let us put

$$f(x) = \begin{cases} f_0(x), & x \in (\alpha, \beta), \\ +\infty, & x \notin (\alpha, \beta). \end{cases}$$

For this function we have $\operatorname{dom} f = (\alpha, \beta)$. Let us verify that it is convex. An arbitrary point of the line segment joining two points $(x_i, z_i) \in \operatorname{epi} f$, $i = 1, 2$, has the form $(\alpha x_1 + (1-\alpha) x_2,\ \alpha z_1 + (1-\alpha) z_2)$, $\alpha \in [0, 1]$. Let us apply Lagrange's formula; this yields

$$\alpha z_1 + (1-\alpha) z_2 - f(\alpha x_1 + (1-\alpha) x_2) \geqslant \alpha f(x_1) + (1-\alpha) f(x_2) - f(\alpha x_1 + (1-\alpha) x_2) =$$
$$= \alpha f'(c_1)(x_1 - (\alpha x_1 + (1-\alpha) x_2)) +$$
$$+ (1-\alpha) f'(c_2)(x_2 - (\alpha x_1 + (1-\alpha) x_2)) = \alpha(1-\alpha)(f'(c_1) - f'(c_2))(x_1 - x_2) \geqslant 0$$

*The arithmetical operations and inequalities involving the improper numbers $+\infty$ and $-\infty$ are defined thus (below $a \in \mathbf{R}$ is arbitrary and $p > 0$):

$$\begin{array}{lll} (+\infty)+(+\infty)=+\infty, & (+\infty)(\pm\infty)=\pm\infty, & +\infty > a, \\ \pm\infty + a = \pm\infty, & (-\infty)(\pm\infty)=\mp\infty, & -\infty < a, \\ (-\infty)+(-\infty)=-\infty, & (\pm\infty)\cdot p = \pm\infty, & +\infty > -\infty, \\ -(+\infty)=-\infty, & (\pm\infty)\cdot 0 = 0, & \\ -(-\infty)=+\infty, & (\pm\infty)\cdot(-p)=\mp\infty. & \end{array}$$

The expressions $(+\infty) - (+\infty)$, $(+\infty) + (-\infty)$, and $(-\infty) - (-\infty)$ are regarded as meaningless; the operations of division of the improper numbers are not defined either.

because we have $x_1 \leqslant c_1 \leqslant \alpha x_1+(1-\alpha) x_2 \leqslant c_2 \leqslant x_2$ and the derivative is nondecreasing by the hypothesis. Consequently, $(\alpha x_1+(1-\alpha) x_2, \alpha z_1+(1-\alpha) z_2) \in \operatorname{epi} f$, and f is convex. ■

Exercise 6. Let a convex function $f: \mathbf{R} \to \overline{\mathbf{R}}$ be finite on an open interval $I = (\alpha, \beta)$. Prove that f is continuous on I and possesses the left-hand and the right-hand derivatives $f'_-(x)$ and $f'_+(x)$ at each point $x \in I$ and that these derivatives are nondecreasing on I and coincide everywhere except, possibly, at not more than a countable number of points.

2) Let U be a function of class C^2 (Section 2.2.5) on an open convex set U of a normed space X, and let its second differential $d^2 f_0 = f''_0(x)[\xi, \xi]$ be nonnegative. As in the foregoing example, let us put

$$f(x)=\begin{cases} f_0(x), & x \in U, \\ +\infty, & x \notin U. \end{cases}$$

The restriction of this function to any straight line in X possesses the properties of the function considered in the foregoing example (prove this!). Therefore, f is convex (to verify the convexity of epi f we can limit ourselves to all the possible sections of this set by vertical two-dimensional planes in $X \times \mathbf{R}$).

These examples yield a great number of concrete convex functions. On $\mathbf{R}$ such are the functions e^x, $\alpha x^2 + bx + c$ for $\alpha \geqslant 0$, $-\log x$ (completed by the value $+\infty$ for $x \leqslant 0$), etc. In the Euclidean space the quadratic function $f(x) = (Ax|x) + (b|x) + c$ is convex, where A is a positive-definite symmetric operator.

3) Indicator Function. This is the function

$$\delta A(x)=\begin{cases} +\infty, & x \notin A, \\ 0, & x \in A. \end{cases}$$

It is clear that $\delta A(x) \in \mathcal{V}^{\circ}(X)$ if and only if $A \in \mathfrak{B}(X)$.

4) Minkowski's Function (Section 2.1.3). Jensen's inequality (4) is a source of many important inequalities used in various branches of mathematics. In particular, applying it (with $\alpha_i = 1/n$) to the functions

$$f_1(x)=\begin{cases} -\log x, & x>0, \\ +\infty, & x \leqslant 0, \end{cases}$$

and

$$f_2(x)=\begin{cases} x \log x, & x>0, \\ 0, & x=0, \\ +\infty, & x<0, \end{cases}$$

we obtain, in the first case, the well-known Cauchy inequality connecting the arithmetic and the geometric means and, in the second case, the important inequality of the probability theory for the entropy H(p) of a distribution $(p_1, \ldots, p_n)$, where $p_i = x_i \Big/ \sum_{i=1}^{n} x_i$, $x_i \geqslant 0$:

$$H(p) = -\sum_{i=1}^{n} p_i \log p_i \leqslant \log n.$$

2.6.2. Convex Sets and Functions in Topological Linear Spaces. Let us suppose now that X is a locally convex topological linear space [KF, Chapter 3, Section 5]. As is known, in this case the conjugate space X* consisting of all continuous linear functionals on X is sufficiently rich [KF, Chapter 4, Sections 1-2], which yields a great number of convex sets and functions.

The simplest convex sets in X are hyperplanes and half-spaces (here $x^* \in X^*$):

$$H(x^*, \alpha) = \{x \mid \langle x^*, x\rangle = \alpha\} \text{ (a hyperplane)},$$
$$\left.\begin{array}{l} H_+(x^*, \alpha) = \{x \mid \langle x^*, x\rangle \leqslant \alpha\} \\ H_-(x^*, \alpha) = \{x \mid \langle x^*, x\rangle \geqslant \alpha\} \end{array}\right\} \text{(closed half-spaces)},$$
$$\left.\begin{array}{l} \mathring{H}_-(x^*, \alpha) = \{x \mid \langle x^*, x\rangle < \alpha\} \\ \mathring{H}_-(x^*, \alpha) = \{x \mid \langle x^*, x\rangle > \alpha\} \end{array}\right\} \text{(open half-spaces)}.$$

Exercise 1. Check that these sets are convex.

Further, by a *convex polyhedron* we shall mean the intersection of a finite collection of closed half-spaces (it should be stressed that such an intersection is not necessarily a bounded set).

Exercise 2. Prove that every convex polytope (Section 2.6.1) in $\mathbf{R}^n$ is a convex polyhedron and that every bounded convex polyhedron is a convex polytope.

Now we present the list of the topological properties of convex sets which we need for our further aims (for the topological terminology see [KF]).

Proposition 1. Let A be a convex set. Then:

a) all the points of the half-open line segment $[x_1, x_2)$, where $x_1 \in \operatorname{int} A$ and $x_2 \in A$ belong to int A;

b) the interior int A and the closure $\overline{A}$ of the set A are convex; if $\operatorname{int} A \neq \emptyset$, then $\overline{A} = \overline{\operatorname{int} A}$.

Proof. a) Let $V \subset A$ be a convex neighborhood of the point x_1. An arbitrary point $x \in [x_1, x_2)$ has the form $x = \alpha x_1 + (1 - \alpha)x_2$, $0 < \alpha \leqslant 1$, and $\alpha V + (1 - \alpha)x_2$ is its neighborhood contained in A, i.e., $x \in \operatorname{int} A$.

From the assertion a) it follows that if $\operatorname{int} A \neq \emptyset$ then $A \subset \overline{\operatorname{int} A}$, whence $\overline{A} \subset \overline{\operatorname{int} A}$. The opposite inclusion is obvious.

b) If $x_i \in \operatorname{int} A$, i = 1, 2, then, according to a), we have $[x_1, x_2] \in \operatorname{int} A$; i.e., int A is convex. Now let $x_i \in \overline{A}$, i = 1, 2. Let us take an arbitrary convex neighborhood V of zero. By the definition of the closure, there exist $x_i' \in (x_i + V) \cap A$, i = 1, 2. For an arbitrary point $x = \alpha x_1 + (1 - \alpha) x_2 \in [x_1, x_2]$ let us put $x' = \alpha x_1' + (1 - \alpha)x_2'$. Then $x' \in A$ and $x' \in \alpha(x_1 + V) + (1 - \alpha)(x_2 + V) = x + V$; i.e., every neighborhood of the point x intersects A, whence $x \in \overline{A}$ and A is convex. ∎

Definition 1. The intersection of all the closed convex sets containing a set A is called the *convex closure* of the set A and is denoted by $\overline{\operatorname{conv}} A$.

It is clear that $\overline{\operatorname{conv}} A$ is convex and closed.

Proposition 2. a) $\overline{\operatorname{conv}} A = A \Leftrightarrow$ the set A is convex and closed;

b) the set $\overline{\operatorname{conv}} A$ coincides with the intersection of all the closed half-spaces containing A;

c) $\overline{\operatorname{conv}} A = \overline{\operatorname{conv} A}$.

Proof. The first assertion follows at once from the definition. The inclusion $\operatorname{conv} A \subset \overline{\operatorname{conv}} A$ is also evident (not every convex set containing A is necessarily closed) and therefore $\overline{\operatorname{conv} A} \subset \overline{\operatorname{conv}} A$. On the other hand, the set $\overline{\operatorname{conv} A}$ is convex [Proposition 1 b)] and closed, and contains A, and therefore, according to the definition, it must contain $\overline{\operatorname{conv}} A$, whence $\overline{\operatorname{conv}} A = \overline{\operatorname{conv} A}$.

Further, let us denote by B the intersection of all the closed half-spaces containing A. Then $\overline{\text{conv}}\, A \subset B$ (not every closed convex set containing A is necessarily a half-space). Let $x_0 \notin \overline{\text{conv}}\, A$. By the second separation theorem (Section 2.1.4), there exists a linear functional $x^* \in X^*$ strongly separating the point x_0 and $\overline{\text{conv}}\, A$:

$$\langle x^*, x_0\rangle > \sup_{x \in \overline{\text{conv}}\, A} \langle x^*, x\rangle = \alpha.$$

It follows that $x_0 \notin H_+(x^*, \alpha) \supset \overline{\text{conv}}\, A \supset A$, and therefore $x_0 \notin B$. Consequently, $\overline{\text{conv}}\, A = B$. ■

Exercise 3. Let us consider the infinite-dimensional ellipsoid $B = \left\{x = (x_1, x_2, \ldots) \,\middle|\, \sum_{n=1}^{\infty} (x_n/a_n)^2 \leqslant 1\right\}$ with semiaxes a_n in Hilbert's space l_2.

a) Prove that the set B is convex and closed.

b) Find the conditions on a_n under which $\text{int}\, B \neq \emptyset$.

c) Let $\text{int}\, B \neq \emptyset$, and let a point $y = (y_1, y_2, \ldots)$ lie on the boundary of the ellipsoid, i.e., $\sum_{n=1}^{\infty} (y_n/a_n)^2 = 1$. Prove that it is possible to draw a hyperplane through y such that B lies on its one side.

d) Is the foregoing assertion always true when $\text{int}\, B = \emptyset$?

The simplest convex sets are half-spaces, and, similarly, the simplest convex functions are *affine functions*

$$a(x) = \langle x^*, x\rangle - b, \quad x^* \in X, \quad b \in \mathbf{R}.$$

Exercise 4. Prove that $a(\cdot)$ is convex; find $\text{epi}\, a$.

Propositions 1 and 2 have their analogues to whose statement we now proceed.

Definition 2. Let $f: X \to \overline{\mathbf{R}}$. The function $\bar{f}$ specified by the condition $\text{epi}\, \bar{f} = \overline{\text{epi}\, f}$ is called the *closure* of the function f; if $f = \bar{f}$, the function f is said to be *closed*. The function $\overline{\text{conv}}\, f$ determined by the condition $\text{epi}\, (\overline{\text{conv}}\, f) = \overline{\text{conv}}\, (\text{epi}\, f)$ is called the *convex closure* of f.

Exercise 5. Verify that $\overline{\text{epi}\, f}$ and $\overline{\text{conv}}\, f$ are epigraphs of certain functions (see Exercise 5 in Section 2.6.1).

It is evident that $\overline{\text{conv}}\, f = f \Leftrightarrow$ f is convex and closed.

We also remind the reader that a function $f: X \to \overline{\mathbf{R}}$ is said to be *lower semicontinuous at a point* x_0 if $\varliminf_{x \to x_0} f(x) \geqslant f(x_0)$, and that it is simply called *lower semicontinuous* if the same holds for any x_0.

Exercise 6. Prove that $f: X \to \overline{\mathbf{R}}$ is lower semicontinuous if and only if the set $\mathscr{L}_c f = \{x \mid f(x) \leqslant c\}$ is closed in X for any $c \in \mathbf{R}$.

Proposition 3. a) For a proper function to be closed it is necessary and sufficient that it be lower semicontinuous.

b) For a closed function f to be continuous on $\text{int dom}\, f$ it is sufficient that f be bounded (above) in a neighborhood U of a point $\bar{x}$ and finite at the point $\bar{x}$.

Moreover, f is proper, $\text{int dom}\, f \neq \emptyset$,

$$\text{int epi}\, f = \{(x, \alpha) \in X \times \mathbf{R} \mid x \in \text{int dom}\, f, \ \alpha > f(x)\} \neq \emptyset,$$

and

$$f(x) = (\overline{\text{conv}}\, f)(x), \quad \forall x \in \text{int dom}\, f.$$

In accordance with Exercise 6, we shall verify the closedness of the set $\mathscr{L}_c f$.

Proof. A) Let f be a closed function. Then epi f is a closed set in $X \times \mathbf{R}$. The hyperplane $H=\{(x, \alpha \mid \alpha=c\}$ is also closed. Therefore, the set $\mathscr{L}_c f=\{x \mid f(x) \leqslant c\}$ is also closed because we have $\operatorname{epi} f \cap H=\{(x, c) \mid x \in \mathscr{L}_c f\}$ and the mapping $x \to (x, c)$ is a homeomorphism.

Conversely, let all $\mathscr{L}_c f$ be closed and $(x_0, \alpha_0) \notin \operatorname{epi} f$. Then we have $f(x_0) > \alpha_0$, and consequently $f(x_0) > \alpha_0 + \varepsilon$ for some $\varepsilon > 0$ and $x_0 \notin \mathscr{L}_{\alpha_0+\varepsilon} f$. The set $\mathscr{L}_{\alpha_0+\varepsilon} f$ being closed, there exists a neighborhood $V \ni x_0$ such that $V \cap \mathscr{L}_{\alpha_0+\varepsilon} f=\varnothing$, i.e., we have $f(x) > \alpha_0+\varepsilon$, $x \in V$. The open set $\{(x, \alpha) \mid x \in V, \alpha < \alpha_0 + \varepsilon\}$ contains (x_0, α_0) and does not intersect epi f. Consequently, the complement of epi f is open and epi f itself is closed.

B) LEMMA. Let X be a locally convex space, let V be a convex neighborhood of a point $\bar{x}$ in X, and let f be a convex function on X taking on a finite value at the point $\bar{x} \in V$ ($f(\bar{x}) \neq -\infty$) and bounded above in V. Then f is continuous at the point $\bar{x}$.

Proof of the Lemma. Performing the translation $x \to x + \bar{x}$ and subtracting the constant $f(\bar{x})$ from f we reduce the situation to the case when $\bar{x}=0$, $0 \in V$, $f(0)=0$, $\sup_{x \in V} f(x) \leqslant C$. Here V can be regarded as a symmetric neighborhood of zero [if otherwise, we would consider $W=V \cap(-V)$]. Let $0 < \alpha < 1$ and $x \in \alpha V$. Then we have $x/\alpha \in V$, and the convexity of f implies

$$f(x)=f((1-\alpha) 0+\alpha x/\alpha) \leqslant (1-\alpha) f(0)+\alpha f(x/\alpha) \leqslant \alpha C.$$

On the other hand, the symmetry of V implies $-x/\alpha \in V$, and using the equality $0 = x/(1 + \alpha) + \alpha/(1 + \alpha)(-x/\alpha)$ and the convexity of f we obtain

$$0=f(0)=f(x/(1+\alpha)+\alpha/(1+\alpha)(-x/\alpha)) \leqslant 1/(1+\alpha) f(x)+\alpha/(1+\alpha) f(-x/\alpha),$$

and hence $f(x) \geqslant -\alpha f(-x/\alpha) \geqslant -\alpha C$.

Thus, if $x \in \alpha V$, then $|f(x)| \leqslant \alpha C$, i.e., f is continuous at zero. ■

COROLLARY. Under the conditions of the lemma, int dom f $\neq \emptyset$.

C) Now we come back to the proof of the assertion of Section b) of Proposition 3. According to the lemma and its corollary, f is continuous at $\bar{x}$, and we have int dom f $\neq \emptyset$.

Let us suppose that $f(y) = -\infty$ for some y, i.e., $(y, \alpha) \in \operatorname{epi} f$ for any $\alpha \in \mathbf{R}$. Since epi f is convex,

$$((1-\lambda) \bar{x}+\lambda y, (1-\lambda) f(\bar{x})+\lambda \alpha) \in \operatorname{epi} f$$

for any $\alpha \in \mathbf{R}$, whence $f((1 - \lambda)\bar{x} + \lambda y) = -\infty$, which contradicts the continuity of f at $\bar{x}$ for $\lambda \downarrow 0$. Hence $f(y) > -\infty$ everywhere, i.e., f is proper.

Now let $y \in \operatorname{int} \operatorname{dom} f$. Let us find $\rho > 1$ such that $z=\bar{x}+\rho(y-\bar{x}) \in \operatorname{int} \operatorname{dom} f$ (this can be done because the intersection of the open set int dom f with the straight line passing through $\bar{x}$ and y is an open interval of that line). The homothetic transformation of G with center at z and the coefficient $(\rho - 1)/\rho$ transforms $\bar{x}$ into y and the neighborhood V into a neighborhood V' of the point y. Further, if $\zeta \in G(V)=V'$, then $\zeta=(\rho-1)/\rho \tilde{x}+z/\rho$, $\tilde{x} \in V$ and

$$f(\zeta) \leqslant \frac{\rho-1}{\rho} f(\tilde{x})+\frac{1}{\rho} f(z) \leqslant \frac{\rho-1}{\rho} C+\frac{1}{\rho} f(z).$$

Consequently, according to the lemma, f is continuous at the point y; i.e., f is continuous on int dom f.

If we have $(x_0, \alpha_0) \in \operatorname{int} \operatorname{epi} f$, then, according to the definition, there is a neighborhood W of the point x_0 and $\varepsilon > 0$ such that

$$\{(x, \alpha) \mid x \in W, |\alpha-\alpha_0| < \varepsilon\} \subset \operatorname{epi} f,$$

whence $x_0 \in \operatorname{int}\operatorname{dom} f$ and $\alpha_0 > f(x_0)$. The opposite inclusion

$$\{(x, \alpha) \mid x \in \operatorname{int}\operatorname{dom} f,\ \alpha > f(x)\} \subset \operatorname{int}\operatorname{epi} f$$

is evident. In particular, $\operatorname{int}\operatorname{epi} f \neq \emptyset$.

Since f is convex, conv (epi f) = epi f. Recalling Definition 2 and Proposition 2c), we write

$$\operatorname{epi}(\overline{\operatorname{conv}} f) = \overline{\operatorname{conv}}(\operatorname{epi} f) = \overline{\operatorname{conv}(\operatorname{epi} f)} = \overline{\operatorname{epi}(f)} \supset \operatorname{epi}(f).$$

Consequently, we always have $(\overline{\operatorname{conv}} f)(x) \leqslant f(x)$. In case f is continuous (or at least lower semicontinuous) at a point x and $f(x) > \alpha$, the same inequality is retained in a neighborhood of the point x, and therefore we have $(x, \alpha) \notin \overline{\operatorname{epi} f} = \operatorname{epi}(\overline{\operatorname{conv}} f)$ and $(\overline{\operatorname{conv}} f)(x) > \alpha$. It follows that $(\overline{\operatorname{conv}} f)(x) = f(x)$. In particular, under the conditions of Section b) of the assertion we have to prove this equality holds for $x \in \operatorname{int}(\operatorname{dom} f)$. ■

Exercise 7. Let f be convex on X, and let $f(\bar{x}) = -\infty$ at a point $\bar{x} \in \operatorname{int}\operatorname{dom} f$. Then $f(x) \equiv -\infty$, $\forall x \in \operatorname{int}\operatorname{dom} f$.

Definition 3. An affine function $a(x) = \langle x^*, x\rangle - b$ is said to be a *supporting function to a function* f if:

a) $a(x) \leqslant f(x)$ for all x;

b) for any $\varepsilon > 0$ there is x such that $a(x) > f(x) - \varepsilon$. In other words,

$$b = \sup_x \{\langle x^*, x\rangle - f(x)\}. \tag{1}$$

Minkowski's Theorem. A proper function f is convex and closed if and only if it is the upper bound of the set of all its supporting affine functions.

Proof. 1) "If." An affine function is convex and closed since its epigraph is a closed subspace. Further, for any family of functions $\{f_\alpha\}$ we have

$$\operatorname{epi}\left\{\sup_\alpha f_\alpha\right\} = \left\{(x, z) \mid z \geqslant \sup_\alpha f_\alpha(x)\right\} = \bigcap_\alpha \{(x, z) \mid z \geqslant f_\alpha(x)\} = \bigcap_\alpha \operatorname{epi} f_\alpha, \tag{2}$$

and if all f_α are closed and convex, then $\operatorname{epi} f_\alpha$ are closed convex sets and their intersection possesses the same property.

2) "Only if." By the hypothesis, B = epi f is a nonempty closed convex set in X × **R**. According to Proposition 2, the set B is the intersection of all the closed subspaces containing it. Since every continuous linear functional on X × **R** has the form $\langle x^*, x\rangle + \lambda z$, $x^* \in X^*$, $\lambda \in \mathbf{R}$ (see Section 2.1.2), a closed half-space is determined by an inequality

$$\langle x^*, x\rangle + \lambda z \leqslant b. \tag{3}$$

Since B = epi f is nonempty and contains, along with its point (x_0, z_0), all the points (x_0, z), $z > z_0$, the set B can be contained in the half-space (3) only if $\lambda \leqslant 0$. For $\lambda = 0$ we shall call the half-space (3) *vertical*. It is evident that

$$\operatorname{epi} f = \{(x, z) \mid f(x) \leqslant z\} \subset \{(x, z) \mid \langle x^*, x\rangle \leqslant b\} \Leftrightarrow$$
$$\Leftrightarrow \operatorname{dom} f \subset \{x \mid \langle x^*, x\rangle \leqslant b\} = H_+(x^*, b) \Rightarrow \operatorname{epi} f \subset H_+(x^*, b) \times \mathbf{R}.$$

For $\lambda < 0$ all the terms of inequality (3) can be divided by $|\lambda|$, and thus we can assume that $\lambda = -1$ and the half-space (3) coincides with the epigraph of the affine function $a(x) = \langle x^*, x\rangle - b$. However,

$$B = \operatorname{epi} f \subset \operatorname{epi} a \Leftrightarrow f(x) \geqslant a(x),\ \forall x,$$

and therefore

$$\operatorname{epi} f = \bigcap_{a \leqslant f} \operatorname{epi} a \cap \bigcap_{\operatorname{dom} f \subset H_+(x^*, b)} [H_+(x^*, b) \times \mathbf{R}]. \tag{4}$$

Now we note that at least one affine function $a_0 \leqslant f$ exists [if otherwise, epi f would contain, along with the point (x_0, z_0), all the points (x_0, z), $z \in \mathbf{R}$, whence $f(x_0) = -\infty$, which contradicts the fact that f is proper]. But we have

$$\operatorname{epi} a_0 \cap H_+(x^*, b) \times \mathbf{R} = \{(x, z) \mid a_0(x) \leqslant z, \langle x^*, x\rangle \leqslant b\} =$$
$$= \bigcap_{\lambda \geqslant 0} \{(x, z) \mid a_0(x) + \lambda(\langle x^*, x\rangle - b) \leqslant z\} = \bigcap_{\lambda \geqslant 0} \operatorname{epi}\{a_0(x) + \lambda(\langle x^*, x\rangle - b)\},$$

and if $a_0(x) \leqslant f(x)$ and $\operatorname{dom} f \subset H_+(x^*, b)$, then

$$f(x) < +\infty \Rightarrow \langle x^*, x\rangle - b \leqslant 0 \Rightarrow a_0(x) + \lambda(\langle x^*, x\rangle - b) \leqslant f(x).$$

Consequently, $\operatorname{epi} a_0 \cap [H_+(x^*, b) \times \mathbf{R}] \subset \bigcap_{a \leqslant f} \operatorname{epi} a$, and hence we can discard all the vertical half-space $H_+(x^*, b) \times \mathbf{R}$ in (4). Thus

$$\operatorname{epi} f = \bigcap_{a \leqslant f} \operatorname{epi} a$$

and, according to (2),

$$f(x) = \sup\{a(x) \mid a(x) \text{ is affine and } \leqslant f(x)\}. \tag{5}$$

It remains to note that in (5) we can confine ourselves to the supporting functions to f since the supremum does not change if of all the functions $a(x) = \langle x^*, x\rangle - b \leqslant f(x)$ with one and the same x^* we retain the one for which b is determined by equality (1), i.e., a supporting function.

Exercise 8. If $f: X \to \overline{\mathbf{R}}$ is convex and closed and $f(\bar{x}) = -\infty$, then $f(x) \equiv -\infty$, $\forall x \in \operatorname{dom} f$.

For smooth functions we can use the following assertion.

Proposition 4. If a function f is convex and Gateaux differentiable at a point x_0, then the function $a(x) = \langle f'_\Gamma(x_0), x - x_0\rangle + f(x_0)$ is its supporting function.

Proof. The condition b) of Definition 3 follows from the equality $a(x_0) = f(x_0)$. Let us suppose that $f(x_1) < a(x_1)$ for some x_1. Then $f(x_1) < a(x_1) + \varepsilon$ for some $\varepsilon > 0$, and therefore for $0 < \alpha < 1$ we have

$$f(x_0 + \alpha(x_1 - x_0)) = f((1-\alpha)x_0 + \alpha x_1) \leqslant$$
$$\leqslant (1-\alpha) f(x_0) + \alpha f(x_1) < (1-\alpha) f(x_0) + \alpha(a(x_1) - \varepsilon) =$$
$$= (1-\alpha) f(x_0) + \alpha(f(x_0) + \langle f'_\Gamma(x_0), x_1 - x_0\rangle - \varepsilon) = f(x_0) + \alpha\langle f'_\Gamma(x_0), x_1 - x_0\rangle - \alpha\varepsilon.$$

Consequently,

$$\langle f'_\Gamma(x_0), x_1 - x_0\rangle = \lim_{\alpha \downarrow 0} \frac{f(x_0 + \alpha(x_1 - x_0)) - f(x_0)}{\alpha} \leqslant \langle f'_\Gamma(x_0), x_1 - x_0\rangle - \varepsilon,$$

which is contradictory. Hence the condition a) is also fulfilled. ■

Now let a function $f: X \to \mathbf{R}$ be Gateaux differentiable at each point $x_0 \in X$. Let us form its Weierstrass function (cf. Section 1.4.4):

$$\mathscr{E}(x, x_0) = f(x) - f(x_0) - \langle f'_\Gamma(x_0), x - x_0\rangle.$$

Proposition 5. For f to be convex it is necessary and sufficient that the inequality $\mathscr{E}(x, x_0) \geqslant 0$, $\forall x, x_0$, hold.

Proof. a) If f is convex, then $\mathscr{E}(x, x_0) \geqslant 0$, $\forall x, x_0$ is implied by Proposition 4 we have just proved.

b) Let $x_0 = \alpha x_1 + (1-\alpha)x_2$, $0 \leqslant \alpha \leqslant 1$. The inequalities $\mathscr{E}(x_1, x_0) \geqslant 0$, $\mathscr{E}(x_2, x_0) \geqslant 0$ mean that the points $(x_1, f(x_1))$ and $(x_2, f(x_2))$ lie in the half-space $\{(x, z) \mid z \geqslant f(x_0) + \langle f'_\Gamma(x_0), x - x_0\rangle\}$. Consequently, the line segment joining them also lies in this half-space, and therefore

$$\alpha f(x_1)+(1-\alpha)f(x_2)\geqslant f(x_0)+\langle f'(x_0),\ \alpha x_1+(1-\alpha)x_2-x_0\rangle=f(x_0). \blacksquare$$

Thus, the fulfillment of Weierstrass' condition of Section 1.4.4 for all the points $(t, x, \dot{x})$ is equivalent to the convexity of the Lagrangian $L(t, x, \dot{x})$ with respect to $\dot{x}$.

2.6.3. Legendre–Young–Fenchel Transform. The Fenchel–Moreau Theorem. Let X be again a locally convex topological linear space, and let X* be its conjugate space. Let us take an arbitrary function $f: X \to \overline{\mathbf{R}}$ and investigate in detail the set of its supporting functions. It is natural to assume that the function f is proper because, according to the definition, $f(x) \geqslant a(x) > -\infty$ everywhere if f possesses at least one supporting function, and the case $f(x) \equiv +\infty$ is trivial: any affine function is a supporting function.

Definition. The function on X* specified by the equality

$$f^*(p)=\sup_{x\in X}\{\langle p,\ x\rangle-f(x)\} \tag{1}$$

is called the *conjugate function to* f or the *Young–Fenchel transform* (sometimes the *Legendre transform*) *of* f, and the function

$$f^{**}(x)=\sup_{p\in X^*}\{\langle p,\ x\rangle-f^*(p)\} \tag{2}$$

is called the *second conjugate to* f.

It is seen from equality (1) of Section 2.6.2 that the set of the supporting functions to f and the set of those $p\in X^*$ for which f*(p) is finite are in one-to-one correspondence:

$$(p,\ f^*(p)\neq\pm\infty)\leftrightarrow a_p(x)=\langle p,\ x\rangle-f^*(p). \tag{3}$$

Remark. Despite the seeming symmetry of formulas (1) and (2), we have $f^{**} \neq (f^*)^*$. The matter is that, by definition, $(f^*)^*$ must be regarded on $(X^*)^*$ and not on X. Only for reflexive spaces are $(f^*)^*$ and f^{**} identified in a natural way.

Proposition 1. 1) Functions f* and f** are convex and closed.

2) For any $x\in X$ and $p\in X^*$ there holds Young's inequality

$$f(x)+f^*(p)\geqslant\langle p,\ x\rangle. \tag{4}$$

3) $f^{**}(x) \leqslant f(x)$, $\forall x$.

4) If $f(x) \leqslant g(x)$, then $f^*(p) \geqslant g^*(p)$ and $f^{**}(x) \leqslant g^{**}(x)$.

Proof. 1) It is seen from 1) and 2) that each of the functions f* and f** is obtained as the upper bound of a family of affine functions, and therefore [according to formula (2) of Section 2.6.2] the epigraphs of these functions are intersections of closed subspaces and thus are convex and closed.

2) It follows from (1) that there must be $f^*(p) \geqslant \langle p, x\rangle - f(x)$ for any $x\in X$ and $p\in X^*$, which is equivalent to (4).

3) As has already been proved, $f(x) \geqslant \langle p, x\rangle - f^*(p)$, whence

$$f(x)\geqslant\sup_p\{\langle p,\ x\rangle-f^*(p)\}=f^{**}(x).$$

4) This assertion is an obvious consequence of the definitions. ■

Examples. 1) Let $f(x) = \langle x^*, x\rangle - b$ be an affine function. Such a function f(x) has only one supporting function, which is f(x) itself. Therefore,

$$f^*(p)=\sup_x\{\langle p,\ x\rangle-\langle x_0^*,\ x\rangle+b\}=\begin{cases}+\infty, & p\neq x_0^*,\\ b, & p=x_0^*,\end{cases} \tag{5}$$

and

$$f^{**}(x)=\sup_p\{\langle p, x\rangle - f^*(p)\} = \langle x_0^*, x\rangle - b = f(x). \quad (6)$$

2) As in the example 1) of Section 2.6.1, let

$$f(x)=\begin{cases} f_0(x), & \alpha<x<\beta, \\ +\infty, & x\notin(\alpha,\beta),\end{cases}$$

and let $f_0'(x)$ be continuous and monotone increasing on (α, β). Let us denote

$$A=\lim_{x\to\alpha+0} f_0'(x), \quad B=\lim_{x\to\beta-0} f_0'(x).$$

Given any $p\in(A, B)$, the equality $p = f'(x_0)$ holds for some $x_0\in(\alpha,\beta)$. Since $f(x)$ is a convex function, Proposition 4 of Section 2.6.2 implies that

$$a(x)=f(x_0)+f'(x_0)(x-x_0)=px-(px_0-f(x_0))$$

is a supporting function, and hence the equalities

$$f^*(p)=px_0-f_0(x_0), \quad p=f_0'(x_0) \quad (7)$$

specify $f^*(p)$ parametrically in (A, B) [in the classical mathematical analysis the function f^* specified by equalities (7) is called the *Legendre transform* of f_0].

Exercises. 1. What can be said about $f^*(p)$ for $p\notin(A, B)$ in the above example?

2. For the following functions f on **R** indicate dom f and epi f, verify the convexity and the closedness, and compute f^* and f^{**}:

a) $ax^2+bx+c,\ a\geq 0$, b) $|x|+|x-a|$,

c) $||x|-1|$, d) $\delta[\alpha,\beta](x)=\begin{cases}0, & \alpha\leqslant x\leqslant\beta,\\ \infty, & x\notin[\alpha,\beta],\end{cases}$

e) $\begin{cases}-\sqrt{1+x^2}, & |x|\leqslant 1,\\ +\infty, & |x|>1,\end{cases}$ f) $\begin{cases}+\sqrt{1-x^2}, & |x|<1,\\ +\infty, & |x|\geqslant 1,\end{cases}$

g) e^x, h) $\begin{cases}-\log x, & x>0,\\ 0, & x\leqslant 0,\end{cases}$

i) $\begin{cases}x\log x, & x>0,\\ 0, & x=0,\\ +\infty, & x<0,\end{cases}$ j) $\begin{cases}x\log x, & x>0,\\ +\infty, & x\leqslant 0,\end{cases}$

k) $\begin{cases}\frac{x^a}{a}, & x\geqslant 0,\\ +\infty, & x<0\end{cases}\ (a\geqslant 1)$, l) $\begin{cases}\frac{x^a}{a}, & x\geqslant 0,\\ 0, & x<0\end{cases}\ (a\geqslant 1)$,

m) $|x|^a/a\ \ (a\geqslant 1)$.

3) The example below will be used in Section 3.3.

Proposition 2. Let $f:\mathbf{R}^s\to\mathbf{R}$ be the function determined by the equality $f(x) = f(x_1,\dots,x_s) = \max\{x_1,\dots,x_s\}$.

Then

$$f^*(p)=f^*(p_1,\dots,p_s)=\begin{cases}0 & \text{if } p_i\geqslant 0,\ \sum_{i=1}^s p_i=1,\\ +\infty & \text{in all the other cases.}\end{cases}$$

Proof. Since f is the maximum of a finite collection of linear (and hence convex and continuous) functions, the convexity and the continuity of f are obvious.

If $p_i\geqslant 0,\ \sum_{i=1}^s p_i=1$, then $\langle p, x\rangle \leqslant \max\{x_1,\dots,x_s\}$, and therefore

$$f^*(p)=\sup_x\{\langle p, x\rangle-\max\{x_1,\dots,x_s\}\}=0.$$

On the other hand, if $f^*(p) < +\infty$, then there must be $f^*(p) = 0$ [by virtue of the homogeneity of the functions $x \to \langle p, x\rangle$ and $x \to \max\{x,\ldots,x_s\}$], i.e., for such p the inequality

$$\sum_{i=1}^{s} p_i x_i - \max\{x_1, \ldots, x_s\} \leqslant 0, \ \forall x,$$

holds. Substituting $x_j = 0$, $j \neq i$, $x_i = \xi$ into this inequality, we obtain $p_i\xi \leqslant \max\{\xi, 0\}$, whence $p_i \geqslant 0$. Putting $x_i = a$, $i = 1,\ldots,s$, in the same inequality, we obtain

$$a \sum_{i=1}^{s} p_i \leqslant a, \ \forall a \in \mathbf{R} \Leftrightarrow \sum_{i=1}^{s} p_i = 1. \ \blacksquare$$

4) The conjugate function to the norm in a normed space.

Proposition 3. Let X be a normed space, and let $N(x) = \|x\|$. Then N^* coincides with the indicator function δB^* of the unit ball of the conjugate space X^*, and $N^{**}(x) = N(x)$.

Proof. If $\|p\| > 1$, then $\langle p, x_0\rangle > \|x_0\|$ for some x_0, and then

$$f^*(p) = \sup_x \{\langle p, x\rangle - \|x\|\} \geqslant \sup_{\alpha \geqslant 0} \{\langle p, \alpha x_0\rangle - \alpha\|x_0\|\} = +\infty.$$

In case $\|p\| \leqslant 1$, we have $\langle p, x\rangle \leqslant \|x\|$ and

$$f^*(p) = \sup_x (\langle p, x\rangle - \|x\|) = 0.$$

Further,

$$f^{**}(x) = \sup_p \{\langle p, x\rangle - f^*(p)\} = \sup_{\|p\| \leqslant 1} \{\langle p, x\rangle\} = \|x\| = N(x). \ \blacksquare$$

5) Let $X = \mathbf{R}$, $f(x) = 1/(1 + x^2)$. Then, as can easily be seen,

$$f^*(p) = \begin{cases} 0, & p = 0, \\ +\infty, & p \neq 0, \end{cases} \quad f^{**}(x) \equiv 0,$$

and hence the equality $f^{**}(x) = f(x)$, which was observed in the examples 1) and 4), does not hold here.

The next theorem, which is one of the most important in convex analysis, shows that the coincidence of f and f^{**} is by far not accidental.

Fenchel–Moreau Theorem. Let X be a locally convex topological linear space, and let $f: X \to \bar{\mathbf{R}}$ be a function greater than $-\infty$ everywhere. Then

a) $f^{**}(x) \equiv f(x)$ if and only if f is convex and closed.

b) $f^{**}(x) = \sup\{a(x) \mid a(x)$ is affine and $\leqslant f(x)\}$.

c) If there exists at least one affine function $a(x) \leqslant f(x)$ [equivalent conditions are: $f^*(p) \not\equiv +\infty$ or $f^{**}(x) > -\infty$ everywhere], then $f^{**}(x)$ is the greatest of the convex functions not exceeding $f(x)$, i.e., $f^{**} = \operatorname{conv} f$.

d) $(f^{**})^* = f^*$.

Proof. A) "Only if" follows from Section 1) of Proposition 1. Let us suppose now that $a(x) \leqslant f(x)$ is an affine function. In the example 1) we establish that $a^{**}(x) = a(x)$. Using Proposition 1 again, we write

$$a(x) = a^{**}(x) \leqslant f^{**}(x) \leqslant f(x).$$

Therefore

$$\sup\{a(x) \mid a(x) \text{ is affine and} \leqslant f(x)\} \leqslant f^{**}(x) \leqslant f(x), \tag{8}$$

whence, first of all, the "if" part of a) follows because when f is convex and closed, the left-hand and the right-hand sides of (8) coincide (for a proper function f this is implied by Minkowski's theorem, and if $f \equiv +\infty$ then the left-hand side is also equal to $+\infty$). Hence a) is proved.

B) For an arbitrary f the following three cases are possible. If there is no affine function $a(x) \leqslant f(x)$, then $f^*(p) = \sup_x \{\langle p, x\rangle - f(x)\} = +\infty$ for all p, and hence $f^{**}(x) = -\infty$. However, in this case the left-hand side of (8) is also equal to $-\infty$ ($\sup \emptyset = -\infty$ by definition), i.e., b) takes place.

In case there exists an affine function $a(x) \leqslant f(x)$, relation (8) implies that $f^{**}(x) > -\infty$ everywhere. There remain the following two possibilities. If $f^{**}(x) \equiv +\infty$, then $f(x) \equiv +\infty$ and the two functions are equal to the left-hand side of (8) because here the affine function can be arbitrary. In case f^{**} is proper, Minkowski's theorem implies

$$f^{**}(x) = \sup \{a(x) \mid a(x) \text{ is supporting to } f^{**}(x)\}.$$

Comparing this with (8) we see that $\sup \{a(x) \mid a(x)$ is affine and $\leqslant f(x)\} \leqslant f^{**}(x) = \sup \{a(x) \mid a(x)$ is supporting to $f^{**}(x)\} \leqslant \sup \{a(x) \mid a(x)$ is affine and $\leqslant f(x)\}$. Thus, b) holds for these cases as well.

C) Further, let us suppose that there exists an affine function $a(x) \leqslant f(x)$, and let $g(x) \leqslant f(x)$ be an arbitrary closed convex function. Let us put

$$h(x) = \sup \{g(x),\ f^{**}(x)\}.$$

This function is closed and convex, and $f(x) \geqslant h(x) > -\infty$ everywhere. Consequently,

$$h(x) = h^{**}(x) \leqslant f^{**}(x) \leqslant h(x),$$

and therefore $h(x) = f^{**}(x)$ and $g(x) \leqslant f^{**}(x)$, which proves c).

D) Since $f^{**}(x) \leqslant f(x)$, we have $(f^{**})^*(p) \geqslant f^*(p)$. On the other hand, by virtue of Young's inequality, we have $f(x) \geqslant a(x) = \langle p, x\rangle - f^*(p)$, whence, by virtue of (8),

$$f^{**}(x) \geqslant a(x) \text{ and } (f^{**})^*(p) = \sup_x (\langle p, x\rangle - f^{**}(x)) \leqslant \sup_x (\langle p, x\rangle - a(x)) = f^*(p)$$

for a finite $f^*(p)$. The case $f^*(p) = +\infty$ is trivial, and for $f^*(p) = -\infty$, we obtain

$$f(x) = f^{**}(x) \equiv +\infty \text{ and } (f^{**})^*(p) = -\infty = f^*(p). \quad \blacksquare$$

Exercise 3. Let f be convex and bounded above in a neighborhood of a point. Then $f^{**}(x) = f(x),\ \forall x \in \operatorname{int} \operatorname{dom} f$.

2.6.4. Subdifferential. Moreau–Rockafellar Theorem. Dubovitskii–Milyutin theorem. Let X be a locally convex topological linear space, and let f be a function on X ($f: X \to \overline{\mathbf{R}}$).

Definition. The subset of X^* consisting of those elements $x^* \in X^*$ for which the inequality

$$f(x) - f(x_0) \geqslant \langle x^*,\ x - x_0\rangle, \quad \forall x \in X, \tag{1}$$

holds is said to be the *subdifferential* of f at the point x_0. The subdifferential of the function f at the point x_0 is denoted by $\partial f(x_0)$.

Hence $\partial f(x_0)$ is the set of the "slopes" of the affine functions $a(x) = \langle x^*, x\rangle - b$ *supporting to* f *at the point* x_0, i.e., such that the supremum in the formula (1) of Section 2.6.2 is attained at x_0:

$$b = \langle x^*,\ x_0\rangle - f(x_0). \tag{2}$$

Proposition 1. The subdifferential $\partial f(x_0)$ is a convex (possibly empty) set in X^*.

Proof. Let $x_i^* \in \partial f(x_0)$, i = 1, 2. Then, by definition, $f(x)-f(x_0) \geqslant \langle x_1^*, x-x_0\rangle$, $f(x)-f(x_0) \geqslant \langle x_2^*, x-x_0\rangle$. Multiplying the first inequality by α and the second by $1-\alpha$, where $\alpha \in [0, 1]$, and adding them together, we obtain

$$f(x)-f(x_0) \geqslant \langle \alpha x_1^* + (1-\alpha) x_2^*, x-x_0\rangle, \quad \forall x. \ \blacksquare$$

Examples. 1) Subdifferentials of Certain Functions Defined on the Straight Line and on the Plane. Let the reader check that:

a) $f_1(x)=|x|, \quad (x \in \mathbf{R}) \Rightarrow \partial f_1(x) = \begin{cases} [-1, 1], & x=0, \\ \operatorname{sign} x, & x \neq 0. \end{cases}$

b) $f_2(x)=\sqrt{x_1^2+x_2^2}, \quad (x \in \mathbf{R}^2) \Rightarrow \partial f_2(x) = \begin{cases} \{y \mid \sqrt{y_1^2+y_2^2} \leqslant 1\}, & x=0, \\ \dfrac{x}{|x|} = \left(\dfrac{x_1}{\sqrt{x_1^2+x_2^2}}, \dfrac{x_2}{\sqrt{x_1^2+x_2^2}}\right), & x \neq 0. \end{cases}$

c) $f_3(x) = \max(|x_1|, |x_2|), \ (x \in \mathbf{R}^2) \Rightarrow$

$$\Rightarrow \partial f_3(x) = \begin{cases} \{y \mid |y_1|+|y_2| \leqslant 1\}, & x=0, \\ (\operatorname{sgn} x_1, 0), & |x_1|>|x_2|, \\ (0, \operatorname{sgn} x_2), & |x_2|>|x_1|, \\ \operatorname{conv}((\operatorname{sgn} x_1, 0), (0, \operatorname{sgn} x_2)), & |x_1|=|x_2|=z, z \neq 0. \end{cases}$$

2) Subdifferential of the Norm in a Normed Space.

Proposition 2. Let X be a normed space, let X^* be its conjugate space, and let $N(x) = \|x\|$. Then

$$\partial N(x) = \begin{cases} B^*, \text{ where } B^* \text{ is the unit ball of the conjugate space if } x = 0, \\ \{x^* \in X^* \mid \|x^*\|=1, \ \langle x^*, x\rangle = \|x\|\}, \text{ if } \quad x \neq 0. \end{cases}$$

Proof. As was indicated above [relation (3) of Section 2.6.3], the set of the supporting functions to f and the set $\{x^* \in X^*, \ f^*(x^*) \neq \pm\infty\}$ are in one-to-one correspondence, and, moreover,

$$a_{x^*}(x) = \langle x^*, x\rangle - f^*(x^*).$$

According to Proposition 2 of Section 2.6.3,

$$N^*(x^*) = \delta B^*(x^*) = \begin{cases} 0, & \|x^*\| \leqslant 1, \\ +\infty, & \|x^*\| > 1, \end{cases}$$

and hence the functions $\langle x^*, x\rangle$, $\|x^*\| \leqslant 1$, are supporting to N. Equality (2) goes into

$$0 = \langle x^*, x_0\rangle - \|x_0\|, \tag{3}$$

and hence any function $x \to \langle x^*, x\rangle$, $\|x^*\| \leqslant 1$, is supporting to N at the point $x_0 = 0$, and $\partial N(0) = B^*$. In case $x_0 \neq 0$, the set of $x^* \in B^*$ for which (3) holds coincides with the one indicated in the statement of the proposition. ■

In the above examples the subdifferential existed at each point. Of course, for nonconvex functions (e.g., for $-|x|$) it may exist at none of the points. However, convex functions may have no subdifferentials either even at points of their effective domains.

Here is the simplest example:

$$f(x) = \begin{cases} -\sqrt{1-x^2}, & |x| \leqslant 1, \\ +\infty, & |x| > 1. \end{cases}$$

The subdifferential is empty at the points $x_1 = +1$ and $x_2 = -1$. A sufficient condition for the existence of the subdifferential will be established later (see the corollary of the Moreau–Rockafellar theorem).

In convex analysis the theorem proved below is an analogue of the theorem on the linear property of the differential.

Moreau–Rockafellar Theorem. Let f_i, $i = 1,\dots,n$, be proper convex functions on X. Then

$$\partial\left(\sum_{i=1}^{n} f_i\right)(x) \supseteq \sum_{i=1}^{n} \partial f_i(x). \tag{4}$$

If all the functions except possibly one are continuous at a point $\bar{x}$, and if this function is finite at $\bar{x}$, the equality

$$\partial\left(\sum_{i=1}^{n} f_i\right)(x) = \sum_{i=1}^{n} \partial f_i(x) \tag{5}$$

holds at any point x.

Proof. We shall confine ourselves to the case n = 2; the case of a greater number of summands is considered by induction. The inclusion $\partial f_1(x) + \partial f_2(x) \subset \partial(f_1 + f_2)(x)$ follows at once from the definition of the subdifferential.

Let $x_0^* \in \partial(f_1 + f_2)(x_0)$. Without loss of generality, we can assume that $x_0 = 0$, $x_0^* = 0$, $f_i(0) = 0$, $i = 1, 2$. Indeed, if these relations are not fulfilled, we can consider the functions

$$g_1(x) = f_1(x_0 + x) - f_1(x_0) - \langle x_0^*, x\rangle$$

and

$$g_2(x) = f_2(x_0 + x) - f_2(x_0),$$

instead of the functions f_1 and f_2. Thus, let $0 \in \partial(f_1 + f_2)(0)$. According to (1), this means that

$$\min_x (f_1(x) + f_2(x)) = f_1(0) + f_2(0) = 0.$$

Let $\bar{x}$ be the point at which f_1 is continuous and f_2 is finite. Let us consider the two convex sets

$$C_1 = \{(x, \alpha) \mid \alpha > f_1(x),\ x \in \operatorname{int} \operatorname{dom} f_1\} = \operatorname{int} \operatorname{epi} f_1$$

and

$$C_2 = \{(x, \alpha) \mid -\alpha \geqslant f_2(x)\}.$$

It is clear that the sets C_1 and C_2 are convex and nonempty $((\bar{x}, -f_2(\bar{x})) \in C_2)$, C_1 is open, and $C_1 \neq \emptyset$ by virtue of Proposition 3b) of Section 2.6.2 because the function f_1 is continuous at the point $\bar{x}$ and therefore it is bounded in a neighborhood of $\bar{x}$, and we have $C_1 \cap C_2 = \varnothing$. Indeed, if $(\alpha_1, x_1) \in C_1 \cap C_2$, then $f_1(x_1) < \alpha \leqslant -f_2(x_1)$; i.e.,

$$0 = \min_x (f_1(x) + f_2(x)) \leqslant f_1(x_1) + f_2(x_1) < 0,$$

which is impossible. By the first separation theorem (see Section 2.1.4), C_1 and C_2 are separated by a nonzero linear functional (x_1^*, β):

$$\inf_{(x, \alpha) \in C_2} (\langle x_1^*, x\rangle + \beta\alpha) \geqslant \sup_{(x, \alpha) \in C_1} (\langle x_1^*, x\rangle + \beta\alpha). \tag{6}$$

It is clear that $\beta \leqslant 0$ because, if otherwise, the supremum would be equal to $+\infty$. If we assume that $\beta = 0$, then

$$\sup_{x \in \operatorname{int} \operatorname{dom} f_1} \langle x_1^*, x\rangle \leqslant \inf_{x \in \operatorname{dom} f_2} \langle x_1^*, x\rangle.$$

Since the maximum of a linear function cannot be attained at an interior point, there must be

$$\langle x_1^*, \bar{x}\rangle < \sup_{x\in \text{int dom } f_1} \langle x_1^*, x\rangle \leqslant \inf_{x\in \text{dom } f_2} \langle x_1^*, x\rangle \leqslant \langle x_1^*, \bar{x}\rangle.$$

This contradiction shows that $\beta \neq 0$. Consequently, all the terms in (6) can be divided by $|\beta|$, and then denoting $\tilde{x}_1^* = |\beta|^{-1}x_1^*$ and using the fact that $\text{epi}\, f_1 \subseteq \overline{\text{int epi}\, f_1} = \overline{C}_1$ (Proposition 1 of Section 2.6.2), we obtain

$$\sup_x \{\langle \tilde{x}_1^*, x\rangle - f_1(x)\} = \sup_{\substack{x\in \text{dom } f_1 \\ \alpha \geqslant f_1(x)}} \{\langle \tilde{x}_1^*, x\rangle - \alpha\} =$$

$$= \sup_{(x,\alpha)\in \overline{C}_1} \{\langle \tilde{x}_1^*, x\rangle - \alpha\} \leqslant \inf_{(x,\alpha)\in C_2} \{\langle \tilde{x}_1^*, x\rangle - \alpha\} = \inf_x \{\langle \tilde{x}_1^*, x\rangle + f_2(x)\}.$$

The values of the functions in the braces coincide and are equal to zero for $x = 0$ $[f_1(0) = f_2(0) = 0]$. Hence

$$f_1(x) - f_1(0) \geqslant \langle \tilde{x}_1^*, x\rangle, \quad f_2(x) - f_2(0) \geqslant \langle -\tilde{x}_1^*, x\rangle;$$

i.e.,

$$\tilde{x}_1^* \in \partial f_1(0), \quad -\tilde{x}_1^* \in \partial f_2(0)$$

and

$$0 \in \partial f_1(0) + \partial f_2(0). \ \blacksquare$$

COROLLARY 1. Let f be a convex function continuous at a point x_0. Then $\partial f(x_0) \neq \emptyset$.

Proof. Let us consider the function

$$f + \delta\{x_0\} = \begin{cases} f(x_0), & x = x_0, \\ \infty, & x \neq x_0. \end{cases}$$

The functions f and $\delta\{x_0\}$ satisfy the conditions of the Moreau–Rockafellar theorem. The equality

$$X^* = \partial(f + \delta\{x_0\})(x_0) = \partial f(x_0) + \partial\delta\{x_0\}(x_0) = \partial f(x_0) + X^*$$

means that $\partial f(x_0) \neq \emptyset$. ■

The subdifferential $\partial f(x_0)$ of a convex function continuous at a point x_0 is in fact a compact convex set in the weak* topology.

Let K be a cone (Section 2.6.1). The cone K^* consisting of those elements x^* for which $\langle x^*, x\rangle \geqslant 0$, $\forall x \in K$, is said to be the *cone conjugate to* K. If $0 \in K$, then definition (1) immediately implies the equality $\partial\delta K(0) = -K^*$.

COROLLARY 2. Let $K_1,\ldots,K_n$ be open convex cones whose intersection is nonempty. Then

$$\left(\bigcap_{i=1}^n K_i\right)^* = \sum_{i=1}^n K_i^*.$$

Indeed, let us add the origin to K_i and consider the functions δK_i. Applying the Moreau–Rockafellar theorem to them (all δK_i are continuous at a nonzero point $\bar{x} \in \bigcap_{i=1}^n K_i$), we obtain

$$\left(\bigcap_{i=1}^n K_i\right)^* = -\partial\delta\left(\bigcap_{i=1}^n K_i\right)(0) = -\partial\left(\sum_{i=1}^n \delta K_i\right)(0) = -\sum_{i=1}^n \partial\delta K_i(0) = \sum_{i=1}^n K_i^*.$$

COROLLARY 3. Dubovitskii–Milyutin Theorem on the Intersection of Cones. For convex cones $K_1,\ldots,K_n$, K_{n+1}, among which the first n are open, not to intersect it is necessary and sufficient that there be functionals $x_i^* \in K_i^*$, $i = 1,\ldots,n+1$, not all zero, such that $\sum_{i=1}^{n+1} x_i^* = 0$.

Proof. "Necessary." Without loss of generality, we can assume that $K=\bigcap_{i=1}^{n} K_i \neq \varnothing$. Then K is an open cone which, by the hypothesis, does not intersect K_{n+1}. By the first separation theorem, there is a nonzero linear functional $y^* \in X^*$ separating K and K_{n+1}:

$$\inf_{x \in K} \langle y^*, x\rangle \geqslant \sup_{x \in K_{n+1}} \langle y^*, x\rangle.$$

The last relation means that $y^* \in K^*$, while $(-1)y^* \in K^*_{n+1}$ (x = 0 is a limit point both for K and for K_{n+1}). By Corollary 2, we can represent y* as a sum $y^*=\sum_{i=1}^{n} x_i^*$, $x_i^* \in K_i^*$, $1 \leqslant i \leqslant n$; denoting $(-1)y^*$ by x^*_{n+1}, we obtain what is required.

"Sufficient." Let $x_i^* \in K_i^*$ be not all zero, and let $\sum_{i=1}^{n+1} x_i^*=0$. Let $x \in \bigcap_{i=1}^{n+1} K_i$, $x \neq 0$, and $x^*_{i_0} \neq 0$, $1 \leqslant i_0 \leqslant n$. Then $x \in \operatorname{int} K_{i_0} \Rightarrow \langle x^*_{i_0}, x\rangle > 0$, and hence $0=\langle \sum x_i^*, x\rangle > 0$. This contradiction proves the theorem. ■

Dubovitskii–Milyutin Theorem on the Subdifferential of the Maximum. Let $f_1,\ldots,f_n$ be convex functions on a locally convex topological space X continuous at a point x_0; let $f(x) = \max\{f_1(x),\ldots,f_n(x)\}$, and let $I = \{i_1,\ldots,i_s\}$ be a set of indices such that

$$f_i(x_0)\begin{cases} =f(x_0) & \text{for } i \in I, \\ <f(x_0) & \text{for } i \notin I. \end{cases}$$

Then

$$\partial f(x_0)=\operatorname{conv}\{\partial f_{i_1}(x_0) \cup \cdots \cup \partial f_{i_s}(x_0)\}=$$
$$=\left\{x^* \mid x^*=\sum_{k=1}^{s} \lambda_k x_k^*,\ x_k^* \in \partial f_{i_k}(x_0),\ \lambda_k \geqslant 0,\ \sum_{k=1}^{s} \lambda_k=1\right\}.$$

Proof. A) If $i \in I$ and $x^* \in \partial f_i(x_0)$, then

$$f(x)-f(x_0)=f(x)-f_i(x_0) \geqslant f_i(x)-f_i(x_0) \geqslant$$
$$\geqslant \langle x^*, x-x_0\rangle \Rightarrow x^* \in \partial f(x_0) \Rightarrow \partial f(x_0) \supset \bigcup_{i \in I} \partial f_i(x_0) \Rightarrow \partial f(x_0) \supset \operatorname{conv}\left\{\bigcup_{i \in I} \partial f_i(x_0)\right\},$$

where the last implication is a consequence of the convexity of $\partial f(x_0)$ (Proposition 1).

The opposite inclusion will be proved by induction on the number of functions. For n = 1 the assertion is obvious. In what follows we shall suppose that it is true for n − 1 functions.

B) LEMMA. If a function f(x) is convex and the inequality

$$\varphi(x)=f(x)-f(x_0)-\langle x^*, x-x_0\rangle \geqslant 0 \tag{7}$$

holds in an open convex set V and if $\varphi(\bar{x})=0$ for some $\bar{x} \in V$, then $x^* \in \partial f(x_0)$.

Proof. Let us show that inequality (7) holds for all x; in accordance with the definition of the subdifferential, this will imply the assertion of the lemma.

Let $\tilde{x}$ be arbitrary. The convexity of f implies the convexity of the function φ, and if $\varphi(\tilde{x})<0$, then

$$\varphi((1-\alpha)\bar{x}+\alpha\tilde{x}) \leqslant (1-\alpha)\varphi(\bar{x})+\alpha\varphi(\tilde{x})=\alpha\varphi(\tilde{x})<0$$

for any $\alpha \in (0, 1)$. For a sufficiently small α we have $(1-\alpha)\bar{x}+\alpha\tilde{x} \in V$; we have thus arrived at a contradiction. Consequently, $\varphi(x) \geqslant 0$ for all x. ■

C) Let $x^* \in \partial f(x_0)$. We shall show that in this case there are two possibilities: either x_0 is a solution of the following linear programming problem:

$$\varphi_0(x) = f_{i_1}(x) - f(x_0) - \langle x^*, x - x_0\rangle \to \inf, \tag{8}$$
$$\varphi_i(x) = f_i(x) - f(x_0) - \langle x^*, x - x_0\rangle \leqslant 0, \quad i \neq i_1, \tag{9}$$

or the assertion of the theorem is true. Indeed, if the first possibility does not take place, then

$$\varphi_0(\bar{x}) < 0 = \varphi_0(x_0), \quad \varphi_i(\bar{x}) \leqslant 0, \quad i \neq i_1. \tag{10}$$

for some $\bar{x} \in \operatorname{int} \operatorname{dom} \varphi_0$. By the continuity, the inequality $\varphi_0(x) < 0$ is retained in a convex neighborhood $V \ni \bar{x}$; i.e.,

$$f_{i_1}(x) - f(x_0) < \langle x^*, x - x_0\rangle, \quad x \in V. \tag{11}$$

Now let us take into consideration that $x^* \in \partial f(x_0)$, and consequently the inequality

$$\max\{f_1(x), \ldots, f_n(x)\} - f(x_0) \geqslant \langle x^*, x - x_0\rangle$$

must hold. Comparing it with (11), we conclude that for $x \in V$ the inequality

$$\varphi(x) = \max\{f_i(x) \mid i \neq i_1\} - f(x_0) - \langle x^*, x - x_0\rangle \geqslant 0 \tag{12}$$

holds. At the same time, according to (10),

$$\varphi(\bar{x}) = \max\{f_i(\bar{x}) \mid i \neq i_1\} - f(x_0) - \langle x^*, \bar{x} - x_0\rangle \leqslant 0,$$

and hence $\varphi(\bar{x}) = 0$, and, by the lemma, $x^* \in \partial \tilde{f}(x_0)$, where

$$\tilde{f}(x) = \max\{f_i(x) \mid i \neq i_1\}.$$

Since $\tilde{f}$ is the maximum of $n - 1$ functions, the induction hypothesis implies that

$$x^* \in \operatorname{conv}\left\{\bigcup_{k \geqslant 2} \partial f_{i_k}(x_0)\right\} \subset \operatorname{conv}\left\{\bigcup_{k \geqslant 1} \partial f_{i_k}(x_0)\right\}.$$

D) Knowing that x_0 is a solution of problem (8), (9) we can make use of the Kuhn–Tucker theorem (Section 1.3.3); attention should be paid to the remark made after the proof of that theorem: although, by virtue of Proposition 3 of Section 2.6.2, the functions we deal with are continuous on int dom, they can be equal to $+\infty$ outside it.

According to the Kuhn–Tucker theorem, there exists Lagrange multipliers λ_0, $\lambda_i \geqslant 0$, $i \neq i_1$, not all zero, for which x_0 is a point of minimum of the Lagrange function

$$\mathscr{L}(x; \lambda) = \lambda_0 \varphi_0(x) + \sum_{i \neq i_1} \lambda_i \varphi_i(x).$$

Moreover, the conditions of complementary slackness must hold according to which $\lambda_i = 0$ if $\varphi_i(x_0) = f_i(x_0) - f(x_0) \neq 0$, i.e., if $i \notin I$. Denoting λ_0 as λ_{i_1} and φ_0 as φ_{i_1}, we obtain $\mathscr{L}(x, \lambda) = \sum_{i \in I} \lambda_i \varphi_i(x)$, and since the Lagrange multipliers are determined to within a positive factor, we can assume that $\sum_{i \in I} \lambda_i = 1$.

Now we have

$$\mathscr{L}(x, \lambda) \geqslant \mathscr{L}(x_0, \lambda) \Leftrightarrow \sum_{i \in I} \lambda_i \varphi(x) \geqslant 0 \Leftrightarrow$$
$$\Leftrightarrow \sum_{i \in I} \lambda_i f_i(x) - \left(\sum_{i \in I} \lambda_i\right) f(x_0) - \left(\sum_{i \in I} \lambda_i\right) \langle x^*, x - x_0\rangle \geqslant 0 \Leftrightarrow x^* \subset \partial\left(\sum_{i \in I} \lambda_i f_i(\cdot)\right)(x_0).$$

Applying the Moreau–Rockafellar theorem and taking into account that the definition of the subdifferential implies the evident equality

$$\partial(\lambda f)(x_0) = \lambda \partial f(x_0)$$

holding for $\lambda > 0$, we see that

$$x^* = \sum_{i \in I} \lambda_i x_i^*, \quad x_i^* \in \partial f_i(x_0),$$

$$\lambda_i \geqslant 0, \quad \sum_{i \in I} \lambda_i = 1.$$

Chapter 3

THE LAGRANGE PRINCIPLE FOR SMOOTH PROBLEMS WITH CONSTRAINTS

In the present chapter our primary aim is to prove the Lagrange principle for smooth problems with equality and inequality constraints and to derive sufficient conditions for an extremum in such problems. The general results of this chapter will be applied in Chapter 4 to problems of the classical calculus of variations and optimal control problems. Since a part of the material was already presented at an elementary level in Chapter 1, it is advisable to look through the corresponding passages in Sections 1.3 and 1.4.

3.1. Elementary Problems

3.1.1. Elementary Problems without Constraints. Let X be a topological space, let U be a neighborhood in X, and let $f:U \to \mathbf{R}$. The problem

$$f(x) \to \text{extr}$$

is called an *elementary (extremal) problem without constraints*. In the case when X is a normed linear space and f is smooth in a certain sense, the problem $(\mathfrak{z})$ is called an *elementary smooth problem*; if X is a topological linear space and f is a convex function, the minimum problem $(\mathfrak{z})$ is called an *elementary convex problem*.

The definition of a local extremum for the problem $(\mathfrak{z})$ was stated in Section 1.2.1. If $\hat{x}$ yields a local minimum (or a maximum or an extremum) in the problem $(\mathfrak{z})$ we use the abbreviated notation $\hat{x} \in \text{locmin}\,\mathfrak{z}$ (or $\hat{x} \in \text{locmax}\,\mathfrak{z}$, or $\hat{x} \in \text{locextr}\,\mathfrak{z}$). For the definition of the various terms related to the notion of smoothness, see Section 2.2.

We shall begin with the simplest one-dimensional case when $X = \mathbf{R}$.

LEMMA. Let the function $f:U \to \mathbf{R}$ be defined on an open interval $U \subseteq \mathbf{R}$ containing a point $\hat{x}$.

a) If f possesses the right-hand and the left-hand derivatives at the point $\hat{x}$ and $\hat{x} \in \text{locmin}\,\mathfrak{z}$ $(\text{locmax}\,\mathfrak{z})$, then

$$f'(\hat{x}+0) \geqslant 0, \quad f'(\hat{x}-0) \leqslant 0 \quad (f'(\hat{x}+0) \leqslant 0, \ f'(\hat{x}-0) \geqslant 0). \tag{1}$$

If (1) involves strict inequalities, then $\hat{x} \in \text{locmin}\,\mathfrak{z}$ $(\text{locmax}\,\mathfrak{z})$.

Further, let the function f be k times differentiable at the point $\hat{x}$.

b) If $\hat{x} \in \text{locmin}\,\mathfrak{z}$ $(\text{locmax}\,\mathfrak{z})$, then either $f^{(i)}(\hat{x}) = 0$, $1 \leqslant i \leqslant k$, or there exists s, $1 \leqslant s \leqslant [k/2]$, such that

$$f'(\hat{x}) = \ldots = f^{(2s-1)}(\hat{x}) = 0, \quad f^{(2s)}(\hat{x}) > 0 \qquad (f^{(2s)}(\hat{x}) < 0). \tag{2}$$

c) If there is s, $1 \leqslant s \leqslant [k/2]$, such that the relations (2) hold, then $\hat{x} \in \operatorname{locmin} \mathfrak{z}$ ($\operatorname{locmax} \mathfrak{z}$).

For k = 1 the assertion b) is known in mathematical analysis as the Fermat theorem [$\hat{x} \in \operatorname{locextr} \mathfrak{z} \Rightarrow f'(\hat{x}) = 0$]. We proved this theorem in Section 1.4.1.

Proof. The assertion a) follows at once from the definitions of one-sided derivatives and a local extremum. For k = 1 the assertion b) (the Fermat theorem) follows from a) if we take into account that the existence of the derivative implies the equalities $f'(\hat{x}) = f'(\hat{x} + 0) = f'(\hat{x} - 0)$. Further, let k > 1; for definiteness, let us consider a minimum problem. By Taylor's formula,

$$f(\hat{x} + h) = \sum_{j=0}^{k} \frac{f^{(j)}(\hat{x})}{j!} h^j + o(h^k). \tag{3}$$

Let $f'(\hat{x}) = \ldots = f^{(m-1)}(\hat{x}) = 0$, $1 < m \leqslant k$, $f^{(m)}(\hat{x}) \neq 0$. Here there are two possibilities: m may be odd or even. For an odd m let us put $\varphi(\xi) = f(\hat{x} + \sqrt[m]{\xi})$, $\xi \in \mathbf{R}$. Then from (3) it follows that $\varphi \in D^1(0)$ and $\varphi'(0) = f^{(m)}(\hat{x})/m! \neq 0$, whereas the Fermat theorem implies that $\varphi'(0) = 0$. This contradiction shows that m must be even. For an even m we put $\psi(\xi) = f(\hat{x} + \sqrt[m]{\xi})$, $\xi \geqslant 0$ and obtain from (1) the inequality $\psi'(+0) = f^{(m)}(\hat{x})/m! > 0$ which proves the assertion b). The assertion c) immediately follows from Taylor's formula $f(\hat{x} + h) = f(\hat{x}) + f^{(2s)}(\hat{x})h^{2s}/(2s)! + o(h^{2s})$ which shows that if $f^{(2s)}(\hat{x}) > 0$, then $\hat{x}$ and if $f^{(2s)}(\hat{x}) < 0$, then $\hat{x} \in \operatorname{locmax} \mathfrak{z}$. ■

The assertions that follow are almost immediate consequences of the corresponding definitions.

THEOREM 1 (a First-Order Necessary Condition for an Extremum). Let X in ($\mathfrak{z}$) be a normed linear space, and let the function f possess the derivative in a direction h at a point $\hat{x} \in U$.

If $\hat{x} \in \operatorname{locmin} \mathfrak{z}$ ($\hat{x} \in \operatorname{locmax} \mathfrak{z}$), then

$$f'(\hat{x}; h) \geqslant 0 \qquad (f'(\hat{x}; h) \leqslant 0). \tag{4}$$

COROLLARY 1. Let f possess the first variation (the first Lagrange variation) at a point $\hat{x}$. If $\hat{x} \in \operatorname{locmin} \mathfrak{z}$ ($\hat{x} \in \operatorname{locmax} \mathfrak{z}$), then

$$\delta_+ f(\hat{x}, h) \geqslant 0, \quad \forall h \in X \qquad (\delta_+ f(\hat{x}, h) \leqslant 0, \quad \forall h \in X)$$
$$(\delta f(\hat{x}, h) = 0, \quad \forall h \in X). \tag{5}$$

COROLLARY 2. Let f possess the Fréchet (the Gateaux) derivative at a point $\hat{x}$. If $\hat{x} \in \operatorname{locextr} \mathfrak{z}$, then

$$f'(\hat{x}) = 0 \quad (f'_\Gamma(\hat{x}) = 0). \tag{6}$$

The assertion of Corollary 2 is also called the Fermat theorem. The points $\hat{x}$ at which the equality (6) is fulfilled are called *stationary points* of the problem ($\mathfrak{z}$).

THEOREM 2 (a Second-Order Necessary Condition for an Extremum). Let X in ($\mathfrak{z}$) be a normed linear space, and let f possess the second Lagrange variation at a point $\hat{x} \in U$. If $\hat{x} \in \operatorname{locmin} \mathfrak{z}$ ($\hat{x} \in \operatorname{locmax} \mathfrak{z}$), then the following conditions hold:

$$\delta f(\hat{x}, h) = 0, \quad \forall h \in X, \tag{7}$$

$$\delta^2 f(\hat{x}, h) \geqslant 0, \quad \forall h \in X \quad (\delta^2 f(\hat{x}, h) \leqslant 0, \quad \forall h \in X). \tag{8}$$

Proof. Equality (7) is contained in Corollary 1. Inequalities (8) follow at once from the definition of the second Lagrange variation and the assertion b) of the lemma. ■

COROLLARY 3. Let f possess the second Fréchet derivative at a point $\hat{x}$. If $\hat{x} \in \operatorname{locmin} \mathfrak{z}$ ($\hat{x} \in \operatorname{locmax} \mathfrak{z}$), then the following relations hold:

$$f'(\hat{x}) = 0, \tag{9}$$

$$f''(\hat{x}) \geqslant 0 \qquad (f''(\hat{x}) \leqslant 0). \tag{10}$$

(These inequalities mean that the quadratic form $d^2f = f''(\hat{x})[h, h]$ is nonnegative (nonpositive).)

Proof. Using Taylor's formula (see Section 2.2.5), we write

$$\varphi(\alpha) = f(\hat{x} + \alpha h) = f(\hat{x}) + \alpha f'(\hat{x})[h] + \frac{\alpha^2}{2} f''(\hat{x})[h, h] + o(\alpha^2),$$

whence

$$\delta f(\hat{x}, h) = \varphi'(0) = f'(\hat{x})[h],$$
$$\delta^2 f(\hat{x}, h) = \varphi''(0) = f''(\hat{x})[h, h],$$

and it remains to refer to (7) and (8). ■

Corollary 2 asserts that the points of local extremum are stationary points. As was mentioned in Section 1.3.1, the converse is not true. To obtain sufficient conditions we have to take into consideration higher-order derivatives as has already been done in the lemma.

THEOREM 3 (a Second-Order Sufficient Condition for an Extremum). Let X in ($\mathfrak{z}$) be a normed space, and let f possess the second Fréchet derivative. If the conditions

$$f'(\hat{x}) = 0, \tag{11}$$

$$f''(\hat{x})[h, h] \geqslant \alpha \|h\|^2, \quad \forall h \in X \ (f''(\hat{x})[h, h] \leqslant -\alpha \|h\|^2, \ \forall h \in X) \tag{12}$$

are fulfilled for some $\alpha > 0$, then $x \in \operatorname{locmin} \mathfrak{z}$ ($x \in \operatorname{locmax} \mathfrak{z}$).

Proof. If the first of inequalities (12) holds, then using Taylor's formula once again we obtain

$$f(\hat{x} + h) = f(\hat{x}) + f'(\hat{x})[h] + \frac{f''(\hat{x})}{2}[h, h] + o(\|h\|^2) \geqslant f(\hat{x}) + \frac{\alpha}{2}\|h\|^2 + o\|h\|^2) > f(\hat{x}),$$

provided that $h \neq 0$ and $\|h\|$ is sufficiently small. Consequently, $\hat{x} \in \operatorname{locmin} \mathfrak{z}$. The case of a maximum is considered in a similar way. ■

The condition (12) is called the condition of *strict positivity (negativity) of the second differential of* f.

Remarks. 1. For the finite-dimensional case when $X = \mathbf{R}^n$ the assertions of Theorems 1-3 are well known in mathematical analysis. In this case the Fermat theorem means the following:

$$f'(\hat{x}) = 0 \Leftrightarrow \partial f(\hat{x})/\partial x_1 = \ldots = \partial f(\hat{x})/\partial x_n = 0. \tag{13}$$

From the second-order sufficient condition it follows that in the finite-dimensional extremum problem the assertion $\hat{x} \in \operatorname{locmin} \mathfrak{z}$ implies [besides (13)] that the matrix $\left(\frac{\partial^2 f(\hat{x})}{\partial x_i \partial x_j}\right)$ is positive-semidefinite:

$$\sum_{i, j=1}^{n} \frac{\partial^2 f(\hat{x})}{\partial x_i \partial x_j} h_i h_j \geqslant 0, \quad \forall h \in \mathbf{R}^n.$$

As can easily be seen, the condition that the matrix $\left(\frac{\partial^2 f(\hat{x})}{\partial x_i \partial x_j}\right)$ is positive-definite guarantees the strict positivity of the second differential (and hence it is a sufficient condition for a minimum at a stationary point). This is not the case for infinite-dimensional spaces.

Example. Let $X = l_2$, and let $f(x) = \sum_{j=1}^{\infty} ((x_j^2/n^3) - x_j^4)$. Then $\hat{x} = 0$ is a stationary point, $f'(0) = 0$, and the second differential of f at zero

$$f''(0)[h, h] = \sum_{j=1}^{\infty} h_j^2/n^3$$

is positive-definite. At the same time, $0 \notin \operatorname{locmin} \mathfrak{z}$ because the functional f assumes negative values on the sequence $\{e_n/n\}_{n=1}^{\infty}$ (e_i is the i-th base vector of the space l_2): $f(e_n/n) = (1/n^5) - (1/n^4) < 0$ while the sequence itself tends to zero.

2. In the one-dimensional case the lemma with which this section begins gives in a certain sense an exhaustive answer to the question whether the given point is a point of local minimum of an infinitely differentiable function f: with the exception of the trivial case when all the derivatives vanish at the given point the lemma answers this question. Such an exhaustive procedure is unlikely to exist even in the case of functions of two variables.

3. Inequalities (12) impose severe restrictions on the structure of the normed space X. If the first of them holds, the symmetric bilinear functions $(\xi | \eta) = f''(\hat{x})[\xi, \eta]$ determines a scalar product in X, and the inequalities

$$\|f''(\hat{x})\| \|\xi\|^2 \geqslant (\xi | \xi) = f''(\hat{x})[\xi, \xi] \geqslant \alpha \|\xi\|^2$$

show that the norm generated by this scalar product is equivalent to the original norm. Consequently, X is linear homeomorphic to a Euclidean (pre-Hilbert) space. In particular, inequalities (12) cannot hold in the spaces C and C^1.

3.1.2. Elementary Linear Programming Problem. In Section 1.2.6 we presented the general description of the class of linear programming problems. Here we shall consider the simplest problem of this class. It will be of interest to us because its analysis will lead us to very important conditions for extremum, namely, the conditions of *correspondence and concordance of signs* and the conditions of *complementary slackness*.

Let $X = \mathbf{R}^n$, and let $a \in \mathbf{R}^{n*}$ be an element of the conjugate space: $a = (a_1, \ldots, a_n)$. We shall consider the problem

$$ax \longrightarrow \inf(\sup); \quad \Leftrightarrow \sum_{i=1}^{n} a_i x_i \longrightarrow \inf(\sup). \qquad (\mathfrak{z})$$
$$x_i \gtreqless 0.$$

Here the symbol $\gtreqless$ means that the corresponding constraint may involve one of the signs $\geqslant$, $\leqslant$, and =. The problem $(\mathfrak{z})$ will be called an *elementary linear programming problem.*

We shall say that for a vector a the condition of *correspondence of signs* (with the constraints $x \gtreqless 0$) is fulfilled if $a_i \geqslant 0$ for the constraint $x_i \geqslant 0$, $a_i \leqslant 0$ for the constraint $x_i \leqslant 0$, and a_i may have an arbitrary sign for the equality constraint $x_i = 0$. In the case when these inequalities are "reversed," i.e., if $a_i \geqslant 0$ for the constraint $x_i \leqslant 0$, $a_i \leq 0$ for the constraint $x_i \geqslant 0$ (and again a_i may be of an arbitrary sign for $x_i = 0$), we shall say that the condition of *concordance of signs* is fulfilled. In the former case we write $a_i \gtreqless 0$ and in the latter case $a_i \lesseqgtr 0$.

THEOREM (a Necessary and Sufficient Condition for an Extremum in an Elementary Linear Programming Problem). For a point $\hat{x} \in \mathbf{R}^n$ to yield an absolute minimum (maximum) in the problem $(\mathfrak{z})$, it is necessary and sufficient that there hold:

a) the conditions of correspondence (concordance) of signs

$$a_i \gtrless 0 \qquad (a_i \lesseqgtr 0) \tag{1}$$

and

b) the conditions of complementary slackness

$$a_i \hat{x}_i = 0, \quad i = 1, \ldots, n. \tag{2}$$

Proof. Let us suppose that $(\mathfrak{z})$ is a minimization problem and that the i_0-th constraint has the form $x_{i_0} \geqslant 0$. If we assume that $a_{i_0} < 0$, then the infimum in the problem $(\mathfrak{z})$ is equal to $-\infty$ because x_{i_0} may assume arbitrarily large positive values. Therefore, the condition $a_{i_0} \geqslant 0$ is necessary for the value of the problem (the sought-for inf) to be finite. Now if we assume that $a_{i_0}\hat{x}_{i_0} \neq 0$, then the condition $\hat{x}_{i_0} \geqslant 0$ and the inequality $a_{i_0} \geqslant 0$ we have proved imply that there must be in fact $a_{i_0}\hat{x}_{i_0} > 0$. However, then the point $\hat{x}$ does not yield a minimum in this problem because $a\tilde{x} < a\hat{x}$ on the admissible vector $\tilde{x} = (\hat{x}_1, \ldots, \hat{x}_{i_0-1}, 0, \hat{x}_{i_0+1}, \ldots, \hat{x}_n)$. The sufficiency of conditions (1), (2) is evident: if, for definiteness, $(\mathfrak{z})$ is again a minimization problem and the conditions of correspondence of signs and complementary slackness are fulfilled, then we obtain $ax \geqslant 0$ for any admissible vector $x \in \mathbf{R}^n$, whereas, by virtue of (2), we have $a\hat{x} = 0$ for $\hat{x}$. ■

Exercise. Derive this theorem from the Kuhn–Tucker theorem.

3.1.3. The Bolza Problem. Let Δ be a finite closed interval on the real line. Let us consider the Banach space

$$\Xi = C^1(\Delta, \mathbf{R}^n) \times \mathbf{R} \times \mathbf{R}$$

consisting of the elements $\xi = (x(\cdot), t_0, t_1)$. For this space we shall consider the problem

$$\mathscr{B}(\xi) = \mathscr{B}(x(\cdot), t_0, t_1) = \int_{t_0}^{t_1} L(t, x(t), \dot{x}(t))\,dt + l(t_0, x(t_0), t_1, x(t_1)) \rightarrow \text{extr}. \tag{$\mathfrak{z}$}$$

Here the functions $(t, x, \dot{x}) \rightarrow L(t, x, \dot{x})$ and $(t_0, x_0, t_1, x_1) \rightarrow l(t_0, x_0, t_1, x_1)$ in $(\mathfrak{z})$ are defined in some open subsets V and W of the spaces $\mathbf{R} \times \mathbf{R}^n \times \mathbf{R}^n$ and $\mathbf{R} \times \mathbf{R}^n \times \mathbf{R} \times \mathbf{R}^n$, respectively, and $t_0, t_1 \in \Delta$. We shall suppose that the two functions are at least continuous. The problem $(\mathfrak{z})$ is called the *Bolza problem with nonfixed time* and the functional in $(\mathfrak{z})$ is called the *Bolza functional*. In Section 1.4.2 we considered a special case of this problem when $n = 1$ (the case $n > 1$ was only briefly mentioned) and the instants t_0 and t_1 were fixed.

LEMMA. Let the functions L and l and their partial derivatives L_x, $L_{\dot{x}}$, l_{t_i}, and l_{x_i} $(i = 0, 1)$ be continuous in the domains V and W, respectively. Then the functional $\mathscr{B}$ is continuously (Fréchet) differentiable on the following open subset of the space Ξ:

$$\mathscr{U} = \{\xi = (x(\cdot), t_0, t_1) \mid x(\cdot) \in C^1(\Delta, \mathbf{R}^n);\ t \in \Delta,$$
$$(t, x(t), \dot{x}(t)) \in V;\ (t_0, x(t_0), t_1, x(t_1)) \in W,\ t_0, t_1 \in \operatorname{int}\Delta\},$$

and, moreover,

$$\mathscr{B}'(\hat{x}(\cdot), \hat{t}_0, \hat{t}_1)[h(\cdot), \tau_0, \tau_1] = \int_{\hat{t}_0}^{\hat{t}_1} (\hat{L}_x(t)h(t) +$$
$$+ \hat{L}_{\dot{x}}(t)\dot{h}(t))\,dt + \hat{L}(\hat{t}_1)\tau_1 - \hat{L}(\hat{t}_0)\tau_0 + \hat{l}_{t_0}\tau_0 + \hat{l}_{t_1}\tau_1 + \hat{l}_{x_0}(h(\hat{t}_0) + \dot{\hat{x}}(\hat{t}_0)\tau_0) + \hat{l}_{x_1}(h(\hat{t}_1) + \dot{\hat{x}}(\hat{t}_0)\tau_1), \tag{1}$$

where

$$\hat{L}(t) = L(t, \hat{x}(t), \dot{\hat{x}}(t)),\ \hat{L}_x(t) = L_x(t, \hat{x}(t), \dot{\hat{x}}(t)),$$
$$\hat{L}_{\dot{x}}(t) = L_{\dot{x}}(t, \hat{x}(t), \dot{\hat{x}}(t)), \quad \hat{l}_{t_i} = l_{t_i}(\hat{t}_0, \hat{x}(\hat{t}_0), \hat{t}_1, \hat{x}(\hat{t}_1)),$$
$$\hat{l}_{x_i} = l_{x_i}(\hat{t}_0, \hat{x}(\hat{t}_0), \hat{t}_1, \hat{x}(\hat{t}_1)), \quad i = 0, 1.$$

Proof. The functional $\mathscr{B}$ is a sum of an integral functional and an endpoint functional. Their continuous differentiability was proved in Sections 2.4.2 and 2.4.3. [The endpoint functional $\xi \to l(t_0, x(t_0), t_1, x(t_1))$ is a special case of an operator of boundary conditions.] To obtain (1) it remains to make use of formula (9) of Section 2.4.2 and the formula (3) of Section 2.4.3. ■

According to this lemma, $(\mathfrak{z})$ is an elementary smooth problem without constraints. Therefore, we can apply to it the necessary and sufficient conditions for an extremum of Section 3.1.1. Then the assertion of the Fermat theorem is interpreted as the following theorem:

THEOREM 1 (a First-Order Necessary Condition for an Extremum in the Bolza Problem). Let the conditions of the lemma be fulfilled. If an element $\hat{\xi} = (\hat{x}(\cdot), \hat{t}_0, \hat{t}_1) \in \Xi$ belongs to the set $\mathcal{U}$ and yields a local minimum in the problem $(\mathfrak{z})$, then

1) the stationary conditions with respect to $x(\cdot)$ hold:

a) the Euler equation:

$$-\frac{d}{dt}\hat{L}_{\dot{x}}(t) + \hat{L}_x(t) = 0 \tag{2}$$

and

b) the transversality conditions with respect to x:

$$\hat{L}_{\dot{x}}(\hat{t}_i) = (-1)^i \hat{l}_{x_i}, \quad i = 0, 1; \tag{3}$$

2) the transversality conditions with respect to t hold:

$$\hat{H}(\hat{t}_i) = (-1)^{i+1} \hat{l}_{t_i}, \quad i = 0, 1, \tag{4}$$

where

$$\hat{H}(t) = \hat{L}_{\dot{x}}(t)\dot{\hat{x}}(t) - \hat{L}(t).$$

Relations (2) and (3) were already encountered in Section 1.4.2, whereas the transversality conditions with respect to t have appeared for the first time, and later in Chapter 4 we shall constantly deal with them.

Proof. As was already mentioned, the problem $(\mathfrak{z})$ is an elementary smooth problem, and consequently the Fermat theorem is applicable to it (Corollary 2 of Section 3.1.1) according to which

$$\hat{\xi} \in \operatorname{locextr} \mathfrak{z} \Rightarrow \mathscr{B}'(\hat{\xi}) = 0 \Leftrightarrow \mathscr{B}'(\hat{\xi})[\eta] = 0, \tag{5}$$
$$\forall \eta = (h(\cdot), \tau_0, \tau_1) \in \Xi.$$

Now taking into account (1), we obtain from (5) the relation

$$0 = \mathscr{B}'(\hat{\xi})[\eta] = \int_{\hat{t}_0}^{\hat{t}_1} (\hat{L}_x(t) h(t) + \hat{L}_{\dot{x}}(t) \dot{h}(t))\, dt + \alpha_0 h(\hat{t}_0) + \alpha_1 h(\hat{t}_1) + \beta_0 \tau_0 + \beta_1 \tau_1, \tag{6}$$

where, for brevity, the following notation has been introduced:

$$\alpha_i = \hat{l}_{x_i}, \quad \beta_i = \hat{l}_{t_i} + \hat{l}_{x_i}\dot{\hat{x}}(\hat{t}_i) + (-1)^{i+1}\hat{L}(\hat{t}_i), \quad i = 0, 1. \tag{7}$$

The relation $\mathscr{B}'(\hat{\xi})[\eta] = 0$ holds for any element $\eta = (h(\cdot), \tau_0, \tau_1)$. Let us first consider the elements of the form $\eta = (h(\cdot), 0, 0)$, $h \in C^1(\Delta, \mathbf{R}^n)$, $h(t_i) = 0$, $i = 1, 2$. Then (6) implies that

$$\int_{\hat{t}_0}^{\hat{t}_1} (\hat{L}_x(t) h(t) + \hat{L}_{\dot{x}}(t) \dot{h}(t))\, dt = 0 \tag{8}$$

for any vector function $h(\cdot) \in C^1([\hat{t}_0, \hat{t}_1], \mathbf{R}^n)$ such that $h(\hat{t}_0) = h(\hat{t}_1) = 0$. The situation of this kind was already dealt with in Section 1.4.1. From the DuBois–Reymond's lemma proved there we readily obtain relation (2) [DuBois–Reymond's lemma was proved in Section 1.4.2 for scalar functions; here it should be applied consecutively to each of the components of the vector function $h(\cdot)$; putting $h_1(\cdot) = \ldots = h_{i-1}(\cdot) = h_{i+1}(\cdot) = \ldots = h_n(\cdot) = 0$ we reduce (8) to an analogous scalar relation with one component $h_i(\cdot)$ and obtain the i-th equation of system (2)]. Incidentally, the continuous differentiability of $\hat{L}_{\dot{x}}(t)$ has also been proved. Therefore, we can perform integration by parts in the expression (6), and thus [taking into account (2)] we obtain

$$0 = \mathscr{B}'(\hat{\xi})[\eta] = (\alpha_0 - \hat{L}_{\dot{x}}(\hat{t}_0)) h(t_0) + (\alpha_1 + \hat{L}_{\dot{x}}(\hat{t}_1)) h(t_1) + \beta_0\tau_0 + \beta_1\tau_1 \tag{9}$$

for arbitrary vectors $h(t_0)$ and $h_1(t_1)$ and numbers τ_0 and τ_1. Equalities (9) and (7) immediately imply relations (3) and (4). ■

As is seen, the proof of the theorem is virtually the same as in Chapter 1. The only distinction is that earlier we used primitive methods for computing the derivatives while here we have used the general theorems of differential calculus established in Sections 2.2 and 2.4.

3.1.4. Elementary Optimal Control Problem. Let $\mathfrak{U}$ be a topological space, and let $\varphi: [t_0, t_1] \times \mathfrak{U} \to \mathbf{R}$. We shall consider the problem

$$f(u(\cdot)) = \int_{t_0}^{t_1} \varphi(t, u(t))\, dt \to \inf \tag{$\mathfrak{z}$}$$

in the space $KC([t_0, t_1], \mathfrak{U})$ of piecewise-continuous functions $u(\cdot): [t_0, t_1] \to \mathfrak{U}$ with values belonging to $\mathfrak{U}$. We shall assume that the function φ is continuous in $[t_0, t_1] \times \mathfrak{U}$. The problem ($\mathfrak{z}$) will be called an *elementary optimal control problem*.

THEOREM (a Necessary and Sufficient Condition for a Minimum in an Elementary Optimal Control Problem). Let a function $\hat{u}(\cdot) \in KC([t_0, t_1], \mathfrak{U})$ yield an absolute minimum in the problem ($\mathfrak{z}$). Then for every point of continuity of the function $\hat{u}(\cdot)$, the condition

$$\min_{u \in \mathfrak{U}} f(t, u) = f(t, \hat{u}(t)) \tag{1}$$

holds.

Proof. Let us suppose that relation (1) does not hold and there exist a point τ [at which $\hat{u}(\cdot)$ is continuous] and an element $v \in \mathfrak{U}$ such that $f(\tau, v) < f(\tau, \hat{u}(\tau))$. Since the functions $t \to f(t, v)$ and $t \to f(t, \hat{u}(t))$ are continuous, there is a closed interval $\Delta = [\tau - \delta, \tau + \delta]$ such that $f(t, v) < f(t, \hat{u}(t))$ for $t \in \Delta$. Let us put $\tilde{u}(t) = \hat{u}(t)$ for $t \notin \Delta$ and $\tilde{u}(t) = v$ for $t \in \Delta$. Then we obtain $f(\tilde{u}(\cdot)) < f(\hat{u}(\cdot))$, which contradicts the minimality of $\hat{u}(\cdot)$. The sufficiency is obvious.

3.1.5. Lagrange Principle for Problems with Equality and Inequality Constraints. In Chapter 1 we already discussed the principle to which there correspond necessary conditions for problems with constraints. Now after we have separated out some "elementary problems" it is possible to sum up certain results.

The extremal problems we dealt with were formalized in such a way that the constraints were divided into two groups the first of which had the form of equalities. With respect to the constraints of the first group the

Lagrange function was formed. Then the problem was mentally posed on the corresponding extremum of the Lagrange function with respect to the second group of constraints, i.e., the group which did not take part in the formation of the Lagrange function. Then it turned out that either the problem thus stated was itself elementary or the "partial" problems obtained by fixing all the unknowns except one were elementary. The collection of the necessary conditions for these elementary problems formed the sought-for set of necessary conditions for the extremum. Let the reader verify that all the necessary conditions which were discussed in Chapter 1 and all the necessary conditions which will be spoken of in Chapter 4 correspond to the described procedure. The problems with inequality constraints make an exception: they involve some additional necessary conditions. These problems will be treated in the next section; here we shall show that they can also be included in the scheme of the "Lagrange principle" we have described.

For definiteness, let us consider a minimization problem

$$f_0(x) \to \inf; \quad F(x)=0, \quad f_i(x) \lessgtr 0, \quad i=1, \ldots, m, \qquad (\mathfrak{z})$$

in which besides equality constraints there are also inequality constraints [X and Y in $(\mathfrak{z})$ are topological linear spaces; $f_i: X \to \mathbf{R}$ and $F: X \to Y$]. The introduction of the new variables u_i brings the problem $(\mathfrak{z})$ to the form

$$f_0(x) \to \inf; \quad F(x)=0, \quad f_i(x)+u_i=0, \\ u_i \gtrless 0, \quad i=1, \ldots, m. \qquad (\tilde{\mathfrak{z}})$$

Let us divide the constraints into two groups the first of which consists of equalities [$F(x) = 0$ and $f_i(x) + u_i = 0$] and the other consists of the constraints of the form $u_i \gtrless 0$. Let us form the Lagrange function for the problem $(\tilde{\mathfrak{z}})$, ignoring the constraints of the second group:

$$\tilde{\mathcal{L}}(x, u, y^*, \lambda, \lambda_0)=\lambda_0 f_0(x)+\sum_{i=1}^{m} \lambda_i (f_i(x)+u_i)+\langle y^*, F(x)\rangle, \quad \lambda=(\lambda_1, \ldots, \lambda_m).$$

As to the sign of the multiplier λ_0, we assume that $\lambda_0 \geqslant 0$ in a minimum problem and $\lambda_0 \leqslant 0$ in a maximum problem.

The problem of minimizing the Lagrange function (for fixed values of the multipliers)

$$\tilde{\mathcal{L}}(x, u, \hat{y}^*, \hat{\lambda}, \hat{\lambda}_0) \to \inf$$

involves two groups of variables: x and u. If the variables $u = \hat{u}$ are fixed, we obtain an elementary problem without constraints (a smooth or a convex problem, etc.), and for this problem we can write the required necessary condition for an extremum; as can easily be seen, $\hat{u}_i$ are not involved in this condition. If the variables $x = \hat{x}$ are fixed, we obtain an elementary linear programming problem. The conditions for an extremum written in accordance with Section 3.1.2 yield the conditions of *correspondence of signs* for the Lagrange multipliers $\hat{\lambda}$:

$$\hat{\lambda}_i \gtrless 0 \qquad (1)$$

and the conditions of *complementary slackness:*

$$\hat{\lambda}_i \hat{u}_i = 0 \Leftrightarrow \hat{\lambda}_i f_i(\hat{x}) = 0. \qquad (2)$$

In what follows we shall permanently use this procedure, but if there are inequality constraints we shall form the "shortened" Lagrange function at the very beginning: $\mathcal{L}(x, y^*, \lambda, \lambda_0)=\sum_{i=0}^{m} \lambda_i f_i(x)+\langle y^*, F(x)\rangle$. It is this very function that will be called the *Lagrange function of the problem* $(\mathfrak{z})$. It should be borne in mind, however, that to the conditions for the extremum of this function with respect to x we must add the relations following from (1), (2), namely, the conditions of *concordance of signs:*

$$f_i(x) \leqslant 0 \Rightarrow \hat{\lambda}_i \geqslant 0 \tag{1'}$$

and the conditions of *complementary slackness:*

$$\hat{\lambda}_i f_i(\hat{x}) = 0.$$

Now it should be noted that the Lagrange principle (like, for example, the Fermat theorem for a smooth extremal problem) yields only necessary conditions for an extremum, i.e., separates out a set of "suspected objects" but does not prove their "guilt." Therefore, for the complete solution of the problem we must either have at our disposal a set of sufficient conditions for an extremum or be sure that the solution exists.

Accordingly, in the first case we can "make an examination": each of the objects separated out by the necessary conditions is tested for the sufficiency. Finding among these objects the one satisfying the sufficient conditions, we complete the solution of the problem. In the second case the sought-for solution (whose existence is known in advance or is proved) must be among the suspected objects satisfying the necessary conditions for an extremum. On computing the values of the functional for all these objects, we choose as the solution the object for which the value of the functional is extremal.

Of course, we cannot give preference to either of these approaches. Usually sufficient conditions do not coincide with necessary conditions and therefore there can remain "suspects" whose "guilt" cannot be proved [for example, the theorems of Section 3.1.1 give no answer to the question about the existence of an extremum at a point $\hat{x}$ for a "planar" function $f(x)$ for which $f^{(k)}(\hat{x}) = 0$ for all k]. On the other hand, even if it is known that the solution exists, difficulties may arise when the necessary conditions separate out an infinite (or simply a very large) number of the suspected objects.

For many questions concerning the existence of the solution it is possible to rely on a perfected version of the classical Weierstrass theorem (this theorem was mentioned in Section 1.6.1): the existence is a consequence of the compactness of the set of the admissible elements and the semicontinuity of the functional.

Definition. A function $f: X \to \mathbf{R}$ defined on a topological space X is said to be *lower semicontinuous at a point* x_0 if $\underline{\lim}_{x \to x_0} f(x) \geqslant f(x_0)$, and it is simply called *lower semicontinuous* if it is lower semicontinuous at each point (cf. Section 2.6.2).

The Weierstrass Theorem. A lower semicontinuous function $f: X \to \mathbf{R}$ attains its minimum on every countable compact subset of the topological space X.

Proof. Let $A \subset X$ be a countably compact subset, and let S be the value of the problem

$$f(x) \to \inf, \quad x \in A; \tag{3}$$

i.e.,

$$S = \inf_{x \in A} f(x).$$

According to the definition of the infimum, it is possible to choose a minimizing sequence for problem (3), i.e., a sequence of points $x_n \in A$, such that $f(x_n) \to S$. By the definition of countable compactness, x_n contains a subsequence x_{n_k} convergent to a point $\hat{x} \in A$. Since f is semicontinuous, we have

$$f(\hat{x}) \leqslant \lim_{k\to\infty} f(x_{n_k}) = \lim_{n\to\infty} f(x_n) = S,$$

and, on the other hand, since f(x) cannot be less than the value of the problem (3), there must be $f(\hat{x}) = S$, i.e., $\hat{x}$ is a point of minimum. ■

COROLLARY. Let f be lower semicontinuous on a topological space X. If an α-level set $\mathscr{L}_\alpha f = \{x \mid f(x) \leqslant \alpha\}$ of the function f is nonempty and countably compact, the function f attains its minimum on X.

3.2. Lagrange Principle for Smooth Problems with Equality and Inequality Constraints

3.2.1. Statement of the Theorem.

Let us consider the extremal problem

$$f_0(x) \to \text{extr}; \quad F(x) = 0, \quad f_i(x) \lesseqgtr 0, \quad i = 1, \ldots, m. \tag{1}$$

The symbol $f_i(x) \lesseqgtr 0$ means that the i-th constraint has either the form $f_i(x) = 0$ or $f_i(x) \leqslant 0$ or $f_i(x) \geqslant 0$.

The problems of the type (1) where X and Y are normed spaces, f_i are smooth functions on X, and F is a smooth mapping from X into Y are called *smooth extremal problems with equality and inequality constraints*.

By the *Lagrange function* of the problem (1) we mean the function

$$\mathscr{L}(x, y^*, \lambda, \lambda_0) = \sum_{i=0}^{m} \lambda_i f_i(x) + \langle y^*, F(x)\rangle, \tag{2}$$

where $\lambda = (\lambda_1, \ldots, \lambda_m) \in \mathbf{R}^{m*}$, $\lambda_0 \in \mathbf{R}$, $y^* \in Y^*$ are the *Lagrange multipliers*.

THEOREM (Lagrange Multiplier Rule for Smooth Problems with Equality and Inequality Constraints). Let X and Y be Banach spaces, let U be an open set in X, and let the functions $f_i: U \to \mathbf{R}$, $i = 0, 1, \ldots, m$, and the mapping $F: U \to Y$ be strictly differentiable at a point $\hat{x}$.

If $\hat{x}$ yields a local extremum in problem (1) and if

$$\text{the image } \operatorname{Im} F'(\hat{x}) \text{ is a closed subspace in } Y, \tag{3}$$

then there are Lagrange multipliers $\hat{y}^*$, $\hat{\lambda}$, $\hat{\lambda}_0$ for which there hold:

a) the stationarity condition for the Lagrange function with respect to x:

$$\mathscr{L}_x(\hat{x}, \hat{y}^*, \hat{\lambda}, \hat{\lambda}_0) = 0; \tag{4}$$

b) the condition of concordance of signs: $\hat{\lambda}^* \geqslant 0$ in the case of a minimum problem and $\hat{\lambda}_0 \leqslant 0$ in the case of a maximum problem, and

$$\hat{\lambda}_i \gtreqless 0, \quad i = 1, \ldots, m; \tag{5}$$

c) the conditions of complementary slackness:

$$\hat{\lambda}_i f_i(\hat{x}) = 0, \quad i = 1, \ldots, m. \tag{6}$$

We once again remind the reader that the notation $\hat{\lambda}_i \gtreqless 0$ means the following: if $f_i(x) \geqslant 0$ in problem (1), then $\hat{\lambda}_i \leqslant 0$; if $f_i(x) \leqslant 0$, then $\hat{\lambda}_i \geqslant 0$; and, finally, if $f_i(x) = 0$, then the sign of $\hat{\lambda}_i$ may be arbitrary.

The assertion that there exist Lagrange multipliers satisfying the set of the conditions a)-c) is in exact agreement with the general principle of constraint removal: this very principle was spoken of in the last subsection of Section 3.1. Therefore, the result stated above is also called the *Lagrange principle for smooth problems with equality and inequality constraints*.

The proof of the general theorem is based, in the first place, on the implicit function theorem of Section 2.3 and, in the second place, on the Kuhn–Tucker theorem, i.e. ultimately on the finite-dimensional separation theorem. The Kuhn–Tucker theorem was proved in Section 1.3, and this is in fact the only significant reference to the first chapter in the present part of the book. The reference to the Kuhn–Tucker theorem is connected with the presence of inequalities in (1), and this complicates the proof to a certain extent. As to the case when there are no inequalities, it is quite simple and at the same time meaningful and instructive. Therefore, we shall consider it separately although the result concerning this case follows automatically from the general result. In the study of the next section we recommend that the reader compare the infinite-dimensional version of the proof with the finite-dimensional version considered in Section 1.3.2.

3.2.2. Lagrange Multiplier Rule for Smooth Problems with Equality Constraints.

THEOREM (Lagrange Principle for Smooth Problems with Equality Constraints).

a) Let X and Y be Banach spaces, let U be an open set in X, and let $f: U \to \mathbf{R}$ and $F: U \to Y$ be a function and a mapping strictly differentiable at a point $\hat{x}$.

If the point $\hat{x}$ is a point of local extremum in the problem

$$f(x) \to \text{extr}, \quad F(x) = 0 \tag{1}$$

and if the image $\operatorname{Im} F'(x)$ is a closed subspace of Y, then there are Lagrange multipliers $\hat{\lambda}_0 \in \mathbf{R}$ and $\hat{y}^* \in Y^*$ for which the stationarity condition for the Lagrange function holds:

$$\mathscr{L}_x(\hat{x}, \hat{y}^*, \hat{\lambda}_0) = 0 \Leftrightarrow \langle \hat{\lambda}_0 f'(\hat{x}), x \rangle + \langle \hat{y}^*, F'(\hat{x})[x] \rangle = 0, \forall x. \tag{2}$$

b) If the condition of regularity holds for the mapping F, i.e., if $\operatorname{Im} F'(\hat{x})$ coincides with the whole space Y, then the multiplier $\hat{\lambda}_0$ is nonzero.

Proof. We shall carry out the proof for a minimum problem. Let us define the mapping

$$\mathscr{F}(x) = (f(x) - f(\hat{x}), F(x)), \quad \mathscr{F}: U \to \mathbf{R} \times Y.$$

Obviously, the mapping $\mathscr{F}$ is strictly differentiable at $\hat{x}$, and $\mathscr{F}'(\hat{x}) = (f'(\hat{x}), F'(\hat{x}))$.

There are two possible cases here: the image $\operatorname{Im} \mathscr{F}'(\hat{x})$ may either coincide with the space $\mathbf{R} \times Y$ or be different from it.

A) We shall begin with the case $\operatorname{Im} \mathscr{F}'(\hat{x}) \neq \mathbf{R} \times Y$. Let us apply the lemma on the closed image (Section 2.1.6) to the mapping $\mathscr{F}'(\hat{x})$. By the hypothesis, the image $\operatorname{Im} F'(x)$ is closed, and the image of $f'(\hat{x})(\operatorname{Ker} F'(\hat{x}))$ is either $\{0\}$ or $\mathbf{R}$, i.e., a closed subset of $\mathbf{R}$. Hence, by that lemma, the image $\operatorname{Im} \mathscr{F}'(\hat{x})$ is closed in $\mathbf{R} \times Y$. Since $\operatorname{Im} \mathscr{F}'(\hat{x})$ does not coincide with $\mathbf{R} \times Y$, it is a closed proper subspace. Further, according to the lemma on the nontriviality of the annihilator (Section 2.1.4), there exist a number $\hat{\lambda}_0$ and an element $\hat{y}^*$ (see Section 2.1.5 where the general form of a linear functional in a product space was discussed), not vanishing simultaneously ($|\hat{\lambda}_0| + \|\hat{y}^*\| \neq 0$), such that

$$\langle \hat{\lambda}_0 f'(\hat{x}), x \rangle + \langle \hat{y}^*, F'(\hat{x})[x] \rangle = 0, \quad \forall x.$$

We see that this relation coincides with (2).

B) Now let $\operatorname{Im} \mathscr{F}'(\hat{x}) = \mathbf{R} \times Y$. Then we can apply the implicit function theorem (see Sections 2.3.1-2.3.3).

According to this theorem,† there exist a constant $K > 0$, a neighborhood $\mathcal{U}$ of the point (0, 0) in the space $\mathbf{R} \times Y$, and a mapping $\varphi: \mathcal{U} \to X$ such that

$$\mathcal{F}(\varphi(\alpha, y)) = (\alpha, y) \text{ and } \|\varphi(\alpha, y) - \hat{x}\| \leqslant K \|\mathcal{F}(\hat{x}) - (\alpha, y)\|. \tag{3}$$

Now we put $x(\varepsilon) = \varphi(-\varepsilon, 0) = \varphi(z(\varepsilon))$. Then (3) yields

$$\mathcal{F}(x(\varepsilon)) = z(\varepsilon) \Leftrightarrow f(x(\varepsilon)) - f(\hat{x}) = -\varepsilon \quad F(x(\varepsilon)) = 0, \tag{4}$$

$$\|x(\varepsilon) - \hat{x}\| = \|\varphi(z(\varepsilon)) - x\| \leqslant K \|(0, 0) - (-\varepsilon, 0)\| = K|\varepsilon|. \tag{5}$$

From (4) and (5) it follows that $x(\varepsilon)$ is an admissible element in problem (1) which is arbitrarily close to $\hat{x}$, and at the same time $f(x(\varepsilon)) < f(\hat{x})$, i.e., $\hat{x} \notin \operatorname{locmin}(1)$. This contradiction shows that the equality $\operatorname{Im} \mathcal{F}'(\hat{x}) = \mathbf{R} \times Y$ is impossible. Hence the assertion a) of the theorem is true.

C) Let F be regular at the point $\hat{x}$, and let $\hat{\lambda}_0 = 0$. Then $\hat{y}^* \neq 0$, and relation (2) takes the following form: $\langle y^*, F'(\hat{x})[x] \rangle = 0, \ \forall x$. Let us choose an element $\tilde{y}$ such that $\langle \hat{y}^*, \tilde{y} \rangle \neq 0$ (this is possible because $\hat{y}^* \neq 0$) and find an element $\tilde{x}$ such that $F'(\hat{x})[\tilde{x}] = \tilde{y}$ [it exists by virtue of the equality $\operatorname{Im} F'(\hat{x})X = Y$]. Then $0 \neq \langle \hat{y}^*, \tilde{y} \rangle = \langle \hat{y}^*, F'(\hat{x})[\tilde{x}] \rangle = 0$. This contradiction proves the second assertion. ■

Remark. We constructed the above proof according to the same scheme as the one used in the proof of the finite-dimensional Lagrange multiplier rule (Section 1.3.2), and it turned out to be as simple and concise as the latter. However, this required preliminary work because we used three facts of functional analysis: the theorem on the nontriviality of the annihilator (i.e., in fact the Hahn–Banach theorem), the lemma on the closed image, and the implicit function theorem. To complete the general theory, i.e., to prove the Lagrange multiplier rule for problems with inequality constraints, we shall only use, besides these three facts, the Kuhn–Tucker theorem, which is a direct consequence of the finite-dimensional separation theorem (see Section F) of the proof of the theorem in Section 3.2.4. It is remarkable that such modest means concerning general mathematical structures yield a result from which, as its immediate consequence, in the next chapter, with the aid of some simple auxiliary facts which are also of general character (Riesz' representation theorem for linear functionals in the space C and the standard theorems on differential equations), significant concrete results are derived: necessary conditions for an extremum in problems of the classical calculus of variations and optimal control problems.

3.2.3. Reduction of the Problem.

Before proving the theorem stated in Section 3.2.1, we shall perform a transformation of its conditions. First we *reduce the problem to a minimum problem* by replacing, if necessary, f_0 by $(-1)f_0$. If among the constraints of the form $f_i \leqslant 0$ or $f_i \geqslant 0$ there are such that $f_i(\hat{x}) < 0$ or $f_i(\hat{x}) > 0$, respectively, we discard them since they are insignificant from the local point of view because they hold at all the points in a neighborhood of the point $\hat{x}$. Further, if (1) in Section 3.2.1 involves the relations $f_i(x) \geqslant 0$, $f_i(\hat{x}) = 0$, we *replace them by the inequalities* $(-1)f_i(x) \leqslant 0$. As to the equalities $f_i(x) = 0$, we *add*

†Let us compare the notation of the theorem of Section 2.3.1 with that of the present section. The topological space X in the implicit function theorem (i.f.) consists here of one point $\{x_0\}$, Y(i.f.) $\Leftrightarrow$ X, Z(i.f.) $\Leftrightarrow$ $\mathbf{R} \times Y$, $\Psi(x, y) = \Psi(x_0, y)$(i.f.) $\Leftrightarrow \mathcal{F}(x)$, Λ(i.f.) $\Leftrightarrow \mathcal{F}'(x)$, y_0(i.f.) $\Leftrightarrow \hat{x}$, z_0(i.f.) $\Leftrightarrow 0 \in \mathbf{R} \times Y$. Condition 1) of the theorem becomes trivial, condition 2) holds due to the strict differentiability of $\mathcal{F}$ because

$$\Psi(x_0, y') - \Psi(x_0, y'') - \Lambda(y' - y'') \Leftrightarrow \mathcal{F}(x') - \mathcal{F}(x'') - \mathcal{F}'(\hat{x})[x' - x''],$$

and, finally, condition 3) holds since $\operatorname{Im} \mathcal{F}'(\hat{x}) = \mathbf{R} \times Y$.

them to the equality $F(x) = 0$. After that we renumber all the inequalities specifying the constraints to obtain the following problem equivalent to (1):

$$\tilde{f}_0(x) \to \inf; \quad \tilde{F}(x)=0, \quad \tilde{f}_j(x) \leqslant 0, \quad j=1, \ldots, \tilde{m}, \tag{1'}$$

where $\tilde{F}(x)=(F(x), \tilde{f}_{\tilde{m}+1}(x), \ldots, \tilde{f}_\mu(x))$, and for every j, $0 \leqslant j \leqslant \mu$, there exists $\varepsilon_j = +1$ or $\varepsilon_j = -1$ such that

$$\tilde{f}_j(x) \equiv \varepsilon_j f_{i_j}(x). \tag{2'}$$

The element $\hat{x}$ yields a local minimum in the problem (1'), and, moreover, $\tilde{f}_j(\hat{x}) = 0$, $j = 1,\ldots,\tilde{m}$.

Now we shall show that all the conditions of the theorem hold for problem (1'). Indeed, X and $Y \times \mathbf{R}^{\mu-\tilde{m}}$ are Banach spaces. The functions $\tilde{f}_j$ and the mapping $\tilde{F}$ are strictly differentiable at the point $\hat{x}$ (because F and f_j are). It only remains to show that $L = \operatorname{Im}\tilde{F}'(\hat{x})$ is closed in $Y \times \mathbf{R}^{\mu-\tilde{m}}$. We have $\tilde{F}'(\hat{x})=(F'(\hat{x}), \Phi'(\hat{x}))$, where $\Phi'(\hat{x})[x]=(\langle\tilde{f}'_{\tilde{m}+1}(\hat{x}), x\rangle, \ldots, \langle\tilde{f}'_\mu(\hat{x}), x\rangle)$, $\Phi'(\hat{x}): X \to \mathbf{R}^{\mu-\tilde{m}}$. The image $\operatorname{Im}F'(\hat{x})$ is closed by the hypothesis, and the subspace $\Phi'(\hat{x})[\operatorname{Ker}F'(\hat{x})] \subset \mathbf{R}^{\mu-\tilde{m}}$ is closed like every subspace of a finite-dimensional space. Therefore, by the lemma on the closed image (Section 2.1.6), $\operatorname{Im}F'(x)$ is closed in $Y \times \mathbf{R}^{\mu-\tilde{m}}$.

Now let us suppose that the theorem under consideration is true for problem (1') and hence there exist Lagrange multipliers $\tilde{y}^* = (\hat{y}^*, \tilde{\lambda}_{\tilde{m}+1},\ldots, \tilde{\lambda}_\mu)$, $\tilde{\lambda}_j$, $j = 0, 1,\ldots,\tilde{m}$, for which the conditions a) and b) of Section 3.2.1 hold [the conditions c) hold automatically because $\tilde{f}_j(\hat{x}) = 0$, $j = 1,\ldots,\tilde{m}$). By virtue of b), we have $\tilde{\lambda}_j \geqslant 0$, $j = 0, 1,\ldots,m$, and, by virtue of a),

$$\tilde{\mathscr{L}}_x(\hat{x}, \tilde{y}^*, \tilde{\lambda}_1, \ldots, \tilde{\lambda}_{\tilde{m}}, \tilde{\lambda}_0) = \sum_{j=0}^{\tilde{m}} \tilde{\lambda}_j \tilde{f}'_j(\hat{x}) + (\tilde{y}^* \circ \tilde{F})'(\hat{x}) = 0, \tag{3'}$$

where $\tilde{\mathscr{L}} = \sum_{j=0}^{\tilde{m}} \tilde{\lambda}_j \tilde{f}_j(x) + (\tilde{y}^* \circ \tilde{F})(x)$.

Now it remains to put $\hat{\lambda}_0=\tilde{\lambda}_0$, $\hat{\lambda}_{i_j}=\varepsilon_j\tilde{\lambda}_j$, $j=1, \ldots, \tilde{m}$, $\hat{\lambda}_{i_j}=\tilde{\lambda}_j$; $j=\tilde{m}+1, \ldots, \mu$ and $\hat{\lambda}_i = 0$ if $f_i(\hat{x}) \neq 0$. The collection of the numbers $\hat{\lambda}_{ij}$ obviously satisfies the conditions of complementary slackness and the condition of concordance of signs [because in (2') we have $\varepsilon_j = +1$ when the constraint has the form $f_{ij}(x) \leqslant 0$ and $\varepsilon_j = -1$ for $f_{ij} \geqslant 0$]. Moreover, since

$$\tilde{\mathscr{L}}(x, \tilde{y}^*, \tilde{\lambda}_1, \ldots, \tilde{\lambda}_\mu, \tilde{\lambda}_0) = \mathscr{L}(x, \hat{y}^*, \hat{\lambda}_1, \ldots, \hat{\lambda}_m, \hat{\lambda}_0),$$

the stationarity condition is fulfilled or not fulfilled for the two Lagrange functions simultaneously.

Thus, in what follows we can limit ourselves to problem (1'). However, for the sake of simplicity we shall no longer write the tilde (~) above F, f_j, and m.

3.2.4. Proof of the Theorem. Thus, let a problem

$$f_0(x) \to \inf; \quad F(x)=0, \quad f_j(x) \leqslant 0, \quad j=1, \ldots, m, \tag{1}$$

be given, let all the conditions of the theorem of Section 3.2.1 hold, and let $f_j(\hat{x}) = 0$, $j = 1,\ldots,m$. Without loss of generality, we can also assume that $f_0(\hat{x}) = 0$.

We shall split the proof into several parts.

A) The Linear Case. Let us first consider the simplest situation when $f_0 = x^*$ is a linear functional, $f_i \equiv 0$, $i = 1,\ldots,m$, and $F = \Lambda$ is a *surjective continuous linear operator*. The point $\hat{x} = 0$ is the solution of the problem

$$\langle x^*, x\rangle \to \inf; \; \Lambda x=0 \quad (\Lambda \in \mathscr{L}(X, Y),\ \Lambda(X)=Y) \tag{2}$$

if and only if $x^* \in (\operatorname{Ker}\Lambda)^\perp$. By the lemma on the annihilator of the kernel (Section 2.1.7), there exists an element $\hat{y}^* \in Y^*$ such that $x^* + \Lambda^*\hat{y}^* = 0$. The last relation is nothing other than the Lagrange principle for the problem (2):

$$x^* + \Lambda^*\hat{y}^* = 0 \Leftrightarrow \mathscr{L}_x(0, \hat{y}^*, 1) = 0,$$

where $\mathscr{L}(x, y^*, \lambda_0) = \lambda_0\langle x^*, x\rangle + \langle y^*, \Lambda x\rangle = \langle \lambda_0 x^* + \Lambda^* y^*, x\rangle$ is the Lagrange function of problem (2). Hence in this situation the Lagrange principle is true.

B) The Degenerate Case. Let $\operatorname{Im} F'(x)$ be a proper subspace of the space Y. By the lemma on the nontriviality of the annihilator (Section 2.1.4), there is an element $\hat{y}^* \in (\operatorname{Im} F'(\hat{x}))^\perp \Leftrightarrow (\hat{y}^* \circ F')(\hat{x}) = 0$. It remains to put $\hat{\lambda}_i = 0$, $i = 0,\dots,m$ and to verify that the Lagrange principle holds for the degenerate case under consideration as well.

In what follows we assume that F is a regular operator at the point $\hat{x}$, i.e., $\operatorname{Im} F'(\hat{x}) = \tilde{Y}$.

Let us put

$$A_k = \{x \mid \langle f_i'(\hat{x}), x\rangle < 0,\ i=k, k+1, \dots, m,\ F'(\hat{x})[x]=0\}$$

for $0 \leqslant k \leqslant m$.

C) Lemma 1 (Basic). If $\hat{x}$ is a local solution in the problem (1), then the set A_0 is empty. In other words,

$$\max_{0\leqslant i\leqslant m} \langle f_i'(\hat{x}), x\rangle \geqslant 0, \quad \forall x \in \operatorname{Ker} F'(\hat{x}).$$

Proof. Let us suppose that A_0 is nonempty, i.e., there exists an element ξ such that $F'(\hat{x})[\xi] = 0$, $\langle f_i'(\hat{x}), \xi\rangle = \beta_i$, $\beta_i < 0$, $0 \leqslant i \leqslant m$. Then, by Lyusternik's theorem (Section 2.3.5), there exists a mapping $r:[-\alpha, \alpha] \to X$, $\alpha > 0$, such that

$$F(\hat{x} + \lambda\xi + r(\lambda)) \equiv 0, \quad \lambda \in [-\alpha, \alpha], \quad r(\lambda) = o(\lambda). \tag{3}$$

For small $\lambda > 0$ and $i = 0, 1,\dots,m$ we have the inequalities

$$f_i(\hat{x} + \lambda\xi + r(\lambda)) = f_i(\hat{x}) + \lambda\langle f_i'(\hat{x}), \xi\rangle + o(\lambda) = \lambda\beta_i + o(\lambda) < 0. \tag{4_i}$$

Relations (3) and (4_i) ($i = 1,\dots,m$) mean that for small $\lambda > 0$ the element $\hat{x} + \lambda\xi + r(\lambda)$ is an admissible element in problem (1). Then inequality (4_0) contradicts the fact that $\hat{x}$ yields a local minimum in the problem. ■

D) LEMMA 2. If the set A_m is empty, then the Lagrange principle holds for problem (1).

Proof. The emptiness of A_m means that $x = 0$ is a solution of the problem

$$\langle f_m'(\hat{x}), x\rangle \to \inf, \quad F'(\hat{x})[x] = 0.$$

By virtue of Section A), the Lagrange principle holds for this problem and hence it holds for problem (1) as well [to show this it suffices to put $\hat{\lambda}_0 = \dots = \hat{\lambda}_{m-1} = 0$].

Thus, from Sections C) and D) it follows that either the Lagrange principle has already been proved ($A_m = \emptyset$) or there exists k, $0 \leqslant k < m$, such that

$$A_k = \emptyset, \quad A_{k+1} \neq \emptyset.$$

E) LEMMA 3. If relations (5) hold, then zero is a solution of the following linear programming problem:

$$\langle f_k'(\hat{x}), x\rangle \to \inf; \; \langle f_i'(\hat{x}), x\rangle \leqslant 0,$$

$$i=k+1, \ldots, m, \quad F'(\hat{x})[x]=0. \tag{6}$$

Proof. Let us suppose that η is an admissible element in the problem (6) (i.e., $\langle f_i'(\hat{x}), \eta\rangle \leqslant 0$, $i \geqslant k+1$, $F'(\hat{x})[\eta] = 0$) such that $\langle f_k'(\hat{x}), \eta\rangle < 0$. Let ζ be an element belonging to A_{k+1}, i.e., $\langle f_i'(\hat{x}), \zeta\rangle < 0$, $i \geqslant k+1$, $F'(\hat{x})[\zeta] = 0$. Then the element $\eta + \varepsilon\zeta$ belongs to A_k for small $\varepsilon > 0$, which contradicts (5). ∎

F) Completion of the Proof. We apply the Kuhn–Tucker theorem (Section 1.3.3) to problem (6) taking into account that the Slater condition holds for this problem (because A_{k+1} is nonempty). According to this theorem, there are nonnegative numbers $\hat{\lambda}_{k+1}, \ldots, \hat{\lambda}_m$ such that the point zero is a solution of the problem

$$\langle f_k'(\hat{x}), x\rangle + \sum_{i=k+1}^{m} \langle \hat{\lambda}_i f_i'(\hat{x}), x\rangle \to \inf, \quad F'(\hat{x})[x]=0. \tag{7}$$

[The last assertion is nothing other than the minimum principle for problem (6); the Lagrange multiplier in the functional is put to be equal to 1 since the Slater condition is fulfilled.] Now, problem (7) is again the problem mentioned in Section A). The Lagrange principle holds for the latter problem or, which is the same in the present situation, the lemma on the annihilator of the kernel of the operator $F'(\hat{x})$ is applicable. In other words, there exists an element $\hat{y}^*$ for which

$$f_k'(\hat{x}) + \sum_{i=k+1}^{m} \hat{\lambda}_i f_i'(\hat{x}) + (\hat{y}^* \circ F')(\hat{x}) = 0.$$

We see that this is nothing other than the stationarity condition for the Lagrange function if we put $\hat{\lambda}_0 = \ldots \hat{\lambda}_{k-1} = 0$, $\hat{\lambda}_k = 1$. ∎

In the course of the proof we have found that $\hat{\lambda}_0 = 1$ for $A_1 \neq \emptyset$. Let us show that if F is regular [i.e., $\operatorname{Im} F'(\hat{x}) = Y$], then there is no collection of Lagrange multipliers at all for which the stationarity condition for the Lagrange function holds and which contains the multiplier $\hat{\lambda}_0 = 0$. Indeed, for $A_1 \neq \emptyset$ there exists an element h such that $F'(\hat{x})[h] = 0$, $\langle f_i'(\hat{x}), h\rangle < 0$, $i = 1, \ldots, m$. Now let us suppose that there are Lagrange multipliers $\tilde{\lambda} = (\tilde{\lambda}_1, \ldots, \tilde{\lambda}_m)$, $\tilde{y}^*$, not all zero, such that $\mathscr{L}_x(\hat{x}, \tilde{y}^*, \tilde{\lambda}, 0)=0$. Then, by virtue of the inequalities $\tilde{\lambda}_i\langle f_i'(\hat{x}), h\rangle \leqslant 0$, we have

$$0=\mathscr{L}_x(\hat{x}, \tilde{y}^*, \tilde{\lambda}, 0)[h]=\sum_{i=1}^{m} \tilde{\lambda}_i \langle f_i'(\hat{x}), h\rangle + \langle \tilde{y}^*, F'(\hat{x})[h]\rangle =$$

$$= \sum_{i=1}^{m} \tilde{\lambda}_i \langle f_i'(\hat{x}), h\rangle \Rightarrow \tilde{\lambda}_1 = \ldots = \tilde{\lambda}_m = 0 \Rightarrow \tilde{y}^* \neq 0,$$

and therefore

$$0=\mathscr{L}_x(\hat{x}, \tilde{y}^*, \tilde{\lambda}, 0)[x]=\langle \tilde{y}^*, F'(\hat{x})[x]\rangle, \quad \forall x \Rightarrow \operatorname{Im} F'(\hat{x}) \neq Y,$$

which contradicts the hypothesis.

Let us state what has been said as a separate proposition.

Proposition. For the assertion of the theorem of Section 3.2.1 to involve the inequality $\hat{\lambda}_0 \neq 0$ it suffices to add to its conditions that $\operatorname{Im} F'(x) = Y$ and that there exists an element $h \in \operatorname{Ker} F'(x)$ for which $\langle f_i'(\hat{x}), h\rangle < 0$, $i = 1, \ldots, m$.

The additional assumptions which have been mentioned here will be called the *strengthened regularity conditions for problem* (1).

The statement of the Lagrange principle we have proved includes the important condition (and the only condition besides the requirement of smoothness and the requirement that X and Y be Banach spaces) of the closedness of the image $\operatorname{Im} F'(\hat{x})$. It should be noted that without conditions of this type the Lagrange principle may not hold.

First of all, if the requirement that the operator $\Lambda \in \mathscr{L}(X, Y)$ be surjective (X and Y are Banach spaces) is dropped, the formula $(\operatorname{Ker} \Lambda)^{\perp} = \operatorname{Im} \Lambda^*$ may prove to be wrong or, more precisely, $\operatorname{Im} \Lambda^*$ may turn out to be a proper subspace of $(\operatorname{Ker} \Lambda)^{\perp}$. For instance, if $X=Y=l_2$, $x=(x_1, \ldots, x_n, \ldots) \in l_2$, $\Lambda x=(x_1, x_2/2, \ldots, x_n/n, \ldots)$, then $\operatorname{Ker} \Lambda = \{0\}$, and hence $(\operatorname{Ker}\Lambda)^{\perp} = l_2$, whereas $\operatorname{Im} \Lambda = \operatorname{Im} \Lambda^* \neq l_2$ [for instance, the element $y = (1, 1/2, \ldots, 1/n, \ldots)$ belongs to l_2 while the solution of the equation $\Lambda x = \hat{y}$ obviously does not exist]. Now we can give an example of a problem for which the Lagrange principle does not hold.

Example. Let X and Y be Banach spaces, and let an operator $\Lambda \in \mathscr{L}(X, Y)$ be such that $\operatorname{Ker} \Lambda^* = \{0\}$, while $\operatorname{Im} \Lambda^*$ is a proper subspace of $(\operatorname{Ker} \Lambda)^{\perp}$. Now we choose $x^* \in (\operatorname{Ker} \Lambda)^{\perp} \setminus \operatorname{Im} \Lambda^*$ and consider the problem

$$\langle x^*, x\rangle \to \inf; \ \Lambda x = 0.$$

The Lagrange principle does not hold for this problem. Indeed, $\hat{x} = 0$ is a point of minimum, and if there were $\hat{\lambda}_0$ and $\hat{y}^* \in Y^*$ such that

$$\mathscr{L}_x(0, \hat{y}^*, \hat{\lambda}_0)[h] = 0, \quad \forall h \in X \Leftrightarrow \hat{\lambda}_0 \langle x^*, h\rangle + \langle \hat{y}^*, \Lambda h\rangle = 0, \quad \forall h \in X,$$

then we would have $\hat{\lambda}_0 = 0$ (because, if otherwise, then $x^* \in \operatorname{Im} \Lambda^*$), and consequently $\Lambda^* \hat{y}^* = 0 \Rightarrow \hat{y}^* = 0$.

Exercise.† Let $Z=Y=l_2$, $z=(z_1, \ldots, z_n, \ldots) \in l_2$, $\Lambda z=(z_1, z_2/2, \ldots, z_n/n, \ldots)$, $\hat{y} \notin \operatorname{Im} \Lambda$, $X=\mathbf{R} \times Z$, $f(x)=f(\alpha, z)=\alpha$, $F(x)=F(\alpha, z)=\Lambda z+\alpha^2 \hat{y}$. Show that the Lagrange principle does not hold for the problem $f(x) \to \inf$; $F(x) = 0$.

3.3*. Lagrange Principle and the Duality in Convex Programming Problems

3.3.1. Kuhn–Tucker Theorem (Subdifferential Form).

The Lagrange principle for convex programming problems (the Kuhn–Tucker theorem) was already proved in Section 1.3.3. In the present section the "subdifferential" form of this theorem is given, and its connection with some other notions of convex analysis is elucidated.

Let X and Y be Banach spaces, let $\Lambda: X \to Y$ be a continuous linear operator, let $f_i: X \to \overline{\mathbf{R}}$, $i = 0, 1, \ldots, m$, be convex functions, let $a = (a_1, \ldots, a_m) \in \mathbf{R}^m$, $b \in Y$, and let A be a convex set in X. We shall consider the following convex programming problem:

$$f_0(x) \to \inf; \ f_i(x) \leqslant a_i, \ i=1, \ldots, m, \ \Lambda x=b, \quad x \in A. \tag{$\mathfrak{z}$}$$

The set $\{x \mid \Lambda x = b\}$ will be denoted by B. The Lagrange function of the problem ($\mathfrak{z}$) is

$$\mathscr{L}(x, y^*, \lambda, \lambda_0) = \lambda_0 f_0(x) + \sum_{i=1}^{m} \lambda_i (f_i(x) - a_i) + \langle y^*, \Lambda x - b\rangle.$$

Proposition. Let $\hat{x}$ be a point of absolute minimum in the problem ($\mathfrak{z}$). Then $\hat{x}$ is a point of absolute maximum in the elementary problem

$$f(x) = \max(f_0(x) - f_0(\hat{x}), f_1(x) - a_1, \ldots, f_m(x) - a_m) + \delta(A \cap B)(x) \to \inf, \tag{$\mathfrak{z}'$}$$

where $\delta(A \cap B)$ is the indicator function of the set $A \cap B$.

Indeed, if there exists an element $\tilde{x}$ for which $f(\tilde{x}) < 0$, then this implies that, firstly, $\tilde{x} \in A$, $\tilde{x} \in B (\Leftrightarrow \Lambda \tilde{x} = b)$, secondly, $f_i(\tilde{x}) < a_i$, $i = 1, \ldots, m$ (i.e., $\tilde{x}$ is an admissible element of the problem), and, thirdly, $f_0(\tilde{x}) < f_0(\hat{x})$, which contradicts the hypothesis. ■

THEOREM (Subdifferential Form of the Kuhn–Tucker Theorem). Let the functions f_i, $i = 0, 1, \ldots, m$, in ($\mathfrak{z}$) be continuous at a point $\hat{x} \in A \cap B$,

†This exercise was suggested by the fourth-year student V. V. Uspenskii.

which yields an absolute minimum in the problem. Then there are numbers $\hat{\lambda}_i \geq 0$ such that $\sum_{i=0}^{m} \hat{\lambda}_i = 1$, $\hat{\lambda}_i(f_i(\hat{x}) - a_i) = 0$, $i \geq 1$, and an element $\hat{x}^* \in \partial\delta(A \cap B)(\hat{x})$ for which

$$0 \in \sum_{i=0}^{m} \hat{\lambda}_i \partial f_i(\hat{x}) + \hat{x}^*. \tag{1}$$

Proof. By the hypothesis, $\hat{x}$ yields an absolute minimum in the elementary problem $(\mathfrak{z}')$. According to Theorem 1 of Section 3.1.1, we have $0 \in \partial f(\hat{x})$. The function $g(x) = \max(f_0(x) - f_0(\hat{x}), f_1(x) - a_1, \ldots, f_m(x) - a_m)$ is convex and continuous at the point $\hat{x} \in A \cap B$ [i.e., $\hat{x}$ belongs to $\operatorname{dom}\delta(A \cap B)$], and hence, by the Moreau–Rockafellar theorem (Section 2.6.4), $\partial f(\hat{x}) = \partial g(\hat{x}) + \partial\delta(A \cap B)(\hat{x})$. Finally, the Dubovitskii–Milyutin theorem (Section 2.6.4) implies $\partial g(\hat{x}) = \operatorname{conv}(\partial f_{i_1}(\hat{x}) \cup \ldots \cup \partial f_{i_s}(\hat{x}))$, where i_j are those and only those indices for which $f_i(\hat{x}) = g(\hat{x})$. Thus, there exist two elements $\hat{\xi}^* \in \partial g(\hat{x})$ and $\hat{x}^* \in \partial\delta(A \cap B)(\hat{x})$ for which

$$\hat{\xi}^* + \hat{x}^* = 0,\ \hat{\xi}^* \in \operatorname{conv}(\partial f_{i_1}(\hat{x}) \cup \ldots \cup \partial f_{i_s}(\hat{x})) \Leftrightarrow \hat{\xi}^* =$$
$$= \sum_{j=1}^{s} \hat{\lambda}_{i_j} \hat{x}^*_{i_j},\ \hat{x}^*_{i_j} \in \partial f_{i_j}(\hat{x}),\ \sum_{j=1}^{s} \hat{\lambda}_{i_j} = 1,\ \hat{\lambda}_{i_j} \geq 0.$$

It remains to put $\hat{\lambda}_i = 0$, $i \notin \{i_1, \ldots, i_s\}$. ■

Exercise. Let $B = \{x \mid \Lambda x = b\}$. Prove that if $x_0 \in B$, then $\partial\delta(B)(x_0) = (\operatorname{Ker}\Lambda)^{\perp}$.

COROLLARY (Lagrange Principle for Convex Programming Problems with Equality and Inequality Constraints). Let $A = X$ in the conditions of the theorem, and let the image of X under the mapping Λ be closed in Y. Then there are Lagrange multipliers $\hat{\lambda}_0 \in \mathbf{R}$, $\hat{\lambda} \in \mathbf{R}^{m*}$, $\hat{y}^* \in Y^*$ such that

$$\hat{\lambda}_i \geq 0, \quad i \geq 0, \quad \hat{\lambda}_i(f_i(\hat{x}) - a_i) = 0, \quad i \geq 1,$$
$$\min_x \mathscr{L}(x, \hat{y}^*, \hat{\lambda}, \hat{\lambda}_0) = \mathscr{L}(\hat{x}, \hat{y}^*, \hat{\lambda}, \hat{\lambda}_0). \tag{2}$$

Indeed, if $\operatorname{Im}\Lambda$ is a proper subspace in Y, then we can put $\hat{\lambda}_i = \hat{\lambda}_0 = 0$, $\hat{y}^* \in (\operatorname{Im}\Lambda)^{\perp}$. In case Λ is a surjective operator, we have $\partial\delta B(\hat{x}) = (\operatorname{Ker}\Lambda)^{\perp} = \operatorname{Im}\Lambda^*$. Then, according to (1), $0 \in \partial_x \mathscr{L}(\hat{x}, \hat{y}^*, \hat{\lambda}, \hat{\lambda}_0) = \sum_{i=0}^{m} \hat{\lambda}_i \partial f_i(\hat{x}) + \operatorname{Im}\Lambda^*$, and hence, by the definition of the subdifferential,

$$\mathscr{L}(x, \hat{y}^*, \hat{\lambda}, \hat{\lambda}_0) - \mathscr{L}(\hat{x}, \hat{y}^*, \hat{\lambda}, \hat{\lambda}_0) - \langle 0, x - \hat{x} \rangle \geq 0 \Leftrightarrow$$
$$\Leftrightarrow \mathscr{L}(x, \hat{y}^*, \hat{\lambda}, \hat{\lambda}_0) \geq \mathscr{L}(\hat{x}, \hat{y}^*, \hat{\lambda}, \hat{\lambda}_0),\ \forall x. \ ■$$

3.3.2. Perturbation Method and the Duality Theorem. The convex programming problem $(\mathfrak{z})$ was considered in the foregoing section as an individual problem. However, for many reasons it is natural and useful to consider families of problems of this kind. Let us fix f_i, Λ, a_i, A, and b and include the problem $(\mathfrak{z})$ in the family

$$f_0(x) \to \inf;\ f_i(x) + \alpha_i \leq a_i,\ i = 1, \ldots, m,$$
$$\Lambda x + \eta = b, \quad x \in A. \tag{$\mathfrak{z}(\alpha, \eta)$}$$

(Of course, we could simply assume a_i and b to be variable parameters of the family, but the above introduction of the parameters α_i, η leads to more beautiful formulas.) The family of problems $\{(\mathfrak{z}(\alpha, \eta))\}$ will be called a *perturbation* of the problem $(\mathfrak{z}) = (\mathfrak{z}(0, 0))$.

As was already seen in Section 3.3.1, the problem $(\mathfrak{z})$ is reducible to an elementary problem. Here, in a somewhat different manner, we shall perform an analogous reduction of the family $\mathfrak{z}(\alpha, \eta)$. Namely, let us denote

$$f(x; \alpha, \eta) = \begin{cases} f_0(x) & \text{if } f_i(x) + \alpha_i \leqslant a_i,\ \Lambda x + \eta = b,\ x \in A, \\ +\infty & \text{in all the other cases.} \end{cases} \tag{1}$$

Then the family $(\mathfrak{z}(\alpha, \eta))$ can be written in the form of an elementary problem:

$$f(x; \alpha, \eta) \to \inf \ (\text{with respect to } x \in X). \tag{2}$$

In what follows when speaking about the problem $(\mathfrak{z}(\alpha, \eta))$, we shall not distinguish between the original statement of the problem and (2). For the indicated constraints the value of this problem, i.e., $\inf f_0$, is a function of $\alpha = (\alpha_1, \ldots, \alpha_m)$ and η which will be denoted by S ($S: \mathbf{R}^m \times Y \to \overline{\mathbf{R}}$) and will sometimes be called the *S-function*:

$$S(\alpha, \eta) = \inf_x f(x; \alpha, \eta) = \inf_{x \in A,\ f_i(x) + \alpha_i \leqslant a_i,\ \Lambda x + \eta = b} f_0(x). \tag{3}$$

LEMMA. Let $F(x, z)$ be a convex function on the product of two linear spaces X and Z. Then the function

$$S(z) = \inf_x F(x, z)$$

is convex on Z.

Proof. Let $(z_i, t_i) \in \operatorname{epi} S$, $i = 1, 2$, and $\lambda \in [0, 1]$. Then $S(z_i) \leqslant t_i$, $i = 1, 2$, and for every $\varepsilon > 0$ there exist (x_i, z_i) such that $F(x_i, z_i) < t_i + \varepsilon$, $i = 1, 2$. By the convexity of F, it follows that

$$F(\lambda x_1 + (1-\lambda) x_2, \lambda z_1 + (1-\lambda) z_2) \leqslant$$
$$\leqslant \lambda F(x_1, z_1) + (1-\lambda) F(x_2, z_2) < \lambda (t_1 + \varepsilon) + (1-\lambda)(t_2 + \varepsilon) = \lambda t_1 + (1-\lambda) t_2 + \varepsilon.$$

Since $\varepsilon > 0$ is arbitrary,

$$F(\lambda x_1 + (1-\lambda) x_2, \lambda z_1 + (1-\lambda) z_2) \leqslant \lambda t_1 + (1-\lambda) t_2 \Rightarrow S(\lambda z_1 + (1-\lambda) z_2) \leqslant$$
$$\leqslant \lambda t_1 + (1-\lambda) t_2 \Rightarrow (\lambda z_1 + (1-\lambda) z_2, \lambda t_1 + (1-\lambda) t_2) \in \operatorname{epi} S. \ \blacksquare$$

Proposition 1. Let X and Y be linear spaces, let $A \subset X$ be a convex set, and let $\Lambda: X \to Y$ be a linear mapping.

If the functions $f_i: X \to \overline{\mathbf{R}}$, $i = 0, 1, \ldots, m$, are convex, the function $f(x; \alpha, \eta)$ determined by equality (1) is convex on $X \times \mathbf{R}^m \times Y$.

Proof. The sets

$$\begin{aligned} M_0 &= \{(x, \alpha, \eta, t) \mid (x, t) \in \operatorname{epi} f_0\}, \\ M_i &= \{(x, \alpha, \eta, t) \mid f_i(x) + \alpha_i \leqslant a_i\}, \\ M_A &= \{(x, \alpha, \eta, t) \mid x \in A\}, \\ M_\Lambda &= \{(x, \alpha, \eta, t) \mid \Lambda x + \eta = b\} \end{aligned}$$

are convex in $X \times \mathbf{R}^m \times Y \times \mathbf{R}$. Indeed, we obtain (changing, when necessary, the order of the factors): $M_0 = \operatorname{epi} f_0 \times \mathbf{R}^m \times Y$, $M_A = A \times \mathbf{R}^m \times Y \times \mathbf{R}$, $M_\Lambda = \tilde{\Lambda}^{-1}(b) \times \mathbf{R}^m \times \mathbf{R}$, where $\tilde{\Lambda}: (x, y) \to \Lambda x + y$ is a linear mapping from $X \times Y$ into Y and $M_i = \sigma \operatorname{epi}\{f_i - a_i\} \times \mathbf{R}^{m-1} \times Y \times \mathbf{R}$, where $\sigma: (x, t) \to (x, -t)$ is a symmetry transformation in $X \times \mathbf{R}$. It follows that all these sets are convex, and it remains to note that

$$\operatorname{epi} f = \bigcap_{i=0}^{m} M_i \cap M_A \cap M_\Lambda. \ \blacksquare$$

COROLLARY 1. The S-function of the problem $(\mathfrak{z}(\alpha, \eta))$ is convex on $\mathbf{R}^m \times Y$.

As was already seen in Section 2.6, convexity makes it possible to associate with various objects (functions and sets) the objects dual to them in the conjugate spaces. The same applies to the convex programming problems. In what follows we shall assume that X and Y are locally convex spaces, X* and Y* are their conjugate spaces, and $\lambda=(\lambda_1, \ldots, \lambda_m)\in \mathbf{R}^{m*}$.

Definition 1. The family of extremal problems

$$g(x^*; \lambda, \eta^*)\to \sup \text{ (with respect to } \lambda\in \mathbf{R}^{m*}, \eta^*\in Y^*), \qquad (\mathfrak{z}^*(x^*))$$

where $(-1)g(x^*; \lambda, \eta^*) = f^*(x^*; \lambda, \eta^*)$ is the Young–Fenchel transform (Section 2.6.3) of the function $(x, \alpha, \eta)\to f(x, \alpha, \eta)$, determined by equality (1) is said to be *dual* to the family $\mathfrak{z}(\alpha, \eta)\Leftrightarrow(2)$.

The problem $\mathfrak{z}^*=\mathfrak{z}^*(0)$ is said to be *dual* to the problem $\mathfrak{z}=\mathfrak{z}(0, 0)$ [with respect to the family of perturbations $\mathfrak{z}(\alpha, \eta)$].

The value of the problem $\mathfrak{z}^*(x^*)$ will be denoted

$$\Sigma(x^*) = \sup_{(\lambda, \eta^*)} g(x^*; \lambda, \eta^*). \tag{4}$$

The function f* being convex, its opposite function g is concave. The dual problems could be called concave programming problem, but we would rather do without this new term.

Definition 2. By the *Lagrange function* of the pair of dual families $\mathfrak{z}(\alpha, \eta)$ and $\mathfrak{z}^*(x^*)$ or the *extended Lagrange function* of the problem $(\mathfrak{z})$, we shall mean the function $\tilde{\mathscr{L}}: X\times\mathbf{R}^{m*}\times Y^*\to\mathbf{R}$ determined by the relations

$$\tilde{\mathscr{L}}(x; \lambda, \eta^*) = \begin{cases} f_0(x)+\sum_{i=1}^m \lambda_i(f_i(x)-a_i)+\langle\eta^*, \Lambda x-b\rangle, \\ x\in A, \ \lambda\in\mathbf{R}_+^{m*}, \\ -\infty, \ x\in A, \ \lambda\notin\mathbf{R}_+^{m*}, \\ +\infty, \ x\notin A. \end{cases} \tag{5}$$

This function is convex with respect to x for fixed (λ, η^*), and the opposite function $-\tilde{\mathscr{L}}$ is convex with respect to (λ, η^*) for fixed x.

It should be noted that

$$\tilde{\mathscr{L}}(x; \lambda, \eta^*)=\mathscr{L}(x, \eta^*, \lambda, 1), \quad x\in A, \quad \lambda\in\mathbf{R}_+^{m*}, \tag{6}$$

where $\mathscr{L}$ is the Lagrange function of the problem $(\mathfrak{z})$ defined in Section 3.3.1.

Proposition 2. The pair of dual families $\mathfrak{z}(\alpha, \eta)$ and $\mathfrak{z}^*(x^*)$ is uniquely determined by their Lagrange function since

$$f(x; \alpha, \eta)=(-\tilde{\mathscr{L}})^{*(2)}, \quad g(x^*; \lambda, \eta^*)=-\tilde{\mathscr{L}}^{*(1)}, \tag{7}$$

where *(1) and *(2) denote the Young–Fenchel transforms with respect to the arguments x and (λ, η^*), respectively.

Proof. The definition of the Young–Fenchel transform implies

$$\begin{aligned} -g(x^*; \lambda, \eta^*) &= f^*(x^*; \lambda, \eta^*) = \\ &= \sup_{(x, \alpha, \eta)} \{\langle x^*, x\rangle+\lambda\alpha+\langle\eta^*, \eta\rangle-f(x; \lambda, \eta)\} = \\ &= \sup_{\substack{x\in A, \ \Lambda x+\eta=b \\ f_i(x)+\alpha_i\leqslant a_i}} \{\langle x^*, x\rangle+\lambda\alpha+\langle\eta^*, \eta\rangle-f_0(x)\} = \end{aligned}$$

$$= \begin{cases} \sup_{x\in A}\left\{\langle x^*, x\rangle-f_0(x)-\sum_{i=1}^m\lambda_i(f_i(x)-a_i)-\langle\eta^*, \Lambda x-b\rangle\right\}, \\ \lambda\in\mathbf{R}_+^{m*} \\ +\infty, \ \lambda\notin\mathbf{R}_+^{m*} \end{cases} = \tilde{\mathscr{L}}^{*(1)}(x^*; \lambda, \eta^*),$$

which proves the second of equalities (7). Incidentally, we have derived the useful relation

$$g(x^*;\lambda,\eta^*)=\begin{cases}\inf\limits_{x\in A}(\tilde{\mathscr{L}}(x;\ \lambda,\ \eta^*)-\langle x^*,\ x\rangle),\ \lambda\in\mathbf{R}_+^{m*},\\ -\infty,\quad \lambda\notin\mathbf{R}_+^{m*}.\end{cases}\tag{8}$$

On the other hand,†

$$(-\tilde{\mathscr{L}})^{*(2)}=\sup_{(\lambda,\ \eta^*)}(\lambda\alpha+\langle\eta^*,\ \eta\rangle+\tilde{\mathscr{L}}(x;\ \lambda,\ \eta^*))=\begin{cases}+\infty,\ x\notin A,\\ \sup\limits_{\lambda\geqslant 0,\ \eta^*}(\lambda\alpha+\langle\eta^*,\ \eta\rangle+f_0(x)+\\ +\sum\limits_{i=1}^{m}\lambda_i(f_i(x)-a_i)+\langle\eta^*,\Lambda x-b\rangle)\end{cases}=$$

$$=\begin{cases}+\infty\ \text{if}\ x\notin A\ \text{or}\ \Lambda x-b\neq-\eta\ \text{or}\ f_i(x)-a_i>-\alpha_i\\ \qquad\text{for some}\ i,\\ f_0(x)\ \text{in all the other cases.}\end{cases}=f(x;\ \alpha,\ \eta).\ \blacksquare$$

COROLLARY 2. The conjugate function to the S-function of the problem $\mathfrak{z}(\alpha,\ \eta)$ has the form

$$S^*(\lambda,\ \eta^*)=-g(0;\ \lambda,\ \eta^*)=\begin{cases}-\inf\limits_{x\in A}\mathscr{L}(x,\eta^*,\lambda,1),\ \lambda\in\mathbf{R}_+^{m*},\\ +\infty,\ \lambda\notin\mathbf{R}_+^{m*}.\end{cases}\tag{9}$$

Proof. By definition,

$$S^*(\lambda,\ \eta^*)=\sup_{(\alpha,\ \eta)}(\lambda\alpha+\langle\eta^*,\ \eta\rangle-\inf_x f(x;\ \alpha,\ \eta))=$$
$$=\sup_{(\alpha,\ \eta,\ x)}(\lambda\alpha+\langle\eta^*,\ \eta\rangle-f(x;\ \alpha,\ \eta))=-g(0;\ \lambda,\ \eta^*),$$

after which (9) follows from (6) and (8).

Hence, the problem $\mathfrak{z}^*=\mathfrak{z}^*(0)$ dual to the problem $\mathfrak{z}=\mathfrak{z}(0,\ 0)$ can also be stated thus:

$$-S^*(\lambda,\ \eta^*)=\inf\mathscr{L}(x,\ \eta^*,\ \lambda,\ 1)\dashrightarrow\sup;\qquad \lambda\in\mathbf{R}_+^{m*},\ \eta^*\in Y^*.\tag{10}$$

Remark. Generally speaking, the definition of the dual problem depends on the family of perturbations in which the original problem $(\mathfrak{z})$ is included. Nevertheless, the equalities specifying the extended Lagrange function and equality (7) show that in a sense the families $\{(\mathfrak{z}(\alpha,\ \eta))\}$ and $\{(\mathfrak{z}^*(x^*))\}$ correspond to the problem $(\mathfrak{z})$ in a natural way, and hence the problem $(\mathfrak{z}^*)$ equivalent to (10) is the natural dual problem of $(\mathfrak{z})$ in the same sense.

Exercise. Show that if the functions f_i and the set A are convex and closed and if $f_0(x)$ does not assume the value $-\infty$ on the set of the elements x admissible in the problem $(\mathfrak{z})$, then the family dual to the family $\{(\mathfrak{z}^*(x^*))\}$ (see the footnote to the proof of Proposition 2) coincides with $\{(\mathfrak{z}(\alpha,\ \eta))\}$, and thus $(\mathfrak{z})$ and $(\mathfrak{z}^*)$ form a pair of dual problems.

Duality Theorem for Convex Programming Problems. Let us suppose that the S-function of the family $\{(\mathfrak{z}(\alpha,\ \eta))\}$ is continuous at the point (0, 0). Then

$$S(\alpha,\ \eta)=\sup_{\lambda\geqslant 0,\ \eta^*\in Y^*}(\lambda\alpha+\langle\eta^*,\ \eta\rangle+\inf_{x\in A}\mathscr{L}(x,\ \eta^*,\ \lambda,\ 1)\tag{11}$$

†As in Section 2.6.3, when computing the conjugate function to a function defined on the conjugate space (on $\mathbf{R}^{m*}\times Y^*$ in the case under consideration), we regard the result as a function on the original space and not on the second conjugate space.

for any $(\alpha, \eta) \in \operatorname{int}(\operatorname{dom} S)$, and the values of the problem $(\mathfrak{z})$ and its dual problem $(\mathfrak{z}^*)$ are equal:

$$\inf_{\substack{x \in A \\ \Lambda x = b \\ f_i(x) \leqslant a_i,\ i=1,\ldots,m}} f_0(x) = S(0, 0) = \Sigma(0) =$$

$$= \sup_{\substack{\lambda \geqslant 0 \\ \eta^* \in Y^*}} \left\{ \inf_{x \in A} \left\{ f_0(x) + \sum_{i=1}^{m} \lambda_i (f_i(x) - a_i) + \langle \eta^*, \Lambda x - b \rangle \right\} \right\}. \tag{12}$$

Proof. By Corollary 1, the function S is convex, and its continuity at one point implies (see Proposition 3 of Section 2.6.2) its continuity and the equality $S(\alpha, \eta) = (\overline{\operatorname{conv} S})(\alpha, \eta)$ for all $(\alpha, \eta) \in \operatorname{int}(\operatorname{dom} S)$. By the Fenchel–Moreau theorem (Section 2.6.3) and Corollary 2,

$$S(\alpha, \eta) = S^{**}(\alpha, \eta) = \sup_{(\lambda, \eta^*)} (\lambda\alpha + \langle \eta^*, \eta \rangle - S^*(\lambda, \eta^*)) =$$

$$= \sup_{\substack{\lambda \geqslant 0 \\ \eta^* \in Y^*}} (\lambda\alpha + \langle \eta^*, \eta \rangle + \inf_{x \in A} \mathscr{L}(x, \eta^*, \lambda, 1)),$$

which proves (11). Substituting $\alpha = 0$ and $\eta = 0$, we obtain (12). ■

COROLLARY 3 (Minimax Theorem). Under the conditions of the duality theorem, the relation

$$\sup_{\substack{\lambda \geqslant 0 \\ \eta^* \in Y^*}} \inf_{x \in A} \mathscr{L}(x, \eta^*, \lambda, 1) = \inf_{x \in A} \sup_{\substack{\lambda \geqslant 0 \\ \eta^* \in Y^*}} \mathscr{L}(x, \eta^*, \lambda, 1) \tag{13}$$

holds.

Proof. The left-hand side of (13) is equal to $\Sigma(0)$ and, by virtue of (12), it coincides with S(0, 0). As to the right-hand side, the formula (7) implies that it is equal to

$$\inf_{x \in A} \sup_{\substack{\lambda \geqslant 0 \\ \eta^* \in Y^*}} \mathscr{L}(x, \eta^*, \lambda, 1) = \inf_{x \in A} \sup_{(\lambda, \eta^*)} \tilde{\mathscr{L}}(x; \lambda, \eta^*) =$$

$$= \inf_{x \in A} ((-\tilde{\mathscr{L}})^{*(2)}(x; 0, 0)) = \inf_{x \in A} f(x; 0, 0)$$

and thus it also coincides with S(0, 0).

3.3.3. Linear Programming: the Existence Theorem and the Duality Theorem. Let X be a linear space, let X' be the space algebraically conjugate to X (i.e., the collection of *all* the linear functionals on X), and let K be a polyhedral cone, i.e. the intersection of a finite number of half-spaces $H_j = \{\langle x_j', x \rangle \leqslant 0, x_j' \in X',\ \ j = 1, \ldots, s\}$ (the condition that the cone is polyhedral is sometimes dropped).

The extremal problems of the type

$$\langle x_0', x \rangle \to \inf;\ f_i(x) = \langle x_i', x \rangle \geqslant b_i,\ i = 1, \ldots, m,\ x \in K, \tag{1}$$

form a separate class of *linear programming problems*.

In this section we shall consider the *finite-dimensional case* when $X = \mathbf{R}^n$, $X' = \mathbf{R}^{n*}$, and $K = \mathbf{R}^n_+$. In this case the problem (1) can be rewritten thus†:

$$cx \to \inf;\quad Ax \geqslant b,\quad x \geqslant 0. \tag{2}$$

†We remind the reader that for finite-dimensional spaces, we have two types of notation:

$$\langle c, x \rangle = cx = \sum_{i=1}^{n} c_i x_i,\ c \in \mathbf{R}^{n*},\ x \in \mathbf{R}^n.$$

where $c \in \mathbf{R}^{n*}$, $A=(a_{ij})$, $j = 1,\ldots,n$, $i = 1,\ldots,m$, is a matrix specifying a linear operator from $\mathbf{R}^n$ into $\mathbf{R}^m$, and $b \in \mathbf{R}^m$. For finite-dimensional vectors z' and z the symbol $z' \geqslant z$ means that all the coordinates of the vector z do not exceed the corresponding coordinates of the vector z'. Problem (2) will be called a *finite-dimensional linear programming problem.* This problem is a special case of the general convex programming problem. However, it involves a still more special structure: here *all the functions are linear and the cone is polyhedral.* As usual, we shall agree that if the problem (2) is inconsistent, i.e., has no admissible elements, its value is assumed to be equal to $+\infty$.

The Existence Theorem. If the set of admissible elements of the finite-dimensional linear programming problem (2) is nonempty and its value is finite, then the problem possesses a solution.

Proof. The proof of the theorem is based on a simple fact of finite-dimensional geometry. We remind the reader that the conic hull cone C of a set $C \subset X$ is determined by the equality

$$\operatorname{cone} C=\left\{x \in X \,\middle|\, x=\sum_{i=1}^{s} \lambda_i x_i,\ \lambda_i \geqslant 0,\ x_i \in C\right\}$$

and is a convex set (Proposition 1 of Section 2.6.1).

We shall say that the cone cone C is generated by the set C.

A) The Lemma on the Closedness of a Finitely Generated Cone in a Finite-Dimensional Space. Every cone in a finite-dimensional space generated by a finite set of points is closed.

Proof. Let $K=\operatorname{cone} C \subset \mathbf{R}^N$, $C=\{z_1, \ldots, z_k\}$, and $z_i \in \mathbf{R}^N$. We shall prove the lemma by induction.

1) k = 1. Then $K=\operatorname{cone}\{z_1\}=\{x \mid x=\lambda_1 z_1,\ \lambda_1 \geqslant 0\}$ is a closed half-line.

2) Now let us suppose that the lemma is true for $k = s - 1$. We shall prove it for $k = s$. There are two possible cases here:

a) the cone K contains the vectors $-z_1,\ldots,-z_s$. Then K is a subspace of a finite-dimensional space and thus is a closed subset;

b) at least one of the vectors $(-z_i)$, $i = 1,\ldots,s$, say $(-z_s)$, does not belong to the cone K. Let us denote $K_1 = \operatorname{cone}\{z_1,\ldots,z_{s-1}\}$. By the induction hypothesis, the cone K_1 is closed. Let a vector z belong to the closure of the cone K, i.e., $z=\lim_n \xi_n;\ \xi_n \in K,\ n \geqslant 1$. According to the definition,

$$\xi_n \in K=\operatorname{cone}\{z_1, \ldots, z_s\} \Leftrightarrow \xi_n=\sum_{i=1}^{s} \lambda_{in} z_i=\zeta_n+\lambda_{sn} z_s,$$

where $\zeta_n \in K_1$ and $\lambda_{sn} \geqslant 0$.

If we assume that $\lambda_{sn} \to \infty$ for $n \to \infty$, then

$$-z_s=\lim_{n\to\infty} \frac{\zeta_n-\xi_n}{\lambda_{sn}}=\lim_{n\to\infty} \frac{\zeta_n}{\lambda_{sn}} \in K_1$$

(we remind the reader that $\xi_n \to z$, and therefore $\xi_n/\lambda_{sn} \to 0$, and $\zeta_n/\lambda_{sn} \in K_1$ since $\zeta_n \in K_1$), i.e., $-z_s \in K_1$ because K_1 is closed, which contradicts the hypothesis. Hence λ_{sn} does not tend to infinity, and there is a subsequence λ_{sn_k} convergent to a number $\lambda_{s0} \geqslant 0$. In this case

$$\zeta_{n_k}=\xi_{n_k}-\lambda_{sn_k} z_s \longrightarrow z-\lambda_{s0} z_s=\tilde{z}.$$

Since K_1 is closed, we have $\tilde{z} \in K_1$, and consequently

$$z=\tilde{z}+\lambda_{s0} z_s \in K. \blacksquare$$

B) Now we shall prove the existence theorem.

Let us consider the set K in the space $\mathbf{R}^{m+1}$ formed of those vectors (α, z), $\alpha \in \mathbf{R}$, $z \in \mathbf{R}^m$, for each of which there is at least one vector $\tilde{x} \in \mathbf{R}^n_+$ such that $c\tilde{x} \leqslant \alpha$, $A\tilde{x} \geqslant z$.

The set K is a cone because if $(\alpha, z) \in K$, then $c\tilde{x} \leqslant \alpha$, $A\tilde{x} \geqslant z$ for some $\tilde{x} \in \mathbf{R}^n_+$. Therefore, for any $t \geqslant 0$ we have $t\tilde{x} \in \mathbf{R}^n_+$, $c(t\tilde{x}) \leqslant t\alpha$, $A(t\tilde{x}) \geqslant tz$, i.e., $t(\alpha, z) \in K$. The convexity of K is proved in a similar way.

Let us show that the cone K is generated by a finite set of vectors belonging to $\mathbf{R}^{m+1}$:

$$\zeta_1 = (c_1, a_{11}, \ldots, a_{m1}),\ \zeta_2 = (c_2, a_{12}, \ldots, a_{m2}), \ldots, \zeta_n = (c_n, a_{1n}, \ldots, a_{mn}),$$

$$\zeta_{n+1} = (1, 0, \ldots, 0),\ \zeta_{n+2} = (0, -1, 0, \ldots, 0), \ldots, \zeta_{n+m+1} = (0, \ldots, 0, -1).$$

First of all, we have $\zeta_i \in K$, and hence $\{\zeta_1, \ldots, \zeta_{n+m+1}\} \subset K$. Indeed, if $1 \leqslant i \leqslant n$, then the standard base vector e_i should be taken as $\tilde{x}$; for the other values of i we must take $\tilde{x} = 0$. Now let $\zeta = (\alpha, z) = (\alpha, z_1, \ldots, z_m) \in K$. Then for some $x \in \mathbf{R}^n$, $x = (x_1, \ldots, x_n)$, such that $x_1 \geqslant 0, \ldots, x_n \geqslant 0$ and some $\beta_0 \geqslant 0$, $\beta_j \geqslant 0$ the relations

$$\sum_{j=1}^{n} c_j x_j + \beta_0 = \alpha, \quad \sum_{j=1}^{n} a_{ij} x_j - \beta_j = z_i, \quad i = 1, \ldots, m,$$

hold. But this exactly means that

$$\zeta = (\alpha, z) = \sum_{j=1}^{n} x_j \zeta_j + \beta_0 \zeta_{n+1} + \sum_{i=1}^{m} \beta_i \zeta_{n+i+1},$$
$$x_j \geqslant 0, \quad \beta_0 \geqslant 0, \quad \beta_i \geqslant 0,$$

i.e.,

$$\zeta \in \operatorname{cone}\{\zeta_1, \ldots, \zeta_{n+m+1}\}.$$

According to the lemma proved in the foregoing section, the cone K is closed. By the condition of the theorem, the problem (2) has an admissible element $\tilde{x}$, and the value of the problem is $\hat{\alpha} > -\infty$. From the definition of K it follows that the point $(\tilde{\alpha}, b)$, where $\tilde{\alpha} = c\tilde{x}$, belongs to K. Moreover, we obviously have $\tilde{\alpha} \geqslant \hat{\alpha}$. Consequently, the set $\mathfrak{A} = \{\alpha \in \mathbf{R} \mid (\alpha, b) \in K\}$ is nonempty and $\hat{\alpha} = \inf\{\alpha \mid \alpha \in \mathfrak{A}\}$; i.e., $(\hat{\alpha}, b)$ belongs to the closure of the cone K, and hence to K itself. Therefore, there exists an element $\hat{x} \geqslant 0$ for which $A\hat{x} \geqslant b$ and $c\hat{x} \leqslant \hat{\alpha} \Rightarrow c\hat{x} = \hat{\alpha}$, i.e., $\hat{x}$ is a solution of the problem. ■

Now we proceed to the investigation of the dual problem. Here the duality theorem assumes a more complete form than in the foregoing section since the S-function of the problem turns out to be closed due to its specific structure.

In accordance with formulas (5) and (6) of Section 3.3.2, the extended Lagrange function is of the form

$$\tilde{\mathscr{L}}(x; \lambda) = \begin{cases} cx + \lambda(b - Ax), & x \in \mathbf{R}^n_+, \quad \lambda \in \mathbf{R}^{m*}_+, \\ -\infty, & x \in \mathbf{R}^n_+, \quad \lambda \notin \mathbf{R}^{m*}_+, \\ +\infty, & x \notin \mathbf{R}^n_+. \end{cases}$$

From this expression we find the family of perturbations of the problem (2) and the dual family. According to (7) in Section 3.3.2,

$$f(x; \alpha) = (-\tilde{\mathscr{L}})^{*(2)} = \sup_{\lambda} (\lambda\alpha + \mathscr{L}(x, \lambda)) =$$

$$= \begin{cases} +\infty, & x \notin \mathbf{R}^n_+, \\ \sup\limits_{\lambda \geqslant 0} (\lambda\alpha + cx + \lambda(b - Ax)), & x \in \mathbf{R}^n_+ \end{cases} = \begin{cases} cx, & x \geqslant 0, \ Ax \geqslant b + \alpha, \\ +\infty & \text{in all the other cases.} \end{cases}$$

Denoting $z = b + \alpha$, we obtain the perturbation of problem (2):

$$cx \to \inf, \quad x \geqslant 0, \quad Ax \geqslant z. \tag{3}$$

Similarly, for the determination of the dual problem, we have to calculate

$$g(0, \lambda) = (-\tilde{\mathscr{L}}^{*(1)})(0, \lambda) = -\sup_x (-\tilde{\mathscr{L}}(x, \lambda)) =$$

$$= \begin{cases} -\sup\limits_{x \geqslant 0} (-cx - \lambda(b - Ax)), & \lambda \geqslant 0, \\ -\infty, & \lambda \notin \mathbf{R}_+^{m*}, \end{cases} = \begin{cases} \lambda b, & \lambda \geqslant 0, \ \lambda A \leqslant c, \\ -\infty & \text{in all the other cases.} \end{cases}$$

This leads us to the problem

$$\lambda b \to \sup; \quad \lambda A \leqslant c, \quad \lambda \geqslant 0, \tag{4}$$

dual to problem (2). It possesses the same structure as problem (2), and it can easily be seen that if a dual problem is constructed proceeding from problem (4) in accordance with the same rule with the aid of which problem (4) was obtained from problem (2), we shall come back to the problem (2). Therefore, it makes sense to speak about a *pair of dual linear programming problems*.

The Duality Theorem. For a pair of dual linear programming problems the following alternative holds: either the values of the problems are finite and equal and solutions exist in both the problems or in one of the problems the set of admissible elements is empty.

In the first case elements $\hat{x} \in \mathbf{R}^n$ and $\hat{\lambda} \in \mathbf{R}^{m*}$ are solutions of problems (2) and (4), respectively, if and only if they are admissible in these problems and satisfy one of the two relations

$$c\hat{x} = \hat{\lambda} b \tag{5}$$

and,

$$\hat{\lambda}(A\hat{x} - b) = (\hat{\lambda} A - c)\hat{x}. \tag{6}$$

In the second case one of the problems is inconsistent and the other is either inconsistent (i.e., its set of admissible elements is empty) or has an infinite value.

Proof. A) The Construction of the S-Function. Let us again consider the same cone K as in the existence theorem. If $(\alpha, z) \in K$ and $\beta \geqslant \alpha$, then $(\beta, z) \in K$, and therefore K is the epigraph of the function

$$S(z) = \inf \{\alpha \mid (\alpha, z) \in K\}. \tag{7}$$

From the proof of the existence theorem it follows that inf is attained and S(z) is the value of problem (3); consequently, formula (7) specifies the S-function of problem (2). Since K is a convex and closed cone, the S-function of problem (2) is closed and convex.

B) Let us compute the function S* conjugate to the function S. According to the definition,

$$S^*(\lambda) = \sup_z (\lambda z - S(z)) = \sup_z \{\lambda z - \inf_x \{cx \mid x \in \mathbf{R}_+^n, \ Ax \geqslant z\}\} =$$

$$= \sup_{(z, x)} \{\lambda z - cx \mid x \in \mathbf{R}_+^n, \ z \in \mathbf{R}^m, \ z \leqslant Ax\}.$$

It is evident that $\sup\{\lambda z \mid z \in \mathbf{R}^m, \ z \leqslant Ax\} = \lambda Ax < \infty$ if and only if $\lambda \geqslant 0$.

Thus,

$$S^*(\lambda) = \begin{cases} \sup\limits_{x \geqslant 0} (\lambda A - c)x & \text{if} \quad \lambda \in \mathbf{R}_+^{m*}, \\ +\infty & \text{if} \quad \lambda \notin \mathbf{R}_+^{m*} \end{cases} =$$

$$= \begin{cases} 0 & \text{if } \lambda A \leqslant c, \ \lambda \geqslant 0, \\ +\infty & \text{in all the other cases.} \end{cases}$$

Therefore,

$$S^{**}(z) = \sup \{\lambda z \mid \lambda A \leqslant c,\ \lambda \geqslant 0\}. \tag{8}$$

In particular, it follows that $S^{**}(b)$ is the value of the dual problem (4).

C) Completion of the Proof. The function S cannot be identically equal to $+\infty$ because $S(0) \leqslant 0$ since zero is an admissible element in the problems $cx \to \inf$, $Ax \leqslant 0$, $x \geqslant 0$; consequently, $\operatorname{dom} S \neq \emptyset$.

Here one of the following two cases takes place:

1) $S(z) > -\infty$, $\forall z$ or

2) there exists $\tilde{z}$ for which $S(\tilde{z}) = -\infty$.

The case 1) in its turns splits into two subcases: 1a) $b \in \operatorname{dom} S$ and 1b) $b \notin \operatorname{dom} S$. If the subcase 1a) takes place, then S is a proper function, and the value of the problem is finite. By virtue of the closedness of S, the Fenchel—Moreau theorem (Section 2.6.3) implies that $S^{**}(b) = S(b)$, and now it is seen from (8) that the dual problem (4) has the same value as the primal problem (2) and, in particular, it is consistent. By virtue of the existence theorem, solutions exist in both the problems. If the subcase 1b) takes place, then $S(b) = +\infty$; i.e., problem (2) is inconsistent. The Fenchel—Moreau theorem again implies $S^{**}(b) = S(b)$, and hence problem (4) is consistent but its value is infinite.

If the case 2) takes place, we have $S(z) = -\infty$ for all $z \in \operatorname{dom} S$ (see Exercise 8 in Section 2.6.2). According to the definition, $S^*(\lambda) \equiv +\infty$ and $S^{**}(z) \equiv -\infty$. In particular, $S^{**}(b) = -\infty$, i.e., problem (4) is inconsistent.

If $b \in \operatorname{dom} S$, then problem (2) is consistent and its value is infinite: $S(b) = -\infty$. In case $b \notin \operatorname{dom} S$, we have $S(b) = +\infty$, and problem (2) is inconsistent. The proof of the alternative is completed.

Let us come back to the subcase 1a) in which, as was proved, solutions of problems (2) and (4) exist. We denote them by $\hat{x}$ and $\hat{\lambda}$, respectively. It is proved that the values of the problems are equal: $c\hat{x} = \hat{\lambda}b$, i.e., (5) holds, and therefore

$$\hat{\lambda}(A\hat{x} - b) = \hat{\lambda}A\hat{x} - \hat{\lambda}b = \hat{\lambda}A\hat{x} - c\hat{x} = (\hat{\lambda}A - c)\hat{x},$$

i.e., (6) holds. Further, if x and λ are admissible elements (i.e., $x \geqslant 0$, $\lambda \geqslant 0$, $\lambda A \leqslant c$, $Ax \geqslant b$), then

$$cx \geqslant \lambda Ax \geqslant \lambda b.$$

Therefore, if $c\hat{x} = \hat{\lambda}b$, then $\hat{x}$ and $\hat{\lambda}$ are solutions of the problems. Further, if $\hat{\lambda}(A\hat{x} - b) = (\hat{\lambda}A - c)\hat{x}$, then $c\hat{x} = \hat{\lambda}b$, i.e., $\hat{x}$ and $\hat{\lambda}$ are solutions of the problems.

Thus, if $\hat{x}$ and $\hat{\lambda}$ are solutions, then (5) and (6) hold; if $\hat{x}$ and $\hat{\lambda}$ are admissible and either (5) or (6) holds, then $\hat{x}$ and $\hat{\lambda}$ are solutions of the problems. The proof of the theorem is completed. ■

Exercise. What does the minimax theorem (Corollary 3 in Section 3.3.2) turn into in the situation under consideration?

3.3.4. Duality Theorem for the Shortest Distance Problem. Hoffman's Lemma and the Minimax Lemma. Let Y be a normed space, and let $B \subset Y$ be a nonempty subset of Y. The quantity

$$S_B(\eta) = \rho(\eta, B) = \inf_{y \in B} \|y - \eta\| \tag{1}$$

is called the *distance from the point* η *to the set* B. The investigation of the function $S_B(\eta)$ is one of the basic problems in the approximation theory (see [84]). If B is a convex set, we obtain a convex programming problem. The duality theorem proved below has many applications in mathematical analysis.

Thus, we shall suppose that B is a convex set. (In what follows it is fixed, and we drop the subscript B in the notation of the function S.) Then the function $\eta \to S(\eta)$ is the S-function of the following problem:

$$\|z\| \to \inf; \quad y - z = \eta, \quad y \in B. \tag{2}$$

Let us reduce (2) to the standard form of a family of convex programming problems [see ($\mathfrak{z}$ (α, η)) in Section 3.3.1]. To this end we put $X = Y \times Y$, $x = (y, z)$, $f_0(x) = \|z\|$, $\Lambda x = z - y$, Λ: $X \to Y$, $A = \{x = (y, z) \mid y \in B\}$, and then problem (2) assumes the form

$$f_0(x) \to \inf; \quad \Lambda x + \eta = 0, \quad x \in A. \tag{3}$$

It follows that function S is convex (Corollary 1 in Section 3.3.2).

For our further aims we need the following important geometrical notion.

Definition. Given a set $B \subset Y$, function $sB: Y^* \to \overline{\mathbf{R}}$ determined by the equality

$$sB(y^*) = \sup_{y \in B} \langle y^*, y \rangle$$

is called the *support function* of the set B.

Duality Theorem for the Shortest Distance Problem. The quantity $S(\eta)$ admits of the following dual representation:

$$S(\eta) = \rho(\eta, B) = \sup \{\langle y^*, y \rangle - sB(y^*) \mid \|y^*\| \leqslant 1\}. \tag{4}$$

Proof. It is obvious that $S(\eta) \geqslant 0$, $\forall \eta$. On the other hand, we have $S(\eta) \leqslant \|y_0 - \eta\|$, where y_0 is an arbitrary point belonging to B. Therefore, S is bounded both from above and from below, and consequently it is continuous throughout the space Y (Proposition 3 of Section 2.6.2). Now we can apply the duality theorem proved in Section 3.3.2. Since there are no inequality constraints in (2), the Lagrange function has the form

$$\mathcal{L}(x, y^*, 1) = \|z\| + \langle y^*, z - y \rangle.$$

From the formula (11) of Section 3.3.2 we obtain

$$S(\eta) = \sup_{y^*} \Big(\langle y^*, \eta \rangle + \inf_{\substack{y \in B \\ z \in Y}} (\|z\| + \langle y^*, z - y \rangle) \Big) =$$
$$= \sup_{y^*} \Big\{ \Big(\langle y^*, \eta \rangle - \sup_{y \in B} \langle y^*, y \rangle \Big) - \sup_{z \in Y} (\langle -y^*, z \rangle - \|z\|) \Big\} =$$
$$= \sup_{y^*} \{ \langle y^*, \eta \rangle - sB(y^*) - N^*(-y^*) \},$$

where $N(z) = \|z\|$, and $N^*(z) = 0$ for $\|z^*\| \leqslant 1$ and $N^*(z) = +\infty$ for $\|z^*\| > 1$ (Proposition 3 of Section 2.6.3), whence follows (4). ■

For our further aims we need the following generalization of Corollary 2 of Section 2.6.4:

The Lemma on the Conjugate Cone. Let X and Y be Banach spaces, let $\Lambda: X \to Y$ be a surjective linear operator from X into Y, let $x_1^*, \ldots, x_s^*$ be elements of the conjugate space X^*, and let

$$K = \{x \mid \langle x_i^*, x \rangle \leqslant 0, \ i = 1, \ldots, s, \ \Lambda x = 0\}$$

be a cone in X.

Then every element x_0^* belonging to the conjugate cone

$$K^* = \{x^* \mid \langle x^*, x \rangle \geqslant 0, \ x \in K\}$$

admits of the representation of the form

$$-x_0^* = \sum_{i=1}^{s} \lambda_i x_i^* + \Lambda^* y^*$$

for some $\lambda_i \geqslant 0$ and $y^* \in Y^*$.

Proof. A) Let us denote $L = \bigcap_{i=0}^{s} \operatorname{Ker} x_i^*$, $Z = \operatorname{Ker}\Lambda/L$, and let $\pi:\operatorname{Ker}\Lambda \to Z$ be the natural projection operator transforming $x \in \operatorname{Ker}\Lambda$ into the class $\pi(x) \in Z$ containing it. Let us show that $\dim Z \leqslant s + 1$. Indeed, let $z_0, \ldots, z_{s+1}$ be arbitrary elements of Z, $z_i = \pi(x_i)$. The homogeneous linear system of $s + 1$ equations

$$\sum_{j=0}^{s+1} \langle x_i^*, x_j\rangle \lambda_j = \left\langle x_i^*, \sum_{j=0}^{s+1} \lambda_j x_j \right\rangle = 0, \quad i = 0, 1, \ldots, s,$$

with $s + 2$ unknowns possesses a nonzero solution $\hat{\lambda}_0, \ldots, \hat{\lambda}_{s+1}$. Hence

$$\hat{x} = \sum_{j=0}^{s+1} \hat{\lambda}_j x_j \in \operatorname{Ker} x_i^*, \quad i = 0, \ldots, s, \Rightarrow \hat{x} \in L \Rightarrow 0 = \pi(\hat{x}) = \sum_{j=0}^{s+1} \hat{\lambda}_j \pi(x_j) = \sum_{j=0}^{s+1} \hat{\lambda}_j z_j.$$

Consequently, any $s + 2$ elements of Z are linearly dependent and $d = \dim Z \leqslant s + 1$.

B) Choosing a basis $f_1, \ldots, f_d$ in Z we establish the standard isomorphism between Z and $\mathbf{R}^d$ and between Z^* and $\mathbf{R}^{d*}$:

$$z = \sum_{i=1}^{d} \zeta_i f_i \mapsto (\zeta_1, \ldots, \zeta_d),$$

$$\langle z^*, z\rangle = \sum_{i=1}^{d} \langle z^*, f_i\rangle \zeta_i \Rightarrow z^* \mapsto (\zeta_1^*, \ldots, \zeta_d^*) = (\langle z^*, f_1\rangle, \ldots, \langle z^*, f_d\rangle).$$

Further, let us take the functionals $z_i^* \in Z^*$, $i = 0, 1, \ldots, s$, determined by the equalities

$$\langle z_i^*, \pi(x)\rangle = \langle x_i^*, x\rangle$$

and consider the convex cone

$$\tilde{K} = \operatorname{cone}\{z_1^*, \ldots, z_s^*\} = \left\{ z^* \,\middle|\, \sum_{i=1}^{s} \lambda_i z_i^*, \ \lambda_i \geqslant 0 \right\}$$

in Z^*. Since Z^* is finite-dimensional and the cone $\tilde{K}$ is finitely generated, the lemma of Section 3.3.3 implies that $\tilde{K}$ is closed.

Let us suppose that $-z_0^* \notin \tilde{K}$. Then, by the second separation theorem (Section 2.1.4), there exists a linear functional $l \in (Z^*)^*$ strongly separating $\{-z_0^*\}$ and $\tilde{K}$:

$$\langle l, -z_0^*\rangle > \sup\{\langle l, z^*\rangle \mid z^* \in \tilde{K}\} = \sup\left\{ \sum_{i=1}^{s} \lambda_i \langle l, z_i^*\rangle \mid \lambda_i \geqslant 0 \right\}.$$

Therefore, there must necessarily be $\langle l, z_i^*\rangle \leqslant 0$ (because, if otherwise, then $\sup = +\infty$), $0 = \sup\{\langle l, z^*\rangle \mid z^* \in \tilde{K}\}$, and $\langle l, z^*\rangle < 0$.

Now we note that l, like every linear functional on the finite-dimensional space Z^*, is specified by a linear form:

$$\langle l, z^*\rangle = \sum_{i=1}^{d} a_i \zeta_i^* = \sum_{i=1}^{d} a_i \langle z^*, f_i\rangle = \left\langle z^*, \sum_{i=1}^{d} a_i f_i \right\rangle = \langle z^*, a\rangle.$$

Choosing a representative $x_0 \in \operatorname{Ker}\Lambda$ of the class $a = \sum_{i=1}^{d} a_i f_i \in Z = \operatorname{Ker}\Lambda/L$, we see that, on the one hand,

$$\langle x_i^*, x_0\rangle = \langle z_i^*, \pi(x_0)\rangle = \langle z_i^*, a\rangle = \langle l, z_i^*\rangle \leqslant 0,$$

and consequently

$$x_0 \in \operatorname{Ker}\Lambda \cap \bigcap_{i=1}^{s} \{x \mid \langle x_i^*, x\rangle \leqslant 0\} = K,$$

and, on the other hand,

$$\langle x_0^*, x_0\rangle = \langle z_0^*, \pi(x_0)\rangle = \langle z_0^*, a\rangle = \langle l, z_0\rangle < 0,$$

and consequently $x_0^* \notin K^*$, which contradicts the condition of the lemma.

Thus, the assumption $-z_0^* \notin \tilde{K}$ leads to a contradiction, and therefore $-z_0^* = \sum_{i=1}^{s} \lambda_i z_i^*$ for some $\lambda_i \geqslant 0$.

C) For an arbitrary $x \in \operatorname{Ker}\Lambda$ we have

$$\left\langle x_0^* + \sum_{i=1}^{s} \lambda_i x_i^*, x \right\rangle = \left\langle z_0^* + \sum_{i=1}^{s} \lambda_i z_i^*, \pi(x) \right\rangle = 0.$$

Therefore, $x_0^* + \sum_{i=1}^{s} \lambda_i x_i^* \in (\operatorname{Ker}\Lambda)^{\perp}$, and the lemma on the kernel of a regular operator (Section 2.1.7) implies that $x_0^* + \sum_{i=1}^{s} \lambda_i x_i^* = \Lambda^*(-y^*)$ for some $y^* \in Y^*$. ■

Hoffman's Lemma. Let the same conditions be fulfilled as in the lemma on the conjugate cone.

Then for the distance function from a point x to K the inequality

$$\rho(x, K) \leqslant C\left\{\sum_{i=1}^{s} \langle x_i^*, x\rangle_+ + \|\Lambda x\|\right\} \tag{5}$$

holds, where $\langle x_i^*, x\rangle_+$ is equal to $\langle x_i^*, x\rangle$ if $\langle x_i^*, x\rangle \geqslant 0$ and is equal to zero in all the other cases, the constant C being independent of x.

Proof. The set K is the intersection of a finite collection of half-spaces and a subspace and is therefore a convex cone in X. Let us compute its support function sK. Let $x^* \in X^*$. Here one of the following two cases takes place: either there is an element $x_0 \in K$ such that $\langle x^*, x_0\rangle > 0$ or $\langle x^*, x\rangle \leqslant 0$, $\forall x \in K$. In the former case we have $sK(x^*) \geqslant t\langle x^*, x_0\rangle$ $\forall t \in \mathbf{R}_+$, and consequently $sK(x^*) = +\infty$. In the latter case $(-1)x^*$ belongs to the cone conjugate to the cone K, and hence, by the foregoing lemma, we have $x^* = \sum_{i=1}^{s} \lambda_i x_i^* + \Lambda^* y^*$, $\lambda_i \geqslant 0$, $y^* \in Y^*$. Thus

$$sK(x^*) = \begin{cases} 0 & \text{if } \exists \lambda_i \geq 0,\ y^* \in Y^*\colon\ x^* = \sum_{i=1}^{s} \lambda_i x_i^* + \Lambda^* y, \\ +\infty & \text{in all the other cases.} \end{cases}$$

Now, applying formula (4) (the duality theorem) to the problem under consideration, we obtain

$$\rho(x, K) = \sup\left\{\langle x^*, x\rangle \mid x^* = \sum_{i=1}^{s} \lambda_i x_i^* + \Lambda^* y^*,\ \lambda_i \geqslant 0,\ \|x^*\| \leqslant 1\right\}. \tag{6}$$

The subspace $L = \operatorname{lin}\{x_1^*, \ldots, x_s^*\} + \operatorname{Im}\Lambda^*$ is the sum of the closed subspace $L_1 = \operatorname{Im}\Lambda^*$ [because $\operatorname{Im}\Lambda^* = (\operatorname{Ker}\Lambda)^{\perp}$, and an annihilator is always closed] and the finite-dimensional space $L_2 = \{x_1^*, \ldots, x_s^*\}$. Therefore, L is closed in X (prove this!) and consequently it is a Banach space. The operator $\Lambda_1\colon \mathbf{R}^s \times Y^* \to L$, $\Lambda_1(\lambda, y) = \sum_{i=1}^{s} \lambda_i x_i^* + \Lambda^* y^*$ is linear and continuous, and it maps $\mathbf{R}^s \times Y^*$ onto the Banach space L. By the lemma on the right inverse mapping (Section 2.1.5), there exists a mapping $M_1\colon L \to \mathbf{R}^s \times Y^*$ such that $\Lambda_1 \circ M_1 = I_L$, $\|M_1 x^*\| \leqslant C\|x^*\|$. Therefore, if $\|x^*\| \leqslant 1$, then $\|M_1 x^*\|_{\mathbf{R}^s \times Y^*} = \sum_{i=1}^{s} |\lambda_i| + \|y^*\| \leqslant C$.

Hence we can assume that $0 \leqslant \lambda_i \leqslant C$, $\|y^*\| \leqslant C$ in expression (6), whence

$$S(x)=\rho(x, K) \leqslant \sup\left\{\left\langle \sum_{i=1}^{s} \lambda_i x_i^* + \Lambda^* y^*, x\right\rangle \Big| 0 \leqslant \lambda_i \leqslant C, \|y^*\| \leqslant C\right\} \leqslant$$

$$\leqslant C\left\{\sum_{i=1}^{s} \langle x_i^*, x\rangle_+ + \|\Lambda x\|\right\}. \blacksquare$$

It should be noted that if the cone K is specified only by equalities:

$$K=\{x \mid \langle x_i^*, x\rangle = 0, \Lambda x = 0\}$$

(i.e., it is a subspace), then it can also be specified in terms of inequalities:

$$K=\{x \mid \langle x_i^*, x\rangle \leqslant 0, \langle(-1)x_i^*, x\rangle \leqslant 0, \Lambda x = 0\}.$$

Applying Hoffman's lemma, we see that

$$\rho(x, K) \leqslant C\left\{\sum_{i=1}^{s} |\langle x_i^*, x\rangle| + \|\Lambda x\|\right\} \tag{5'}$$

in this case as well.

The Minimax Lemma. Let X and Y be Banach spaces, let $\Lambda: X \to Y$ be a surjective operator, and let $x_i^* \in X^*$, $i=1, \ldots, s$, $a=(a_1, \ldots, a_s)$. Let us define the function $S: R^s \times Y \to \overline{\mathbf{R}}$ by means of the equality

$$S(a, y) = \inf_{\Lambda x + y = 0} \max_{1 \leqslant i \leqslant s} (a_i + \langle x_i^*, x\rangle). \tag{7}$$

If $\max_{1 \leqslant i \leqslant s} \langle x_i^*, x\rangle \geqslant 0$ for any $x \in \operatorname{Ker}\Lambda$, then the following duality formula holds:

$$S(a, y) = \sup\left\{\sum_{i=1}^{s} \alpha_i a_i + \langle y^*, y\rangle \mid \alpha_i \geqslant 0,\right.$$

$$\left.\sum_{i=1}^{s} \alpha_i = 1, \Lambda^* y^* + \sum_{i=1}^{s} \alpha_i x_i^* = 0\right\}. \tag{8}$$

Moreover, inf in (7) is attained on some element $\hat{x} = \hat{x}(a, y)$ (possibly not unique), and there exist $C > 0$ and $\tilde{C} > 0$ independent of a and y such that

$$\|\hat{x}(a, y)\| \leqslant C\{|S(a, y)| + |a| + \|y\|\} \leqslant \tilde{C}(|a| + \|y\|) \tag{9}$$

for an appropriately chosen $\hat{x}(a, y)$.

Proof. The existence of the minimum in (7), the convexity of S, and formula (8) will be proved by the reduction to the two standard problems considered above.

A) Auxiliary Mappings. We shall use the lemma on the right inverse mapping (Section 2.1.5). According to the lemma, there exists $M: Y \to X$ such that $\Lambda \circ M = I$, $\|M(y)\| \leqslant C_1\|y\|$. Further, let $\varphi: X \to \mathbf{R}^s$ be specified by the equality $\varphi(x) = (\langle x_1^*, x\rangle, \ldots, \langle x_s^*, x\rangle)$. Let us denote $L = \varphi(\operatorname{Ker}\Lambda)$. Applying the same lemma once again, we conclude that there exists a mapping $\mu: L \to \operatorname{Ker}\Lambda$ such that

$$\varphi \circ \mu = I, \|\mu(\xi)\| \leqslant C_2|\xi|.$$

Further, L, like every subspace of $\mathbf{R}^s$, can be determined by a linear system of equations $\sum_{j=1}^{s} b_{lj}\xi_j = 0$, $l=1, \ldots, p$, or with the aid of a mapping $\beta: \mathbf{R}^s \to \mathbf{R}^p$, $\beta\xi = \left(\sum_{j=1}^{s} b_{lj}\xi_j\right)$ by means of the condition $\beta\xi = 0$.

B) The Boundedness of $S(a, b)$. We have

$$\Lambda x+y=0 \Leftrightarrow x=M(-y)+x_0, \quad x_0 \in \operatorname{Ker}\Lambda \Rightarrow a_i+\langle x_i^*, x\rangle = a_i+\langle x_i^*, M(-y)\rangle+\langle x_i^*, x_0\rangle. \tag{10}$$

Therefore, putting $x_0 = 0$ we obtain, on the one hand, the inequality

$$\inf_{\Lambda x+y=0} \max_{1\leqslant i\leqslant s} (a_i+\langle x_i^*, x\rangle) \leqslant \max_{1\leqslant i\leqslant s} \{|a_i|+C_1\|x_i^*\|\|y\|\} \overset{\text{def}}{=} K, \tag{11}$$

and, on the other hand, for $\Lambda x + y = 0$, according to (10),

$$\max_{1\leqslant i\leqslant s} (a_i+\langle x_i^*, x\rangle) \geqslant \min_{1\leqslant i\leqslant s} (a_i-C_1\|x_i^*\|\|y\|) \geqslant -K \tag{12}$$

because, by the hypothesis, $\max\limits_{1\leqslant i\leqslant s} (\langle x_i^*, x_0\rangle) \geqslant 0$ $(x_0 \in \operatorname{Ker}\Lambda)$. Now (11) and (12) imply the estimate

$$|S(a, y)| \leqslant K = \max_{1\leqslant i\leqslant s} \{|a_i|+C_1\|x_i^*\|\|y\|\}. \tag{13}$$

C) The Existence of the Minimum. Denoting $\tilde{a}_i = a_i + \langle x_i^*, M(-y)\rangle + K$ and recalling that

$$x_0 \in \operatorname{Ker}\Lambda \Leftrightarrow (\langle x_1^*, x_0\rangle, \ldots, \langle x_s^*, x_0\rangle) = \varphi(x_0) \in L = \varphi(\operatorname{Ker}\Lambda) = \operatorname{Ker}\beta,$$

we conclude from (10) and (7) that $S(a, y) + K$ is the value of the problem

$$\max(\tilde{a}_i+\xi_i) \to \inf;\ \xi \in L \Leftrightarrow c \to \inf;\ \tilde{a}_i+\xi_i \leqslant c, \beta\xi = 0. \tag{14}$$

By virtue of (13), we have $S(a, y) + K \geqslant 0$, and therefore the condition $c \geqslant 0$ can simply be added to (14).

Problem (14) can be brought to the standard form of a linear programming problem (2) of Section 3.3.3 in $\mathbf{R}^{s+1}$. To this end we denote $z_0 = c$, $z_i = c - \xi_i - \tilde{a}_i$, $z = (z_1,\ldots,z_s)$, $\theta = (1,\ldots,1)$, $\tilde{a} = (\tilde{a}_1,\ldots,\tilde{a}_s)$, $\tilde{z} = (z_0, z)$. The condition $\beta\xi = 0$ is written in the form

$$\beta\xi = 0 \Leftrightarrow \beta(c\theta - z - \tilde{a}) = c\beta\theta - \beta z - \beta\tilde{a} = 0 \Leftrightarrow \tilde{B}\tilde{z} \geqslant \tilde{b},$$

where

$$\tilde{B} = \begin{pmatrix} \beta\theta & -\beta \\ -\beta\theta & \beta \end{pmatrix} \quad \text{and} \quad \tilde{b} = \begin{pmatrix} \beta\tilde{a} \\ -\beta\tilde{a} \end{pmatrix}$$

are a matrix of order $2p \times (s + 1)$ and a $2p$-dimensional vector (we have in fact replaced the equality $\beta\xi = 0$ by the two inequalities $\beta\xi \geqslant 0$ and $\beta\xi \leqslant 0$).

Therefore, problem (14) is equivalent to the problem

$$z_0 \to \inf, \quad \tilde{B}\tilde{z} \geqslant \tilde{b}, \quad \tilde{z} \geqslant 0, \tag{15}$$

having the standard form. This problem is consistent [for example, $(2K, 2K - \tilde{a}_1,\ldots,2K - \tilde{a}_s)$ is an admissible vector to which $\xi = 0 \in L$ corresponds], and its value $S(a, y) + K$ is finite (and even nonnegative). By the existence theorem of Section 3.3.3, problem (15) possesses a solution: $\hat{z} = (\hat{z}_0, \hat{z}) = (S(a, y) + K, \hat{z})$. Consequently, problem (14) possesses the solution $\hat{\xi} = (S(a, y) + K)\theta - \hat{z} - \tilde{a}$, and therefore

$$S(a, y)+K = \max_{1\leqslant i\leqslant s} \{\tilde{a}_i+\hat{\xi}_i\} = \max_{1\leqslant i\leqslant s} \{a_i+\langle x_i^*, M(-y)\rangle+\hat{\xi}_i\}+K$$

and

$$S(a, y) = \max_{1\leqslant i\leqslant s} \{a_i + \langle x_i^*, \hat{x}\rangle\},$$

where, according to (10) and the definition of the mapping $\mu: L \to \operatorname{Ker}\Lambda$, we have

$$\hat{x}=M(-y)+\mu\hat{\xi}. \tag{16}$$

D) Convexity of S and the Duality Theorem. The function $(a, y) \to S(a, y)$ is the S-function of the following convex programming problem:

$$\max(\eta_1, \ldots, \eta_s) \to \inf, \quad -\eta_i + \langle x_i^*, x\rangle + a_i = 0, \quad \Lambda x + y = 0. \tag{17}$$

Putting $Z = \mathbf{R}^s \times X$, $\tilde{Y} = \mathbf{R}^s \times Y$, $z = (\eta, x)$,

$$f_0(z) = \max(\eta_1, \ldots, \eta_s),$$

$$\tilde{\Lambda}z = (\langle x_1^*, x\rangle - \eta_1, \ldots, \langle x_s^*, x\rangle - \eta_s, \Lambda x),$$

we reduce (17) to the standard form [$(\mathfrak{z}(\alpha, \eta))$ in Section 3.3.2]:

$$f_0(z) \to \inf, \quad \tilde{\Lambda}z + (a, y) = 0. \tag{18}$$

Consequently, function $(a, y) \to S(a, y)$ is convex (Corollary 1 of Section 3.3.2). According to (13), it is bounded and therefore continuous throughout the space $\tilde{Y}$ (Proposition 3 of Section 2.6.2). By the duality theorem proved in Section 3.3.2 [we remind the reader that problem (18) involves no inequality constraints],

$$S(a, y) = \sup_{z^*}\{\langle z^*, (a, y)\rangle + \inf_z[\langle z^*, \tilde{\Lambda}z\rangle + \max(\eta_1, \ldots, \eta_s)]\}. \tag{19}$$

However, $z^* \in (\mathbf{R}^s \times Y)^* = \mathbf{R}^{s*} \times Y^*$, i.e., z^* is representable in the form (α, y^*), $\alpha = (\alpha_1, \ldots, \alpha_s) \in \mathbf{R}^{s*}$, $y^* \in Y^*$, and therefore

$$\langle z^*, (a, y)\rangle = \alpha a + \langle y^*, y\rangle,$$

$$\langle z^*, \tilde{\Lambda}z\rangle = \sum_{i=1}^{s} \alpha_i(\langle x_i^*, x\rangle - \eta_i) + \langle y^*, \Lambda x\rangle. \tag{20}$$

According to Proposition 2 of Section 2.6.3, the Young–Fenchel transform of the function $f(\eta) = \max(\eta_1, \ldots, \eta_s)$ has the form

$$f^*(\alpha) = \sup_\eta(\alpha\eta - f(\eta)) = \begin{cases} 0 & \text{for } \alpha_i \geqslant 0, \ \sum_{i=1}^{s}\alpha_i = 1, \\ +\infty & \text{in all the other cases.} \end{cases} \tag{21}$$

Now from (20) and (21) we obtain

$$\inf_z(\max(\eta_1, \ldots, \eta_s) + \langle z^*, \tilde{\Lambda}z\rangle) =$$

$$= -\sup_{(\eta, x)}\left\{\alpha\eta - \max(\eta_1, \ldots, \eta_s) - \left\langle \sum_{i=1}^{s}\alpha_i x_i^* + \Lambda^* y^*, x\right\rangle\right\} =$$

$$= \begin{cases} 0 & \text{for } \alpha_i \geqslant 0, \ \sum_{i=1}^{s}\alpha_i = 1, \ \Lambda^* y^* + \sum_{i=1}^{s}\alpha_i x_i^* = 0, \\ -\infty & \text{in all the other cases.} \end{cases} \tag{22}$$

Substituting (22) into (19), we obtain (8).

E) A Geometrical Lemma. Let L be a subspace of $\mathbf{R}^s$ and let K be a finitely generated cone in $\mathbf{R}^s$. Then there exists $N > 0$ such that

$$\rho(0, (L+a) \cap K) \leqslant N\rho(0, L+a) \leqslant N|a| \tag{23}$$

for any $a \in \mathbf{R}^s$ [this formula remains valid in the case $(L+a) \cap K = \varnothing$ as well if we agree that $\rho(0, \emptyset) = -\infty$].

Proof. a) Let $K = \text{cone}\{x_1, \ldots, x_m\}$. Let us represent every vector as a sum of two components one of which is contained in the subspace L and the other is orthogonal to L:

$$a = b + c, \quad x_i = y_i + z_i, \quad i = 1, \ldots, m, \quad b, y_i \in L, \quad c, z_i \in L^\perp.$$

Then

$$\rho(0,\ (L+a)\cap K)=\inf\{|\xi|\,|\,\xi\in K,\ (\xi-a)\in L\}=$$
$$=\inf\left\{\left|\sum_{i=1}^{m}\lambda_i(y_i+z_i)\right|\,\middle|\,\lambda_i\geqslant 0,\ \sum_{i=1}^{m}\lambda_i(y_i+z_i)-b-c\in L\right\}=$$
$$=\inf\left\{\left|\sum_{i=1}^{m}\lambda_i(y_i+z_i)\right|\,\middle|\,\lambda_i\geqslant 0,\ \sum_{i=1}^{m}\lambda_i z_i=y_i\in L\right\}\leqslant$$
$$\leqslant\inf\left\{\sum_{i=1}^{m}\lambda_i\max_{1\leqslant i\leqslant m}\{|y_i|\}+|c|\,\middle|\,\lambda_i\geqslant 0,\ \sum_{i=1}^{m}\lambda_i z_i=y_i\in L\right\}. \tag{24}$$

On the other hand,

$$\rho(0,\ L+a)=\inf\{|\eta+a|\,|\,\eta\in L\}=\inf\{|\eta'+c|\,|\,\eta'\in L\}=|c|\leqslant\sqrt{|b|^2+|c|^2}=|a|.$$

Since

$$\lambda_i\geqslant 0,\quad \sum_{i=1}^{m}\lambda_i z_i=c\Leftrightarrow c\in\operatorname{cone}\{z_1,\ \ldots,\ z_m\},$$

we see that (23) will follow from (24) if

$$c\in\operatorname{cone}\{z_1,\ \ldots,\ z_m\}\Rightarrow\exists\lambda_i\geqslant 0,$$
$$\sum_{i=1}^{m}\lambda_i z_i=c,\quad \sum_{i=1}^{m}\lambda_i\leqslant N_1|c| \tag{25}$$

for some $N_1 > 0$ [and then we shall have $N=N_1\max\limits_{1\leqslant i\leqslant m}\{|y_i|\}+1$ in (23)].

b) Let $L_1=\operatorname{lin}\{z_1,\ \ldots,\ z_m\}$, $\dim L_1=n$. Let us separate out the various collections of indices $I = \{i_1,\ldots,i_n\}$ for which $\{z_{i_1},\ldots,z_{i_n}\}$ is a basis in L_1. To each of these collections there corresponds a linear mapping $\Lambda_I:\mathbf{R}^n\to L_1$ determined by the formula $\Lambda_I x=\sum_{k=1}^{n}x_k z_{i_k}$. Since $\{z_{i_1},\ldots,z_{i_n}\}$ is a basis in L_1, there exists the inverse mapping.

By Carathéodory's theorem (Section 2.6.1), every vector $c\in\operatorname{cone}\{z_1,\ \ldots,$ $z_m\}$ is a conic combination of not more than n linearly independent vectors z_i:

$$c=\lambda_{i_1}z_{i_1}+\ldots+\lambda_{i_s}z_{i_s},\quad \lambda_{i_k}>0,\quad s\leqslant n.$$

Completing, if necessary, the set $\{z_{i_1},\ldots,z_{i_s}\}$ with some vectors $z_{i_{s+1}},\ldots,$ z_{i_n}, to obtain a basis in L_1 and putting $\lambda_{i_{s+1}}=\ldots=\lambda_{i_n}=0$ we see that

$$c=\Lambda_I\lambda,\quad \lambda=(\lambda_{i_1},\ \ldots,\ \lambda_{i_n}),\quad \lambda_{i_k}\geqslant 0,\quad I=(i_1,\ \ldots,\ i_n).$$

Therefore

$$\sum_{k=1}^{n}\lambda_{i_k}\leqslant n|\lambda|\leqslant n\|\Lambda_I^{-1}\|\,|c|\leqslant n\max_I\{\|\Lambda_I^{-1}\|\}\,|c|.$$

This proves (25) for $N_1=n\max\limits_I\{\|\Lambda_I^{-1}\|\}$.

F) Completion of the Proof of the Minimax Lemma. It only remains to derive estimate (9). To this end we come back to the left problem in (14). An element $\tilde{\xi}=(\tilde{\xi}_1,\ldots,\tilde{\xi}_s)$ is its solution if and only if

$$\tilde{\xi}_i+\tilde{a}_i\leqslant S(a,\ y)+K,\quad i=1,\ \ldots,\ s,\quad \tilde{\xi}\in L \tag{26}$$

[since $S(a,\ b)+K$ is the value of the problem, for at least one index i the equality takes place in (26)]. Recalling the definition of a_i in Section C) and denoting $\hat{a}=(\hat{a}_1,\ldots,\hat{a}_s)$, where

$$\hat{a}_i=\tilde{a}_i-S(a,\ y)-K=a_i+\langle x_i^*,\ \mathrm{M}(-y)\rangle-S(a,\ y), \tag{27}$$

we rewrite (26) in the form $\tilde{\xi}_i + \hat{a}_i \leqslant 0$.

Thus, $\tilde{\xi}$ is a solution of (14) $\Leftrightarrow \tilde{\xi}+\hat{a} \in (L+\hat{a}) \cap (-\mathbf{R}_+^s)$. Therefore,

$$\inf\{|\tilde{\xi}+\hat{a}|\,|\,\tilde{\xi} \text{ is a solution of } (14)\} = \rho(0,\ (L+\hat{a}) \cap (-\mathbf{R}_+^s)).$$

Since $\mathbf{R}_+^s = \text{cone}\{-e_1, \ldots, -e_s\}$ is a finitely generated cone, the geometrical lemma of Section E) implies that the right-hand member of the last relation does not exceed $N\rho(0,\ L+\hat{a}) \leqslant N|a|$. Consequently,

$$\inf\{|\tilde{\xi}+\hat{a}|\,|\,\tilde{\xi} \text{ is a solution of } (14)\} \leqslant N|\hat{a}|,$$

and since $|\tilde{\xi}| \leqslant |\tilde{\xi}+\hat{a}|+|\hat{a}|$, we have

$$\inf\{|\tilde{\xi}|\,|\,\tilde{\xi} \text{ is a solution of } (14)\} \leqslant (N+1)|\hat{a}|.$$

If $\hat{a} \neq 0$, then $(N+1)|\hat{a}| < (N+2)|\hat{a}|$, and among the solutions of problem (14) there is one (we denote it by $\hat{\xi}$) for which

$$|\hat{\xi}| \leqslant (N+2)|\hat{a}|. \tag{28}$$

If $\hat{a} = 0$, then $\hat{\xi} = 0$ is a solution, and inequality (28) again holds.

Using formula (16), we find the element $\hat{x} = \hat{x}(a, y)$ from $\hat{\xi}$. To estimate this element we take into account the inequalities

$$\|M(-y)\| \leqslant C_1\|y\|, \quad \|\mu(\hat{\xi})\| \leqslant C_2|\hat{\xi}| \tag{29}$$

[cf. the definitions of the right inverse mappings M and μ in Section A)] and

$$\begin{aligned} |\hat{a}| = \left(\sum_{i=1}^{s} \hat{a}_i^2\right)^{1/2} &\leqslant \sqrt{s}\max\{|\hat{a}_1|, \ldots, |\hat{a}_s|\} \overset{(27)}{\leqslant} \\ &\leqslant \sqrt{s}\left[\max_{1\leqslant i\leqslant s}\{|a_i|+\|x_i^*\|C_1\|y\|\}+|S(a, y)|\right] \leqslant \\ &\leqslant \sqrt{s}\left[|a|+C_1\max_{1\leqslant i\leqslant s}\|x_i^*\|\|y\|+|S(a, y)|\right], \end{aligned} \tag{30}$$

and also estimate (13):

$$\begin{aligned} \|\hat{x}(a, y)\| &\overset{(16),(29)}{\leqslant} C_1\|y\|+C_2|\hat{\xi}| \overset{(28)}{\leqslant} C_1\|y\|+C_2(N+2)|\hat{a}| \overset{(30)}{\leqslant} \\ &\leqslant C_1(1+C_2(N+2)\sqrt{s}\max_{1\leqslant i\leqslant s}\|x_i^*\|)\|y\| + C_2(N+2)\sqrt{s}(|a|+|S(a, y)|) \overset{(13)}{\leqslant} \\ &\leqslant C_1(1+\sqrt{s}\max_{1\leqslant i\leqslant s}\|x_i^*\|2C_2(N+2))\|y\|+2C_2(N+2)\sqrt{s}|a|. \end{aligned}$$

Therefore, inequality (9) with the constants

$$C = \max\{C_1(1+C_2(N+2)\sqrt{s}\max_{1\leqslant i\leqslant s}\|x_i^*\|,\ C_2(N+2)\sqrt{s}\},$$
$$\tilde{C} = \max\{C_1(1+\sqrt{s}\max_{1\leqslant i\leqslant s}\|x_i^*\|2C_2(N+2)),\ 2C_2(N+2)\sqrt{s}\}$$

holds.

3.4*. Second-Order Necessary Conditions and Sufficient Conditions for Extremum in Smooth Problems

We shall again begin with the case when there are no inequality constraints.

3.4.1. Smooth Problems with Equality Constraints. We shall consider the problem

$$f(x) \to \text{extr}; \quad F(x) = 0. \tag{1}$$

THEOREM 1 (Second-Order Necessary Conditions). Let X and Y be Banach spaces, let U be an open set in X, and let the function $f:U \to \mathbf{R}$ and the mapping $F:U \to Y$ possess the second Fréchet derivatives at a point $\hat{x} \in U$. Let $\mathscr{L}(x, \hat{y}^*, 1) = f(x) + \langle \hat{y}^*, F(x)\rangle$.

If $\hat{x}$ yields a local minimum (maximum) in problem (1) and if F is regular at the point $\hat{x}$ [i.e., $\operatorname{Im} F'(\hat{x}) = Y$], then there exists a Lagrange multiplier $\hat{y}^* \in Y^*$ such that

$$\mathscr{L}_x(\hat{x}, \hat{y}^*, 1) = 0, \tag{2}$$

and for any y* possessing this property the relation

$$\mathscr{L}_{xx}(\hat{x}, \hat{y}^*, 1)[h, h] \geqslant 0 \ (\leqslant 0), \quad \forall h \in \operatorname{Ker} F'(\hat{x}) \tag{3}$$

holds.

Proof. The existence of y* for which equality (2) holds was proved in Section 3.2.2. Further we consider the case $\hat{x} \in \operatorname{locmin}(1)$. Let $h \in \operatorname{Ker} F'(x)$. According to Lyusternik's theorem (Section 2.3.5), we have $h \in T_{\hat{x}}\mathscr{M}$ where $\mathscr{M} = \{x \mid F(x) = 0\}$; i.e., there exists a mapping $r(\cdot): [-\varepsilon, \varepsilon] \to \mathscr{M}$ such that

$$F(\hat{x} + th + r(t)) = 0, \quad r(t) = o(t), \quad t \in [-\varepsilon, \varepsilon]. \tag{4}$$

By virtue of (4), the element $\hat{x} + th + r(t)$ is an admissible element in the problem for $t \in [-\varepsilon, \varepsilon]$, and consequently $f(\hat{x}) \leqslant f(\hat{x} + th + r(t))$ because $\hat{x} \in \operatorname{locmin}(1)$. Therefore,

$$\begin{gathered} f(\hat{x}) \leqslant f(\hat{x} + th + r(t)) = \mathscr{L}(\hat{x} + th + r(t), \hat{y}^*, 1) = \\ = \mathscr{L}(\hat{x}, \hat{y}^*, 1) + \mathscr{L}_x(\hat{x}, \hat{y}^*, 1)[th + r(t)] + \\ + \frac{1}{2}\mathscr{L}_{xx}(\hat{x}, \hat{y}^*, 1)[th + r(t), th + r(t)] + o(t^2) = f(\hat{x}) + \frac{t^2}{2}\mathscr{L}_{xx}(\hat{x}, \hat{y}^*, 1)[h, h] + o(t^2), \end{gathered}$$

whence (3) immediately follows. ■

THEOREM 2 (a Sufficient Condition for a Minimum). Let the conditions of the foregoing theorem be fulfilled and, moreover, let the inequality

$$\mathscr{L}_{xx}(\hat{x}, \hat{y}^*, 1)[h, h] \geqslant 2\alpha\|h\|^2, \quad \forall h \in \operatorname{Ker} F'(\hat{x}) \tag{5}$$

be fulfilled for some $\alpha > 0$. Then $\hat{x}$ is a point of local minimum in problem (1).

Proof. Without loss of generality, we can assume that $f(\hat{x}) = 0$. Let us denote by $B(h_1, h_2)$ the bilinear form $\frac{1}{2}\mathscr{L}_{xx}(\hat{x}, \hat{y}^*, 1)[h_1, h_2]$. By the theorem on mixed derivatives (Section 2.2.5), B is a continuous symmetric bilinear form. Let us choose $\varepsilon > 0$ such that

$$\varphi(\varepsilon) = \alpha(1-\varepsilon)^2 - 2\|B\|(1+\varepsilon)\varepsilon - \|B\|\varepsilon^2 - \frac{\alpha}{2} > 0 \tag{6}$$

[this can be done because $\varphi(0) = \alpha/2 > 0$].

By the hypothesis, the functions F and $\mathscr{L}(\cdot, \hat{y}^*, 1)$ possess the second Fréchet derivatives with respect to x at the point $\hat{x}$. Using Taylor's formula (Theorem 2 of Section 2.2.5) and taking into account that $F(\hat{x}) = 0$, $\mathscr{L}(\hat{x}, \hat{y}^*, 1) = 0$ and $\mathscr{L}_x(\hat{x}, \hat{y}^*, 1) = 0$, we find $C_1 > 0$ and $\delta_1 > 0$ such that the inequalities

$$\begin{gathered} \|F(\hat{x}+h) - F'(\hat{x})[h]\| = \|F(\hat{x}+h) - F(\hat{x}) - F'(\hat{x})[h]\| \leqslant C_1\|h\|^2, \\ |\mathscr{L}(\hat{x}+h, \hat{y}^*, 1) - B(h, h)| \leqslant \frac{\alpha}{2}\|h\|^2 \end{gathered} \tag{7}$$

hold for $\|h\| < \delta_1$. Applying the lemma proved in Section 2.1.5, we construct the right inverse mapping $M: Y \to X$, $F'(\hat{x}) \circ M = I$, $\|M(y)\| \leqslant C\|y\|$ to

$F'(x)$. For $\varepsilon > 0$ chosen earlier we find δ, $0 < \delta < \delta_1$, such that

$$\delta C C_1 < \varepsilon. \tag{8}$$

Now let $\|h\| < \delta$, and let $\hat{x} + h$ be an admissible element in the problem, i.e., $F(\hat{x} + h) = 0$. Let us put $h_2 = M(F'(\hat{x})h)$ and denote by h_1 the difference $h - h_2$. Then from the estimate for $M(y)$ and from (7) and (8) we obtain

$$\|h_2\| \leqslant C\|F'(\hat{x})[h]\| \leqslant CC_1\|h\|^2 < \varepsilon\|h\|, \tag{9}$$

$$F'(\hat{x})[h_1] = F'(\hat{x})[h - h_2] = F'(\hat{x})[h] - F'(\hat{x})M(F'(\hat{x})[h]) = 0.$$

Hence $h_1 \in \operatorname{Ker} F'(\hat{x})$. From (9) it follows that $(1-\varepsilon)\|h\| \leqslant \|h_1\| \leqslant (1+\varepsilon)\|h\|$. As a result, taking into account (6), (5), and (7) we obtain

$$f(\hat{x}+h) = \mathscr{L}(\hat{x}+h, \hat{y}^*, 1) \geqslant B(h, h) - \frac{\alpha}{2}\|h\|^2 = B(h_1+h_2, h_1+h_2) - \frac{\alpha}{2}\|h\|^2 \geqslant$$

$$\geqslant B(h_1, h_1) - 2\|B\|\|h_1\|\|h_2\| - \|B\|\|h_2\|^2 - \frac{\alpha}{2}\|h\|^2 \geqslant$$

$$\geqslant \left(\alpha(1-\varepsilon)^2 - 2\|B\|(1+\varepsilon)\varepsilon - \|B\|\varepsilon^2 - \frac{\alpha}{2}\right)\|h\|^2 > 0,$$

i.e., $\hat{x} \in \operatorname{locmin}(1)$. ■

3.4.2. Second-Order Necessary Conditions for Smooth Problems with Equality and Inequality Constraints. We shall consider the problem

$$f_0(x) \to \inf, \quad F(x) = 0, \quad f_i(x) \leqslant 0, \quad i = 1, \ldots, m. \tag{$\mathfrak{z}$}$$

Changing, if necessary, the signs of the functions we can reduce to ($\mathfrak{z}$) any problem of Section 3.2. The Lagrange function of the problem ($\mathfrak{z}$) has the form

$$\mathscr{L}(x, y^*, \lambda, \lambda_0) = \sum_{i=0}^{m} \lambda_i f_i(x) + \langle y^*, F(x)\rangle.$$

THEOREM. Let X and Y be Banach spaces, let U be an open set in X, let the functions $f_i: U \to R$, $i = 0, 1, \ldots, m$, and the mapping $F: U \to Y$ possess the second Fréchet derivatives at a point $\hat{x} \in U$, and, moreover, let $f_i(\hat{x}) = 0$, $i = 1, \ldots, m$.

If $\hat{x}$ yields a local minimum in the problem ($\mathfrak{z}$), and if F is regular at the point $\hat{x}$ [i.e., $\operatorname{Im} F'(x) = Y$], then

a) the collection D of the Lagrange multipliers $(y^*, \lambda, \lambda_0)$, $y^* \in Y^*$, $\lambda \in \mathbf{R}^{m*}$ such that

$$\lambda_0 \geqslant 0, \quad \lambda_i \geqslant 0, \quad \sum_{i=0}^{m} \lambda_i = 1, \qquad \mathscr{L}_x(\hat{x}, y^*, \lambda, \lambda_0) = 0 \tag{1}$$

is a nonempty convex compact set in $Y^* \times \mathbf{R}^{m*} \times \mathbf{R}$;

b) for any h_0 belonging to the subspace

$$L = \{h \mid \langle f_i'(\hat{x}), h\rangle = 0, \quad i \geqslant 0, \quad F'(\hat{x})[h] = 0\}, \tag{2}$$

there are Lagrange multipliers $(y^*(h_0), \lambda(h_0), \lambda_0(h_0)) \in D$ such that

$$\mathscr{L}_{xx}(\hat{x}, y^*(h_0), \lambda(h_0), \lambda_0(h_0))[h_0, h_0] \geqslant 0. \tag{3}$$

Proof. For the sake of brevity, we shall denote $F'(\hat{x}) = \Lambda$, $f'(\hat{x}) = x_i^*$.

A) Let us consider the mapping $\varphi: \sigma \to X^*$ of the simplex $\sigma = \left\{(\lambda, \lambda_0) \mid \lambda_i \geqslant 0, \sum_{i=0}^{m} \lambda_i = 1\right\}$ determined by the equality $\varphi(\lambda, \lambda_0) = \sum_{i=0}^{m} \lambda_i x_i^*$. Then

$$(y^*, \lambda, \lambda_0) \in D \Leftrightarrow \sum_{i=0}^{m} \lambda_i x_i^* + y^* \circ \Lambda =$$

$$= \varphi(\lambda, \lambda_0) + \Lambda^* y^* = 0 \Rightarrow \varphi(\lambda, \lambda_0) \in \operatorname{Im} \Lambda^* \Rightarrow (\lambda, \lambda_0) \in \sigma_1 = \varphi^{-1}(\operatorname{Im} \Lambda^*).$$

By the hypothesis, $\operatorname{Im} \Lambda = Y$, and consequently (Section 2.1.7) $\operatorname{Im} \Lambda^* = (\operatorname{Ker} \Lambda)^{\perp}$, and therefore $\operatorname{Im} \Lambda^*$ is closed. Consequently, the set σ_1 is closed subset of a compact set and hence it is itself a compact set.

Now we note that $\operatorname{Ker} \Lambda^* = \{0\}$. Indeed,

$$h^* \in \operatorname{Ker} \Lambda^* \Rightarrow \langle \Lambda^* h^*, x \rangle = 0, \ \forall x \Rightarrow \langle h^*, \Lambda x \rangle = 0, \ \forall x \Rightarrow h^* \in (\operatorname{Im} \Lambda)^{\perp} \Rightarrow h^* = 0.$$

Since the closed subspace $\operatorname{Im} \Lambda^*$ of the Banach space is itself a Banach space, the Banach theorem implies that there exists the inverse mapping $\Gamma : \operatorname{Im} \Lambda^* \to Y^*$ to Λ^*. Now we have

$$(y^*, \lambda, \lambda_0) \in D \Leftrightarrow (\lambda, \lambda_0) \in \sigma_1,$$
$$\varphi(\lambda, \lambda_0) + \Lambda^* y^* = 0 \Leftrightarrow (\lambda, \lambda_0) \in \sigma_1,$$
$$y^* = -\Gamma \varphi(\lambda, \lambda_0).$$

Consequently, D is the image of the compact set σ_1 under the continuous mapping $(\lambda, \lambda_0) \to (-\Gamma\varphi(\lambda, \lambda_0), \lambda_1, \lambda_0)$ and hence it is itself a compact set.

From the Lagrange principle proved in Section 3.2 it follows that D is nonempty. The convexity of D is an obvious consequence of conditions (1), and it can also be derived from the convexity of σ and $\operatorname{Im} \Lambda^*$ if we take into account that the mappings considered above are linear. This completes the proof of the assertion a).

B) As in Section 3.3.1, we replace the given problem with inequality constraints by the corresponding problem with equality constraints:

$$f(x) = \max\{f_0(x) - f_0(\hat{x}), f_1(x), \ldots, f_m(x)\} \to \inf; \ F(x) = 0.$$

If $\hat{x} \in \operatorname{locmin} \mathfrak{z}$, then $\hat{x} \in \operatorname{locmin} \mathfrak{z}'$. Indeed,

$$\hat{x} \notin \operatorname{locmin} \mathfrak{z}' \Rightarrow \forall \varepsilon > 0, \exists x_\varepsilon : \|x_\varepsilon - \hat{x}\| < \varepsilon,$$
$$F(x_\varepsilon) = 0, \quad f(x_\varepsilon) < 0 \Rightarrow f_0(x_\varepsilon) < f_0(\hat{x}), \quad f_i(x_\varepsilon) < 0,$$
$$i = 1, \ldots, m, \ F(x_\varepsilon) = 0 \Rightarrow \hat{x} \notin \operatorname{locmin} \mathfrak{z}.$$

Further we investigate the problem $(\mathfrak{z}')$.

C) Let $h_0 \in L$. Since D is compact, there are $(\hat{y}^*, \hat{\lambda}, \hat{\lambda}_0)$ such that

$$\Psi(h_0) = \mathscr{L}_{xx}(\hat{x}, \hat{y}^*, \hat{\lambda}, \hat{\lambda}_0)[h_0, h_0] = \max_{(y^*, \lambda, \lambda_0) \in D} \mathscr{L}_{xx}(\hat{x}, y^*, \lambda, \lambda_0)[h_0, h_0] =$$

$$= \max\left\{ \sum_{i=0}^{m} \lambda_i f_i''(\hat{x})[h_0, h_0] + \langle y^*, F''(\hat{x})[h_0, h_0] \rangle \mid \lambda_i \geqslant 0, \right.$$
$$\left. \sum_{i=0}^{m} \lambda_i = 1, \ \sum_{i=0}^{m} \lambda_i x_i^* + \Lambda^* y^* = 0 \right\}.$$

The assertion b) is equivalent to the inequality $\Psi(h_0) \geqslant 0$.

Let us suppose that $\Psi(h_0) < 0$. Denoting

$$a_i(\lambda) = \frac{\lambda^2}{2} f_i''(\hat{x})[h_0, h_0], \quad i = 0, \ldots, m, \qquad (4)$$
$$y(\lambda) = \frac{\lambda^2}{2} F''(\hat{x})[h_0, h_0],$$

we consider the problem

$$\max_{0 \leqslant i \leqslant m} (a_i(\lambda) + \langle x_i^*, x \rangle) \to \inf; \quad \Lambda x + y(\lambda) = 0. \qquad (5)$$

Let us verify that the minimax lemma of Section 3.3.4 is applicable to problem (5). Indeed,

$$\max_{0 \leqslant i \leqslant m} \langle x_i^*, x \rangle \geqslant 0, \quad \forall x \in \operatorname{Ker} \Lambda,$$

is a necessary condition for $\hat{x}$ to yield a minimum in the problem ($\mathfrak{z}$) (Lemma 1 of Section 3.2.4). Moreover, by the hypothesis, operator Λ is surjective.

By the minimax lemma, there is an element $x_0(\lambda)$ possessing the properties

$$\max_{0 \leqslant i \leqslant m} (a_i(\lambda) + \langle x_i^*, x_0(\lambda) \rangle) = S(a(\lambda), y(\lambda)), \tag{6}$$

where $S(a(\lambda), y(\lambda))$ is the value of problem (5) and

$$\|x_0(\lambda)\| \leqslant C_1 \{\max |a_i(\lambda)| + \|y(\lambda)\|\} \leqslant C\lambda^2. \tag{7}$$

Here, according to formula (8) of Section 3.3.4,

$$S(a(\lambda), y(\lambda)) = \frac{\lambda^2}{2} \max \left\{ \sum_{i=0}^{m} \lambda_i f_i''(\hat{x})[h_0, h_0] + \langle y^*, F''(\hat{x})[h_0, h_0] \rangle \,\middle|\, \lambda_i \geqslant 0, \right.$$
$$\left. \sum_{i=0}^{m} \lambda_i = 1, \quad \sum_{i=0}^{m} \lambda_i x_i^* + \Lambda^* y^* = 0 \right\} = \frac{\lambda^2}{2} \Psi(h_0). \tag{8}$$

Therefore, (7) implies that $\|\lambda h_0 + x_0(\lambda)\| = O(h)$. By Taylor's formula (Section 2.2.5),

$$F(\hat{x} + x_0(\lambda) + \lambda h_0) = F(\hat{x}) + F'(\hat{x})[x_0(\lambda)] + \frac{1}{2} F''(\hat{x})[\lambda h_0 + x_0(\lambda), \lambda h_0 + x_0(\lambda)] + o(\lambda^2) =$$

$$= F''(\hat{x})[\lambda h_0, x_0(\lambda)] + \frac{1}{2} F''(\hat{x})[x_0(\lambda), x_0(\lambda)] + o(\lambda^2) = o(\lambda^2)$$

(we remind the reader that

$$h_0 \in L \subset \operatorname{Ker} F'(\hat{x})$$

and that, by virtue of (5),

$$F'(\hat{x})[x_0(\lambda)] + (\lambda^2/2) F''(\hat{x})[h_0, h_0] = 0.)$$

When proving Lyusternik's theorem in Section 2.3.5, we constructed a mapping $\varphi: U \to X$ of a neighborhood $U \ni \hat{x}$ such that

$$F(x + \varphi(x)) \equiv 0 \text{ and } \|\varphi(x)\| \leqslant K \|F(x)\|.$$

Putting $r(\lambda) = \varphi(\hat{x} + x_0(\lambda) + \lambda h_0)$, we write

$$F(\hat{x} + x_0(\lambda) + \lambda h_0 + r(\lambda)) \equiv 0,$$
$$\|r(\lambda)\| \leqslant K \|F(\hat{x} + x_0(\lambda) + \lambda h_0)\| = o(\lambda^2).$$

Now applying Taylor's formula to f_i and using (6) and (8) we obtain

$$f(\hat{x} + x_0(\lambda) + \lambda h_0 + r(\lambda)) = \max\{(f_0(\hat{x} + x_0(\lambda) + \lambda h_0 + r(\lambda)) - f_0(\hat{x}),$$
$$f_i(\hat{x} + x_0(\lambda) + \lambda h_0 + r(\lambda)), \quad i = 1, \ldots, m\} =$$
$$= \max_{0 \leqslant i \leqslant m} \left(\langle x_i^*, x_0(\lambda) \rangle + \frac{\lambda^2}{2} f_i''(\hat{x})[h_0, h_0] + o(\lambda^2) \right) \leqslant$$
$$\leqslant \max_{0 \leqslant i \leqslant m} \left(\langle x_i^*, x_0(\lambda) \rangle + \frac{\lambda^2}{2} f_i''(\hat{x})[h_0, h_0] \right) + o(\lambda^2) = \frac{\lambda^2}{2} \Psi(h_0) + o(\lambda^2) < 0$$

for small λ^2. Hence $\Psi(h_0) < 0 \Rightarrow \hat{x} \notin \operatorname{locmin} \mathfrak{z}' \Rightarrow \hat{x} \notin \operatorname{locmin} \mathfrak{z}$. We have arrived at a contradiction, which proves the theorem. ■

<u>3.4.3. Sufficient Conditions for an Extremum for Smooth Problems with Equality and Inequality Constraints.</u> As in the foregoing section, we shall investigate the problem

$$f_0(x) \to \inf; \quad F(x) = 0, \quad f_i(x) \leqslant 0. \tag{$\mathfrak{z}$}$$

THEOREM. Let X and Y be Banach spaces, let U be an open set in X, let $\hat{x} \in U$, let the functions $f_i : U \to \mathbf{R}$, $i = 0, 1, \ldots, m$, and the mapping $F: U \to Y$ possess the second Fréchet derivatives at a point $\hat{x}$ [which is admissible in $(\mathfrak{z})$], and let $f_i(\hat{x}) = 0$, $i = 1, \ldots, m$.

Let us suppose that there exist Lagrange multipliers $\hat{\lambda} \in \mathbf{R}^{m*}$, $\hat{y}^* \in Y^*$ and a number $\alpha > 0$ such that $\hat{\lambda}_i > 0$,

$$\mathscr{L}_x(\hat{x}, \hat{y}^*, \hat{\lambda}, 1) = f_0'(\hat{x}) + \sum_{i=1}^{m} \hat{\lambda}_i f_i'(\hat{x}) + F'^*(\hat{x})[\hat{y}^*] = 0 \tag{1}$$

and

$$\mathscr{L}_{xx}(\hat{x}, \hat{y}^*, \hat{\lambda}, 1)[h, h] \geqslant 2\alpha \|h\|^2 \tag{2}$$

for any h belonging to the subspace

$$L = \{h \mid \langle f_i'(\hat{x}), h \rangle = 0, \quad i \geqslant 1, \quad F'(\hat{x})[h] = 0\}. \tag{3}$$

Then $\hat{x}$ yields a local minimum in the problem $(\mathfrak{z})$.

Proof. A) Let $\hat{x} + h$ be an admissible element, i.e.,

$$f_i(\hat{x}+h) \leqslant 0, \quad i \geqslant 1, \quad F(\hat{x}+h) = 0. \tag{4}$$

We shall estimate the quantity $f_0(\hat{x} + h)$ in two different ways. We have

$$f_0(\hat{x}+h) = f_0(\hat{x}+h) + \sum_{i=1}^{m} \hat{\lambda}_i f_i(\hat{x}+h) +$$

$$+ \langle \hat{y}^*, F(\hat{x}+h) \rangle - \sum_{i=1}^{m} \hat{\lambda}_i f_i(\hat{x}+h) = \mathscr{L}(\hat{x}+h, \hat{y}^*, \hat{\lambda}, 1) - \sum_{i=1}^{m} \hat{\lambda}_i f_i(\hat{x}+h). \tag{5}$$

The first estimation method is based on the direct expansion of $f_0(\hat{x} + h)$:

$$f_0(\hat{x}+h) = \mathscr{L}(\hat{x}+h, \hat{y}^*, \hat{\lambda}, 1) - \sum_{i=1}^{m} \hat{\lambda}_i f_i(\hat{x}+h) = \mathscr{L}(\hat{x}, \hat{y}^*, \hat{\lambda}, 1) + \mathscr{L}_x(\hat{x}, \hat{y}^*, \hat{\lambda}, 1)[h] -$$

$$- \sum_{i=1}^{m} \hat{\lambda}_i f_i(\hat{x}) - \sum_{i=1}^{m} \langle \hat{\lambda}_i f_i'(\hat{x}), h \rangle + r_1(h) = f_0(\hat{x}) - \sum_{i=1}^{m} \langle \hat{\lambda}_i f_i'(\hat{x}), h \rangle + r_1(h), \tag{6}$$

where the remainder term $r_1(h)$ is $O(\|h\|^2)$.

The second method is based on the fact that, by virtue of the inequalities $\hat{\lambda}_i > 0$, $f_i(\hat{x}+h) \leqslant 0$, we have $\sum_{i=1}^{m} \hat{\lambda}_i f_i(\hat{x}+h) \leqslant 0$, and consequently

$$f_0(\hat{x}+h) \geqslant \mathscr{L}(\hat{x}+h, \hat{y}^*, \hat{\lambda}, 1) = \mathscr{L}(\hat{x}, \hat{y}^*, \hat{\lambda}, 1) + \mathscr{L}_x(\hat{x}, \hat{y}^*, \hat{\lambda}, 1)[h] +$$

$$+ \frac{1}{2} \mathscr{L}_{xx}(\hat{x}, \hat{y}^*, \hat{\lambda}, 1)[h, h] + r_2(h) = f_0(\hat{x}) + B(h, h) + r_2(h), \tag{7}$$

where B(h, h) denotes the quadratic form $\frac{1}{2} \mathscr{L}_{xx}(\hat{x}, \hat{y}^*, \hat{\lambda}, 1)[h, h]$, and for the remainder term we have $r_2(h) = o(\|h\|^2)$.

B) Let us denote by K the cone consisting of those h for which $\langle f_i'(\hat{x}), h \rangle \leqslant 0$, $i \geqslant 1$, $F'(\hat{x})h = 0$.

By virtue of Hoffman's lemma, an arbitrary h can be represented as a sum $h_1 + h_2$, where $h_1 \in K$ and h_2 satisfies the inequality

$$\|h_2\| \leqslant C_1 \left\{ \sum_{i=1}^{m} \langle f_i'(\hat{x}), h \rangle_+ + \|F'(\hat{x}) h\| \right\}. \tag{8}$$

Further, making use of the remark following Hoffman's lemma (Section 3.3.4) and of formula (5') of Section 3.3.4 we can also represent h_1 as a sum $h_1 = h_1' + h_1''$, where $h_1' \in L$ and h_1'' satisfies the inequality

$$\|h_1''\| \leqslant C_2 \left\{ \sum_{i=1}^{m} \left| \langle f_i'(\hat{x}), h_1 \rangle \right| \right\} = C_2 \left\{ - \sum_{i=1}^{m} \langle f_i'(\hat{x}), h_1'' \rangle \right\}. \tag{9}$$

Here we have used the fact that $F'(\hat{x})[h_1] = 0$ (because $h_1 \in K$) and that

$$|\langle f_i'(\hat{x}), h_1\rangle| = |\langle f_i'(\hat{x}), h_1' + h_1''\rangle| = |\langle f_i'(\hat{x}), h_1''\rangle| = -\langle f_i'(\hat{x}), h_1''\rangle \text{ (because } \langle f_i'(\hat{x}), h_1'\rangle = 0 \text{ and } \langle f_i'(\hat{x}), h_1\rangle \leqslant 0).$$

C) From (4) we obtain

$$\begin{aligned} 0 &= F(\hat{x}+h) - F(\hat{x}) = F'(\hat{x})h + r_0(h), \\ 0 &\geqslant f_i(\hat{x}+h) = \langle f_i'(\hat{x}), h\rangle + \rho_i(h), \quad i = 1, \ldots, m, \end{aligned} \tag{10}$$

where $\|r_0(h)\|$ and $|\rho_i(h)|$ are on the order of $O(\|h\|^2)$. Substituting these estimates into (8), we find that if $\hat{x} + h$ is an admissible element, then

$$\|h_2\| \leqslant C_1\left\{\sum_{i=1}^{m} |\rho_i(h)| + \|r_0(h)\|\right\}. \tag{11}$$

Now we fix a number $\varepsilon_1 \in (0, 1]$, whose magnitude will be specified later and choose $\delta \in (0, \|B\|^{-1})$ such that the inequality $\|h\| \leqslant \delta$ implies the inequalities

$$\sum_{i=1}^{m} |\rho_i(h)| + \sum_{j=0}^{1} \|r_j(h)\| \leqslant \varepsilon_1 \|h\|, \quad \|r_2(h)\| \leqslant \frac{\alpha}{2}\|h\|^2. \tag{12}$$

Further we choose a number $A > 0$ such that

$$AC_2^{-1}\left(\min_i \hat{\lambda}_i\right) - C_1 \max_i (\hat{\lambda}_i \|f_i'(\hat{x})\|) - 1 \geqslant 0, \tag{13}$$

and, finally, we choose ε_1 so that the inequalities

$$\varepsilon < 1, \ \alpha(1-\varepsilon)^2 - 2\|B\|\varepsilon(1+\varepsilon) - \|B\|\varepsilon^2 - \alpha/2 \geqslant 0 \tag{14}$$

hold for $\varepsilon = (C_1 + A)\varepsilon_1$.

D) Completion of the Proof. Let $\hat{x} + h$ be an admissible element. Let us represent h in the form of a sum as it was done in B): $h = h_1' + h_1'' + h_2$. Here the following two cases are possible: a) $\|h_1''\| > A\varepsilon_1\|h\|$ and b) $\|h_1''\| \leqslant A\varepsilon_1\|h\|$. If the case a) takes place, relation (9) implies

$$A\varepsilon_1\|h\| < \|h_1''\| < C_2\left(-\sum_{i=1}^{m} \langle f_i'(\hat{x}), h_1''\rangle\right) \tag{15}$$

for $\|h\| \leqslant \delta$.

Then, by virtue of (6), (15), (11), (12), and (13), we obtain

$$f_0(\hat{x}+h) \overset{(6)}{=} f_0(\hat{x}) - \sum_{i=1}^{m} \langle \hat{\lambda}_i f_i'(\hat{x}), h_1'' + h_2\rangle + r_1(h) \overset{(15),\,(11),\,(12)}{\geqslant}$$

$$\geqslant f_0(\hat{x}) + AC_2^{-1}\left(\min_i \hat{\lambda}_i\right)\varepsilon_1\|h\| - \max_i (\hat{\lambda}_i\|f_i'(\hat{x})\|) C_1\varepsilon_1\|h\| - \varepsilon_1\|h\| = f_0(\hat{x}) +$$

$$+ \varepsilon_1\|h\|\left(AC_2^{-1}\left(\min_i \hat{\lambda}_i\right) - C_1 \max_i (\hat{\lambda}_i\|f_i'(\hat{x})\|) - 1\right) \overset{(13)}{\geqslant} f_0(\hat{x}).$$

Now let the case b) take place. Then $\|h_1''\| \leqslant A\varepsilon_1\|h\|$, and hence if we denote $h_2' = h_1'' + h_2$, relations (11) and (12) imply the inequality $\|h_2'\| = \|h_1'' + h_2\| \leqslant (A + C_1)\varepsilon_1\|h\| = \varepsilon\|h\|$. Then $h = h_1' + h_2'$, where $(1 - \varepsilon)\|h\| \leqslant \|h_1'\| \leqslant (1 + \varepsilon)\|h\|$. Now applying these inequalities and also (7), (2), and (12), we obtain

$$f_0(\hat{x}+h) \overset{(7)}{\geqslant} f_0(\hat{x}) + B(h, h) + r_2(h) = f_0(\hat{x}) + B(h_1' + h_2', h_1' + h_2') + r_2(h) =$$

$$= f_0(\hat{x}) + B(h_1', h_1') + 2B(h_1', h_2') + B(h_2', h_2') + r_2(h) \overset{(2),\,(12)}{\geqslant}$$

$$\geqslant f_0(\hat{x}) + \left(\alpha(1-\varepsilon)^2 - 2\|B\|\varepsilon(1+\varepsilon) - \|B\|\varepsilon^2 - \frac{\alpha}{2}\right)\|h\|^2 \geqslant f_0(\hat{x}).$$

The theorem is proved.

3.5. Application of the Theory to Algebra and Mathematical Analysis

In this section we gather several examples in which the solution of an extremal problem is the key to the proof of the theoretical result. The fundamental theorem of algebra, the theorem on the orthogonal complement, and Hilbert's theorem belong to the most important theorems of the university course of mathematics. Sylvester's theorem and the Gram determinants are popular tools used by mathematicians. As to Section 3.5.5, it contains a more exquisite material which is included because of the great importance of the finiteness of the index of a quadratic form in the classical calculus of variations and the Morse theory.

3.5.1. Fundamental Theorem of Algebra

THEOREM. Every polynomial with complex coefficients of degree not less than unity has at least one complex root.

Proof. Let $p(z) = a_0 + a_1 z + \ldots + a_n z^n$ be a polynomial of degree $n \geqslant 1$, where $a_n \neq 0$. Let us consider the elementary problem

$$f(z) = |p(z)|^2 \to \inf. \quad (\mathfrak{z})$$

LEMMA. There exists a solution of the problem $(\mathfrak{z})$.

Indeed,

$$f(z) = |p(z)|^2 = \left|\sum_{k=0}^{n} a_k z^k\right|^2 \geqslant \left(|a_n||z|^n - \sum_{k=0}^{n-1} |a_k||z|^k\right)^2$$

$$= |a_n|^2 |z|^{2n}(1 + O(1/|z|)) \to \infty$$

for $z \to \infty$. It follows that *all the level sets of* f *are compact*, and now the assertion of the lemma follows from the continuity of the function f and the corollary of Weierstrass' theorem (Section 3.1.7).

Let $\hat{z}$ be a solution of the problem $(\mathfrak{z})$. Without loss of generality, we can assume that $\hat{z} = 0$ [if otherwise, we would consider the polynomial $g(z) = p(z - \hat{z})$]. Thus,

$$f(0) \leqslant f(z) = |a_0 + a_1 z + \ldots + a_n z^n|^2, \qquad \forall z \in \mathbf{C}.$$

If $a_0 = 0$, then the point zero is a root of the polynomial, and the theorem has been proved. Let $a_0 \neq 0$, and let s be the index such that $a_1 = \ldots = a_{s-1} = 0$, $a_s \neq 0$. Now we fix $\zeta = e^{i\theta}$ and consider the function of one variable $\varphi(t) = f(t\zeta)$. By the hypothesis, the point zero yields a minimum for this function. We have

$$\varphi^{(k)}(0) = \frac{d^k}{dt^k}\varphi|_{t=0} = \frac{d^k}{dt^k}[(a_0 + a_s t^s e^{is\theta} + O(t^{s+1}))(\bar{a}_0 + \bar{a}_s t^s e^{-is\theta} + O(t^{s+1}))]|_{t=0}$$

$$= \begin{cases} 0, \ k = 1, \ldots, s-1, \\ (2s!)\operatorname{Re}(\bar{a}_0 a_s e^{is\theta}) = (2s)!|a_0||a_s|\cos(s\theta + \gamma), & k = s. \end{cases}$$

Since $s \geqslant 1$, function $\theta \to \cos(s\theta + \gamma)$ assumes both positive and negative values. Hence, by virtue of the lemma of Section 3.1.1, function φ has no minimum at zero if ζ is chosen so that $\varphi^{(s)}(0) < 0$. This contradiction shows that $a_0 = 0$. ■

The first attempts to state the fundamental theorem of algebra were made by A. Girard (in 1629) and R. Descartes (in 1637). R. Descartes wrote: "You must know that every equation may have as many roots as its dimension. However, it sometimes happens that some of these roots are false or less than nothing." The first proof of the theorem is known to have been given by K. F. Gauss (in 1799). The idea of the proof presented above was in fact suggested in the works by J. D'Alembert (in 1746-1748).

3.5.2. Sylvester's Theorem. As we have already seen several times in the present chapter, second-order conditions are connected, in the case of a minimum, with the nonnegativity (a necessary condition) or the positivity (a sufficient condition) of a quadratic form. In the finite-dimensional case the positivity of a quadratic form is established with the aid of the well-known Sylvester theorem. Here we shall show that the theorem itself can be derived from Fermat's theorem.

We remind the reader that a symmetric matrix $A_m = (a_{ij})$; $i, j = 1,\ldots, m$, $a_{ij} = a_{ji}$, is said to be positive-definite if the quadratic form $x^TAx = \sum_{i,j=1}^{m} a_{ij}x_ix_j = Q_m(x)$ corresponding to it is positive for any nonzero vector $x \in \mathbf{R}^m$. The determinants $\det A_k$, where $A_k = (a_{ij})$, $i, j = 1,\ldots,k$, $1 \leqslant k \leqslant m$, are called the *principal minors of the matrix* A_m.

Sylvester's Theorem. For a matrix to be positive-definite it is necessary and sufficient that its principal minors be positive.

Proof. A) For first-order matrices the assertion of the theorem is trivial. We shall suppose that the theorem is true for the matrices of order $n - 1$ ($n \geqslant 2$) and then prove it for the matrices of the n-th order. Let A_n be a matrix of order n, and let A_{n-1} be its submatrix (of order $n - 1$) obtained from the matrix A_n by deleting the last row and the last column of the latter.

For an arbitrary $\xi = (\xi_1,\ldots,\xi_{m-1}) \in \mathbf{R}^{m-1}$ we denote $\tilde{\xi} = (\xi_1,\ldots,\xi_{m-1}, 0) \in \mathbf{R}^m$. Then $Q_{n-1}(\xi) = Q_n(\tilde{\xi})$, and if A_n is positive-definite, the matrix A_{n-1} is also positive-definite and consequently $\det A_k > 0$; $k = 1,\ldots,n - 1$.

Thus, we have to prove the following assertion:

LEMMA 1. Let $\det A_k > 0$, $k = 1,\ldots,n - 1$. The matrix A_n is positive-definite if and only if $\det A_n > 0$.

B) Let $x = (x_1,\ldots,x_n) \in \mathbf{R}^n$, where $x_n \neq 0$. Let us put

$$y = (y_1, \ldots, y_{n-1}) = (x_1/x_n, \ldots, x_{n-1}/x_n) \in \mathbf{R}^{n-1}, \qquad a = (a_{1n}, \ldots, a_{n-1,n}).$$

It is clear that

$$Q_n(x) = x^TA_nx = (y^TA_{n-1}y + 2a^Ty + a_{nn})x_n^2. \tag{1}$$

Let us consider the following elementary extremal problem:

$$f(y) = y^TA_{n-1}y + 2a^Ty + a_{nn} \to \inf. \tag{2}$$

LEMMA 2. Under the conditions of Lemma 1, problem (2) has a solution.

Proof of Lemma 2. By assumption, $\det A_k > 0$, $k = 1,\ldots,n - 1$. Since the determinant of a matrix is continuous with respect to the elements of the matrix, for the principal minors of the matrix $A_{n-1}(\varepsilon) = A_{n-1} - \varepsilon I$ (where I is the unit matrix) the same inequalities $\det A_k(\varepsilon) > 0$, $k = 1,\ldots,n - 1$, are retained when $\varepsilon > 0$ is sufficiently small. By the induction hypothesis, it follows that the matrix $A_{n-1}(\varepsilon)$ is positive-definite and the inequality

$$Q_{n-1}(y) = y^TA_{n-1}y = y^T(A_{n-1} - \varepsilon I)y + \varepsilon|y|^2 \geqslant \varepsilon|y|^2$$

holds, whence, taking into account the Cauchy–Bunyakovskii inequality, we obtain

$$|f(y)| = |y^TA_{n-1}y + 2a^Ty + a_{nn}| \geqslant \varepsilon|y|^2 - 2|a||y| - |a_{nn}| \to \infty$$

for $|y| \to \infty$. It remains to refer to the corollary of Weierstrass' theorem (Section 3.1.7).

C) Proof of Lemma 1. Let $\hat{y}$ be a solution of the problem (2). Applying Fermat's theorem, we obtain

$$0=f'(\hat{y})[h]=2h^T(A_{n-1}\hat{y}+a), \forall h \in \mathbf{R}^{n-1}$$
$$\Leftrightarrow A_{n-1}\hat{y}+a=0 \Leftrightarrow \hat{y}=-A_{n-1}^{-1}a,$$

whence

$$f(\hat{y})=a_{nn}-a^TA_{n-1}^{-1}a.$$

Further, expanding $\det A_n$ in the minors of its last column and then expanding all the minors except one (in the resultant formula) in the minors of their last rows we arrive at the equality

$$\det A_{nn}=a_{nn}\det A_{n-1}-\sum_{i,k=1}^{n-1}\alpha_{ik}a_{ni}a_{kn}$$
$$=\det A_{n-1}\left\{a_{nn}-\sum_{i,k=1}^{n-1}\frac{\alpha_{ik}}{\det A_{n-1}}a_{ni}a_{kn}\right\}=\det A_{n-1}\{a_{nn}-a^TA_{n-1}^{-1}a\}=\det A_{n-1}f(\hat{y}).$$

Here α_{ik} is the algebraic adjunct of the element a_{ik} of the matrix A_{n-1} (by virtue of the symmetry of the matrix, $\alpha_{ik} = \alpha_{ki}$), and we have used the well-known formula for the elements of the inverse matrix.

According to (1) and (2), we have $Q_n(x) = f(y)x_n^2$ for $x_n \neq 0$, and, in particular, $f(\hat{y}) = Q_n(\hat{x})$, where $\hat{x} = (\hat{y}_1,\ldots,1) \neq 0$. Consequently, if A_n is positive-definite, then $f(\hat{y}) = Q_n(\hat{x}) > 0$, and, according to (3), $\det A_n > 0$. Conversely, if $\det A_n > 0$, then $f(\hat{y}) > 0$ and $Q_n(x) = f(y)x_n^2 \geqslant f(\hat{y})x_n^2 > 0$ for $x_n \neq 0$. If $x_n = 0$, then $x = (\xi_1,\ldots,\xi_{n-1}, 0) = \tilde{\xi}$, and, by the induction hypothesis, $Q_n(x) = Q_{n-1}(\xi) > 0$ for $x \neq 0$. This proves Lemma 1 together with Sylvester's theorem.

Exercise. Give an example in which $\det A_k \geqslant 0$, $k = 1,\ldots,n$, but $Q_n(x) < 0$ for some $x \neq 0$.

3.5.3. Distance from a Point to a Subspace. Theorem on the Orthogonal Complement. Gram Determinants.

Let X be a real Hilbert space with a scalar product $(\cdot|\cdot)$, and let L be a subspace of X. An element x is said to be *orthogonal* to L if $(x|y) = 0$ for any $y \in L$. The collection of all the vectors orthogonal to the given subspace L forms a subspace which is called the *orthogonal complement* to the subspace L and is denoted by $L^\perp$. In the foundation of the theory of Hilbert spaces lies the following theorem:

Theorem on the Orthogonal Complement. Let L be a closed subspace of a Hilbert space X. Then X is representable as the orthogonal direct decomposition $L \oplus L^\perp$; in other words, for any $\hat{x} \in X$ there exist uniquely determined elements $\hat{y} \in L$ and $\hat{z} \in L^\perp$ such that $\hat{x} = \hat{y} + \hat{z}$.

The *proof* of this theorem is based on the consideration of the problem on the shortest distance from a point $\hat{x}$ to an affine manifold L:

$$\|\hat{x}-y\|\to\inf, \qquad y\in L, \tag{1}$$

which was already studied in Section 3.3.4. Here using specific features of Hilbert spaces, we can obtain a more extensive result.

LEMMA. If L is closed, then there exists a solution of the problem (1).

Proof. A) First we shall solve this problem for a straight line $l = \{x|x = \xi + t\eta\}$, $\eta \neq 0$. Let $\varphi(t) = (\hat{x} - \xi - t\eta|\hat{x} - \xi - t\eta)$. Function $t \to \varphi(t)$ is quadratic and has a single minimum because the coefficient in t^2 is positive. By Fermat's theorem, $\hat{t} \in \operatorname{abs\,min} \varphi \Leftrightarrow \varphi'(\hat{t}) = 0 \Leftrightarrow (\eta|\hat{x} - \xi - \hat{t}\eta) = 0$. It readily follows that

$$\rho^2(\hat{x}, l)=\|\hat{x}-\hat{z}\|^2=\|\hat{x}-\xi\|^2-(\hat{x}-\xi|e)^2, e=\eta/\|\eta\|, \hat{z}=\xi+\hat{t}\eta, \tag{2}$$

and that $\hat{x} - \hat{z}$ is orthogonal to η. The point $\hat{z}$ is the foot of the perpendicular dropped from $\hat{x}$ to l.

B) Now we shall prove the lemma for the general case. If $\hat{x} \in L$, then the lemma has been proved. If $\hat{x} \notin L$, then, by virtue of the closedness of L, the value ρ of problem (1) is positive. Let $\{y_n\}$ be a minimizing sequence of problem (1), i.e., $y_n \in L$ and $\|\hat{x} - y_n\| \to \rho$. Without loss of generality, we can assume that the sequence $\{\|\hat{x} - y_n\|\}$ is monotone. Let us show that it is fundamental. Let $0 < (\|\hat{x} - y_n\|^2 - \rho^2)^{1/2} < \varepsilon$. Let us draw a straight line l through the point y_n and y_{n+k}, $k \geqslant 1$, and drop, as above, a perpendicular with the foot $\hat{z}$ from $\hat{x}$ to l. The farther the foot of an inclined line lies from the foot of the perpendicular, the greater the length of the former is, and therefore

$$\|\hat{x} - y_n\| \geqslant \|\hat{x} - y_{n+k}\| \Rightarrow \|\hat{z} - y_n\| > \|\hat{z} - y_{n+k}\|$$

$$\Rightarrow \|y_n - y_{n+k}\| \leqslant \|\hat{z} - y_{n+k}\| + \|\hat{y}_n - \hat{z}\| \leqslant 2\|\hat{z} - y_n\| = 2(\|\hat{x} - y_n\|^2 - \|\hat{z} - \hat{x}\|^2)^{1/2} < 2\varepsilon,$$

whence it follows that $\{y_n\}$ is fundamental.

Since X is complete and L is closed, the sequence $\{y_n\}$ is convergent to an element $\hat{y}$. By virtue of the continuity of the norm, $\|\hat{x} - \hat{y}\| = \lim_{n\to\infty} \|\hat{x} - y_n\| = \rho$.

Now the theorem can be proved quite simply. If $\hat{y}$ is a solution of problem (1), then the function $\varphi(t) = \|\hat{x} - \hat{y} - tz\|^2$ has a minimum at zero for any vector $z \in L$. Applying Fermat's theorem once again $[\varphi'(0) = 0]$, we readily conclude that $(\hat{x} - \hat{y} | z) = 0$, i.e., $\hat{x} - \hat{y} \in L^{\perp}$. If there exist two representations $\hat{x} = y_1 + z_1 = y_2 + z_2$, $y_i \in L$, $z_i \in L^{\perp}$; $i = 1, 2$, then $y_1 - y_2 = z_2 - z_1 \Rightarrow (y_1 - y_2 | y_1 - y_2) = 0 \Rightarrow y_1 = y_2$, $z_1 = z_2$. ■

Remark. Of course, following what was said in Section 3.1.1, we could have proved the existence of the solution of problem (1) using the lower semicontinuity (in the weak topology) of the function $x \to N(x) = \|x\|$ and the weak compactness of the level set $\{x | N(x) \leqslant 1\}$, i.e., of the unit ball in the space X. However, it should be taken into account that the fact of the weak compactness is based on the isometric isomorphism between X and X*, and this isomorphism is proved with the aid of the theorem on the orthogonal complement. (For the theorem on the weak compactness of the unit ball in the space conjugate to a separable normed space, see [KF, p. 202]. The function N is convex and continuous, and therefore it is closed. By Minkowski's theorem (Section 2.6.2), this function is the upper bound of affine functions which are continuous in the weak topology, and therefore it is lower semicontinuous in the same topology.)

Now let us find the distance from a point to a hyperplane. Let a hyperplane H be specified by an equation $H = \{y | (y|e) = \alpha\}$, $\|e\| = 1$. The square of the distance from a point $\hat{x}$ to the hyperplane H is the value of the smooth problem $\|\hat{x} - y\|^2 = (\hat{x} - y | \hat{x} - y) \to \inf$; $(y|e) = \alpha$, which possesses a solution $\hat{y}$ according to the lemma proved above (it is clear that $\|\hat{x} - y\|$ and $\|\hat{x} - y\|^2$ attain their minima simultaneously).

Applying the Lagrange multiplier rule and omitting trivial details, we obtain

$$\mathscr{L} = \tfrac{1}{2}(\hat{x} - y | \hat{x} - y) + \lambda(y|e), \qquad \mathscr{L}_y = 0 \Rightarrow \hat{y} = \hat{x} + \hat{\lambda} e,$$

$$\hat{\lambda} = \alpha - (\hat{x}|e), \qquad \rho(\hat{x}, H) = \|\hat{x} - \hat{y}\| = |\hat{\lambda} e| = |\alpha - (\hat{x}|e)|.$$

Finally, we shall present the well-known explicit formula for the distance from a point x_0 of a Hilbert space X to a subspace L_n, which is the linear hull of n linearly independent vectors $x_1, \ldots, x_n$. The square of this distance is the value of the following finite-dimensional elementary problem:

$$f(\alpha)=\left|x_0+\sum_{i=1}^{n}\alpha_i x_i\right|^2\to\inf. \tag{4}$$

The existence and the uniqueness of the solution are obvious here, and it is also possible to refer to the theorem on the orthogonal complement. However, writing $f(\alpha)$ in the form

$$f(\alpha)=\sum_{i,j=1}^{n}(x_i|x_j)\alpha_i\alpha_j+2\sum_{i=1}^{n}(x_0|x_i)\alpha_i+(x_0,x_0),$$

we see that problem (4) coincides with the problem considered in Lemma 1 of the foregoing section if the matrix A_n is replaced by the matrix

$$((x_i|x_j)|i,j=0,1,\dots,n) \tag{5}$$

and the matrix A_{n-1} is replaced by the minor of matrix (5) obtained by deleting the first row and the first column. A matrix of form (5) is called the *Gram matrix* of the system of vectors $\{x_0, x_1,\dots,x_n\}$; the determinant of this matrix is called the *Gram determinant* (the *Gramian*) of the system of vectors and is denoted by $G(x_0, x_1,\dots,x_n)$. If $\hat{\alpha}$ is the solution of the problem (4), then, according to formula (3) of Section 3.5.2,

$$G(x_0,x_1,\dots,x_n)=f(\hat{\alpha})G(x_1,\dots,x_n).$$

Thus,

$$\rho(x_0,L_n)=(G(x_0,x_1,\dots,x_n)/G(x_1,\dots,x_n))^{1/2}. \tag{6}$$

In conclusion, we recommend the reader to solve the infinite-dimensional analogue of the problem of Apollonius mentioned in Chapter 1.

Exercise. Let $\Lambda:X\to X$ be a compact continuous linear operator in a Hilbert space X. Find the distance from a point $\hat{x}$ to the ellipsoid

$$E=\{\xi|\xi=\Lambda y,\quad \|y\|\leqslant 1\}.$$

3.5.4. Reduction of a Quadratic Form to Its Principal Axes.

As is known, every quadratic form $Q(x)$ in a finite-dimensional Euclidean space can be brought to a diagonal form $Q(x)=Q(Uy)=\sum\lambda_i y_i^2$ with the aid of an orthogonal transformation of coordinates $x = Uy$. The vectors $x^{(i)} = Ue^{(i)}$ (where $e^{(i)}$ are the elements of the standard orthonormal basis in the coordinates y) are directed along the "principal axes," i.e., they are stationary points of the function Q on the sphere $\|x\|^2 = 1$.

Without exaggeration we can say that the infinite-dimensional generalization of this fact established by D. Hilbert stimulated first the creation of the theory of Hilbert space and then the development of functional analysis as a whole.

We shall remind the reader of some necessary notions. Let X be a separable real Hilbert space with a scalar product $(\cdot|\cdot)$. A sequence x_n is said to be *weakly convergent to* x if $(y|x_n)\to(y|x)$ for any $y\in X$. A continuous linear operator $\Lambda:X\to X$ is said to be *self-adjoint* if $(x|\Lambda y)\equiv(\Lambda x|y)$ for any $x, y\in X$. The quadratic form $Q(x) = (\Lambda x|x)$ generated by the operator Λ is said to be *weakly continuous* if $x_n\xrightarrow[\text{weakly}]{}x\Rightarrow Q(x_n)\to Q(x)$. Finally, an operator Λ is said to be *compact* if it transforms every bounded set into a relatively compact set. In the case under consideration this is equivalent to the fact that $x_n\xrightarrow[\text{weakly}]{}x\Rightarrow\|\Lambda x_n-\Lambda x\|\to 0$ (D. Hilbert himself used the definition based on the last fact).

Hilbert's Theorem. For a quadratic form $Q(x) = (\Lambda x|x)$, where Λ is a self-adjoint continuous linear operator in a separable Hilbert space X, to be weakly continuous, it is necessary and sufficient that the operator Λ be compact.

Moreover, there exists an orthonormal basis e_n in the space X consisting of eigenvectors of the operator Λ and

$$Q(x)=\sum_{n=1}^{\infty}\lambda_n(x|e_n)^2.$$

It is evident that the assertion about the finite-dimensional quadratic forms stated at the beginning of the present section follows immediately from this theorem since every continuous linear operator in a finite-dimensional space is compact.

Proof. "Sufficient." Let Λ be a compact operator, and let $\{x_n\}$ be a sequence weakly convergent to x. Then, by the Banach–Steinhaus theorem [KF, p. 194], the sequence is strongly bounded, i.e., there exists a number $C > 0$ such that $\|x_n\| \leqslant C$, $\forall n$. It follows that $\|x\| \leqslant C$. By virtue of the compactness of the operator, $\|\Lambda x_n - \Lambda x\| \to 0$. As a result,

$$|Q(x_n)-Q(x)|=|(\Lambda x_n|x_n)-(\Lambda x|x)|\leqslant|(\Lambda x_n|x_n)-(\Lambda x_n|x)|$$

$$+|(\Lambda x_n|x)-(\Lambda x|x)|\leqslant\|\Lambda x_n-\Lambda x\|\,\|x_n\|+\|\Lambda x_n-\Lambda x\|\,\|x\|\leqslant 2C\|\Lambda x_n-\Lambda x\|\to 0$$

for $n \to \infty$.

B) Construction of the Orthonormal Basis. Let the form Q be weakly continuous. Let us consider the following extremal problem:

$$|(\Lambda x|x)|\to\sup,\qquad (x|x)\leqslant 1,\qquad x\in L, \tag{1}$$

where L is a closed subspace.

LEMMA. There exists a solution of problem (1).

Indeed, as was already mentioned in the foregoing section the Hilbert space X is isometrically isomorphic to its conjugate space, and therefore its unit ball $\{x|(x|x) \leqslant 1\}$ is weakly compact.

On the other hand, the decomposition $X = L\oplus L^{\perp}$ proved in the foregoing section implies the equality $L = \{x|(x|y) = 0,\ y \in L^{\perp}\} = (L^{\perp})^{\perp}$, and if $x_n \in L$ is a sequence weakly convergent to x, then $(x|y)=\lim\limits_{n\to\infty}(x_n|y)=0$, $\forall y\in L^{\perp}$, and hence $x \in L$. Consequently, L is weakly closed, and therefore the set of the admissible elements of problem (1) is weakly compact. By assumption, $(\Lambda x|x)$ is weakly continuous. Now the existence of the solution follows from Weierstrass' theorem (Section 3.1.7). ■

Let $\mathfrak{z}_1$ be the value of problem (1) with L = X. First let us suppose that $\mathfrak{z}_1 = 0$. Then $(\Lambda x|x) \equiv 0$, and hence any vector yields an extremum in the elementary problem $(\Lambda x|x) \to \inf$. Applying Fermat's theorem to this problem, we conclude that $\Lambda x \equiv 0$, i.e., any vector x is an eigenvector belonging to the eigenvalue zero, and $Q(x) \equiv 0$. In this case the theorem has been proved.

Now let $\mathfrak{z}_1 \neq 0$, and let e_1 be a solution of problem (1) with L = X.

For definiteness, let $(\Lambda e_1|e_1) < 0$; then $e_1 \neq 0$ is a solution of the problem

$$(\Lambda x|x)\to\inf,\qquad (x|x)\leqslant 1. \tag{2}$$

The Lagrange function of this problem has the form

$$\mathscr{L}(x,\lambda_1,\lambda_0)=\lambda_0(\Lambda x|x)+\lambda_1((x|x)^{-1}),$$

and, according to the Lagrange multiplier rule, there exist $\hat{\lambda}_0 \geqslant 0$, $\hat{\lambda}_1 \geqslant 0$, not vanishing simultaneously, such that

$$\mathscr{L}_x(e_1,\hat{\lambda}_1,\hat{\lambda}_0)=2\hat{\lambda}_0\Lambda e_1+2\hat{\lambda}_1 e_1=0.$$

It can easily be seen that neither $\hat{\lambda}_0$ nor $\hat{\lambda}_1$ can be equal to zero. By the condition of complementary slackness, $\|e_1\| = 1$; denoting $\lambda_1 = -\hat{\lambda}_1/\hat{\lambda}_0$ we obtain $\Lambda e_1 = \lambda_1 e_1$, $\lambda_1 = (\Lambda e_1|e_1)$, and, as a result,

$$\Lambda e_1 = \lambda_1 e_1, \qquad \mathfrak{z}_1 = |\lambda_1|, \qquad \|e_1\| = 1. \tag{3}$$

Now let us suppose that n orthonormal vectors $e_1,\ldots,e_n$ (which are eigenvectors of the operator Λ belonging to eigenvalues $\lambda_1,\ldots,\lambda_n$) have already been constructed. Let us consider the following extremal problem:

$$|(\Lambda x|x)| \to \sup, \qquad (x|x) = 1, \qquad (x|e_i) = 0, \qquad i = 1,\ldots,n. \tag{4}$$

By the lemma, there exists a solution of this problem.

We shall denote by $\mathfrak{z}_{n+1}$ the value of problem (4). Again, if $\mathfrak{z}_{n+1} = 0$, then $\Lambda x = 0$ for all $x \in [\mathrm{lin}\{e_1,\ldots,e_n\}]^{\perp}$. In this case Hilbert's theorem has already been proved:

$$\Lambda x = \sum_{i=1}^{n} \lambda_i (x|e_i) e_i, \qquad Q(x) = \sum_{i=1}^{n} \lambda_i (x|e_i)^2.$$

The first of these equalities obviously implies the compactness of Λ.

Let $\mathfrak{z}_{n+1} \neq 0$, and let e_{n+1} be a solution of problem (4); $\lambda_{n+1} = (\Lambda e_{n+1}|e_{n+1})$, $|\lambda_{n+1}| = \mathfrak{z}_{n+1}$. Then e_{n+1} is a solution of the problem

$$\frac{(\Lambda x|x)}{\lambda_{n+1}} \to \sup; \qquad (x|x) \leqslant 1, \qquad (x|e_i) = 0, \qquad i = 1,\ldots,n, \tag{5}$$

and, by the Lagrange multiplier rule, there are $\hat{\lambda}_0, \hat{\lambda}_1,\ldots,\hat{\lambda}_{n+1}$ such that $\hat{\lambda}_0 \leqslant 0$, $\hat{\lambda}_{n+1} \geqslant 0$, and

$$\hat{\lambda}_0 \Lambda e_{n+1} + \hat{\lambda}_{n+1} e_{n+1} + \sum_{i=1}^{n} \hat{\lambda}_i e_i = 0, \qquad \hat{\lambda}_{n+1}((e_{n+1}|e_{n+1}) - 1) = 0. \tag{6}$$

Multiplying scalarly, in succession, the first of these equalities by $e_1,\ldots,e_n$ and using the orthonormality of the vectors e_i and their orthogonality to e_{n+1}, we obtain

$$\hat{\lambda}_0(\Lambda e_{n+1}|e_i) + \hat{\lambda}_i = 0, \qquad i = 1,\ldots,n,$$

whence

$$\hat{\lambda}_i = -\hat{\lambda}_0(\Lambda e_{n+1}|e_i) = -\hat{\lambda}_0(e_{n+1}|\Lambda e_i) = -\hat{\lambda}_0 \hat{\lambda}_i (e_{n+1}|e_i) = 0$$

for $i = 1,\ldots,n$. It follows that $\hat{\lambda}_0 \neq 0$, $\hat{\lambda}_{n+1} \neq 0$ in (6) and $\|e_{n+1}\| = 1$. Let us denote $\lambda_{n+1} = -\hat{\lambda}_{n+1}/\hat{\lambda}_0$. Then

$$\Lambda e_{n+1} = \lambda_{n+1} e_n, \qquad \|e_{n+1}\| = 1, \qquad |\lambda_{n+1}| = (\Lambda e_{n+1}|e_{n+1}) = \mathfrak{z}_{n+1}. \tag{7}$$

Thus, the inductive construction can be continued. As a result, either $\mathfrak{z}_N = 0$ at a finite step or there is a sequence $\{e_n\}_{n=1}^{\infty}$ of orthonormal eigenvectors of the operator Λ. Moreover, this construction immediately implies that $|\lambda_1| \geqslant |\lambda_2| \geqslant \ldots \geqslant |\lambda_n| \geqslant \ldots$. Indeed, when we pass from n to n + 1 in (4), the set of the admissible elements narrows, and therefore $\mathfrak{z}_n = |\lambda_{n+1}| \geqslant \mathfrak{z}_{n+1} = |\lambda_{n+1}|$. The sequence $\{e_n\}_{n=1}^{\infty}$, $\|e_n\| = 1$, is weakly convergent to zero, and therefore $|\lambda_n| = |(\Lambda e_n|e_n) \to 0$ because the form $(\Lambda x|x)$ is weakly continuous. Let us consider the orthogonal complement $L^{\perp}$ to the subspace L spanned by the vectors $\{e_n\}_{n=1}^{\infty}$ and the problem

$$|(\Lambda x|x)| \to \sup; \qquad (x|x) = 1, \qquad x \in L^{\perp}. \tag{8}$$

By the lemma, there exists a solution of problem (8). For the value of the problem we have $\mathfrak{z}_{\infty} \leqslant \min \mathfrak{z}_n = 0$, i.e., $\mathfrak{z}_{\infty}$ is equal to zero. Consequently, $\Lambda x \equiv 0$ on $L^{\perp}$. Taking a basis $f_1,\ldots,f_s,\ldots$ in the space $L^{\perp}$ we form the basis consisting of the vectors $\{e_i\}_{i=1}^{\infty}$ and $\{f_j\}$. Moreover, if $x = \sum_{k=1}^{\infty} (x|e_k)e_k + \sum_j (x|f_j)f_j$, then $\Lambda x = \sum_{k=1}^{\infty} \lambda_k (x|e_k)e_k$. Let us put $\Lambda_n x = \sum_{k=1}^{n} \lambda_k (x|e_k)e_k$. Then each of the operators Λ_n is compact and

$$\|\Lambda_n - \Lambda\| = \sup\{\|\Lambda_n x - \Lambda x\| \,|\, \|x\| \leq 1\} \leq \max_{k \geq n+1} |\lambda_k| \to 0.$$

Consequently, the operator Λ is the limit of a sequence of compact operators convergent with respect to the norm and therefore Λ itself is compact [KF, p. 241].

Moreover,

$$Q(x) = (\Lambda x|x) = \left(\sum_{k=1}^{\infty} \lambda_k (x|e_k) e_k | x \right) = \sum_{k=1}^{\infty} \lambda_k (x|e_k)^2. \quad \blacksquare$$

COROLLARY. The vectors e_n and $-e_n$ are stationary points of the extremal problem

$$Q(x) = (\Lambda x|x) \to \text{extr}, \qquad (x|x) = 1. \tag{9}$$

If all λ_n are distinct, then this problem has no stationary points other than e_n and $-e_n$.

Proof. By definition, the stationary points of the problem (9) are the points at which the derivative of the Lagrange function

$$\mathcal{L} = \lambda_0 (\Lambda x|x) + \lambda_1 (x|x)$$

vanishes. As before, if $\hat{x}$ is a stationary point, then $\hat{\lambda}_0 \neq 0$ and $\Lambda\hat{x} = \lambda\hat{x}$, where $\lambda = -\hat{\lambda}_1/\hat{\lambda}_0$. Consequently,

$$\lambda (x|e_n) = (\Lambda x|e_n) = (\hat{x}|\Lambda e_n) = \lambda_n (\hat{x}|e_n),$$

and therefore $(\hat{x}|e_n) \neq 0$ only if $\lambda = \lambda_n$. This immediately implies the assertion we intend to prove. $\blacksquare$

In conclusion, let us suppose that the form $Q(x) = (\Lambda x|x)$ is positive-definite (all λ_n are positive) and consider the ellipsoid

$$E = \{x | (\Lambda x|x) = 1\}. \tag{10}$$

For the sake of simplicity, we shall confine ourselves to the case when all λ_n are distinct.

The stationary points $\hat{x}$ of the problem

$$(x|x) \to \text{extr}, \qquad (\Lambda x|x) = 1 \tag{11}$$

specify the directions of the principal axes of the ellipsoid, and $\|\hat{x}\|$ are the magnitudes of its semiaxes. Again, forming the Lagrange function

$$\mathcal{L} = \mu_0 (x|x) + \mu_1 (\Lambda x|x)$$

we find $\hat{\mu}_0\hat{x} + \hat{\mu}_1\Lambda\hat{x} = 0$, whence, as before, $\Lambda\hat{x} = \lambda\hat{x}$. Thus, the directions of the principal axes of the ellipsoid (10) are specified by the eigenvectors of the operator Λ: $\hat{x}^n = \|\hat{x}^{(n)}\| e_n$, $n = 1, 2, \ldots$, and the semiaxes $\|\hat{x}^{(n)}\|$ are equal to $\lambda_n^{-1/2}$.

Exercise. Let Λ be a compact self-adjoint operator. Let us arrange its positive and negative eigenvalues in the decreasing order of their absolute values:

$$\lambda_1^+ \geq \lambda_2^+ \geq \ldots \lambda_n^+ \geq \ldots > 0, \qquad \lambda_1^- \leq \lambda_2^- \leq \ldots \leq \lambda_n^- \leq \ldots < 0,$$

and let e_n^+, e_n^- be the corresponding eigenvectors of the operator Λ. Let us denote by $S_n^+(h_1, \ldots, h_{n-1})$ $(S_n^-(h_1, \ldots, h_{n-1}))$ the value of the extremal problem

$$Q(x) - (\Lambda x|x) \to \sup\,(\inf); \quad (x|h_i) - 0, \; i - 1, \ldots, n \quad 1.$$

Prove that

$$\lambda_n^+ = \min_{\{h_1, \ldots, h_{n-1}\}} S^+(h_1, \ldots, h_{n-1}) \left(\lambda_n^- = \max_{\{h_1, \ldots, h_{n-1}\}} S^-(h_1, \ldots, h_{n-1}) \right),$$

where the minimum (the maximum) is attained for $h_k = e_k^+$ ($h_k = e_k^-$).

3.5.5. Legendre Quadratic Forms. In this section we shall continue the study of extremal properties of quadratic forms in a Hilbert space and prove Hestenes' theorem on the index of a quadratic form, which plays an important role in the classical calculus of variations.

Definition. A quadratic form Q in a Hilbert space is said to be a *Legendre form* if it is weakly lower semicontinuous and if the weak convergence of x_n to x and the convergence of $Q(x_n)$ to $Q(x)$ imply the strong convergence of x_n to x.

The maximum dimension of the subspaces L on which a quadratic from Q is negative-definite, i.e., $Q(x) < 0$ for all $x \in L$, $x \neq 0$, is called the *index* of the form and is denoted ind Q.

Hestenes' Theorem. The index of a Legendre quadratic form is finite.

Proof. Let us consider again the extremal problem

$$(\Lambda x|x) = Q(x) \to \inf, \qquad (x|x) \leqslant 1. \tag{1}$$

The existence of a solution in problem (1) is guaranteed, as before, by Weierstrass' theorem (Section 3.1.7): the ball $\{x|(x|x) \leqslant 1\}$ is weakly compact, and, by the hypothesis, Q is weakly lower semicontinuous. If the value of problem (1) is equal to zero, then $Q(x) \geqslant 0$, $\forall x$, the index of Q is equal to zero, and the theorem has been proved. Let the value of problem (1) be negative. Then, as in the foregoing section [problem (2)], the solution e_1 of problem (1) is an eigenvector:

$$\Lambda e_1 = \lambda_1 e_1, \qquad \|e_1\| = 1, \qquad \lambda_1 < 0.$$

We shall again construct by induction an orthonormal system by solving in succession the extremal problems

$$(\Lambda x|x) = Q(x) \to \inf, \qquad (x|x) = 1, \qquad (x|e_i) = 0, \qquad i = 1, \ldots, n, \tag{2}$$

and finding in this way the eigenvectors:

$$\Lambda e_i = \lambda_i e_i, \qquad (e_i|e_i) = 1, \qquad i = 1, 2, \ldots, \lambda_1 \leqslant \lambda_2 \leqslant \ldots \leqslant 0. \tag{3}$$

Let us show that after a finite number of steps this procedure will stop because the extremal problem (2) will no longer have a nonzero solution. Indeed, let the sequence $\{e_i | i \geqslant 1\}$ be infinite. Then it is weakly convergent to zero. By the lower semicontinuity of Q and the monotonicity of $\{\lambda_i\}$,

$$\lim_{n\to\infty} \lambda_n = \underline{\lim}_{n\to\infty} \lambda_n = \underline{\lim}_{n\to\infty} (\Lambda e_n|e_n) = \underline{\lim}_{n\to\infty} Q(e_n) \geqslant Q(0) = 0.$$

However, $\lambda_n \leqslant 0$, and consequently $\lim \lambda_n = 0$. Therefore, $\{e_n\}$ is weakly convergent to zero, and, moreover, $Q(e_n) = (\Lambda e_n | e_n) = \lambda_n \to 0$. The form Q being Legendre, the norm $\|e_n\|$ must tend to zero, but this is not so because $\|e_n\| = 1$. This contradiction shows that after a finite number of steps N the procedure will stop. The form Q must be nonnegative on the orthogonal complement $L_N^\perp = (\text{lin}\{e_1, \ldots, e_N\})^\perp$.

Let the form Q be negative-definite on a subspace L, and let $\{f_n\}$ be a basis in L (which is infinite in case $\dim L = \infty$). By the theorem on the orthogonal complement (Section 3.5.8), $f_n = g_n + h_n$, where $g_n \in L_N$, $h_n \in L_N^\perp$. If $\dim L$, i.e., the number of the vectors in the basis $\{f_n\}$, exceeds $N = \dim L_N$, then the vectors $g_1, \ldots, g_{N+1}$ are linearly dependent, i.e., $\sum_{n=1}^{N} \alpha_n g_n = 0$, where α_n are not all zero. It follows that

$$\sum_{n=1}^{N+1} \alpha_n f_n = \sum_{n=1}^{N+1} \alpha_n h_n \in L_N^\perp,$$

and therefore $Q\left(\sum_{n=1}^{N+1} \alpha_n f_n\right) \geqslant 0$, whence $\sum_{n=1}^{N+1} \alpha_n f_n = 0$ because the form $Q|_L$ is negative-definite. The last equality is impossible since $\{f_n\}$ is a basis, and, consequently, $\dim L \leqslant N$. By the definition of the index, $\operatorname{ind} Q \leqslant N$. On the other hand, the form Q is negative on L:

$$x \in L_N \Rightarrow x = \sum_{i=1}^{N} (x|e_i)e_i \Rightarrow Q(x) = (\Lambda x|x) = \sum_{i=1}^{N} \lambda_i (x_i|e_i)^2 < 0$$

for $x \neq 0$. Hence, $\operatorname{ind} Q \geqslant N$, i.e., $\operatorname{ind} Q = N$. ■

Chapter 4

THE LAGRANGE PRINCIPLE FOR PROBLEMS OF THE CLASSICAL CALCULUS OF VARIATIONS AND OPTIMAL CONTROL THEORY

The subject of this chapter is clear from its title. Here our primary aim is, on the one hand, to show the similarity between the two versions of the theory, which is stressed by the unified system of notation, and, on the other hand, to elucidate the distinction between the classical and the modern statements of the problem. We begin with necessary conditions for the so-called *Lagrange problem* to which many other problems of the classical calculus of variations can be reduced. Then we derive the Pontryagin maximum principle, which is one of the most important means of the modern theory of optimal control problems. The rest of the chapter is devoted to some more special classes of problems and to the derivation of consequences of the general theory. Sufficient conditions for an extremum are treated less thoroughly, and we confine ourselves to some particular situations.

4.1. Lagrange Principle for the Lagrange Problem

4.1.1. Statement of the Problem and the Formulation of the Theorem.

Let us fix a closed interval $\Delta=[\alpha, \beta]\subset\mathbf{R}$ and consider the Banach space

$$\Xi=C^1(\Delta, \mathbf{R}^n)\times C(\Delta, \mathbf{R}^r)\times\mathbf{R}\times\mathbf{R}$$

consisting of the elements $\xi = (x(\cdot), u(\cdot), t_0, t_1)$.

For this space we shall consider the problem

$$\mathscr{B}_0(x(\cdot), u(\cdot), t_0, t_1) = \int_{t_0}^{t_1} f_0(t, x(t), u(t))\,dt+\psi_0(t_0, x(t_0), t_1, x(t_1))\rightarrow \text{extr}, \tag{1}$$

$$\dot{x}=\varphi(t, x(t), u(t)), \tag{2}$$

$$\mathscr{B}_i(x(\cdot), u(\cdot), t_0, t_1)=\int_{t_0}^{t_1} f_i(t, x(t), u(t))\,dt+\psi_i(t_0, x(t_0), t_1, x(t_1))\lesseqgtr 0. \tag{3}$$

Here and henceforth, $f_i\colon V\rightarrow\mathbf{R}$, $\psi_i\colon W\rightarrow\mathbf{R}$, $i=0, 1, \ldots m$; $\varphi\colon V\rightarrow\mathbf{R}^n$ in (1)-(3) are given functions, where V and W are open sets in the spaces $\mathbf{R}\times\mathbf{R}^n\times\mathbf{R}^r$ and $\mathbf{R}\times\mathbf{R}^n\times\mathbf{R}\times\mathbf{R}^n$, respectively. All the enumerated functions are assumed to be at least continuous. The symbol $\lesseqgtr$ is used in the same sense as in Section 3.2.

Problem (1)-(3) will be referred to as the *Lagrange problem in Pontryagin's form*. Some special cases of this problem were discussed in Sections 1.4 and 1.5 and in Section 3.1. Functionals of the type of $\mathscr{B}_i$, involving both integral and endpoint parts were called Bolza functionals (see Sections 1.4.2 and 3.1.3). If the endpoint part in a constraint $\mathcal{B}_i \gtreqless 0$ is a constant, i.e., if the constraint takes the form $\int_{t_0}^{t_1} f_i(t, x, u)\,dt \gtreqless a_i$, then, following L. Euler, we call it an *isoperimetric* constraint. Conversely, if there is no integral term, the constraint assumes the form $\psi_i(t_0, x(t_0), t_1, x(t_1)) \gtreqless 0$ and is called a *boundary condition.*

Constraint (2) is called a *differential constraint.* This type of a differential constraint in the solved form with respect to $\dot{x}$ is a characteristic feature of Pontryagin's form of the problem. J. L. Lagrange specified a differential constraint by an equation $\psi(t, x, \dot{x}) = 0$ (this was briefly mentioned in Section 1.5.1).

Finally, in contrast to Chapter 1, the time interval $[t_0, t_1]$ in problem (1)-(3) is not assumed to be fixed.

A quadruple $\xi = (x(\cdot), u(\cdot), t_0, t_1) \in \Xi$ will be called a *controlled process* in problem (1)-(3) if $t_0, t_1 \in \operatorname{int}\Delta$, $(t, x(t), u(t)) \in V$ for $t \in \Delta$ and if the differential constraint holds everywhere on $[t_0, t_1]$, i.e., $\dot{x}(t) = \varphi(t, x(t), u(t))$. Further, such a quadruple is called an *admissible controlled process* if it is a controlled process and, moreover, conditions (3) hold.

A quadruple $\hat{\xi} = (\hat{x}(\cdot), \hat{u}(\cdot), \hat{t}_0, \hat{t}_1)$ is called an *optimal process in the weak sense* or is said to yield a *weak extremum* in problem (1)-(3) if it yields a local extremum in the space Ξ, i.e., if there is $\varepsilon > 0$ such that for any admissible controlled process ξ satisfying the condition $\|\xi - \hat{\xi}\|_\Xi < \varepsilon$ there holds one of the following two inequalities: $\mathcal{B}_0(\xi) \geqslant \mathcal{B}_0(\hat{\xi})$ in the case of a minimum and $\mathcal{B}_0(\xi) \leqslant \mathcal{B}_0(\hat{\xi})$ in the case of a maximum.

By the *Lagrange function* of problem (1)-(3) we shall mean the function

$$\mathscr{L}(x(\cdot), u(\cdot), t_0, t_1; p(\cdot), \lambda, \lambda_0) = \int_{t_0}^{t_1} L\,dt + l, \tag{4}$$

where

$$\lambda_0 \in \mathbf{R}, \quad p(\cdot) \in C^1(\Delta, \mathbf{R}^{n*}), \quad \lambda = (\lambda_1, \ldots, \lambda_m) \in \mathbf{R}^{m*}$$

are *Lagrange multipliers*,

$$L(t, x, \dot{x}, u) = \sum_{i=0}^{m} \lambda_i f_i(t, x, u) + p(t)(\dot{x} - \varphi(t, x, u)) \tag{5}$$

is the *Lagrangian* or the *integrand*, and

$$l(t_0, x_0, t_1, x_1) = \sum_{i=0}^{m} \lambda_i \psi_i(t_0, x_0, t_1, x_1) \tag{6}$$

is an *endpoint functional.*

As before, here and in the following sections we shall use the following abbreviated notation:

$$\hat{L}_{\dot{x}}(t) = L_{\dot{x}}(t, \hat{x}(t), \dot{\hat{x}}(t), \hat{u}(t)),$$
$$\hat{L}_x(t) = L_x(t, \hat{x}(t), \dot{\hat{x}}(t), \hat{u}(t)),$$
$$\hat{L}_u(t) = L_u(t, \hat{x}(t), \dot{\hat{x}}(t), \hat{u}(t)),$$
$$\hat{l}_{x_k} = l_{x_k}(\hat{t}_0, \hat{x}(t_0), \hat{t}_1, \hat{x}(t_1)),$$
$$\hat{\mathscr{L}}_{t_k} = \mathscr{L}_{t_k}(\hat{x}(\cdot), \hat{u}(\cdot), \hat{t}_0, \hat{t}_1; \hat{p}(\cdot), \hat{\lambda}, \hat{\lambda}_0), \text{ etc.}$$

Euler–Lagrange Theorem. Let the functions $f_i: V \to \mathbf{R}$, $i = 0, 1, \ldots, m$, $\varphi: V \to \mathbf{R}^n$ and their partial derivatives with respect to x and u be continuous in the open set V in the space $\mathbf{R} \times \mathbf{R}^n \times \mathbf{R}^r$, let the functions

$\psi_i\colon W \to \mathbf{R}$, $i = 0, 1, \ldots, m$, be continuously differentiable in the open set W in the space $\mathbf{R} \times \mathbf{R}^n \times \mathbf{R} \times \mathbf{R}^n$, and let $\hat{x}(\cdot) \in C^1(\Delta, \mathbf{R}^n)$, $\hat{u}(\cdot) \in C(\Delta, \mathbf{R}^r)$, $\hat{t}_0$, $\hat{t}_1 \in \operatorname{int}\Delta$ be such that

$$(t, \hat{x}(t), \hat{u}(t)) \in V, \quad t \in \Delta, \quad (\hat{t}_0, \hat{x}(\hat{t}_0), \hat{t}_1, \hat{x}(\hat{t}_1)) \in W.$$

If $\hat{\xi} = (\hat{x}(\cdot), \hat{u}(\cdot), \hat{t}_0, \hat{t}_1)$ is an optimal process in the weak sense for problem (1)-(3), then there are Lagrange multipliers $\hat{\lambda}_0 \geqslant 0$ (in the case of a minimum problem) or $\hat{\lambda}_0 \leqslant 0$ (in the case of a maximum problem), $\hat{p}(\cdot) \in C^1(\Delta, \mathbf{R}^{n*})$ and $\hat{\lambda} = (\hat{\lambda}_1, \ldots, \hat{\lambda}_m)$, not all zero, such that:

a) the stationarity conditions hold for the Lagrange function:

with respect to $x(\cdot)$ $(\hat{\mathscr{L}}_{x(\cdot)} = 0)$:

$$-\frac{d}{dt}\hat{L}_{\dot{x}}(t) + \hat{L}_x(t) \equiv 0, \tag{7}$$

$$\hat{L}_{\dot{x}}(\hat{t}_k) = (-1)^k \hat{l}_{x_k}, \quad k = 0, 1; \tag{8}$$

with respect to $u(\cdot)$ $(\hat{\mathscr{L}}_{u(\cdot)} = 0)$:

$$\hat{L}_u(t) \equiv 0; \tag{9}$$

with respect to t_k:

$$\hat{\mathscr{L}}_{t_k} = 0, \quad k = 0, 1; \tag{10}$$

b) the condition of concordance of signs holds:

$$\hat{\lambda}_i \gtreqless 0, \quad i = 1, \ldots, m; \tag{11}$$

c) the conditions of complementary slackness hold:

$$\hat{\lambda}_i \mathscr{B}_i(\hat{\xi}) = 0, \quad i = 1, \ldots, m. \tag{12}$$

As in Sections 3.1 and 3.2, inequalities (11) mean that $\hat{\lambda}_i \geqslant 0$ if $\mathscr{B}_i(\xi) \leqslant 0$ in condition (3), $\hat{\lambda}_i \leqslant 0$ if $\mathscr{B}_i(\xi) \geqslant 0$ in (3), and $\hat{\lambda}_i$ may have an arbitrary sign if $\mathscr{B}_i(\xi) = 0$.

The assertion on the existence of the Lagrange multipliers satisfying the set of conditions a)-c) will be briefly called the *Lagrange principle for the Lagrange problem* (1)-(3).

Let us show that the assertion of the theorem is in complete agreement with the general Lagrange principle mentioned in Chapter 1 and Section 3.1.5. Indeed, the Lagrange function $\mathscr{L}$ is a function of three arguments: $x(\cdot)$, $u(\cdot)$, and (t_0, t_1). Hence, in accordance with the general Lagrange principle, we have to consider the following three problems:

$$(\alpha)\quad \mathscr{L}(x(\cdot), \hat{u}(\cdot), \hat{t}_0, \hat{t}_1, \hat{p}(\cdot), \hat{\lambda}, \hat{\lambda}_0) \to \text{extr},$$

$$(\beta)\quad \mathscr{L}(\hat{x}(\cdot), u(\cdot), \hat{t}_0, \hat{t}_1; \hat{p}(\cdot), \hat{\lambda}, \hat{\lambda}_0) \to \text{extr},$$

$$(\gamma)\quad \mathscr{L}(\hat{x}(\cdot), \hat{u}(\cdot), t_0, t_1; \hat{p}(\cdot), \hat{\lambda}, \hat{\lambda}_0) \to \text{extr}.$$

Problem (α) is an elementary Bolza problem, and the stationarity conditions (7) and (8) are written in complete accordance with Section 1.4.2 and the theorem of Section 3.1.3. Problem (β) is also a Bolza problem, but it is degenerate since $\dot{u}$ is not involved in the Lagrangian and $u(t_0)$ and $u(t_1)$ are not involved in the endpoint functional. Therefore, the Euler–Lagrange equation turns into (9), and the transversality condition becomes trivial: $0 = 0$. Finally, (γ) is an elementary problem, and equalities (10) simply express the Fermat theorem (see Sections 1.3.1 and 3.1.1).

As we know, the conditions of complementary slackness and the conditions of concordance of signs are also written in accordance with the general Lagrange principle applied to inequality constraints (Section 3.1.5).

Hence, the basic theorem of the present section can be stated thus: if the functions involved in the statement of the problem are sufficiently smooth, then the Lagrange principle holds for a local extremum.

The stationarity conditions can be written in full; according to (5) we obtain:

from (7):

$$\frac{d\hat{p}(t)}{dt} = -\hat{p}(t)\hat{\varphi}_x(t) + \sum_{i=0}^{m} \hat{\lambda}_i \hat{f}_{ix}(t) \tag{7a}$$

[(7a) is sometimes called the *conjugate equation*];

from (9):

$$0 = -\hat{p}(t)\hat{\varphi}_u(t) + \sum_{i=0}^{m} \hat{\lambda}_i \hat{f}_{iu}(t); \tag{9a}$$

from (8):

$$\hat{p}(\hat{t}_k) = (-1)^k \hat{l}_{x_k}, \quad k = 0, 1; \tag{8a}$$

and, finally, from (10):

$$0 = (-1)^{k-1} \hat{L}(\hat{t}_k) + \hat{l}_{t_k} + \hat{l}_{x_k} \dot{\hat{x}}(\hat{t}_k) = (-1)^{k-1} \left[\sum_{i=0}^{m} \hat{\lambda}_i \hat{f}_i(\hat{t}_k) + \hat{p}(\hat{t}_k)(\dot{\hat{x}}(\hat{t}_k) - \hat{\varphi}(\hat{t}_k)) \right] +$$

$$+ \hat{l}_{t_k} + (-1)^k \hat{p}(\hat{t}_k) \dot{\hat{x}}(\hat{t}_k) = (-1)^{k-1} \left[\sum_{i=0}^{m} \hat{\lambda}_i \hat{f}_i(\hat{t}_k) - \hat{p}(\hat{t}_k)\hat{\varphi}(\hat{t}_k) \right] + \hat{l}_{t_k};$$

that is,

$$\hat{p}(\hat{t}_k)\hat{\varphi}(\hat{t}_k) - \sum_{i=0}^{m} \hat{\lambda}_i \hat{f}_i(\hat{t}_k) = (-1)^{k-1} \hat{l}_{t_k}. \tag{10a}$$

The expression

$$H(t, x, p, u) = L_{\dot{x}} \dot{x} - L = p\varphi(t, x, u) - \sum_{i=0}^{m} \hat{\lambda}_i f_i(t, x, u) \tag{13}$$

will be called the *Pontryagin function* of the problem under consideration. Using this function, we can rewrite Eqs. (7a)-(10a) and Eq. (2) thus:

$$\frac{d\hat{x}}{dt} = \hat{H}_p, \quad \frac{d\hat{p}}{dt} = -\hat{H}_x, \quad \hat{H}_u = 0 \tag{14}$$

and

$$\hat{p}(\hat{t}_k) = (-1)^k \hat{l}_{x_k}, \quad \hat{H}(\hat{t}_k) = (-1)^{k-1} \hat{l}_{t_k}, \quad k = 0, 1. \tag{15}$$

Remark. Function (13) is sometimes called the Hamiltonian function. We use the term the "Hamiltonian function" or the "Hamiltonian" in a different sense which is more natural from the viewpoint of classical mechanics; namely, by the Hamiltonian we mean the function

$$\mathscr{H}(t, x, p) = H(t, x, p, u(t, x, p)), \tag{16}$$

where $u(t, x, p)$ is the implicit function determined by the equation $H_u(t, x, p, u) = 0$. When the standard conditions for the applicability of the implicit function theorem hold, Eqs. (14) assume the ordinary form of a canonical Hamiltonian system:

$$\frac{d\hat{x}}{dt} = \mathscr{H}_p, \quad \frac{d\hat{p}}{dt} = -\mathscr{H}_x. \tag{17}$$

It should be noted that the set of the conditions stated in the theorem is complete in a certain sense. Indeed, we have the system of the differential equations (2) and (7a) and the finite equation (9a) for the determination of the unknown functions $(x(\cdot), p(\cdot), u(\cdot))$. Expressing $u(\cdot)$ in terms of $x(\cdot)$ and $p(\cdot)$ with the aid of Eq. (9a) (of course, when it is possible, for instance when the conditions of the implicit function theorem are fulfilled), we obtain a system of $2n$ scalar differential equations [equivalent to system (17)]. Its general solution depends on $2n$ arbitrary constants and also on the Lagrange multipliers λ_i among which there are m independent ones. Adding t_0 and t_1 to them, we obtain altogether $2n + m + 2$ unknowns. For their determination we have the $2n + 2$ transversality conditions (15) and the m conditions of complementary slackness (12).

Thus, the number of the unknowns coincides with that of the equations. It is this fact that was meant when we mentioned the "completeness" of the set of the conditions. What has been said does not, of course, guarantee the solvability of the resultant system of equations.

4.1.2. Reduction of the Lagrange Problem to a Smooth Problem. Let us denote by Y the space $C(\Delta, \mathbf{R}^n)$ and write problem (1)-(3) of Section 4.1.1 in the form

$$\mathscr{B}_0(\xi) \to \text{extr}; \quad \Phi(\xi) = 0, \quad \mathscr{B}_i(\xi) \leqslant 0, \quad i = 1, \ldots, m, \tag{1}$$

where

$$\Phi(\xi)(t) = x(t) - \varphi(t, x(t), u(t)). \tag{2}$$

The mapping Φ and the functionals $\mathscr{B}_i$ are defined in the domain

$$\mathscr{V} = \{(x(\cdot), u(\cdot), t_0, t_1) \mid (t, x(t), u(t)) \in V, \ t \in \Delta; \\ t_0, t_1 \in \operatorname{int} \Delta, \ (t_0, x(t_0), t_1, x(t_1)) \in W\}$$

in the space Ξ.

Problem (1) has the same form as the problem of Section 3.2.1 of the foregoing chapter for which the Lagrange principle was already proved. The hypotheses of the corresponding theorem split into three groups: the conditions that the basic spaces were *Banach* spaces, the conditions of *smoothness* of the mappings, and the condition of the *closedness of the image* of the infinite-dimensional mapping. Let us verify that all these requirements are fulfilled in problem (1).

The spaces Ξ and Y are obviously *Banach* spaces: C^1 and C are the basic examples of Banach spaces in our presentation of the material (see Sections 2.1.1 and 2.1.2).

The *smoothness* (namely, the continuous differentiability in Fréchet's sense) of the functionals $\mathscr{B}_i$, $i = 0, 1, \ldots, m$, follows from the results of Section 2.4.

Indeed, the integral part can be considered in exactly the same way as in Proposition 2 of Section 2.4.2 if we replace $\dot{x}(\cdot)$ by $u(\cdot)$ everywhere, and the differentiability of the endpoint part was proved in Section 2.4.3 (this part is an operator of boundary conditions). According to formula (9) of Section 2.4.2 and formula (3) of Section 2.4.3, for $\hat{\xi} = (\hat{x}(\cdot), \hat{u}(\cdot), \hat{t}_0, \hat{t}_1)$ and $\eta = (h(\cdot), v(\cdot), \tau_0, \tau_1)$, we have

$$\mathscr{B}_i'(\hat{\xi})[\eta] = \int_{t_0}^{t_1} (\hat{f}_{ix}(t) h(t) + \hat{f}_{iu}(t) v(t))\, dt + \\ + \hat{f}_i(\hat{t}_1)\tau_1 - \hat{f}_i(\hat{t}_0)\tau_0 + \hat{l}_{t_0}\tau_0 + \hat{l}_{t_1}\tau_1 + \hat{l}_{x_0} h(\tau_0) + \hat{l}_{x_1} h(\hat{t}_1) + \hat{l}_{x_0}\dot{\hat{x}}(\hat{t}_0)\tau_0 + \hat{l}_{x_1}\dot{\hat{x}}(\hat{t}_1)\tau_1 \tag{3}$$

(cf. the lemma of Section 3.1.3).

Mapping (2) is also continuously Fréchet differentiable (see Proposition 3 of Section 2.4.1). According to the formula (14) of Section 2.4.1, its derivative has the form

$$\Phi'(\hat{\xi})[\eta](t) = \dot{h}(t) - \hat{\varphi}_x(t)h(t) - \hat{\varphi}_u(t)v(t). \quad (4)$$

The continuous Fréchet differentiability of the functionals $\mathcal{B}_i$ and the mapping Φ implies their strict differentiability required in the theorem on the Lagrange principle.

The *closedness of the image* of the mapping Φ' follows from its regularity: the image $\Phi'(\hat{\xi})\Xi$ coincides with Y. Indeed, let us take an arbitrary $y(\cdot) \in Y = C(\Delta, \mathbf{R}^n)$, and put $v(\cdot) = 0$, $\tau_0 = \tau_1 = 0$. The equation

$$\Phi'(\hat{\xi})[h(\cdot), 0, 0, 0](\cdot) = y(\cdot)$$

is equivalent to the linear differential equation $\dot{h}(t) - \hat{\varphi}_x(t)h(t) = y(t)$ with continuous coefficients which possesses a solution $h(\cdot) \in C^1(\Delta, \mathbf{R}^n)$ by virtue of the existence theorem for linear systems (Section 2.5.4). Hence, all the conditions of the theorem of Section 3.2.1 are fulfilled.

For definiteness, let us assume that (1) is a minimum problem. According to the Lagrange principle, if $\hat{\xi} = (\hat{x}(\cdot), \hat{u}(\cdot), \hat{t}_0, \hat{t}_1)$ yields a local minimum for (1), then there are Lagrange multipliers $\hat{\lambda}_0$, $\hat{\lambda} \in \mathbf{R}^{m*}$ and an element $\hat{y}^* \in Y^* = (C(\Delta, \mathbf{R}^n))^*$ such that for the Lagrange function

$$\tilde{\mathcal{L}}(\xi; y^*, \lambda, \lambda_0) = \tilde{\mathcal{L}}(x(\cdot), u(\cdot), t_0, t_1; y^*, \lambda, \lambda_0) =$$
$$= \int_{t_0}^{t_1} \left(\sum_{i=0}^{m} \lambda_i f_i(t, x, u) \right) dt + \sum_{i=0}^{m} \lambda_i \psi_i(t_0, x(t_0), t_1, x(t_1)) + y^* \circ \Phi \quad (5)$$

there hold the conditions of concordance of signs, the conditions of complementary slackness, and the stationarity conditions:

$$\hat{\tilde{\mathcal{L}}}_\xi = 0 \Longleftrightarrow \hat{\tilde{\mathcal{L}}}_{x(\cdot)} = 0, \quad \hat{\tilde{\mathcal{L}}}_{u(\cdot)} = 0, \quad \hat{\tilde{\mathcal{L}}}_{t_k} = 0, \quad k = 0, 1.$$

The three relations forming the right-hand side will be referred to as the *stationarity* conditions with respect to $x(\cdot)$, $u(\cdot)$ and t_k.

However, it should be noted that function (5) is not identical with the Lagrange function (4) of Section 4.1.1. Using Riesz' representation theorem for the general continuous linear functional on the space $C(\Delta, \mathbf{R}^n)$ (Section 2.1.9), we can only assert that

$$\langle y^*, \eta(\cdot) \rangle = \int_\Delta d\nu(t)\, \eta(t),$$

where $\nu(t) = (\nu_1(t), \ldots, \nu_n(t))$ is a row vector formed of canonical functions of bounded variation. Substituting this expression into (5) and taking into account the notation in the formula (6) of Section 4.1.1, we obtain

$$\tilde{\mathcal{L}} = \int_{t_0}^{t_1} \left(\sum_{i=0}^{m} \lambda_i f_i(t, x(t), u(t)) \right) dt +$$
$$+ \int_\Delta d\nu(t) \{\dot{x}(t) - \varphi(t, x(t), u(t))\} + l(t_0, x(t_0), t_1, x(t_1)). \quad (6)$$

Comparing this formula with formulas (4)-(6) of Section 4.1.1, we see that the Lagrange function $\mathcal{L}$ is obtained from $\tilde{\mathcal{L}}$ if the function $\nu(\cdot)$ is absolutely continuous and its derivative is equal to zero outside the closed interval $[t_0, t_1]$ and coincides with $p(\cdot)$ on this interval. Here the situation resembles the one in Section 1.4.1, where the DuBois—Reymond's lemma allowed us to derive Euler's equation without the additional assumption that the function $t \to p(t) = \hat{L}_{\dot{x}}(t)$ was differentiable. Similarly, an appropriate generalization of that lemma will enable us to prove the continuity of the function $\nu(\cdot)$.

4.1.3. Generalized DuBois—Reymond's Lemma. Let a vector function $a(\cdot)\colon [\alpha, \beta] \to \mathbf{R}^{n*}$ be Lebesgue integrable on a closed interval $[\alpha, \beta]$, let a matrix function $b(\cdot)$: $[\alpha, \beta] \to \mathscr{L}(\mathbf{R}^n, \mathbf{R}^n)$ be continuous on $[\alpha, \beta]$, let $\nu(\cdot)$: $[\alpha, \beta] \to \mathbf{R}^{n*}$ be a canonical vector function of bounded variation, let t_k be different points belonging to the open interval (α, β), and let $c_k \in \mathbf{R}^{n*}$, $k = 0, 1, \ldots, p$.

If the equality

$$\int_\alpha^\beta a(t) h(t) dt + \int_\alpha^\beta d\nu(t) \{\dot{h}(t) - b(t) h(t)\} + \sum_{k=0}^{p} c_k h(t_k) = 0 \tag{1}$$

holds for all $h(\cdot) \in C^1([\alpha, \beta], \mathbf{R}^n)$, then function $\nu(\cdot)$ is absolutely continuous; its derivative $p(\cdot)$ is continuous on $[\alpha, \beta]$ except possibly at the points t_k, where it has the limits on the left and on the right, and is absolutely continuous on any interval not containing the points t_k, and, moreover,

$$\dot{p}(t) = a(t) - p(t) b(t) \text{ almost everywhere,} \tag{2}$$

$$p(\alpha) = p(\beta) = 0, \tag{3}$$

$$p(t_k + 0) - p(t_k - 0) = c_k. \tag{4}$$

Proof. Putting $h(t) = \gamma + \int_\alpha^t \eta(s) ds$, we obtain from (1) the relation

$$\left\{\int_\alpha^\beta a(t) dt - \int_\alpha^\beta d\nu(t) b(t) + \sum_{k=0}^{p} c_k\right\} \gamma + \int_\alpha^\beta a(t) \int_\alpha^t \eta(s) ds\, dt +$$
$$+ \int_\alpha^\beta d\nu(t) \eta(t) - \int_\alpha^\beta d\nu(t) b(t) \int_\alpha^t \eta(s) ds + \sum_{k=0}^{p} c_k \int_\alpha^{t_k} \eta(s) ds = 0.$$

Since $\gamma \in \mathbf{R}^n$ and $\eta(\cdot) \in C([\alpha, \beta], \mathbf{R}^n)$ can be chosen arbitrarily and independently, in the first place, we have

$$\int_\alpha^\beta a(t) dt - \int_\alpha^\beta d\nu(t) b(t) + \sum_{k=0}^{p} c_k = 0 \tag{5}$$

[because we can put $\eta(\cdot) = 0$ and take an arbitrary γ] and, in the second place, the equality

$$\int_\alpha^\beta a(t) \int_\alpha^t \eta(s) ds\, dt + \int_\alpha^\beta d\nu(t) \eta(t) - \int_\alpha^\beta d\nu(t) b(t) \int_\alpha^t \eta(s) ds + \sum_{k=0}^{p} c_k \int_\alpha^{t_k} \eta(s) ds = 0 \tag{6}$$

holds for any $\eta(\cdot) \in C([\alpha, \beta], \mathbf{R}^n)$.

Let us change the order of integration in the first and the third terms in accordance with Dirichlet's formula [formula (5) in Section 2.1.9], denote by s the variable of integration in the second term, put

$$\chi_{[\alpha, t]}(s) = \begin{cases} 1, & \alpha \leqslant s \leqslant t, \\ 0, & t < s \leqslant \beta, \end{cases} \tag{7}$$

and use the equality

$$\int_\alpha^\beta f(t) d\mu(t) = \int_\alpha^\beta f(t) \mu'(t) dt \tag{8}$$

(which holds for absolutely continuous functions [KF, p. 359]) to bring equality (6) to the form

$$\int_\alpha^\beta \left(\int_s^\beta a(t) dt\right) \eta(s) ds + \int_\alpha^\beta d\nu(s) \eta(s) - \int_\alpha^\beta \left(\int_s^\beta d\nu(t) b(t)\right) \eta(s) ds +$$
$$+ \int_\alpha^\beta \sum_{k=0}^{p} c_k \chi_{[\alpha, t_k]}(s) \eta(s) ds = \int_\alpha^\beta d\Lambda(s) \eta(s) + \int_\alpha^\beta d\nu(s) \eta(s) = 0, \tag{9}$$

where

$$\Lambda(s)=\int_{\alpha}^{s}\left\{\int_{\sigma}^{\beta} a(t)\,dt-\int_{\sigma}^{\beta} d\nu(t)\,b(t)+\sum_{k=0}^{p} c_k\chi_{[\alpha,\,t_k]}(\sigma)\right\}d\sigma.$$

Function $\Lambda(\cdot)$ is absolutely continuous on $[\alpha, \beta]$, and we have $\Lambda(\alpha) = 0$, and consequently $\Lambda(\cdot)$ is a canonical function of bounded variation. Since $\nu(\cdot)$ is also a canonical function of bounded variation, relation (9) and the uniqueness property in Riesz' theorem (Section 2.1.9) imply the identity $\Lambda(s) + \nu(s) \equiv 0$. Hence, function $\nu(\cdot)$ is absolutely continuous, and its derivative $p(s) = \nu'(s)$ is equal to

$$p(s)=\nu'(s)=-\Lambda'(s)=-\int_{s}^{\beta} a(t)\,dt+\int_{s}^{\beta} p(t)\,b(t)\,dt-\sum_{k=0}^{p} c_k\chi_{[\alpha,\,t_k]}(s)$$

for $s \neq t_k$ [here we again use (8)]. It is seen from this formula that function $p(\cdot)$ is in its turn continuous everywhere except at the points t_k, where (4) holds and that (2) takes place. The equality $p(\beta) = 0$ is evident, and $p(\alpha) = 0$ is equivalent to (5), and hence (3) is also proved. ■

4.1.4. Derivation of Stationarity Conditions. Thus, we have shown that if $\hat{\xi}$ yields a local minimum in problem (1)-(3) of Section 4.1.1, then there are Lagrange multipliers $\hat{\lambda}_0 \geqslant 0$, $\hat{\lambda} \in \mathbf{R}^{m*}$ and $\hat{y}^* \in C(\Delta, \mathbf{R}^n)^*$ such that there hold the conditions of concordance of signs and the conditions of complementary slackness for λ, and the stationarity conditions:

$$\text{a) } \hat{\tilde{\mathscr{L}}}_{x(\cdot)}=0,\quad \text{b) } \hat{\tilde{\mathscr{L}}}_{u(\cdot)}=0,\quad \text{c) } \hat{\tilde{\mathscr{L}}}_{t_k}=0,\quad k=0,\,1, \tag{1}$$

where $\tilde{\mathscr{L}}$ is determined by equality (1) of Section 4.1.2, and $\hat{\tilde{\mathscr{L}}}_{x(\cdot)}$, $\hat{\tilde{\mathscr{L}}}_{u(\cdot)}$, and $\hat{\tilde{\mathscr{L}}}_{t_k}$ are obtained from $\tilde{\mathscr{L}}_{x(\cdot)}$, $\tilde{\mathscr{L}}_{u(\cdot)}$, and $\tilde{\mathscr{L}}_{t_k}$ by substituting $x(\cdot) = \hat{x}(\cdot)$, $u(\cdot) = \hat{u}(\cdot)$, $t_k = \hat{t}_k$, $k = 0, 1$.

Now we pass to the interpretation of conditions (1). When differentiating the function $\tilde{\mathscr{L}}$, we shall take into consideration that the first and the third terms of the formula (6) of Section 4.1.2 form a Bolza functional whose derivative is expressed by formula (3) of that same section, and the second term is the composition of the differential constraint operator and the linear mapping $\eta(\cdot)\to\int_{\Delta} d\nu(t)\,\eta(t)$, and therefore its derivative is obtained by differentiating the formula (4) of Section 4.1.2 with respect to $d\nu(t)$.

A) Stationarity with Respect to $x(\cdot)$. Differentiating $\tilde{\mathscr{L}}$ with respect to $x(\cdot)$ we conclude, in accordance with (1a), that for all $h(\cdot)\in C^1(\Delta, \mathbf{R}^n)$ the equality

$$0=\hat{\tilde{\mathscr{L}}}_{x(\cdot)}[h(\cdot)]=\int_{\hat{t}_0}^{\hat{t}_1}\sum_{i=0}^{m}\hat{\lambda}_i\hat{f}_{ix}(t)\,h(t)\,dt+\int_{\Delta} d\nu(t)\,\{\dot{h}(t)-\hat{\varphi}_x(t)\,h(t)\}+\hat{l}_{x_0}h(\hat{t}_0)+\hat{l}_{x_1}h(\hat{t}_1) \tag{2}$$

holds. Let us begin with the case $\hat{t}_1 > \hat{t}_0$. According to the generalized DuBois—Reymond's lemma, it follows from (2) that the function $\hat{\nu}(\cdot)$ is absolutely continuous on Δ and its derivative $p(\cdot) = \hat{\nu}'(\cdot)$ is continuous on Δ everywhere except possibly at the points $\hat{t}_0$ and $\hat{t}_1$, where it has limits on the left and on the right, and relations (2)-(4) of Section 4.1.3 hold. For the situation under consideration, these relations mean that

$$dp/dt=-p(t)\,\hat{\varphi}_x(t)+\chi_{[\hat{t}_0,\,\hat{t}_1]}(t)\sum_{i=0}^{m}\hat{\lambda}_i\hat{f}_{ix}(t) \tag{3}$$

for $t \neq \hat{t}_0, \hat{t}_1$, $p(\alpha) = p(\beta) = 0$, where α and β are the end points of Δ and that

$$p(\hat{t}_0+0)-p(\hat{t}_0-0)=\hat{l}_{x_0}, \quad p(\hat{t}_1+0)-p(\hat{t}_1-0)=\hat{l}_{x_1} \tag{4}$$

(here $\chi_{[\hat{t}_0,\hat{t}_1]}$ is the characteristic function equal to 1 on the closed interval $[\hat{t}_0, \hat{t}_1]$ and to 0 outside it). It follows from Eq. (3) that $p(\cdot)$ satisfies the homogeneous linear differential equation on the half-open intervals $[\alpha, \hat{t}_0)$ and $(\hat{t}_1, \beta]$. Since it vanishes at the end points α and β of these intervals, the uniqueness theorem of Section 2.5.3 implies that $p(t) \equiv 0$ on $\Delta \setminus [\hat{t}_0, \hat{t}_1]$ and $p(\hat{t}_0 - 0) = p(\hat{t}_1 + 0) = 0$.

Now let us denote by $\hat{p}(\cdot)$ the solution of Eq. (7a) of Section 4.1.1 coinciding with $p(\cdot)$ on the interval $(\hat{t}_0, \hat{t}_1)$ [Eq. (7a) of Section 4.1.1 and Eq. (3) are identical in this interval]. Since (7a) is a linear equation, the solution $\hat{p}(\cdot)$ is defined on the whole interval Δ and belongs to $C^1(\Delta, \mathbf{R}^{n*})$ (Section 2.5.4). We have $\hat{p}(\hat{t}_0) = p(\hat{t}_0 + 0)$ and $\hat{p}(\hat{t}_1) = p(\hat{t}_1 - 0)$ for this solution, and since $p(\hat{t}_0 - 0) = 0$ and $p(\hat{t}_1 + 0) = 0$, relation (4) turns into condition (8a) of Section 4.1.1. Hence, stationarity conditions (7) and (8) [or (7a) and (8a)] of Section 4.1.1 hold for $\hat{p}(\cdot)$.

For $\hat{t}_1 < \hat{t}_0$ we obtain, instead of (3), the equation

$$dp/dt = -p(t)\hat{\varphi}_x(t) - \chi_{[\hat{t}_1, \hat{t}_0]}(t) \sum_{i=0}^{m} \hat{\lambda}_i \hat{f}_{ix}(t)$$

whose solution $p(\cdot)$ vanishes on $[\alpha, \hat{t}_1)$ and $(\hat{t}_0, \beta]$. In this case we denote by $\hat{p}(\cdot)$ the solution of Eq. (7a) of Section 4.1.1, which coincides with $(-1)p(\cdot)$ on $(\hat{t}_1, \hat{t}_0)$, and then conditions (7) and (8) of Section 4.1.1 hold.

Finally, for $\hat{t}_1 = \hat{t}_0$ we can again make use of the generalized DuBois–Reymond's lemma, but in this case we have $\hat{\nu}'(t) = p(t) \equiv 0$ everywhere on Δ except possibly the point $\hat{t}_0 = \hat{t}_1$ at which the condition $p(\hat{t}_1 + 0) - p(\hat{t}_1 - 0) = \hat{l}_{x_0} + \hat{l}_{x_1} = 0$ must be fulfilled. It follows that $\hat{l}_{x_0} + \hat{l}_{x_1} = 0$. Taking as $\hat{p}(\cdot)$ the solution of Eq. (7a) of Section 4.1.1 for which $\hat{p}(\hat{t}_0) = \hat{l}_{x_0}$, we again see that conditions (7) and (8) of Section 4.1.1 hold.

B) Stationarity with Respect to $u(\cdot)$. Differentiating $\bar{\mathscr{L}}$ with respect to $u(\cdot)$ and taking into account that the function $\nu(\cdot)$ is absolutely continuous and its derivative $p(\cdot) = \nu'(\cdot)$ is nonzero only in the interval $(\hat{t}_0, \hat{t}_1)$ and coincides with $\hat{p}(\cdot)$ in this interval, we obtain the equality

$$0 = \hat{\bar{\mathscr{L}}}_u[v(\cdot)] = \int_{\hat{t}_0}^{\hat{t}_1} \left\{ \sum_{i=0}^{m} \hat{\lambda}_i \hat{f}_{iu}(t) - \hat{p}(t)\hat{\varphi}_u(t) \right\} v(t)\,dt.$$

Since this equality holds for all $v(\cdot) \in C(\Delta, \mathbf{R}^r)$, we again refer to the uniqueness in Riesz's theorem (Section 2.1.9) to obtain the equation

$$\sum_{i=0}^{m} \hat{\lambda}_i \hat{f}_{iu}(t) - \hat{p}(t)\hat{\varphi}_u(t) = 0,$$

which coincides with Eq. (9a) and consequently with the condition (9a) of Section 4.1.1.

C) Stationarity with Respect to t_k. Since

$$\bar{\mathscr{L}}(\hat{x}(\cdot), \hat{u}(\cdot), t_0, t_1; \ldots) - \mathscr{L}(\hat{x}(\cdot), \hat{u}(\cdot), t_0, t_1; \ldots) =$$

$$= \int_{\Delta} d\nu(t)\{\dot{\hat{x}}(t) - \varphi(t, \hat{x}(t), \hat{u}(t))\} - \int_{t_0}^{t_1} \hat{p}(t)\{\dot{\hat{x}}(t) - \varphi(t, \hat{x}(t), \hat{u}(t))\}\,dt \equiv 0,$$

the derivatives of these two functions with respect to t_k at the point $\hat{\xi} = (\hat{x}(\cdot), \hat{u}(\cdot), \hat{t}_0, \hat{t}_1)$ coincide, and since the equality $\hat{\bar{\mathscr{L}}}_{t_k} = 0$ takes place, condition (10) of Section 4.1.1 also holds.

4.1.5. Problem with Higher-Order Derivatives. The Euler—Poisson Equation. Let us again fix a closed interval $\Delta\subset\mathbf{R}$. We shall consider the space

$$\Xi_s = C^s(\Delta,\ \mathbf{R}^n)\times\mathbf{R}\times\mathbf{R}$$

whose elements are triples $\xi = (x(\cdot),\ t_0,\ t_1)$, where the functions $x(\cdot)\colon \Delta\to\mathbf{R}^n$ are continuously differentiable up to the order s inclusive, and $t_0,\ t_1\in\operatorname{int}\Delta$. When speaking about a *problem with higher-order derivatives* (in the classical calculus of variations), we shall mean the extremal problem

$$\mathscr{P}_0(\xi)\to\text{extr};\quad \mathscr{P}_i(\xi)\leqslant 0,\quad i=1,\ \ldots,\ m, \tag{1}$$

in the space Ξ_s where

$$\mathscr{P}_i(\xi)=\int_{t_0}^{t_1} f_i(t,\ x(t),\ \ldots,\ x^{(s)}(t))\,dt+ \\ +\psi_i(t_0,\ x(t_0),\ \ldots,\ x^{(s-1)}(t_0),\ t_1,\ x(t_1),\ \ldots,\ x^{(s-1)}(t_1)). \tag{2}$$

Here in (1) and (2) we have $f_i\colon V\to\mathbf{R}$, $\psi_i\colon W\to\mathbf{R}$, $i = 0, 1,\ldots,m$, where V and W are open sets in the space $\mathbf{R}\times(\mathbf{R}^n)^{s+1}$ and $\mathbf{R}\times(\mathbf{R}^n)^s\times\mathbf{R}\times(\mathbf{R}^n)^s$, respectively; functions f_i and ψ_i are at least continuous. A triple $\xi = (x(\cdot),\ t_0,\ t_1)\in\Xi_s$ is *admissible* in problem (1), (2) if $(t,\ x(t),\ \ldots,\ x^{(s)}(t))\in V$, $t\in\Delta$ and

$$(t_0,\ x(t_0),\ \ldots,\ x^{(s-1)}(t_0),\ t_1,\ x(t_1),\ \ldots,\ x^{(s-1)}(t_1))\in W.$$

An admissible triple $\hat{\xi} = (\hat{x}(\cdot),\ \hat{t}_0,\ \hat{t}_1)$ is called a *(local) solution* of problem (1), (2) if it yields a local extremum for the functional $\mathscr{P}_0$ in the space Ξ_s, i.e., if there exists $\varepsilon > 0$ such that for all admissible triples $\xi = (x(\cdot),\ t_0,\ t_1)$ satisfying the conditions

$$\|\xi-\hat{\xi}\|_{\Xi_s}<\varepsilon\Leftrightarrow|t_k-\hat{t}_k|<\varepsilon,\quad k=0,\ 1;\ \|x-\hat{x}(\cdot)\|_{C^s(\Delta,\ R^n)}<\varepsilon,$$

there holds one of the two inequalities: $\mathscr{P}_0(\xi)\geqslant\mathscr{P}_0(\hat{\xi})$ in the case of a minimum and $\mathscr{P}_0(\xi)\leqslant\mathscr{P}_0(\hat{\xi})$ in this case of a maximum.

Let us show that the problem with higher-order derivatives (1), (2) is reducible to the Lagrange problem. To this end we denote

$$\boldsymbol{x}=(x_1,\ \ldots,\ x_s)\in(\mathbf{R}^n)^s, \\ \boldsymbol{x}(\cdot)=(x(\cdot),\ \dot{x}(\cdot),\ \ldots,\ x^{(s-1)}(\cdot)), \\ u(\cdot)=x^{(s)}(\cdot),\quad \boldsymbol{p}(\cdot)=(p_1(\cdot),\ \ldots,\ p_s(\cdot)).$$

Then problem (1), (2) takes the following standard form of the problem considered in Section 4.1.1:

$$\int_{t_0}^{t_1} f_0(t,\ \boldsymbol{x}(t),\ u(t))\,dt+\psi_0(t_0,\ \boldsymbol{x}(t_0),\ t_1,\ \boldsymbol{x}(t_1))\to\text{extr}, \tag{1'}$$

$$\int_{t_0}^{t_1} f_i(t,\ \boldsymbol{x}(t),\ u(t))\,dt+\psi_i(t_0,\ \boldsymbol{x}(t_0),\ t_1,\ \boldsymbol{x}(t_1))\leqslant 0, \\ i=1,\ \ldots,\ m, \tag{2'}$$

$$\dot{x}_j=x_{j+1},\quad j=1,\ \ldots,\ s-1,\quad \dot{x}_s=u \tag{3}$$

with the Lagrange function

$$\mathscr{L}(\boldsymbol{x}(\cdot),\ u(\cdot),\ t_0,\ t_1;\ \boldsymbol{p}(\cdot),\ \lambda,\ \lambda_0)=\int_{t_0}^{t_1} L\,dt+l, \tag{4}$$

where

$$L(t, \boldsymbol{x}, \dot{\boldsymbol{x}}, u)=\sum_{i=0}^{m} \lambda_i f_i(t, \boldsymbol{x}, u)+\sum_{j=1}^{s-1} p_j(t)(\dot{x}_j-x_{j+1})+p_s(t)(\dot{x}_s-u) \tag{5}$$

and

$$\begin{aligned} &l(t_0, x_0, \dot{x}_0, \ldots, x_0^{(s-1)}, t_1, x_1, \dot{x}_1, \ldots, x_1^{(s-1)})= \\ &=\sum_{i=0}^{m} \lambda_i \psi_i(t_0, x_0, \dot{x}_0, \ldots, x_0^{(s-1)}, t_1, x_1, \dot{x}_1, \ldots, x_1^{(s-1)}). \end{aligned} \tag{6}$$

Restating the theorem of Section 4.1.2 for this case, we arrive at the following necessary condition for an extremum:

THEOREM. Let the functions $f_i: V \to \mathbf{R}$, $i = 0, 1, \ldots, m$, and their partial derivatives with respect to $x, \ldots, x^{(s)}$ be continuous in the open set $V \subset \mathbf{R}\times(\mathbf{R}^n)^{s+1}$, let the functions $\psi_i: W \to \mathbf{R}$, $i = 0, 1, \ldots, m$, be continuously differentiable in the open set $W\subset\mathbf{R}\times(\mathbf{R}^n)^s\times\mathbf{R}\times(\mathbf{R}^n)^s$, and let $\hat{x}(\cdot)\in C^s(\Delta, \mathbf{R}^n)$, $\hat{t}_0, \hat{t}_1\in \operatorname{int}\Delta$ be such that

$$\begin{gathered} (t, \hat{x}(t), \ldots, \hat{x}^{(s)}(t))\in V, \quad t\in\Delta; \\ (\hat{t}_0, \hat{x}(\hat{t}_0), \ldots, \hat{x}^{(s-1)}(\hat{t}_0), \hat{t}_1, \hat{x}(\hat{t}_1), \ldots, \hat{x}^{(s-1)}(\hat{t}_1))\in W. \end{gathered}$$

If $\hat{\xi} = (\hat{x}(\cdot), \hat{t}_0, \hat{t}_1)$ is a local solution of problem (1), (2), then there are Lagrange multipliers: $\hat{\lambda}_0 \geqslant 0$ (in the case of a minimum problem) or $\hat{\lambda}_0 \leqslant 0$ (in the case of a maximum problem), $\hat{p}(\cdot)=(\hat{p}_1(\cdot), \ldots, \hat{p}_s(\cdot))\in C^1(\Delta, (\mathbf{R}^n)^{s*})$ and $\hat{\lambda} = (\hat{\lambda}_1, \ldots, \hat{\lambda}_n)$, not all zero, such that:

a) the stationarity conditions for the Lagrange function hold:

with respect to $x(\cdot)$ ($\hat{\mathscr{L}}_{x(\cdot)}=0$, $\hat{\mathscr{L}}_{u(\cdot)}=0$) :

$$\left.\begin{aligned} &\dot{\hat{p}}_1(t)=f_x(t, \hat{x}(t), \ldots, \hat{x}^{(s)}(t)), \\ &\dot{\hat{p}}_j(t)=f_{x^{(j-1)}}(t, \hat{x}(t), \ldots, \hat{x}^{(s)}(t))-\hat{p}_{j-1}(t), \\ &\qquad\qquad j=2, \ldots, s, \\ &0=f_{x^{(s)}}(t, \hat{x}(t), \ldots, \hat{x}^{(s)}(t))-\hat{p}_s(t), \end{aligned}\right\} \tag{7}$$

$$\hat{p}_j(\hat{t}_k)=(-1)^k \hat{l}_{x_k^{(j-1)}}, \quad k=0, 1; \tag{8}$$

with respect to t_k ($\hat{\mathscr{L}}_{t_k}=0$) :

$$\hat{H}(\hat{t}_k)=(-1)^{k+1}\hat{l}_{t_k}, \quad k=0, 1; \tag{9}$$

b) the condition of concordance of signs holds:

$$\hat{\lambda}_i \gtrless 0, \quad i=1, \ldots, m; \tag{10}$$

c) the conditions of complementary slackness hold:

$$\hat{\lambda}_i \mathscr{F}_i(\hat{\xi})=0, \quad i=1, \ldots, m. \tag{11}$$

Here

$$f(t, x, \dot{x}, \ldots, x^{(s)})=\sum_{i=0}^{m} \hat{\lambda}_i f_i(t, x, \ldots, x^{(s)}), \tag{12}$$

$$\hat{H}(t)=\sum_{j=1}^{s} \hat{p}_j(t)\hat{x}^{(j)}(t)-f(t, \hat{x}(t), \ldots, \hat{x}^{(s)}(t)), \tag{13}$$

and, as usual,

$$\hat{l}_{x_k^{(j)}}=l_{x_k^{(j)}}(\hat{t}_0, \hat{x}(\hat{t}_0), \ldots, \hat{x}^{(s-1)}(\hat{t}_0), \hat{t}_1, \hat{x}(\hat{t}_1), \ldots, \hat{x}^{(s-1)}(\hat{t}_1)),$$

etc. and inequalities (10) mean that $\hat{\lambda}_i \geqslant 0$ if $\mathscr{F}_i(\xi)\leqslant 0$ in problem (1), $\hat{\lambda}_i \leqslant 0$ if $\mathscr{F}_i(\xi)\geqslant 0$ in (1), and $\hat{\lambda}_0$ may have an arbitrary sign if $\mathscr{F}_i(\xi)=0$ in (1).

The detailed calculations bringing conditions (7)-(12) of Section 4.1.1 to the analogous conditions of the theorem under consideration are left to the reader as an exercise.

For the special case of a problem with higher-order derivatives in which the endpoint terms in (1') and (2') are constants and the end points are fixed, the assertion of the theorem can be expressed in a simpler form. In this case the transversality conditions (8) and (9) are dropped and the system (7) can be written in the form of one equation

$$\hat{f}_x(t) - \frac{d}{dt}\left(\hat{f}_{\dot{x}}(t) - \frac{d}{dt}\left(\hat{f}_{\ddot{x}} - \ldots - \frac{d}{dt}\hat{f}_{x^{(s)}}(t)\right)\right) = 0. \qquad (14)$$

Finally, if we assume that the required derivatives exist, Eq. (14) can be written in the form of the Euler—Poisson equation

$$\hat{f}_x(t) - \frac{d}{dt}\hat{f}_{\dot{x}} + \frac{d^2}{dt^2}\hat{f}_{\ddot{x}}(t) - \ldots - (-1)^s \frac{d^s}{dt^s}\hat{f}_{x^{(s)}}(t) = 0. \qquad (15)$$

The general solution of this equation of the 2s-th order involves 2s arbitrary constants and also depends on the numbers $\lambda_0, \ldots, \lambda_m$ one of which can be chosen arbitrarily. We have altogether 2s + m unknowns. For their determination we use 2s boundary conditions (for the fixed end points), and m more conditions are obtained from the conditions of complementary slackness and inequalities (1) and (10) (check that altogether we can derive from them exactly m equalities). It can similarly be verified that in the general case the number of the equations the theorem yields also coincides with the number of the unknowns.

4.2. Pontryagin Maximum Principle

4.2.1. Statement of the Optimal Control Problem.

The extremal problem to which this section is devoted almost coincides in its form with the Lagrange problem (1)-(3) of Section 4.1.1. Here we deal with the same functional, the same differential constraint, and the same inequality constraints as before. The distribution of the new problem is that in its formulation a constraint on the control is separated out explicitly; this constraint has the form $u(t) \in \mathfrak{U}$, where $\mathfrak{U}$ is a topological space.

At first glance the new formulation does not provide anything new and even seems less general. Indeed, in the Lagrange problem we dealt with admissible processes which satisfied the condition $(t, x(t), u(t)) \in V$. In the special case, taking $V = G \times \mathfrak{U}$, we split this condition into two: $(t, x(t)) \in G$ and $u(t) \in \mathfrak{U}$.

However, the division of the variables into phase coordinates and controls and the separation of the constraints imposed on the controls proved very significant. It led to the appearance of a new branch of the theory of extremal problems which very soon became very popular among applied scientists due to its applications to practical problems and allowed us to look at the classical theory from a new viewpoint.

Now, what caused all this? In the first place, it was the necessity to investigate technical problems. The separation of phase coordinates and controls and their connection by means of a differential equation is a usual model of a process whose development obeys the laws of nature (the differential equation!) and which is acted upon by the man controlling this process according to his purposes and trying to make it in a sense optimal. It is clear now that the constraints of the form $u(t) \in \mathfrak{U}$ imposed on the controls are connected with the limitation of the possibility of influencing the process (say, with the limitation of the range of the angle of rotation of the rudder of a guided vehicle).

If $\mathfrak{U}$ is assumed to be an open set in $\mathbf{R}^r$, we obtain nothing new indeed. However, in technical and other applied problems the set $\mathfrak{U}$ is not necessarily open. The control can often be even discrete (switching on—switching off). There naturally appear constraints for which $\mathfrak{U}$ is a closed set: the importance of such a situation was already demonstrated in the solution of Newton's problem (Section 1.6.2). Even the simplest optimal control problems involve constraints of the form $|u(t)| \leqslant 1$ (see the time-optimal problem in Section 1.6.3). In the last case the set of the admissible controls is a closed ball in the space $L_\infty(\Delta)$ of complex structure, and in the typical cases the optimal control lies on the boundary of this ball [from what was said in Section 1.6.3 it is seen that the problem would have no solution if the constraint $|u(t)| < 1$ were imposed because the minimum time corresponds to a control for which $|u(t)| = 1$ almost everywhere], and this boundary is highly nonsmooth and "multifaceted" or, more precisely, "infinite-faceted." All this hampers the application of the ordinary methods of differential calculus, and smoothness conditions with respect to the control often turn out to be unnatural. That is why we shall no longer require that the derivatives f_{iu}, φ_u, etc. exist, and even the set $\mathfrak{U}$ of the possible values of the control will be regarded as an arbitrary topological space which does not necessarily possess a linear structure in the general case. The absence of the derivatives with respect to u is the second distinction of the new problem from the old one. In order to stress this fact we shall denote the functionals involved in the formulation of the problems by a new symbol, although they almost coincide in their form with Bolza functionals.

Finally, the last distinction is that the control is no longer supposed to be continuous. Basically, this is connected with the fact that the problem often has no solution in the class of continuous controls (for instance, this is the case for most problems in which $\mathfrak{U}$ is discrete or for the time-optimal problem of Section 1.6.3). It should be noted, however, that the passage to "polygonal extremals" (which corresponds to piecewise-continuous controls) is also made in the classical calculus of variations (see Section 1.4.3). As to the very possibility of the free choice of the control within the set $\mathfrak{U}$ [it was already used when we considered needle-like variations in the derivation of Weierstrass' conditions in Section 1.4.4 and in the proof of the maximum principle for the simplest situation (Section 1.5.4)], it generates "latent convexity" with respect to the controls, with which the form of the basic condition, i.e., the maximum principle [see (10), (11) in Section 4.2.2], is connected: it resembles the corresponding condition in the convex programming problems. Here this connection will not be revealed in full. This will be done in the next section for a particular version of the general problem.

Thus, let us consider the extremal problem

$$\mathcal{J}_0(x(\cdot), u(\cdot), t_0, t_1) = \int_{t_0}^{t_1} f_0(t, x(t), u(t))\,dt + \psi_0(t_0, x(t_0), t_1, x(t_1)) \to \inf, \tag{1}$$

$$\dot{x} = \varphi(t, x(t), u(t)), \quad u(t) \in \mathfrak{U}, \tag{2}$$

$$\mathcal{J}_i(x(\cdot), u(\cdot), t_0, t_1) = \int_{t_0}^{t_1} f_i(t, x(t), u(t))\,dt + \\ + \psi_i(t_0, x(t_0), t_1, x(t_1)) \lesseqgtr 0, \quad i = 1, 2, \ldots, m. \tag{3}$$

Here f_i: $G \times \mathfrak{U} \to \mathbf{R}$, $\psi_i: W \to \mathbf{R}$, $i = 0, 1, \ldots, m$, φ: $G \times \mathfrak{U} \to \mathbf{R}^n$, where G and W are open sets in the spaces $\mathbf{R} \times \mathbf{R}^n$ and $\mathbf{R} \times \mathbf{R}^n \times \mathbf{R} \times \mathbf{R}^n$, respectively, and $\mathfrak{U}$ is an arbitrary topological space. The symbol $\lesseqgtr$ has the same meaning as in Sections 3.2 and 4.1. All the functions f_i, ψ_j, and φ are assumed to be at least continuous.

A quadruple $(x(\cdot), u(\cdot), t_0, t_1)$ will be called a *controlled process* in problem (1)-(3) if:

a) the *control* $u(\cdot)$: $[t_0, t_1] \to \mathfrak{U}$ is a piecewise-continuous function. Its values at the points of discontinuity are insignificant; for definiteness, we shall assume that $u(\cdot)$ is continuous on the right for $t_0 \leqslant t < t_1$ and on the left at the point t_1;

b) the *phase trajectory* $x(\cdot):[t_0, t_1] \to \mathbf{R}^n$ is continuous and its graph lies in G:

$$\Gamma = \{(t, x(t)) \mid t_0 \leqslant t \leqslant t_1\} \subset G.$$

c) the function $x(\cdot)$ satisfies the differential equation

$$\dot{x}(t) = \varphi(t, x(t), u(t))$$

for all $t \in [t_0, t_1]$ except possibly at the points of discontinuity of the control $u(\cdot)$ [in this case we say that $x(\cdot)$ corresponds to the control $u(\cdot)$; it can easily be seen that $x(\cdot)$ possesses the left-hand and the right-hand derivatives at the points of discontinuity of the control].

A controlled process is said to be *admissible* if the conditions (3) are satisfied.

An admissible controlled process $(\hat{x}(\cdot), \hat{u}(\cdot), \hat{t}_0, \hat{t}_1)$ is said to be (locally) *optimal* if there is $\varepsilon > 0$ such that for every admissible controlled process $(x(\cdot), u(\cdot), t_0, t_1)$ such that

$$|t_k - \hat{t}_k| < \varepsilon, \; k = 0, 1 \text{ and } |x(t) - \hat{x}(t)| < \varepsilon \quad \text{for all } t \in [t_0, t_1] \cap [\hat{t}_0, \hat{t}_1] \tag{4}$$

the inequality

$$\mathcal{J}_0(x(\cdot), u(\cdot), t_0, t_1) \geqslant \mathcal{J}_0(\hat{x}(\cdot), \hat{u}(\cdot), \hat{t}_0, \hat{t}_1) \tag{5}$$

holds.

It should be noted that here there is a change in comparison with the Lagrange problem: we no longer require that the derivatives $\dot{x}(t)$ and $\dot{\hat{x}}(t)$ be close to each other. In the classical calculus of variations this corresponds to the passage from a weak extremum to a strong one (see Section 1.4.3).

It is instructive to compare the Lagrange problem and the optimal control problem for the case when $\mathfrak{U} = \mathbf{R}^r$ and $V = G \times \mathbf{R}^r$ in Section 4.1.1. Under these assumptions:

Every admissible controlled process $(x(\cdot), u(\cdot), t_0, t_1)$ of the Lagrange problem is an admissible controlled process for the optimal control problem as well (more precisely, it becomes such for the restriction of $u(\cdot)$ to $[t_0, t_1]$).

This means that the problem has been extended (cf. Section 1.4.3). Consequently, the value of the optimal control problem is not greater than that of the Lagrange problem ($\inf \mathcal{J}_0 \leqslant \inf \mathcal{B}_0$). However, for some natural assumptions about the functions f_i, ψ_i, and φ the two values coincide (for the trivial case when $\dot{x} = u$ this was proved in the "rounding the corners" lemma in Section 1.4.3).

If $\inf \mathcal{J}_0 < \inf \mathcal{B}_0$, then, as can easily be seen, the solutions of the two problems exist or do not exist independently. In case the values of the problem are equal, there takes place one of the following three possibilities:

a) The Lagrange problem possesses a solution: $\inf \mathcal{B}_0 = \mathcal{B}_0(\hat{x}(\cdot), \hat{u}(\cdot), \hat{t}_0, \hat{t}_1)$. Then the same quadruple is a solution of the optimal control problem because

$$\mathcal{J}_0(x(\cdot), u(\cdot), t_0, t_1) \geqslant \mathcal{B}_0(\hat{x}(\cdot), \hat{u}(\cdot), \hat{t}_0, \hat{t}_1) = \inf \mathcal{B}_0 = \inf \mathcal{J}_0.$$

b) The Lagrange problem has no solution while the optimal control problem has: $\inf \mathcal{J}_0$ is attained on a piecewise-smooth curve $\hat{x}(\cdot)$. Such a situation was observed in the example (6) of Section 1.4.3 (for $\dot{x} = u$ the passage from the class C^1 to the class KC^1 exactly corresponds to the passage to the optimal control problem).

c) The two problems have no solutions: the infimum of the functional is not attained even when its domain is extended. This is the case in the Bolza example (Section 1.4.3).

In the foregoing argument it was tacitly assumed that the minimization was considered with respect to the whole class of the admissible controlled processes. However, usually extremal problems are considered from a local viewpoint, which, by the way, is reflected in the definitions we stated. In Section 4.1.1 an admissible controlled process $(\hat{x}(\cdot), \hat{u}(\cdot), \hat{t}_0, \hat{t}_1)$ was called optimal (in the weak sense) when it yielded a minimum for the functional $\mathcal{B}_0$ in a C^1-neighborhood with respect to $x(\cdot)$ and in a C-neighborhood with respect to $u(\cdot)$, whereas in the present definition of optimality we assume that the minimum of $\mathcal{J}_0$ takes place in a wider neighborhood described by inequalities (4). This is in fact a C-neighborhood with respect to $x(\cdot)$, and there are no constraints on the controls at all. As was already mentioned, in the classical situation this would be a strong minimum. Therefore, a local solution of the Lagrange problem may not be an optimal process for the optimal control problem. In case the control $\hat{u}(\cdot)$ in an optimal process $(\hat{x}(\cdot), \hat{u}(\cdot), \hat{t}_0, \hat{t}_1)$ is continuous, this process is a local solution of the Lagrange problem as well under the ordinary assumptions about the functions involved in the problem.

Conditions (3) are not the most general constraints which we have to consider in optimal control problems. For instance, we do not consider at all the so-called *phase constraints* of the form

$$(t, x(t)) \in D, \quad t_0 \leqslant t \leqslant t_1, \tag{6}$$

where D is a subset in $\mathbf{R} \times \mathbf{R}^n$. Of course, if D is open, then replacing G by $G \cap D$ in the definition of a controlled process and diminishing, if necessary, ε in (4) we automatically take into account constraints (6). Therefore, of primary interest is the case when D is not an open set and the optimal phase trajectory x(t) reaches the boundary of D for some t (see Fig. 37; here the arc length is minimized, and the shaded part of the plane does not belong to D). This leads to some essentially new facts, in particular it becomes necessary to introduce a generalized function in the Euler–Lagrange equation, namely, the derivative of a measure which is not absolutely continuous (or to replace this differential equation by an integral equation with Stieltjes' integral with respect to such a measure).

<u>4.2.2. Formulation of the Maximum Principle. The Lagrange Principle for the Optimal Control Problem.</u> Let us again consider the problem

$$\mathcal{J}_0(x(\cdot), u(\cdot), t_0, t_1) = \int_{t_0}^{t_1} f_0(t, x(t), u(t))\,dt + \psi_0(t_0, x(t_0), t_1, x(t_1)) \to \inf, \tag{1}$$

$$\dot{x} = \varphi(t, x(t), u(t)), \quad u(t) \in \mathfrak{U}, \tag{2}$$

$$\mathcal{J}_i(x(\cdot), u(\cdot), t_0, t_1) = \int_{t_0}^{t_1} f_i(t, x(t), u(t))\,dt + \psi_i(t_0, x(t_0), t_1, x(t_1)) \lesseqgtr 0, \quad i = 1, 2, \ldots, m. \tag{3}$$

As in Section 4.1.1, by the Lagrange function of this problem we mean the function

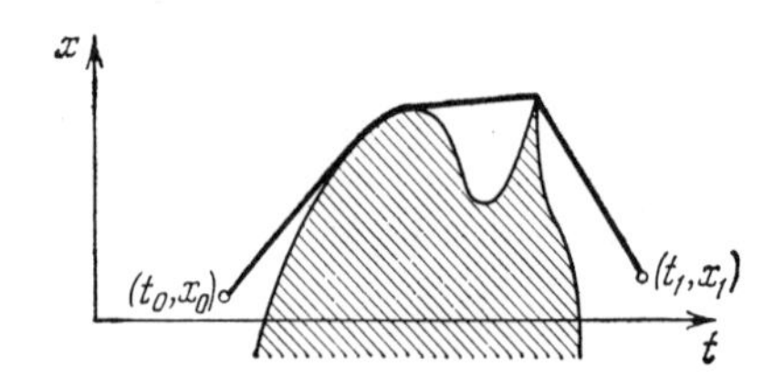

Fig. 37

$$\mathscr{L}(x(\cdot),\ u(\cdot),\ t_0,\ t_1;\ p(\cdot),\ \lambda,\ \lambda_0)=\int_{t_0}^{t_1} L\,dt+l, \tag{4}$$

where

$$\lambda_0 \in \mathbf{R},\ \lambda=(\lambda_1,\ \ldots,\ \lambda_m)\in \mathbf{R}^{m*},$$

$$L(t,\ x,\ \dot{x},\ u)=\sum_{i=0}^{m} \lambda_i f_i(t,\ x,\ u)+p(t)(\dot{x}-\varphi(t,\ x,\ u)), \tag{5}$$

$$l(t_0,\ x_0,\ t_1,\ x_1)=\sum_{i=0}^{m} \lambda_i \psi_i(t_0,\ x_0,\ t_1,\ x_1). \tag{6}$$

The function $p(\cdot):[t_0,\ t_1] \to \mathbf{R}^{n*}$ is assumed to be continuous. When necessary, we shall suppose that the function $p(\cdot)$ is defined on a wider interval than $[t_0,\ t_1]$; to this end we extend it arbitrarily outside that interval with the preservation of the continuity. Function $p(\cdot)$ and the numbers λ_i and λ_0 are called the *Lagrange multipliers* of problem (1)-(3); function (5) is called the *Lagrangian*, and (6) is called the *endpoint functional*. Finally, the function

$$H(t,\ x,\ u,\ p)=L_{\dot{x}}\dot{x}-L=p\varphi(t,\ x,\ u)-\sum_{i=0}^{m} \lambda_i f_i(t,\ x,\ u) \tag{7}$$

will be called the *Pontryagin function*. As usual, we shall use the notation

$$\hat{L}_x(t)=L_x(t,\ \hat{x}(t),\ \dot{\hat{x}}(t),\ \hat{u}(t)),\ \hat{\varphi}(t)=\varphi(t,\ \hat{x}(t),\ \hat{u}(t)),$$
$$\hat{l}_{x_k}=l_{x_k}(\hat{t}_0,\ \hat{x}(\hat{t}_0),\ \hat{t}_1,\ \hat{x}(\hat{t}_1)),$$

etc.

THEOREM (Pontryagin Maximum Principle). Let G be an open set in the space $\mathbf{R} \times \mathbf{R}^n$, let W be an open set in the space $\mathbf{R} \times \mathbf{R}^n \times \mathbf{R} \times \mathbf{R}^n$, and let $\mathfrak{U}$ be an arbitrary topological space. Let the functions f_i: $G\times\mathfrak{U}\to\mathbf{R}$, $i = 0, 1,\ldots,m$, and φ: $G\times\mathfrak{U}\to\mathbf{R}^n$ and their partial derivatives with respect to x be continuous in $G\times\mathfrak{U}$, and let the functions ψ_i: $W\to\mathbf{R}$, $i = 1,\ldots,m$, be continuously differentiable in W.

If $(\hat{x}(\cdot),\ \hat{u}(\cdot),\ \hat{t}_0,\ \hat{t}_1)$ is an optimal process for the problem (1)-(3), then there are Lagrange multipliers

$$\hat{\lambda}_0\geqslant 0,\ \hat{p}(\cdot),\quad \hat{\lambda}=(\hat{\lambda}_1,\ \ldots,\ \hat{\lambda}_m),$$

not all zero, such that:

a) the stationarity conditions and the minimum principle for the Lagrange function hold:

the stationarity condition with respect to $x(\cdot)$ $(\hat{\mathscr{L}}_{x(\cdot)}=0)$:

$$-\frac{d}{dt}\hat{L}_{\dot{x}}(t)+\hat{L}_x(t)=0, \tag{8}$$

$$\hat{L}_{\dot{x}}(\hat{t}_k)=(-1)^k\hat{l}_{x_k},\quad k=0,\ 1; \tag{9}$$

the minimum principle in the Lagrange form with respect to $u(\cdot)$:

$$\hat{L}(t) \equiv L(t, \hat{x}(t), \dot{\hat{x}}(t), \hat{u}(t)) \equiv \min_{v \in \mathfrak{U}} L(t, \hat{x}(t), \dot{\hat{x}}(t), v), \tag{10}$$

or the maximum principle in the Hamiltonian (Pontryagin) form

$$\hat{H}(t) \equiv H(t, \hat{x}(t), \hat{u}(t), \hat{p}(t)) \equiv \max_{v \in \mathfrak{U}} H(t, \hat{x}(t), v, \hat{p}(t)), \tag{11}$$

the function $\hat{H}(t)$ being continuous on the closed interval $[\hat{t}_0, \hat{t}_1]$;

the stationarity conditions with respect to t_k:

$$\hat{\mathscr{L}}_{t_k} = 0, \quad k = 0, 1; \tag{12}$$

b) the condition of concordance of signs holds:

$$\lambda_i \gtrless 0; \tag{13}$$

c) the conditions of complementary slackness hold:

$$\hat{\lambda}_i \mathfrak{J}_i(\hat{x}(\cdot), \hat{u}(\cdot), \hat{t}_0, \hat{t}_1) = 0, \quad i = 1, 2, \ldots, m \tag{14}$$

[as in the foregoing section, inequalities (13) mean that $\hat{\lambda}_i \geqslant 0$ if $\mathfrak{J}_i \leqslant 0$ in condition (3), $\hat{\lambda}_i \leqslant 0$ if $\mathfrak{J}_i \geqslant 0$ in condition (3), and λ_i may have an arbitrary sign if $\mathfrak{J}_i = 0$].

The assertion about the existence of the Lagrange multipliers satisfying the set of the conditions a)-c) will be briefly called the *Lagrange principle for the optimal control problem* (1)-(3) or the *Pontryagin maximum principle*. As before, this assertion is in complete agreement with the general Lagrange principle stated in Section 3.1.5. Since the Lagrange function $\mathscr{L}$ is a function of the three arguments $x(\cdot)$, $u(\cdot)$, and (t_0, t_1), we have to consider the following three problems:

$$\begin{aligned}
&(\alpha) \quad \mathscr{L}(x(\cdot), \hat{u}(\cdot), \hat{t}_0, \hat{t}_1; \hat{p}(\cdot), \hat{\lambda}, \hat{\lambda}_0) \to \inf, \\
&(\beta) \quad \mathscr{L}(\hat{x}(\cdot), u(\cdot), \hat{t}_0, \hat{t}_1; \hat{p}(\cdot), \hat{\lambda}, \hat{\lambda}_0) \to \inf, \\
&(\gamma) \quad \mathscr{L}(\hat{x}(\cdot), \hat{u}(\cdot), t_0, t_1; \hat{p}(\cdot), \hat{\lambda}, \hat{\lambda}_0) \to \inf.
\end{aligned}$$

Here problems (α) and (γ) are the same as in Section 4.1.1. Therefore, the corresponding stationarity conditions with respect to $x(\cdot)$ and t_k must also have the same form as in the Lagrange problem, and this is really so. As to problem (β), this is an elementary optimal control problem of Section 3.1.4, and the minimum principle (10) corresponds exactly to the condition (1) of that subsection. Thus, the above formulation of the theorem does in fact agree with the general Lagrange principle.

The stationarity conditions (8), (9), and (12) are usually written in a different form: (8) is written as the Euler–Lagrange equation (also called the adjoint equation):

$$d\hat{p}(t)/dt = -\hat{p}(t)\hat{\varphi}_x(t) + \sum_{i=0}^{m} \hat{\lambda}_i \hat{f}_{ix}(t); \tag{8a}$$

and (9) and (12) are written as the *transversality conditions*

$$\hat{p}(\hat{t}_1) + \hat{l}_{x_1} = 0, \quad -\hat{p}(\hat{t}_0) + \hat{l}_{x_0} = 0 \tag{9a}$$

and

$$-\hat{H}(\hat{t}_1) + \hat{l}_{t_1} = 0, \quad \hat{H}(\hat{t}_0) + \hat{l}_{t_0} = 0. \tag{12a}$$

The corresponding transformations are carried out in the same way as in Section 4.1.1 and we do not have to repeat them. At the points of discontinuity of the control the right-hand derivative $\hat{p}'_+(\cdot)$ should be taken in Eq. (8a) (the left-hand derivative also exists). The condition of continuity of $\hat{p}(\cdot)$ and $\hat{H}(\cdot)$ at the points of discontinuity of the control [i.e., at the corner points of $x(\cdot)$] is referred to as the *Weierstrass–Erdmann condition*.

Here it is also possible to carry out the analysis of the "completeness" of the set of the necessary conditions given by the theorem we have stated. However, it does not in fact differ from what was done for the Lagrange problem. The distinction is that in the case under consideration instead of finding $u(\cdot)$ as a function of $x(\cdot)$ and $p(\cdot)$ from the condition $L_u = 0$ we must determine $u(\cdot)$ using the relation (10).

4.2.3. Needlelike Variations. As in the simplest situation considered in Section 1.5.4, the basic method used in the proof of the maximum principle which will be presented later consists in replacing the optimal control problem under consideration by a finite-dimensional extremal problem or, more precisely, by a set of such problems. To this end we shall embed the optimal process $(\hat{x}(\cdot), \hat{u}(\cdot), \hat{t}_0, \hat{t}_1)$ in a special family of controlled processes, namely, a "packet" of needlelike variations, and consider the restriction of problem (1)-(3) of Section 4.2.2 to this family.

The family of processes we need depends on the following parameters:

the initial data (t_0, x_0) and the terminal instant t_1;

a set $\bar{\alpha} = (\alpha_1,\ldots,\alpha_N)$ where all $\alpha_i \in \mathbf{R}$ are sufficiently small; for brevity we denote $\alpha = \Sigma\alpha_i$;

a set $\bar{\tau} = (\tau_1,\ldots,\tau_N)$, where $\hat{t}_0 < \tau_1 \leqslant \tau_2 \leqslant \ldots \leqslant \tau_N < \hat{t}_1$, the points τ_i $(i = 1,\ldots,N)$ containing all the points of discontinuity of the optimal control $\hat{u}(\cdot)$;

a set $\bar{v} = (v_1,\ldots,v_N)$, where $v_i \in \mathfrak{U}$.

In what follows $\bar{\tau}$ and $\bar{v}$ are fixed while the parameters $(t_1, t_0, x_0, \bar{\alpha})$ change, and we shall differentiate various functions with respect to them.

Let us begin with $N = 1$. An elementary needlelike variation or, simply, an "elementary needle" corresponding to a pair (τ, v) is determined by the equalities

$$u(t;\alpha,\tau,v)=\begin{cases}\hat{u}(t), & t\notin[\tau,\tau+\alpha),\\ v, & t\in[\tau,\tau+\alpha).\end{cases} \tag{1}$$

Accordingly, a needlelike variation $x(t; \alpha, \tau, v)$ of the phase trajectory $\hat{x}(t)$ is defined as the solution of Cauchy's problem

$$\begin{aligned}\dot{x}&=\varphi(t,x,u(t;\alpha,\tau,v)),\\ x(\tau)&=\hat{x}(\tau).\end{aligned} \tag{2}$$

To within an insignificant distinction, we considered such a variation in Section 1.5.4 (we took the half-open interval $[\tau - \lambda, \tau)$ instead of $[\tau, \tau + \alpha)$. It was found that the function $x(t; \alpha, \tau, v)$ possessed the right-hand derivative with respect to α for $\alpha = 0$ and that for $t \geqslant \tau$ this derivative $y(t) = x_\alpha(t; 0, \tau, v)$ was the solution of the variational equation

$$\dot{y}=\hat{\varphi}_x(t)\,y \tag{3}$$

with the initial condition

$$y(\tau)=\varphi(\tau,\hat{x}(\tau),v)-\varphi(\tau,\hat{x}(\tau),\hat{u}(\tau)). \tag{4}$$

As in Section 2.5.4, here and in the remaining part of the present section we shall denote by $\Omega(t, \tau)$ the principal matrix solution of Eq. (3). Moreover, for the sake of brevity we introduce a special symbol for the right-hand side of equality (4):

$$\Delta\hat{\varphi}(\tau,v)=\varphi(\tau,\hat{x}(\tau),v)-\varphi(\tau,\hat{x}(\tau),\hat{u}(\tau)). \tag{5}$$

Using this notation, we write

$$y(t)=x_\alpha(t;\ 0,\ \tau,\ v)=\Omega(t,\ \tau)\,\Delta\hat{\varphi}(\tau,\ v). \tag{6}$$

In Section 1.5.4 we managed to limit ourselves to the consideration of one "needle" (a needlelike variation). Here we shall need a "packet" of such variations because the problem under consideration is more complex and we need a greater number of free parameters. A packet involves an arbitrary finite number of needles with parameters (τ_i, v_i, α_i), $i = 1,\ldots, N$. Here it is important that some of these needles may only differ in the values v_i of the control for the same values of τ_i. Therefore, the needles can no longer be specified by formulas of the form (1) because different needles with the same τ_i will overlap. To avoid this we shall shift the interval of action of the i-th needle by a distance of $i\alpha=i\sum_{j=1}^{N}\alpha_j$; i.e., we shall take the interval $\Delta_i = [\tau_i + i\alpha,\ \tau_i + i\alpha + \alpha_i]$ (it is evident that if $\alpha_j > 0$ and α_j are sufficiently small, then Δ_i do not overlap).

According to the assumptions made in Section 4.2.1, the optimal control $\hat{u}(\cdot)$ is continuous from the left at the point $\hat{t}_1$ and continuous from the right at the point $\hat{t}_0$. Let us extend $\hat{u}(\cdot)$ outside the closed interval $[\hat{t}_0,\ \hat{t}_1]$ with the preservation of the continuity [for instance, by putting $\hat{u}(t) \equiv \hat{u}(t_0)$ for $t < t_0$ and $\hat{u}(t) \equiv u(t_1)$ for $t > t_1$]; in what follows we shall not stipulate this again.

Let $\alpha_i > 0$ for all $i = 1,\ldots,N$. We define a needlelike variation of the control $\hat{u}(\cdot)$ by means of the formulas

$$u(t;\ \bar{\alpha},\ \bar{\tau},\ \bar{v})=\begin{cases}\hat{u}(t), & t\in(\hat{t}_0-\delta,\ \hat{t}_1+\delta)\setminus\bigcup_{j=1}^{N}\Delta_i,\\ v_i, & t\in\Delta_i=\left[\tau_i+i\sum_{j=1}^{N}\alpha_j,\ \tau_i+i\sum_{j=1}^{N}\alpha_j+\alpha_i\right).\end{cases} \tag{7}$$

The corresponding family of phase trajectories $x(t;\ t_0,\ x_0,\ \bar{\alpha},\ \bar{\tau},\ \bar{v})$ is defined as the solution of Cauchy's problem

$$\dot{x}=\varphi(t,\ x,\ u(t;\ \bar{\alpha},\ \bar{\tau},\ \bar{v})),\quad x(t_0)=x_0. \tag{8}$$

For brevity, we shall write $\hat{x}_0 = \hat{x}(\hat{t}_0)$.

<u>The Lemma on the Packet of Needles.</u> 1) For sufficiently small $\varepsilon_0 > 0$ and $\delta > 0$, the solution of Cauchy's problem (8) such that

$$|t_0-\hat{t}_0|<\varepsilon_0,\ |x_0-\hat{x}_0|<\varepsilon_0,\quad 0<\alpha_j<\varepsilon_0, \tag{9}$$

is defined for $\hat{t}_0 - \delta \leqslant t \leqslant \hat{t}_1 + \delta$.

2) If $t_0 \to \hat{t}_0$, $x_0 \to \hat{x}_0$, and $\alpha_j \downarrow 0$, then $x(t;\ t_0,\ x_0,\ \bar{\alpha},\ \bar{\tau},\ \bar{v}) \to \hat{x}(t)$ uniformly on the closed interval $[t_0,\ t_1]$.

3) The mapping $(t_1,\ t_0,\ x_0,\ \bar{\alpha}) \to x(t_1;\ t_0,\ x_0,\ \bar{\alpha},\ \bar{\tau},\ \bar{v})$ can be extended to a continuously differentiable mapping defined in a neighborhood of the point $(\hat{t}_1,\ \hat{t}_0,\ \hat{x}_0,\ \bar{0})$, and, moreover,

$$\partial\hat{x}/\partial t_1=x_{t_1}(\hat{t}_1;\ \hat{t}_0,\ \hat{x}_0,\ \bar{0},\ \bar{\tau},\ \bar{v})=\varphi(\hat{t}_1,\ \hat{x}(\hat{t}_1),\ \hat{u}(\hat{t}_1))=\hat{\varphi}(\hat{t}_1), \tag{10}$$

$$\partial\hat{x}/\partial t_0=x_{t_0}(\hat{t}_1;\ \hat{t}_0,\ \hat{x}_0,\ \bar{0},\ \bar{\tau},\ \bar{v})=-\Omega(\hat{t}_1,\ \hat{t}_0)\,\varphi(\hat{t}_0,\ \hat{x}_0,\ \hat{u}(\hat{t}_0))=-\Omega(\hat{t}_1,\ \hat{t}_0)\,\hat{\varphi}(\hat{t}_0), \tag{11}$$

$$\partial\hat{x}/\partial x_0=x_{x_0}(\hat{t}_1;\ \hat{t}_0,\ \hat{x}_0,\ \bar{0},\ \bar{\tau},\ \bar{v})=\Omega(\hat{t}_1,\ \hat{t}_0), \tag{12}$$

$$\partial\hat{x}/\partial\alpha_k=x_{\alpha_k}(\hat{t}_1;\ \hat{t}_0,\ \hat{x}_0,\ \bar{0},\ \bar{\tau},\ \bar{v})=\Omega(\hat{t}_1,\ \hat{t}_0)\,\Delta\hat{\varphi}(\tau_k,\ v_k), \tag{13}$$

where $\Omega(t,\ \tau)$ is the principal matrix solution of the variational system (3) and $\Delta\hat{\varphi}(\tau,\ v)$ is determined by the formula (5).

The proof of this lemma will be presented in Section 4.2.6. It is clear that the lemma guarantees the differentiability of the endpoint terms involved in the conditions of the problem posed in Section 4.2.2. As to the integral terms, we shall devote a separate lemma to them as in Section 1.5.4.

The Lemma on the Integral Functionals. Let the functions $f_i(t, x, u)$ satisfy the same conditions as in the formulation of the theorem of Section 4.2.2.

If $u(t, \bar{\alpha}, \bar{\tau}, \bar{v})$ is defined for $\alpha_i > 0$ by the formulas (7) and $x(t; t_0, x_0, \bar{\alpha}, \bar{\tau}, \bar{v})$ is the solution of Cauchy's problem (8), then the functions

$$F_i(t_1, t_0, x_0, \bar{\alpha}) = \int_{t_0}^{t_1} f_i(t, x(t; t_0, x_0, \bar{\alpha}, \bar{\tau}, \bar{v}), u(t; \bar{\alpha}, \bar{\tau}, \bar{v}))\, dt,$$
$$i = 0, 1, \ldots, m,$$

can be extended to continuously differentiable functions in a neighborhood of the point $(\hat{t}_1, \hat{t}_0, \hat{x}_0, \bar{0})$ so that

$$\partial\hat{F}_i/\partial t_1 = F_{it_1}(\hat{t}_1, \hat{t}_0, \hat{x}_0, \bar{0}) = f_i(\hat{t}_1, \hat{x}(\hat{t}_1), \hat{u}(t_1)) = \hat{f}_i(\hat{t}_1), \tag{14}$$

$$\partial\hat{F}_i/\partial t_0 = F_{it_0}(\hat{t}_1, \hat{t}_0, \hat{x}_0, \bar{0}) =$$
$$= -f_i(\hat{t}_0, \hat{x}_0, \hat{u}(\hat{t}_0)) + p_{0i}(\hat{t}_0)\varphi(\hat{t}_0, \hat{x}_0, \hat{u}(\hat{t}_0)) = -\hat{f}_i(\hat{t}_0) + p_{0i}(\hat{t}_0)\hat{\varphi}(\hat{t}_0), \tag{15}$$

$$\partial\hat{F}_i/\partial x_0 = F_{ix_0}(\hat{t}_1, \hat{t}_0, \hat{x}_0, \bar{0}) = -p_{0i}(\hat{t}_0), \tag{16}$$

$$\partial\hat{F}_i/\partial\alpha_k = F_{i\alpha_k}(\hat{t}_1, \hat{t}_0, \hat{x}_0, \bar{0}) = \Delta\hat{f}_i(\tau_k, v_k) - p_{0i}(\tau_k)\Delta\hat{\varphi}(\tau_k, v_k), \tag{17}$$

where $p_{0i}(\tau)$ is the solution of Cauchy's problem

$$dp_{0i}(\tau)/d\tau = -p_{0i}(\tau)\varphi_x(\tau, \hat{x}(\tau), \hat{u}(\tau)) + f_{ix}(\tau, \hat{x}(\tau), \hat{u}(\tau)),$$
$$p_{0i}(\hat{t}_1) = 0 \tag{18}$$

for the homogeneous linear system adjoint to the variational system (3), and $\Delta\hat{f}_i(\tau, v)$ is determined by a formula analogous to (5). The proof of this lemma will be presented in Section 4.2.7.

4.2.4. Reduction to the Finite-Dimensional Problem. The graph of the optimal phase trajectory $\hat{x}(t)$ lies in the domain G and, by virtue of the second assertion of the lemma of Section 4.2.3 (on the packet of needles), the graph of the solution $x(t; t_0, x_0, \bar{\alpha}, \bar{\tau}, \bar{v})$ of Cauchy's problem (8) of that same section also lies in this domain for $t \in [t_0, t_1]$ if for a sufficiently small ε_0 inequalities (9) of Section 4.2.3 hold. Hence the quadruple $(x(\cdot; t_0, x_0, \bar{\alpha}, \bar{\tau}, \bar{v}), u(\cdot; \bar{\alpha}, \bar{\tau}, \bar{v}), t_0, t_1)$ is a controlled process for problem (1)-(3) of Section 4.2.2.

Let us put

$$\Psi_i(t_1, t_0, x_0, \bar{\alpha}) = \psi_i(t_0, x_0, t_1, x(t_1; t_0, x_0, \bar{\alpha}, \bar{\tau}, \bar{v})), \tag{1}$$

$$F_i(t_1, t_0, x_0, \bar{\alpha}) = \int_{t_0}^{t_1} f_i(t, x(t; t_0, x_0, \bar{\alpha}, \bar{\tau}, \bar{v}), u(t; \bar{\alpha}, \bar{\tau}, \bar{v}))\, dt,$$
$$i = 0, 1, \ldots, m. \tag{2}$$

By virtue of the third assertion of the lemma on the packet of needles and according to the composition theorem, the functions (1) are continuously differentiable in a neighborhood of the point $(\hat{t}_1, \hat{t}_0, \hat{x}_0, \bar{0})$. The same property of functions (2) follows from the lemma on the integral functionals of Section 4.2.3.

Now we can consider the finite-dimensional extremal problem, which is the restriction of problem (1)-(3) of Section 4.2.2 to the family of controlled processes we have constructed:

$$I_0(t_1, t_0, x_0, \bar{\alpha}) = F_0(t_1, t_0, x_0, \bar{\alpha}) + \Psi_0(t_1, t_0, x_0, \bar{\alpha}) \to \inf, \tag{3}$$

$$I_i(t_1, t_0, x_0, \bar{\alpha}) = F_i(t_1, t_0, x_0, \bar{\alpha}) + \Psi_i(t_1, t_0, x_0, \bar{\alpha}) \leq 0, \quad i = 1, 2, \ldots, m, \tag{4}$$

$$\alpha_k \geq 0, \quad k = 1, 2, \ldots, N. \tag{5}$$

The point $(\hat{t}_1, \hat{t}_0, \hat{x}_0, \bar{0})$ is a local solution of this problem. Indeed, if the constraints (4) hold, then the quadruple $(x(\cdot; t_0, x_0, \bar{\alpha}, \bar{\tau}, \bar{v}), u(\cdot; \bar{\alpha}, \bar{\tau}, \bar{v})$ is an admissible controlled process for problem (1)-(3) of Section 4.2.2. If ε_0 in inequalities (9) of Section 4.2.3 is sufficiently small then, by virtue of the second assertion of the lemma on the packet of needles, inequalities (4) of Section 4.2.1 hold. Consequently, inequality (5) of Section 4.2.1 hold; whence

$$I_0(t_1, t_0, x_0, \bar{\alpha}) = \mathcal{J}_0(x(\cdot; t_0, x_0, \bar{\alpha}, \bar{\tau}, \bar{v}), u(\cdot, \bar{\alpha}, \bar{\tau}, \bar{v}), t_0, t_1) \geq$$
$$\geq \mathcal{J}_0(\hat{x}(\cdot), \hat{u}(\cdot), \hat{t}_0, \hat{t}_1) = I_0(\hat{t}_1, \hat{t}_0, \hat{x}_0, \bar{0}),$$

which means that $(\hat{t}_1, \hat{t}_0, \hat{x}_0, \bar{0})$ is a local solution of problem (3)-(5).

Applying the Lagrange multiplier rule (Section 3.2) to problem (3)-(5), we arrive at the following assertion:

LEMMA (Lagrange Multiplier Rule for the Auxiliary Finite-Dimensional Problem). There exist Lagrange multipliers

$$\hat{\lambda}_0 \geq 0, \; \hat{\lambda} = (\hat{\lambda}_1, \ldots, \hat{\lambda}_m) \in \mathbf{R}^{m*}, \quad \hat{\mu} = (\mu_1, \ldots, \mu_N), \tag{6}$$

not all zero, such that for the Lagrange function

$$\Lambda(t_1, t_0, x_0, \bar{\alpha}; \lambda, \mu, \lambda_0) = \sum_{i=0}^{m} \lambda_i (F_i + \Psi_i) + \sum_{k=1}^{N} \mu_k \alpha_k \tag{7}$$

the following conditions hold:

1) the stationarity conditions

$$\hat{\Lambda}_{t_1} = \sum_{i=0}^{m} \hat{\lambda}_i \hat{F}_{it_1} + \sum_{i=0}^{m} \hat{\lambda}_i \hat{\Psi}_{it_1} = 0, \tag{8}$$

$$\hat{\Lambda}_{t_0} = \sum_{i=0}^{m} \hat{\lambda}_i \hat{F}_{it_0} + \sum_{i=0}^{m} \hat{\lambda}_i \hat{\Psi}_{it_0} = 0, \tag{9}$$

$$\hat{\Lambda}_{x_0} = \sum_{i=0}^{m} \hat{\lambda}_i \hat{F}_{ix_0} + \sum_{i=0}^{m} \hat{\lambda}_i \hat{\Psi}_{ix_0} = 0, \tag{10}$$

$$\hat{\Lambda}_{\alpha_k} = \sum_{i=0}^{m} \hat{\lambda}_i \hat{F}_{i\alpha_k} + \sum_{i=0}^{m} \hat{\lambda}_i \hat{\Psi}_{i\alpha_k} + \mu_k = 0, \tag{11}$$

where the notation $\hat{\Lambda}_{\alpha_k} = \Lambda_{\alpha_k}(\hat{t}_1, \hat{t}_0, \hat{x}_0, \bar{0}; \hat{\lambda}, \hat{\mu}, \hat{\lambda}_0)$, etc. is used;

2) the conditions of concordance of signs

$$\hat{\lambda}_i \geqq 0, \quad \hat{\mu}_k \leq 0; \quad i = 1, 2, \ldots, m; \quad k = 1, 2, \ldots, N; \tag{12}$$

3) the conditions of complementary slackness

$$\hat{\lambda}_i [F_i(\hat{t}_1, \hat{t}_0, \hat{x}_0, \bar{0}) + \Psi_i(\hat{t}_1, \hat{t}_0, \hat{x}_0, \bar{0})] = 0, \quad i = 1, 2, \ldots, m. \tag{13}$$

4.2.5. Proof of the Maximum Principle. With the exception of one passage which, to a certain extent, is not trivial, the remaining part of the proof will be devoted to the interpretation of conditions (8)-(13) of Section 4.2.4. The maximum principle itself (or, in the Lagrange form, the minimum principle), i.e., relations (11) [or (10)] of Section 4.2.2, is then obtained from equalities (11) of Section 4.2.4, i.e., from the conditions $\hat{\Lambda}_{\alpha_k} = 0$. Each of those conditions corresponds to one of the needles included in the "packet," and if this needle is specified by a pair (τ_k, v_k), we obtain the inequality $H(\tau_k, \hat{x}(\tau_k), v_k, \cdot) \leqslant H(\tau_k, \hat{x}(\tau_k), \hat{u}(\tau_k), \cdot)$ from this condition. Hence, to derive the maximum principle we have to consider the needles corresponding to all the possible pairs (τ, v) while the "packet" contains an arbitrary but finite number of needles. This difficulty is overcome with the aid of a simple topological consideration based on the compactness. Using the lemma on the "centered system of sets," we can choose "universal" Lagrange multipliers applicable to the whole set of needles.

A) Existence of the "Universal" Lagrange Multipliers. Let us study in greater detail conditions (8)-(13) of Section 4.2.4 into which we shall introduce the values of the derivatives of the functions F_i and Ψ_i calculated with the aid of the two lemmas of Section 4.2.3; particular attention will be paid to the terms involving the parameters (τ_k, v_k) of the needles.

The derivatives F_i are given in the lemma on the integral functionals, and denoting

$$A_i(\tau, v)=f_i(\tau, \hat{x}(\tau), v)-f_i(\tau, \hat{x}(\tau), \hat{u}(\tau))-$$
$$-p_{0i}(\tau)[\varphi(\tau, \hat{x}(\tau), v)-\varphi(\tau, \hat{x}(\tau), \hat{u}(\tau)]=\Delta\hat{f}(\tau, v)-p_{0i}(\tau)\Delta\hat{\varphi}(\tau, v), \tag{1}$$

we see that

$$\hat{F}_{i\alpha_k}=A_i(\tau_k, v_k),$$

and none of the derivatives $\hat{F}_{it_0}$, $\hat{F}_{it_1}$, and $\hat{F}_{ix_0}$ involves the parameters of the needles, and hence these derivatives have the same values for any "packet of needles" of Section 4.2.3. Similarly, the derivatives of the functions

$$\Psi_i(t_1, t_0, x_0, \bar{\alpha})=\psi_i(t_0, x_0, t_1, x(t_1;\, t_0, x_0, \bar{\alpha}, \bar{\tau}, \bar{v}))$$

are calculated with the aid of the composition theorem and the lemma on the packet of needles, the parameters of the needles being involved only in $\hat{\Psi}_{i\alpha_k}$. If we denote

$$B_i(\tau, v)=\frac{\partial\psi_i}{\partial x_1}(\hat{t}_0, \hat{x}(\hat{t}_0), \hat{t}_1, \hat{x}(\hat{t}_1))\Omega(\hat{t}_1, \tau)[\varphi(\tau, \hat{x}(\tau), v)-$$
$$-\varphi(\tau, \hat{x}(\tau), \hat{u}(\tau))]=\hat{\psi}_{ix_1}\Omega(\hat{t}_1, \tau)\Delta\hat{\varphi}(\tau, v), \tag{2}$$

then

$$\hat{\Psi}_{i\alpha_k}=B_i(\tau_k, v_k),$$

and the values of the derivatives $\hat{\Psi}_{it_0}$, $\hat{\Psi}_{it_1}$, $\hat{\Psi}_{ix_0}$ are the same for any "packet of needles."

Now we have

$$\hat{\Lambda}_{\alpha_k}=\sum_{i=0}^{m}\hat{\lambda}_i\hat{F}_{i\alpha_k}+\sum_{i=0}^{m}\hat{\lambda}_i\hat{\Psi}_{i\alpha_k}+\hat{\mu}_k=\sum_{i=0}^{m}\hat{\lambda}_i(A_i(\tau_k, v_k)+B_i(\tau_k, v_k))+\hat{\mu}_k.$$

According to conditions (12) and (11) of Section 4.2.4, we have $\hat{\mu}_k \leqslant 0$ and $\hat{\Lambda}_{\alpha_k} = 0$, and therefore the inequalities

$$\sum_{i=0}^{m}\hat{\lambda}_i[A_i(\tau_k, v_k)+B_i(\tau_k, v_k)]\geqslant 0 \tag{3}$$

must hold.

Now let us consider the following sets in the space $\mathbf{R}^{(m+1)*}$:

$$S=\{(\lambda, \lambda_0)\mid \sum_{i=0}^{m}\lambda_i^2=1\},$$

$$K(\tau, v)=\{(\lambda, \lambda_0)| \sum_{i=0}^{m} \lambda_i (A_i(\tau, v)+B_i(\tau, v))\geqslant 0\},$$

$$T_l=\{(\lambda, \lambda_0)| \sum_{i=0}^{m} \lambda_i (\hat{F}_{it_l}+\hat{\Psi}_{it_l})=0\}, \quad l=0, 1,$$

$$X=\{(\lambda, \lambda_0)| \sum_{i=0}^{m} \lambda_i (\hat{F}_{ix_0}+\hat{\Psi}_{ix_0})=0\},$$

and, finally, the set Z consisting of those (λ, λ_0) for which $\lambda_0 \geqslant 0$ and λ_i, $i = 1,\ldots,m$, satisfy the conditions of concordance of signs and the conditions of complementary slackness (12) and (13) of Section 4.2.4 which are identical with the conditions (13) and (14) of Section 4.2.2.

All these sets are closed, and, moreover, the sphere S is compact since it is a bounded closed subset of the finite-dimensional space $\mathbf{R}^{(m+1)*}$.

The stationarity conditions (8)-(10) of Section 4.2.4 correspond to the inclusions $(\hat{\lambda}, \hat{\lambda}_0)\in T_1, T_0, X$, and the conditions of concordance of signs and the conditions of complementary slackness correspond to the inclusion $(\hat{\lambda}, \hat{\lambda}_0)\in Z$ (all these conditions are the same for all packets of needles). Conditions (11) of Section 4.2.4, which depend on the needles, hold when $(\hat{\lambda}, \hat{\lambda}_0)\in K(\tau_k, v_k)$, $k = 1,\ldots,N$. Finally, since the Lagrange multipliers are determined to within a proportionality factor and are not all zero, they can be normalized so that $(\hat{\lambda}, \hat{\lambda}_0)\in S$. Therefore, the assertion of the lemma of the foregoing section means that

$$S\cap T_0\cap T_1\cap X\cap Z\cap \bigcap_{k=1}^{N} K(\tau_k, v_k)\neq\varnothing. \tag{4}$$

Let us consider the following system of closed subsets of the compact set S:

$$\tilde{K}(\tau, v)=S\cap T_0\cap T_1\cap X\cap Z\cap K(\tau, v), \quad \tau\in[\hat{t}_0, \hat{t}_1], \quad v\in\mathfrak{U}. \tag{5}$$

Lemma on the Centered System of Sets. The intersection of the system of sets $\tilde{K}(\tau, v)$ is nonempty.

Proof. According to (4), the intersection of any finite collection of sets belonging to the system $\tilde{K}(\tau, v)$ is nonempty. Such a system is said to be *centered*. By the well-known theorem [KF, p. 99], the intersection of a centered system of subsets of a compact set is nonempty. ■

Thus, there exist

$$(\hat{\lambda}, \hat{\lambda}_0)\in \bigcap_{\substack{\tau\in[\hat{t}_0, \hat{t}_1], \\ v\in\mathfrak{U}}} \tilde{K}(\tau, v)=S\cap T_0\cap T_1\cap X\cap Z\cap \bigcap_{\substack{\tau\in[\hat{t}_0, \hat{t}_1] \\ v\in\mathfrak{U}}} K(\tau, v). \tag{6}$$

These are the sought-for "universal" Lagrange multipliers. Indeed, the multipliers $\hat{\lambda}_i$ are not all zero $((\hat{\lambda}, \hat{\lambda}_0)\in S)$, satisfy the conditions of concordance of signs and the conditions of complementary slackness $((\hat{\lambda}, \hat{\lambda}_0)\in Z)$, satisfy the conditions $\hat{\Lambda}_{t_l} = 0$, $l = 0, 1$, and $\hat{\Lambda}_{x_0}=0$ $((\hat{\lambda}, \hat{\lambda}_0)\in T_0\cap T_1\cap X)$, and, finally,

$$\sum_{i=0}^{m} \hat{\lambda}_i [A_i(\tau, v)+B_i(\tau, v)]\geqslant 0 \tag{7}$$

for all $\tau\in[\hat{t}_0, \hat{t}_1]$, $v\in\mathfrak{U}$.

B) Derivation of the Maximum Principle. Substituting (1) and (2) into inequality (7), we bring it to the form

$$0\leqslant\sum_{i=0}^{m}\hat{\lambda}_i[\Delta\hat{f}_i(\tau, v)-p_{0i}(\tau)\Delta\hat{\varphi}(\tau, v)]+\sum_{i=0}^{m}\hat{\lambda}_i\frac{\partial\hat{\psi}_i}{\partial x_1}\Omega(\hat{t}_1, \tau)[\Delta\hat{\varphi}(\tau, v)]=$$

$$=f(\tau, \hat{x}(\tau), v)-f(\tau, \hat{x}(\tau), \hat{u}(\tau))-\hat{p}(\tau)[\varphi(\tau, \hat{x}(\tau), v)-\varphi(\tau, \hat{x}(\tau), \hat{u}(\tau))]=$$

$$=-H(\tau, \hat{x}(\tau), v, \hat{p}(\tau))+H(\tau, \hat{x}(\tau), \hat{u}(\tau), \hat{p}(\tau)), \tag{8}$$

where

$$f(\tau, x, u)=\sum_{i=0}^{m}\hat{\lambda}_i f_i(\tau, x, u), \tag{9}$$

$$\hat{p}(\tau)=\sum_{i=0}^{m}\hat{\lambda}_i p_{0i}(\tau)-\sum_{i=0}^{m}\hat{\lambda}_i\frac{\partial\hat{\psi}_i}{\partial x_1}\Omega(\hat{t}_1, \tau) \tag{10}$$

is a solution of the Euler–Lagrange equation (8a) of Section 4.2.2 (as will be seen later), and

$$H(\tau, x, u, p)=p\varphi(\tau, x, u)-f(\tau, x, u) \tag{11}$$

is the Pontryagin function [cf. (7) in Section 4.2.2].

Inequality (8), which holds for all $\tau\in[\hat{t}_0, \hat{t}_1]$ and $v\in\mathfrak{U}$, is equivalent to the maximum principle (11) of Section 4.2.2:

$$H(\tau, \hat{x}(\tau), \hat{u}(\tau), \hat{p}(\tau))=\max H(\tau, \hat{x}(\tau), v, \hat{p}(\tau)). \tag{12}$$

C) Derivation of the Stationarity Conditions with Respect to $x(\cdot)$. Differentiating (10) and taking into account that $p_{0i}(\tau)$ satisfy Eqs. (18) of Section 4.2.4 and

$$\frac{\partial\Omega(t, \tau)}{\partial\tau}=-\Omega(t, \tau)\hat{\varphi}_x(\tau)$$

(see the theorem of Section 2.5.4), we obtain

$$d\hat{p}(\tau)/d\tau=\sum_{i=0}^{m}\hat{\lambda}_i\{-p_{0i}(\tau)\hat{\varphi}_x(\tau)+\hat{f}_{ix}(\tau)\}-$$

$$-\sum_{i=0}^{m}\hat{\lambda}_i\frac{\partial\hat{\psi}_i}{\partial x_1}\Omega(\hat{t}_1, \tau)\hat{\varphi}_x(\tau)=-\hat{p}(\tau)\hat{\varphi}_x(\tau)+\sum_{i=0}^{m}\hat{\lambda}_i\hat{f}_{ix}(\tau)=-\hat{p}(\tau)\hat{\varphi}_x(\tau)+\hat{f}_x(\tau);$$

i.e., $\hat{p}(\cdot)$ is a solution of the Euler–Lagrange equation (8a) of Section 4.2.2.

Further, (10) directly implies that

$$\hat{p}(\hat{t}_1)=\sum_{i=0}^{m}\hat{\lambda}_i p_{0i}(\hat{t}_1)-\sum_{i=0}^{m}\hat{\lambda}_i\frac{\partial\hat{\psi}_i}{\partial x_1}\Omega(\hat{t}_1, \hat{t}_1)=-\sum_{i=0}^{m}\hat{\lambda}_i\frac{\partial\hat{\psi}_i}{\partial x_1}=-\frac{\partial\hat{l}}{\partial x_1}, \tag{13}$$

since $p_{0i}(\hat{t}_1) = 0$ by virtue of formula (18) of Section 4.2.3, and $\Omega(t, t) \equiv E$. Here, as in formula (6) of Section 4.2.2, the expression

$$l(t_0, x_0, t_1, x_1)=\sum_{i=0}^{m}\hat{\lambda}_i\psi_i(t_0, x_0, t_1, x_1) \tag{14}$$

is an endpoint functional, and equality (13) we have obtained is equivalent to the first equality (9a) of Section 4.2.2.

Finally, the condition $\hat{\Lambda}_{x_0} = 0$, i.e., relation (10) of Section 4.2.4, after the values of the derivatives (12) and (16) given in Section 4.2.3 have been substituted into it, yields

$$0=\hat{\Lambda}_{x_0}=\sum_{i=0}^{m}\hat{\lambda}_i\hat{F}_{ix_0}+\sum_{i=0}^{m}\hat{\lambda}_i\hat{\Psi}_{ix_0}=-\sum_{i=0}^{m}\hat{\lambda}_i p_{0i}(\hat{t}_0)+\sum_{i=0}^{m}\hat{\lambda}_i\left(\hat{\psi}_{ix_0}+\hat{\psi}_{ix_1}\frac{\partial\hat{x}}{\partial x_0}\right)=$$

$$= -\sum_{i=0}^{m} \hat{\lambda}_i p_{0i}(\hat{t}_0) + \sum_{i=0}^{m} \hat{\lambda}_i \hat{\psi}_{ix_0} + \sum_{i=0}^{m} \hat{\lambda}_i \hat{\psi}_{ix_1} \Omega(\hat{t}_1, \hat{t}_0) = -\hat{p}(\hat{t}_0) + \hat{l}_{x_0},$$

and hence the second transversality condition (9a) of Section 4.2.2 also holds.

D) Derivation of the Stationarity Conditions with Respect to t_i. After the values of the derivatives expressed by formulas (10), (11), (14), and (15) of Section 4.2.3 have been substituted into the conditions $\hat{\Lambda}_{t_0} = \hat{\Lambda}_{t_1} = 0$ [conditions (8) and (9) of Section 4.2.4], we take into account the notation (9)-(11) and equality (13) (which has already been proved) and obtain

$$0 = \hat{\Lambda}_{t_1} = \sum_{i=0}^{m} \hat{\lambda}_i \hat{F}_{it_1} + \sum_{i=0}^{m} \hat{\lambda}_i \hat{\Psi}_{it_1} = \sum_{i=0}^{m} \hat{\lambda}_i \hat{f}_i(\hat{t}_1) + \sum_{i=0}^{m} \hat{\lambda}_i \left[\hat{\psi}_{it_1} + \hat{\psi}_{ix_1} \frac{\partial \hat{x}}{\partial t_1} \right] =$$

$$= \hat{f}(\hat{t}_1) + \hat{l}_{t_1} + \hat{l}_{x_1} \hat{\varphi}(\hat{t}_1) = \hat{f}(\hat{t}_1) - \hat{p}(\hat{t}_1) \hat{\varphi}(\hat{t}_1) + \hat{l}_{t_1} = -\hat{H}(t_1) + \hat{l}_{t_1}$$

and

$$0 = \hat{\Lambda}_{t_0} = \sum_{i=0}^{m} \hat{\lambda}_i \hat{F}_{it_0} + \sum_{i=0}^{m} \hat{\lambda}_i \hat{\Psi}_{it_0} =$$

$$= \sum_{i=0}^{m} \hat{\lambda}_i [-\hat{f}_i(\hat{t}_0) + p_{0i}(\hat{t}_0) \hat{\varphi}(\hat{t}_0)] + \sum_{i=0}^{m} \hat{\lambda}_i \left[\hat{\psi}_{it_0} + \hat{\psi}_{ix_1} \frac{\partial \hat{x}}{\partial t_0} \right] =$$

$$= -\hat{f}(\hat{t}_0) + \sum_{i=0}^{m} \hat{\lambda}_i p_{0i}(\hat{t}_0) \hat{\varphi}(\hat{t}_0) + \hat{l}_{t_0} + \hat{l}_{x_1} [-\Omega(\hat{t}_1, \hat{t}_0) \hat{\varphi}(\hat{t}_0)] =$$

$$= -\hat{f}(\hat{t}_0) + \hat{p}(\hat{t}_0) \hat{\varphi}(\hat{t}_0) + \hat{l}_{t_0} = \hat{H}(\hat{t}_0) + \hat{l}_{t_0}.$$

This proves the two formulas (12a) of Section 4.2.2.

E) Continuity of the Function $\hat{H}(\cdot)$. Since the control $\hat{u}(t)$ is continuous from the right, so is the function

$$\hat{H}(t) = H(t, \hat{x}(t), \hat{u}(t), \hat{p}(t)) = -f(t, \hat{x}(t), \hat{u}(t)) + \hat{p}(t) \varphi(t, \hat{x}(t), \hat{u}(t)) =$$

$$= -\sum_{i=0}^{m} \hat{\lambda}_i f_i(t, \hat{x}(t), \hat{u}(t)) + \hat{p}(t) \varphi(t, \hat{x}(t), \hat{u}(t)).$$

Moreover, each of the functions $H(t, \hat{x}(t), v, \hat{p}(t))$ is continuous with respect to t, and passing to the limit in the inequality

$$H(t, \hat{x}(t), v, \hat{p}(\tau)) \leqslant H(t, \hat{x}(t), \hat{u}(t), \hat{p}(t)) = \hat{H}(t), \tag{15}$$

we obtain

$$H(\tau, \hat{x}(\tau), v, \hat{p}(\tau)) \leqslant \lim_{t \to \tau - 0} \hat{H}(t) = \hat{H}(\tau - 0),$$

whence

$$\hat{H}(\tau) = \sup_{v \in \mathrm{U}} H(\tau, \hat{x}(t), v, p(\tau)) \leqslant \hat{H}(\tau - 0). \tag{16}$$

On the other hand,

$$\hat{H}(\tau - 0) = \lim_{t \to \tau - 0} H(t, \hat{x}(t), \hat{u}(t), \hat{p}(t)) =$$

$$= H(\tau, \hat{x}(\tau), \hat{u}(\tau - 0), \hat{p}(\tau)) = \lim_{v \to \hat{u}(\tau - 0)} H(\tau, \hat{x}(\tau), v, \hat{p}(\tau)) \leqslant$$

$$\leqslant H(\tau, \hat{x}(t), \hat{u}(\tau), \hat{p}(\tau)) = \hat{H}(\tau)$$

by virtue of the same inequality (15). The last inequality and (16) imply $\hat{H}(\tau - 0) = \hat{H}(\tau)$, i.e., $\hat{H}(t)$ is continuous from the left as well. ■

4.2.6. Proof of the Lemma on the Packet of Needles. We remind the reader that the needlelike variation $x(t; t_0, x_0, \bar{\alpha}, \bar{\tau}, \bar{v})$ of the optimal phase trajectory $\hat{x}(\cdot)$ is the solution of Cauchy's problem

$$\dot{x} = \varphi(t, x, u(t; \bar{\alpha}, \bar{\tau}, \bar{v})), \tag{1}$$

$$x(t_0) = x_0, \tag{2}$$

where the variation $u(t; \bar{\alpha}, \bar{\tau}, \bar{v})$ of the optimal control $u(t)$ is specified for $\alpha_j > 0$ by the formulas

$$u(t; \bar{\alpha}, \bar{\tau}, \bar{v}) = \begin{cases} v_i, & t \in \Delta_i = [\tau_i + i\alpha, \tau_i + i\alpha + \alpha_i), \\ \hat{u}(t), & t \notin \bigcup_i \Delta_i; \ \alpha = \sum_{j=1}^{N} \alpha_j. \end{cases} \tag{3}$$

It is convenient to split the proof of the lemma on the packet of needles (the lemma of Section 4.2.3) into two parts. In the first part, which is stated as a separate assertion, we prove the uniform convergence of the varied phase trajectories to $\hat{x}(\cdot)$. Here it is not the piecewise continuity that plays the key role but the integral convergence of the corresponding right-hand members of the differential constraint [see formula (5) below]. To stress this fact we extend the assertion to the case of measurable variations as well (this is preceded by the statement of an appropriate definition of measurability for the case when the controls assume values belonging to an arbitrary topological space). The second part of the proof of the lemma on the packet of needles is directly connected with the piecewise continuity and is based on the classical theorem on continuous differentiability of solutions with respect to the initial data.

Definition. Let $\mathfrak{U}$ be an arbitrary topological space, and let I be a closed interval on the real line. A mapping $u: I \to \mathfrak{U}$ is said to be *measurable* (in Lebesgue's sense) if there exists a finite or a countable sequence of pairwise disjoint measurable subsets A_n such that:

a) the restriction $u|A_n$ can be extended to a function continuous on the closure $\bar{A}_n$;

b) the set $I \setminus \bigcup_n A_n$ is of Lebesgue's measure zero.

It is evident that every piecewise-continuous mapping is measurable in the sense of this definition (in this case we take the intervals of continuity as the sets A_n). It is also readily seen that if $\mathfrak{U} \subset \mathbf{R}^r$, then a function measurable in the sense of this definition is measurable in the ordinary sense as well [KF, Chapter 5, Section 4].

Indeed, let $u_n(\cdot)$ be a function continuous on $\bar{A}_n$ and coinciding with $u(\cdot)$ on A_n. Putting $u_n(t) = 0$ outside A_n, we obtain a measurable function on $\mathbf{R}$. The characteristic function $\chi_{A_n}(\cdot)$ of the set A_n is also measurable, and since $u(t) = \sum_n \chi_{A_n}(t) u_n(t)$ almost everywhere on I, the function $u(t)$ is also measurable.

Exercise. Prove that if $\mathfrak{U} \subset \mathbf{R}^r$ and $u: I \to \mathfrak{U}$ is a function measurable in the standard Lebesgue sense, then it is also measurable in the sense of the above definition. Hint: use Luzin's C-property [KF, p. 291].

In the formulation of the next lemma, $\mathfrak{A}$ is an arbitrary Hausdorff topological space [KF, p. 95].

LEMMA 1. Let a function φ and its derivative φ_x be continuous in $G \times \mathfrak{U}$, and let $\hat{x}: [\hat{t}_0, \hat{t}_1] \to \mathbf{R}^n$ be a solution of the differential equation

$$\dot{x} = \varphi(t, x, \hat{u}(t)), \tag{4}$$

and let its graph lie in G: $\Gamma = \{(t, \hat{x}(t)) \mid \hat{t}_0 \leqslant t \leqslant \hat{t}_1\} \subset G$.

Further, let $\delta_0 > 0$, and let $u_\alpha: (\hat{t}_0 - \delta_0, \hat{t}_1 + \delta_0) \to \mathfrak{U}$ be a family of measurable controls such that the values of $u_\alpha(t)$ belong to a compact set $\mathcal{K}_0 \subset \mathfrak{U}$ for all t and for all $\alpha \in \mathfrak{A}$.

Finally, let $\hat{u}(t) \equiv u_{\hat{\alpha}}(t)$ for some $\hat{\alpha} \in \mathfrak{A}$, and let

$$\lim_{\alpha \to \hat{\alpha}} \int_{\hat{t}_0 - \delta_0}^{\hat{t}_1 + \delta_0} |\varphi(t, \hat{x}(t), u_\alpha(t)) - \varphi(t, \hat{x}(t), \hat{u}(t))| \, dt = 0. \tag{5}$$

Then for some $\hat{\delta} > 0$ the solution $X_\alpha(t, t_0, x_0)$ of Cauchy's problem

$$\dot{x} = \varphi(t, x, u_\alpha(t)), \quad x(t_0) = x_0$$

is defined on the closed interval $[\hat{t}_0 - \hat{\delta}, \hat{t}_1 - \hat{\delta}]$ for all (t_0, x_0, α) belonging to a neighborhood of the point $(\hat{t}_0, \hat{x}(\hat{t}_0), \hat{\alpha})$ in $\mathbf{R} \times \mathbf{R}^n \times \mathfrak{A}$, and, moreover,

$$X_\alpha(t, t_0, x_0) \to \hat{x}(t)$$

uniformly with respect to $t \in [\hat{t}_0, \hat{t}_1]$ for $t_0 \to \hat{t}_0$, $x_0 \to \hat{x}(\hat{t}_0)$, $\alpha \to \hat{\alpha}$.

Proof. Let us put

$$F_\alpha(t, x) = \varphi(t, x, u_\alpha(t)) \tag{6}$$

and apply Theorem 2 of Section 2.5.5 to the case under consideration.

A) If A_n are those sets which are involved in the definition of measurability applied to $u_\alpha(\cdot)$, then, for a fixed x, the functions $t \to \varphi(t, x, u_\alpha(t))$ are continuous on A_n, and consequently $t \to F_\alpha(t, x)$ are measurable functions and the assumption A) of Section 2.5.1 holds. For a fixed t the functions F_α, along with φ, are differentiable with respect to x, and hence the assumption B) of Section 2.5.1 also holds.

B) For any compact set $\mathcal{K} \subset G$ the functions φ and φ_x are continuous on the compact set $\mathcal{K} \times \mathcal{K}_0$ and consequently they are bounded. Denoting

$$M = \max_{\mathcal{K} \times \mathcal{K}_0} |\varphi(t, x, u)|, \quad M_x = \max_{\mathcal{K} \times \mathcal{K}_0} \|\varphi_x(t, x, u)\|,$$

we see that the condition C') of Theorem 2 of Section 2.5.5 with $\varkappa(t) \equiv M$ and $k(t) \equiv M_x$ holds because $u_\alpha(t) \in \mathcal{K}_0$ by the hypothesis. The condition D) of the same theorem coincides with (5).

Thus, in the situation under consideration Theorem 2 of Section 2.5.5 is applicable, whence follows the required assertion. ■

Proof of the Lemma on the Packet of Needles. A) Let us verify that the conditions of the foregoing lemma hold. Their part related to φ and $\hat{x}(\cdot)$ is involved in the conditions of the theorem on the maximum principle.

Above we already assumed that the control $\hat{u}(\cdot)$ was extended outside the closed interval $[\hat{t}_0, \hat{t}_1]$ so that it is a continuous function for $t \leqslant \hat{t}_0$ and for $t \geqslant \hat{t}_1$. Therefore, we shall regard $\hat{u}(\cdot)$ as a function defined on the closed interval $\Delta = [\hat{t}_0 - \delta_0, \hat{t}_1 + \delta_0]$, $\delta_0 > 0$, its points of discontinuity lying within the open interval $(\hat{t}_0, \hat{t}_1)$.

Let $t^{(1)} < t^{(2)} < \ldots < t^{(s)}$ be these points of discontinuity, and let $t^{(0)} = \hat{t}_0 - \delta_0$ and $t^{s+1} = \hat{t}_1 + \delta_0$. The function

$$u_i(t) = \begin{cases} \hat{u}(t), & t^{(i-1)} \leqslant t < t^{(i)}, \\ \hat{u}(t^{(i)} - 0), & t = t^{(i)} \end{cases}$$

is continuous on the closed interval $I_i = [t^{(i-1)}, t^{(i)}]$, and the image $\mathcal{K}_i$ of this interval under the continuous mapping $u_i\colon I \to \mathfrak{U}$ is compact in $\mathfrak{U}$ since I_i is a compact set. According to formulas (3), the values of the control $u(t; \bar{\alpha}, \bar{\tau}, \bar{v})$ belong to the compact set

$$\mathcal{K}_0 = \bigcup_{i=1}^{s+1} \mathcal{K}_i \cup \{v_1, \ldots, v_N\}.$$

It remains to verify the condition of integral convergence (5). The continuous function φ is bounded on the compact set

$$\mathcal{K}=\{(t, \hat{x}(t), u) \mid t \in \Delta,\ u \in \mathcal{K}_0\}=\Gamma \times \mathcal{K}_0,$$

i.e., $|\varphi| \leqslant M$. For $\alpha_j > 0$ so small that the half-open intervals Δ_j do not intersect we conclude from the formula (3) that

$$\int_{\hat{t}_0-\delta_0}^{\hat{t}_1+\delta_0} |\varphi(t, \hat{x}(t), u(t; \bar{\alpha}, \bar{\tau}, \bar{v}))-\varphi(t, \hat{x}(t), \hat{u}(t))|\, dt=$$

$$=\sum_{j=1}^{N} \int_{\Delta_j} |\varphi(t, \hat{x}(t), v_j)-\varphi(t, \hat{x}(t), \hat{u}(t))|\, dt \leqslant 2M \sum_{j=1}^{N} \alpha_j=2M\alpha \rightarrow 0$$

for $\alpha_j \downarrow 0$, and hence condition (5) holds.

Applying Lemma 1, we see that the first two assertions of the lemma on the packet of needles of Section 4.2.3 hold: for sufficiently small $\varepsilon_0 > 0$ the solution $x(t; t_0, x_0, \bar{\alpha}, \bar{\tau}, \bar{v})$ of Cauchy's problem (1), (2) with (t_0, x_0, α) satisfying inequalities (9) of Section 4.2.3 is defined on $\Delta = [t_0 - \hat{\delta}, t_1 + \hat{\delta}]$ (this is the first assertion), and

$$x(t; t_0, x_0, \bar{\alpha}, \bar{\tau}, \bar{v}) \rightarrow \hat{x}(t)$$

uniformly with respect to $t \in [\hat{t}_0, \hat{t}_1]$ for $t_0 \rightarrow \hat{t}_0$, $x_0 \rightarrow \hat{x}_0$, $\alpha_j \downarrow 0$ (the second assertion).

B) Let us proceed to the consideration of the mapping $(t_1, t_0, x_0, \bar{\alpha}) \rightarrow x(t_1; t_0, \bar{\alpha}, \bar{\tau}, \bar{v})$. As usual, we denote by $X(t, t_0, x_0)$ the solution of Cauchy's problem

$$\dot{x}=\varphi(t, x, \hat{u}(t)), \quad x(t_0)=x_0. \tag{7}$$

By Theorem 1 of Section 2.5.6, function $X(\cdot, \cdot, \cdot)$ is defined and continuous in the domain $(\hat{t}_0 - \delta_0, \hat{t}_1 + \delta_0) \times \hat{G}$, where $\hat{G}$ is a neighborhood of the graph Γ of the solution $\hat{x}(\cdot)$. If the domain is narrowed so that t_0 and t do not pass through the points of discontinuity $t^{(i)}$ of the control $\hat{u}(\cdot)$, then the differential equation (7) satisfies the conditions of the classical theorem on the differentiability of the solution with respect to the initial data (Section 2.5.7), and therefore the function $X(t, t_0, x_0)$ is jointly continuously differentiable with respect to its arguments. In particular, X is continuously differentiable in some neighborhoods of the points $(\hat{t}_k, \hat{t}_k, \hat{x}(\hat{t}_k))$ since, by the hypothesis, the control $\hat{u}(\cdot)$ is continuous in some neighborhoods of the points t_k.

Further, let us denote by $\Xi(\xi, \bar{\alpha}, \bar{\tau}, \bar{v})$ the value at $t = \hat{t}_1$ of the solution of Cauchy's problem for Eq. (1) with the initial condition $x(\hat{t}_0) = \xi$, i.e.,

$$\Xi(\xi, \bar{\alpha}, \bar{\tau}, \bar{v})=x(\hat{t}_1, \hat{t}_0, \xi, \bar{\alpha}, \bar{\tau}, \bar{v}). \tag{8}$$

According to formula (13) of Section 2.5.5 and the formulas (3) specifying the control $u(t, \bar{\alpha}, \bar{\tau}, \bar{v})$, we have

$$x(t_1, t_0, x_0, \bar{\alpha}, \bar{\tau}, \bar{v})=X(t_1, \hat{t}_1, \Xi(X(\hat{t}_0, t_0, x_0), \bar{\alpha}, \bar{\tau}, \bar{v})) \tag{9}$$

[in the interval from t_0 to $\hat{t}_0$ the equation $\dot{x}=\varphi(t, x, u)$ is solved for the control $u = \hat{u}(t)$, from $\hat{t}_0$ and $\hat{t}_1$ it is solved for the control $u = u(t; \bar{\alpha}, \bar{\tau}, \bar{u})$, and, finally, from $\hat{t}_1$ to t_1 it is again solved for the control $u = \hat{u}(t)$]. It is important to note that, like the formula (13) of Section 2.5.5, formula (9) is valid for any location of the points t_i relative to $\hat{t}_i$.

Now if we manage to extend the mapping $(\xi, \bar{\alpha}) \rightarrow \Xi(\xi, \bar{\alpha}, \bar{\tau}, \bar{v})$ to a mapping of class C^1 defined in a neighborhood of the point $(\hat{x}(\hat{t}_0), \bar{0})$ (i.e., if we manage to extend it to negative α_j with the preservation of the continuous differentiability) then, according to formula (9), the mapping $(t_1, t_0, x_0, \bar{\alpha}) \to x(t_1; t_0, x_0, \bar{\alpha}, \bar{\tau}, \bar{v})$ will also be extended to a neighborhood of the point $(\hat{t}_1, \hat{t}_0, x, \bar{0})$, and, by the theorem of Section 2.2.2, it will be continuously differentiable since it is a composition of three continuously differentiable mappings: $(t_0, x_0) \to X(\hat{t}_0, t_0, x_0)$, $(\xi, \bar{\alpha}) \to \Xi(\xi, \bar{\alpha}, \bar{\tau}, \bar{v})$, and $(t_1, \eta_0) \to X(t_1, \hat{t}_1, \eta_0)$.

C) Let $\tau \in (\hat{t}_0, \hat{t}_1)$ and $v \in \mathfrak{U}$ be fixed. We shall denote by $Y_v(t, t_0, y_0)$ the solution of Cauchy's problem

$$\dot{y} = \varphi(t, y, v), \quad y(t_0) = y_0. \tag{10}$$

By virtue of the local existence theorem (Section 2.5.2) and the classical theorem on the differentiability of the solution with respect to the initial data (Section 2.5.7), the solution $Y_v(t, t_0, y_0)$ is defined and continuously differentiable in a neighborhood of the point $(\tau, \tau, \hat{x}(\tau))$.

If τ is a point of discontinuity of the control $\hat{u}(\cdot)$, i.e., $\hat{u}(\tau - 0) \neq \hat{u}(\tau)$, then, along with $\hat{u}(\cdot)$, we shall consider the control

$$\tilde{u}(t) = \begin{cases} \hat{u}(t), & t \geqslant \tau, \\ \hat{u}(\tau), & t < \tau. \end{cases}$$

Since $\tilde{u}(\cdot)$ is continuous in a neighborhood of the point τ, the solution $\tilde{X}(t, t_0, x_0)$ of Cauchy's problem

$$\dot{x} = \varphi(t, x, \tilde{u}(t)), \quad x(t_0) = x_0, \tag{11}$$

is also defined and continuously differentiable in a neighborhood of the point $(\tau, \tau, \hat{x}(\tau))$. In case τ is a point of continuity of $\hat{u}(\cdot)$, the solution $X(t, t_0, x_0)$ itself possesses the same property, and we can put $\tilde{X} \equiv X$ in the further formulas.

Let us add the points $\tau_0 = \hat{t}_0$ and $\tau_{N+1} = \hat{t}_1$ to the set $\tau = (\tau_1, \ldots, \tau_N)$ so that $\hat{t}_0 = \tau_0 < \tau_1 \leqslant \ldots \leqslant \tau_N < \tau_{N+1} = \hat{t}_1$ and consider the composition of mappings

$$\tilde{\Xi}(\xi, \bar{\alpha}, \bar{\tau}, \bar{v}) = P \circ X_N \circ Z_N \circ \ldots \circ X_k \circ Z_k \circ X_{k-1} \circ \ldots \circ Z_1 \circ X_0, \tag{12}$$

where $P(\xi, \bar{\alpha}) = \xi$,

$$X_k(\xi, \bar{\alpha}) = (X(\tau_{k+1}, \tau_k, \xi), \bar{\alpha}) \tag{13}$$

and

$$Z_k(\xi, \bar{\alpha}) = (\tilde{X}(\tau_k, \tau_k + k\alpha + \alpha_k, Y_{v_k}(\tau_k + k\alpha + \alpha_k, \tau_k + k\alpha, \tilde{X}(\tau_k + k\alpha, \tau_k, \xi))), \bar{\alpha}). \tag{14}$$

Since X_k and Z_k are continuously differentiable in a neighborhood of the point $p_k = (\hat{x}(\tau_k), 0)$, and we have $X_k(p_k) = p_{k+1}$, $Z_k(p_k) = p_k$, and P is linear, composition (12) is defined and continuously differentiable in a neighborhood of the point $p_0(\hat{x}(\hat{t}_0), \bar{0}) = (\hat{x}_0, 0)$.

<u>LEMMA 2.</u> If all α_k are positive and sufficiently small, then

$$\tilde{\Xi}(\xi, \bar{\alpha}, \bar{\tau}, \bar{v}) = \Xi(\xi, \bar{\alpha}, \bar{\tau}, \bar{v}) = x(\hat{t}_1, \hat{t}_0, \xi, \bar{\alpha}, \bar{\tau}, \bar{v}). \tag{15}$$

<u>Proof.</u> For the sake of brevity, we denote $x(t) = x(t, t_0, \xi, \bar{\alpha}, \bar{\tau}, \bar{v})$. If all α_k are positive and sufficiently small, then the half-open intervals Δ_i involved in definition (3) are pairwise disjoint and are arranged between $\hat{t}_0$ and $\hat{t}_1$ in the ascending order of their indices.

Let $s_i = \tau_i + i\alpha = \tau_i + i\sum_{j=1}^{N} \alpha_j$ be the left end point of the half-open interval Δ_i, and let $s_0 = \tau_0 = \hat{t}_0$ and $s_{N+1} = \tau_{N+1} = \hat{t}_1$. We shall prove by induction that

$$x(s_k)=X(s_k, \tau_k, \xi_k), \tag{16}$$

where $\xi_0 = \xi$ and

$$\xi_k=P\circ X_{k-1}\circ Z_{k-1}\circ X_{k-2}\circ\ldots\circ Z_1\circ X_0(\xi, \bar{\alpha}),\quad k\geqslant 1, \tag{17}$$

and, in particular, $\xi_1=P\circ X_0(\xi, \bar{\alpha})=X(\tau_1, \tau_0, \xi)$. Since the differential equations in (7) and (1) coincide on the half-open interval $\hat{t}_0 = \tau_0 \leqslant t < s_1$ and $x(\tau_0) = \xi = X(\tau_0, \tau_0, \xi_0)$, we have $x(t) \equiv X(t, \tau_0, \xi_0)$ on that interval. Passing to the limit for $t \to s$ and taking into consideration the identity $X(s_1, \tau_0, \xi_0) = X(s_1, \tau_1, X(\tau_1, \tau_0, \xi_0))$, we obtain (16) for $k = 1$.

Let us suppose that (16) holds for some k. By virtue of (3), to pass from s_k to s_{k+1} along the solution $x(\cdot)$ the following operations must be performed. On the interval from s_k to s_{k+1} we solve Eq. (10) with $v = v_k$ and the initial condition $y_0 = x(s_k)$, which results in $x(s_k + \alpha_k) = Y_{v_k}(s_k + \alpha_k, s_k, x(s_k))$. Further, from $s_k + \alpha_k$, i.e., from the end point of Δ_k, to s_{k+1}, i.e., to the initial point of Δ_{k+1}, we solve Eq. (7) with the initial condition $x_0 = x(s_k + \alpha_k)$, which yields the equality

$$x(s_{k+1})=X(s_{k+1}, s_k+\alpha_k, x(s_k+\alpha_k))=X(s_{k+1}, s_k+\alpha_k, Y_{v_k}(s_k+\alpha_k, s_k, x(s_k)).$$

According to formula (13) of Section 2.5.5, the identity

$$X(t, \tau, X(\tau, t_0, x_0))\equiv X(t, t_0, x_0)$$

holds. Using this identity and equality (16), which holds by the induction hypothesis, we consecutively obtain

$$\begin{aligned}x(s_{k+1})=X(s_{k+1}, \tau_{k+1}, X(\tau_{k+1}, \tau_k, X(\tau_k, s_k+\alpha_k, x(s_k+\alpha_k))))=\\=X(s_{k+1}, \tau_{k+1}, X(\tau_{k+1}, \tau_k, X(\tau_k, s_k+\alpha_k,\\ Y_{v_k}(s_k+\alpha_k, s_k, X(s_k, \tau_k, \xi_k))))).\end{aligned} \tag{18}$$

Now we note that if all α_k are positive, then $\tau_k < s_k < s_k + \alpha_k$, and since Eqs. (7) and (11) coincide for $t \geqslant \tau_k$, we have $X(\tau_k, s_k + \alpha_k, \xi) = \tilde{X}(\tau_k, s_k + \alpha_k, \xi)$ and $X(s_k, \tau_k, \xi_k) = \tilde{X}(s_k, \tau_k, \xi_k)$. Therefore, using first the definitions (13) and (14) and then (17), we obtain

$$\begin{aligned}X(\tau_{k+1}, \tau_k, X(\tau_k, s_k+\alpha_k, Y_{v_k}(s_k+\alpha_k, s_k, X(s_k, \tau_k, \xi_k))))=\\=P\circ X_k\circ Z_k(\xi_k, \bar{\alpha})=P\circ X_k\circ Z_k\circ X_{k-1}\circ\ldots\circ X_0(\xi, \bar{\alpha})=\xi_{k+1}.\end{aligned}$$

It follows that (18) coincides with (16), where k is replaced by k + 1, and hence (16) holds for all $k \geqslant 1$. It remains to note that for $k = N + 1$ relation (16) results in the equalities

$$x(\hat{t}_1)=X(\hat{t}_1, \hat{t}_1, \xi_{N+1})=\xi_{N+1}=P\circ X_N\circ Z_N\circ\ldots\circ X_0(\xi, \bar{\alpha})=\Xi(\xi, \bar{\alpha}, \bar{\tau}, \bar{v})$$

[the last equality follows from (12)], and since, by definition, we have $\hat{x}(t_1) = \Xi(\xi, \bar{\alpha}, \bar{\tau}, \bar{v})$, relation (15) is proved. ■

As was already mentioned, composition (12) is continuously differentiable in a neighborhood of the point $(\hat{x}_0, \bar{0})$. Consequently, function $\Xi(\xi, \bar{\alpha}, \bar{\tau}, \bar{v})$ (which was earlier defined only for $\alpha_j > 0$) admits of a continuously differentiable extension to that neighborhood. According to Section B), it follows that the function $x(t_1, t_0, x_0, \bar{\alpha}, \bar{\tau}, \bar{v})$ admits (for fixed $\bar{\tau}$ and $\bar{v}$) of a continuously differentiable extension to a neighborhood of the point $(\hat{t}_1, \hat{t}_0, \hat{x}_0, \bar{0})$.

It only remains to prove formulas (10)-(13) of Section 4.2.3 for the partial derivatives of this function at the point $(\hat{t}_1, \hat{t}_0, \hat{x}_0, \bar{0})$.

D) First of all we note that

$$x(t, t_0, x_0, \bar{0}, \bar{\tau}, \bar{v})\equiv X(t, t_0, x_0), \tag{19}$$

$$x(t, t_0, \hat{x}(t_0), \overline{0}, \overline{\tau}, \overline{v}) \equiv X(t, t_0, \hat{x}(t_0)) = \hat{x}(t), \tag{20}$$

$$x(\hat{t}_1, \hat{t}_0, x_0, \overline{\alpha}, \overline{\tau}, \overline{v}) \equiv \Xi(x_0, \overline{\alpha}, \overline{\tau}, \overline{v}). \tag{21}$$

According to (19) and (21), the partial derivative with respect to x_0 can be found using the theorem of Section 2.5.6:

$$\frac{\partial \hat{x}}{\partial x_0} = \frac{\partial x}{\partial x_0}(\hat{t}_1, \hat{t}_0, x_0, \overline{0}, \overline{\tau}, \overline{v})\Big|_{x_0=\hat{x}(\hat{t}_0)} = \frac{\partial \Xi(x_0, \overline{0}, \overline{\tau}, \overline{v})}{\partial x_0}\Big|_{x_0=\hat{x}(\hat{t}_0)} =$$

$$= \frac{\partial X(\hat{t}_1, \hat{t}_0, x_0)}{\partial x_0}\Big|_{x_0=\hat{x}(\hat{t}_0)} = \Omega(\hat{t}_1, \hat{t}_0), \tag{22}$$

where $\Omega(t, \tau)$ is the principal matrix solution of the variational equation

$$\dot{z} = \varphi_x(t, \hat{x}(t), \hat{u}(t)) z$$

[in formula (2) of Section 2.5.6 we must put $F(t, x) = \varphi(t, x, \hat{u}(t))$ and take into account (20)]. This proves relation (12) of Section 4.2.3.

Now we shall make use of formula (9) and the fact that, as was said in Section B), the derivatives of the function $X(t, t_0, x_0)$ in the neighborhood of the points $(\hat{t}_k, \hat{t}_k, \hat{x}(t_k))$, $k = 0, 1$, can be calculated with the aid of the classical theorem on the differentiability with respect to the initial data [formulas (3)-(6) in Section 2.5.7 where again $F(t, x) = \varphi(t, x, \hat{u}(t))$]. Let us differentiate (9) taking into account formula (22) which we have already derived and also the fact that, according to (20) and (21),

$$\Xi(X(\hat{t}_0, t_0, \hat{x}(t_0)), \overline{0}, \overline{\tau}, \overline{v}) = \Xi(\hat{x}(\hat{t}_0), \overline{0}, \overline{\tau}, \overline{v}) = x(\hat{t}_1, \hat{t}_0, \hat{x}(\hat{t}_0), \overline{0}, \overline{\tau}, \overline{v}) = \hat{x}(\hat{t}_1).$$

This results in

$$\partial \hat{x}/\partial t_1 = \partial X(\hat{t}_1, \hat{t}_1, \hat{x}(\hat{t}_1))/\partial t_1 = \varphi(\hat{t}_1, \hat{x}(\hat{t}_1), \hat{u}(\hat{t}_1)),$$

$$\partial \hat{x}/\partial t_0 = \partial \Xi(\xi, \overline{0}, \overline{\tau}, \overline{v})/\partial \xi\,|_{\xi=\hat{x}(\hat{t}_0)}\, \partial X(\hat{t}_0, t_0, \hat{x}(\hat{t}_0))/\partial t_0\,|_{t_0=\hat{t}_0} = \Omega(\hat{t}_1, \hat{t}_0)[-\varphi(\hat{t}_0, \hat{x}(\hat{t}_0), \hat{u}(\hat{t}_0))].$$

We have thus proved relations (10) and (11) of Section 4.2.3.

Now let $\alpha_i = 0$, $i \neq k$, and $\alpha_k > 0$. For the sake of brevity, we shall denote

$$x(t, \alpha_k) = x(t, \hat{t}_0, \hat{x}(\hat{t}_0), (0, \ldots, 0, \alpha_k, \ldots, 0), \overline{\tau}, \overline{v}).$$

According to (3), for this choice of $\overline{\alpha}$ the differential equations (1) and (7) coincide in the half-open interval $\hat{t}_0 \leqslant t < \tau_k + k\alpha_k$ and $x(\hat{t}_0, \alpha_k) = \hat{x}(t_0)$; therefore $x(t, \alpha_k) \equiv \hat{x}(t)$ in that interval, and passing to the limit we obtain

$$x(\tau_k + k\alpha_k, \alpha_k) = \hat{x}(\tau_k + k\alpha_k). \tag{23}$$

Further, for $\tau_k + k\alpha_k \leqslant t < \tau_k + k\alpha_k + \alpha_k$ we solve the differential equation (10) with the initial condition (23) and $v = v_k$, which results in

$$x(\tau_k + k\alpha_k + \alpha_k, \alpha_k) = Y_{v_k}(\tau_k + k\alpha_k + \alpha_k, \tau_k + k\alpha_k, \hat{x}(\tau_k + k\alpha_k)). \tag{24}$$

In the remaining interval $\tau_k + (k + 1)\alpha_k \leqslant t \leqslant \hat{t}_1$ we return to Eq. (7) for which, however, the initial condition (24) is taken. We shall split this interval into two by fixing s so that the interval (τ_k, s) does not contain points of discontinuity of $\hat{u}(\cdot)$. We shall assume that the quantity $\alpha_k > 0$ is so small that $\tau_k < \tau_k + (k + 1)\alpha_k < s$. We have

$$x(\hat{t}_1, \alpha_k) = X(\hat{t}_1, s, x(s, \alpha_k)) = X(\hat{t}_1, s, X(s, \tau_k + (k+1)\alpha_k, x(\tau_k + (k+1)\alpha_k, \alpha_k))) =$$

$$= X(\hat{t}_1, s, X(s, \tau_k + (k+1)\alpha_k, Y_{v_k}(\tau_k + (k+1)\alpha_k, \tau_k + k\alpha_k, \hat{x}(\tau_k + k\alpha_k)))).$$

When differentiating the last formula, we must take into account the following equalities:

$$\frac{\partial X}{\partial x_0}(\hat{t}_1, s, x_0)\Big|_{x_0=x(s,\,0)=\hat{x}(s)} = \Omega(\hat{t}_1, s)$$

[see equality (1) of Section 2.5.6; $\hat{t}_1$ and s are fixed];

$$\frac{\partial X}{\partial t_0}(s, t_0, x(\tau_k, 0))\Big|_{t_0=\tau_k} = -\Omega(s, \tau_k)\varphi(\tau_k, \hat{x}(\tau_k), \hat{u}(\tau_k)),$$
$$\frac{\partial X}{\partial x_0}(s, \tau_k, x_0)\Big|_{x_0=x(\tau_k,\,0)=\hat{x}(\tau_k)} = \Omega(s, \tau_k)$$

(see equalities (5) and (6) of Section 2.5.7; the classical theorem is applicable because the control $\hat{u}(\cdot)$ is continuous on the closed interval $[\tau_k, s]$, and we have $\tau_k < \tau_k + (k + 1)\alpha_s < s$);

$$\frac{\partial Y_{v_k}}{\partial t}(t, \tau_k, \hat{x}(\tau_k))\Big|_{t=\tau_k} = \varphi(\tau_k, \hat{x}(\tau_k), v_k),$$
$$\frac{\partial Y_{v_k}}{\partial t_0}(\tau_k, t_0, \hat{x}(\tau_k))\Big|_{t_0=\tau_k} = -\varphi(\tau_k, \hat{x}(\tau_k), v_k),$$
$$\frac{\partial Y_{v_k}}{\partial y_0}(\tau_k, \tau_k, y_0) = E$$

[see formulas (4)-(6) of Section 2.5.7 in which we must put $F(t, x)=\varphi(t, x, v_k)$ and take into account that $\Omega(t, t) = E$ for any t]; and, finally, (7) implies $\dot{\hat{x}}(\tau_k)=\varphi(\tau_k, \hat{x}(\tau_k), \hat{u}(\tau_k))$.

Thus,

$$\begin{aligned}
\frac{\partial \hat{x}}{\partial \alpha_k} &= \frac{\partial x}{\partial \alpha_k}(\hat{t}_1, \alpha_k)\Big|_{\alpha_k=0} = \frac{\partial X}{\partial x_0}(\hat{t}_1, s, x_0)\Big|_{x_0=\hat{x}(s)}\Big\{\frac{\partial X(s, \tau_k, \hat{x}(\tau_k))}{\partial t_0}(k+1) + \\
&+ \frac{\partial X}{\partial x_0}(s, \tau_k, \hat{x}(\tau_k))\Big[\frac{\partial Y}{\partial t}(\tau_k, \tau_k, \hat{x}(\tau_k))(k+1) + \\
&+ \frac{\partial Y}{\partial t_0}(\tau_k, \tau_k, \hat{x}(\tau_k))k + \frac{\partial Y}{\partial y_0}(\tau_k, \tau_k, \hat{x}(\tau_k))\dot{\hat{x}}(\tau_k)k\Big]\Big\} = \\
&= \Omega(\hat{t}_1, s)\{-\Omega(s, \tau_k)\varphi(\tau_k, \hat{x}(\tau_k), \hat{u}(\tau_k))(k+1) + \\
&+ \Omega(s, \tau_k)[\varphi(\tau_k, \hat{x}(\tau_k), v_k)(k+1) - \varphi(\tau_k, \hat{x}(\tau_k), v_k)k + \\
&+ \varphi(\tau_k, \hat{x}(\tau_k), \hat{u}(\tau_k)k]\} = \Omega(\hat{t}_1, \tau_k)[\varphi(\tau_k, \hat{x}(\tau_k), v_k) - \varphi(\tau_k, \hat{x}(\tau_k), \hat{u}(\tau_k))],
\end{aligned}$$

which proves relation (13) of Section 4.2.3 [the equality $\Omega(\hat{t}_1, s)\Omega(s, \tau_k) = \Omega(\hat{t}_1, \tau_k)$ holds by virtue of the basic property of the principal matrix solution expressed by formula (11) of Section 2.5.4].

4.2.7. Proof of the Lemma on the Integral Functionals. Let $\xi = (\xi_0, \xi_1, \ldots, \xi_m)$, $f = (f_0, f_1, \ldots, f_m)$, $F = (F_0, F_1, \ldots, F_m)$ (the functions F_i are defined in the formulation of the lemma of Section 4.2.3), $u(t, \bar{\alpha}, \bar{\tau}, \bar{v})$, and $x(t; t_0, x_0, \bar{\alpha}, \bar{\tau}, \bar{v})$ have the same meaning as in Sections 4.2.3 and 4.2.6.

Let us consider Cauchy's problem for (n + m + 1)-th-order system

$$\dot{\tilde{x}} = \begin{pmatrix}\dot{x}\\ \dot{\xi}\end{pmatrix} = \tilde{\varphi}(t, x, \xi, u(t, \bar{\alpha}, \bar{\tau}, \bar{v})) = \begin{pmatrix}\varphi(t, x, u(t, \bar{\alpha}, \bar{\tau}, \bar{v}))\\ f(t, x, u(t, \bar{\alpha}, \bar{\tau}, \bar{v}))\end{pmatrix}, \tag{1}$$
$$x(t_0) = x_0, \quad \xi(t_0) = 0,$$

where $\alpha_k > 0$. Here the equations for x do not involve ξ and coincide with Eqs. (2) of Section 4.2.6. Therefore, their solution is $x(t; t_0, x_0, \bar{\alpha}, \bar{\tau}, \bar{v})$, whence, as can easily be seen,

$$F(t_1, t_0, x_0, \bar{\alpha}) \equiv \xi(t_1; t_0, x_0, \bar{\alpha}, \bar{\tau}, \bar{v}).$$

Consequently, the possibility of extending the function F to nonpositive α_k and its continuous differentiability in a neighborhood of $(\hat{t}_1, \hat{t}_0, \hat{x}_0, 0)$ follow from the lemma on the packet of needles applied to system (1), and the values of the derivatives are found by means of formulas similar to formulas (10)-(13) of Section 4.2.3:

$$\frac{\partial \hat{\tilde{x}}}{\partial t_1} = \hat{\tilde{\varphi}}(\hat{t}_1), \qquad \frac{\partial \hat{\tilde{x}}}{\partial t_0} = -\tilde{\Omega}(\hat{t}_1, \hat{t}_0)\hat{\tilde{\varphi}}(\hat{t}_0),$$
$$\frac{\partial \hat{\tilde{x}}}{\partial \tilde{x}_0} = \tilde{\Omega}(\hat{t}_1, \hat{t}_0), \qquad \frac{\partial \hat{\tilde{x}}}{\partial \alpha_k} = \tilde{\Omega}(\hat{t}_1, \tau_k)\Delta\hat{\tilde{\varphi}}(\tau_k, v_k), \tag{2}$$

where

$$\tilde{\Omega}(t, \tau) = \begin{pmatrix} \Omega_{11} & \Omega_{12} \\ \Omega_{21} & \Omega_{22} \end{pmatrix}$$

is the principal matrix solution of the variational system

$$\dot{\tilde{z}} = \begin{pmatrix} \dot{z} \\ \dot{\xi} \end{pmatrix} = \hat{\tilde{\varphi}}_{\tilde{x}}(t)\begin{pmatrix} z \\ \xi \end{pmatrix} \tag{3}$$

(here Ω_{ij} are the rectangular blocks of the matrix $\tilde{\Omega}$ of dimensions $n \times n$, $n \times m$, $m \times n$, and $m \times m$, respectively). Since $\tilde{\varphi}$ does not depend on ξ, we conclude from (3) that $\tilde{\Omega}$ is the solution of the following Cauchy problem for the block-matrix differential equation:

$$\frac{d}{dt}\begin{pmatrix} \Omega_{11} & \Omega_{12} \\ \Omega_{21} & \Omega_{22} \end{pmatrix} = \begin{pmatrix} \hat{\varphi}_x(t) & 0 \\ \hat{f}_x(t) & 0 \end{pmatrix}\begin{pmatrix} \Omega_{11} & \Omega_{12} \\ \Omega_{21} & \Omega_{22} \end{pmatrix},$$
$$\begin{pmatrix} \Omega_{11}(\tau, \tau) & \Omega_{12}(\tau, \tau) \\ \Omega_{21}(\tau, \tau) & \Omega_{22}(\tau, \tau) \end{pmatrix} = \begin{pmatrix} E_n & 0 \\ 0 & E_m \end{pmatrix}. \tag{4}$$

Solving consecutively the four matrix equations forming (4), we obtain

$$d\Omega_{11}(t, \tau)/dt = \hat{\varphi}_x(t)\Omega_{11}(t, \tau), \quad \Omega_{11}(\tau, \tau) = E_n,$$

whence it follows that $\Omega_{11}(t, \tau) = \Omega(t, \tau)$, where $\Omega(t, \tau)$ is the principal matrix solution of system (3) of Section 4.2.3. The equations

$$d\Omega_{12}(t, \tau)/dt = \hat{\varphi}_x(t)\Omega_{12}(t, \tau), \quad \Omega_{12}(\tau, \tau) = 0$$

are satisfied by $\Omega_{12}(t, \tau) \equiv 0$, and, by the uniqueness theorem, there can be no other solution. Further, we have

$$d\Omega_{21}(t, \tau)/dt = \hat{f}_x(t)\Omega_{11} = \hat{f}_x(t)\Omega(t, \tau), \quad \Omega_{21}(t, \tau) = 0,$$

whence

$$\Omega_{21}(t, \tau) = \int_\tau^t \hat{f}_x(s)\Omega(s, \tau)\, ds.$$

In particular,

$$-\Omega_{21}(\hat{t}_1, \tau) = p_0(\tau) = (p_{00}(\tau), \ldots, p_{0m}(\tau))$$

is the solution of the Cauchy problem (18) of Section 4.2.3 [see formula (14) of Section 2.5.4; this can also be shown directly by differentiating with respect to τ and using the properties of the principal matrix solution described in the same theorem of Section 2.5.4]. Finally,

$$d\Omega_{22}(t, \tau)/dt = \hat{f}_x(t)\Omega_{12}(t, \tau) = 0, \quad \Omega_{22}(\tau, \tau) = E_m,$$

whence $\Omega_{22}(t, \tau) \equiv E_m$. Hence

$$\tilde{\Omega}(\hat{t}_1, \tau) = \begin{pmatrix} \Omega(\hat{t}_1, \tau) & 0 \\ -p_0(\tau) & E_m \end{pmatrix}.$$

Substituting this into (2) and separating what is associated with $\xi(t_1; t_0, \bar{x}_0, \bar{\alpha}, \bar{\tau}, \bar{v}) = F(t_1, t_0, \bar{x}_0, \bar{\alpha})$, we obtain

$$\partial\hat{F}/\partial t_1 = \partial\hat{\xi}/\partial t_1 = \hat{f}(\hat{t}_1),$$
$$\partial\hat{F}/\partial t_0 = \partial\hat{\xi}/\partial t_0 = -\Omega_{21}(\hat{t}_1, \hat{t}_0)\hat{\varphi}(\hat{t}_0) - \Omega_{22}(\hat{t}_1, \hat{t}_0)\hat{f}(\hat{t}_0) = p_0(\hat{t}_0)\hat{\varphi}(\hat{t}_0) - \hat{f}(\hat{t}_0),$$

$$\partial\hat{F}/\partial x_0 = \partial\hat{\xi}/\partial x_0 = \Omega_{21}(\hat{t}_1, \hat{t}_0) = -p_0(\hat{t}_0),$$
$$\partial\hat{F}/\partial\alpha_k = \partial\hat{\xi}/\partial\alpha_k = \Omega_{21}(\hat{t}_1, \tau_k)\Delta\hat{\varphi}(\tau_k, v_k) +$$
$$+\Omega_{22}(\hat{t}_1, \tau_k)\Delta\hat{f}(\tau_k, v_k) = -p_0(\tau_k)\Delta\hat{\varphi}(\tau_k, v_k) + \Delta\hat{f}(\tau_k, v_k),$$

which coincides with relations (14)-(17) of Section 4.2.3. ■

4.3*. Optimal Control Problems Linear with Respect to Phase Coordinates

As is seen from the title, this section is devoted to a special class of optimal control problems. This class is rather important for applications, but this is not the only reason why we are interested in it. The reason is that problems with a linear structure with respect to the phase coordinates allow us to demonstrate most vividly one of the most important ideas of the whole theory. We mean the "latent convexity" which is always present in optimal control problems. Ultimately, it is this factor that makes it possible to write the necessary condition in the form of the "maximum principle," i.e., in the form characteristic of convex programming problems. In contrast to the foregoing section, where we wanted to stress the connection between the theory of optimal control problems and the general theory of smooth extremal problems, here emphasis will be laid upon the connection between optimal control problems and convex analysis.

It should also be noted that here the necessary conditions for an extremum almost coincide with the sufficient conditions (which is characteristic of convex programming problems). Finally, in the present section we shall consider not only piecewise-continuous controls but also measurable ones, which is important for revealing the latent convexity.

4.3.1. Reduction of the Optimal Control Problem Linear with Respect to the Phase Coordinates to the Lyapunov-Type Problem. Let $\Delta = [t_0, t_1]$ be a fixed closed interval of the real line, let $a_i: \Delta \to \mathbf{R}^{n*}$, $i = 0, 1, \ldots, m$, and $A: \Delta \to \mathscr{L}(\mathbf{R}^n, \mathbf{R}^n)$ be integrable vector and matrix functions, respectively, let $\mathfrak{U}$ be a topological space, let $f_i: \Delta\times\mathfrak{U} \to \mathbf{R}$, $i = 0, 1, \ldots, m$, and $F(\cdot, \cdot): \Delta\times\mathfrak{U} \to \mathbf{R}^n$ be continuous functions, let γ_{0i}, γ_{1i}, $i = 1, \ldots, m$, be elements of $\mathbf{R}^{n*}$, and let c_i, $i = 1, 2, \ldots, m$, be some numbers.

The extremal problem

$$J_0(x(\cdot), u(\cdot)) = \int_\Delta (a_0(t)x(t) + f_0(t, u(t)))\,dt + \gamma_{00}x(t_0) + \gamma_{10}x(t_1) \to \inf,$$
$$\dot{x} = A(t)x + F(t, u(t)), \quad u(t) \in \mathfrak{U}, \tag{1}$$
$$J_i(x(\cdot), u(\cdot)) = \int_\Delta (a_i(t)x(t) + f_i(t, u(t)))\,dt + \gamma_{0i}x(t_0) + \gamma_{1i}x(t_1) \lesseqgtr c_i,$$

which involves linearly the quantities associated with the phase trajectory x, is called an *optimal control problem linear with respect to the phase coordinates*. Of course, it can be regarded as a special case of the general optimal control problem stated in Section 4.2. However, here we shall consider a wider set of admissible processes.

Let us denote by $\mathcal{U}$ the set of measurable mappings (in the sense of the definition of Section 4.2.6) $u: \Delta \to \mathfrak{U}$ such that the functions $t \to f_i(t, u(t))$ and $t \to F(t, u(t))$ are integrable. A pair $(x(\cdot), u(\cdot))$ is called an *admissible process* if $x(\cdot): \Delta \to \mathbf{R}^n$ is an absolutely continuous vector function (see Section 2.1.8), $u(\cdot) \in \mathcal{U}$,

$$\dot{x}(t) = A(t)x(t) + F(t, u(t))$$

almost everywhere, and the inequalities $J_i(x(\cdot), u(\cdot)) \lesseqgtr c_i$ hold. An admissible process $(\hat{x}(\cdot), \hat{u}(\cdot))$ is said to be *optimal* if there exists $\varepsilon > 0$

such that for any admissible process $(x(\cdot), u(\cdot))$ satisfying the condition $\|x(\cdot)-\hat{x}(\cdot)\|_{C(\Delta, \mathbf{R}^n)} < \varepsilon$ the inequality

$$J_0(x(\cdot), u(\cdot)) \geqslant J_0(\hat{x}(\cdot), \hat{u}(\cdot))$$

holds.

We shall show that the linear structure makes it possible to transform problem (1) to a form in which $x(\cdot)$ and $u(\cdot)$ are, in a sense, separated (and, in particular, there is no differential constraint).

Let $\Omega(\cdot, \cdot)$ be the principal matrix solution (Section 2.5.4) of the homogeneous linear system

$$\dot{x} = A(t)x. \tag{2}$$

According to formula (13) of Section 2.5.4,

$$x(t) = \Omega(t, t_0)x(t_0) + \int_{t_0}^{t} \Omega(t, \tau)F(\tau, u(\tau))\,d\tau. \tag{3}$$

Let us transform the functionals of problem (1) by substituting (3):

$$J_i(x(\cdot), u(\cdot)) = \int_{t_0}^{t_1}\left\{a_i(t)\left[\Omega(t, t_0)x(t_0) + \int_{t_0}^{t}\Omega(t, \tau)F(\tau, u(\tau))\,d\tau\right]\right\}dt +$$

$$+\int_{t_0}^{t_1} f_i(t, u(t))\,dt + \gamma_{0i}x(t_0) + \gamma_{1i}\left[\Omega(t_1, t_0)x(t_0) + \int_{t_0}^{t_1}\Omega(t_1, \tau)F(\tau, u(\tau))\,d\tau\right] =$$

$$= \int_{t_0}^{t_1}\int_{t_0}^{t} a_i(t)\Omega(t, \tau)F(\tau, u(\tau))\,d\tau\,dt + \int_{t_0}^{t_1} f_i(\tau, u(\tau))\,d\tau +$$

$$+\left\{\int_{t_0}^{t_1} a_i(t)\Omega(t, t_0)\,dt + \gamma_{0i} + \gamma_{1i}\Omega(t_1, t_0)\right\}x(t_0) +$$

$$+\gamma_{1i}\int_{t_0}^{t_1}\Omega(t_1, \tau)F(\tau, u(\tau))\,d\tau = \int_{t_0}^{t_1}\left\{\int_{\tau}^{t_1} a_i(t)\Omega(t, \tau)\,dt\,F(\tau, u(\tau)) +\right.$$

$$\left.+f_i(\tau, u(\tau)) + \gamma_{1i}\Omega(t_1, \tau)F(\tau, u(\tau))\right\}d\tau +$$

$$+\left\{\int_{t_0}^{t_1} a_i(t)\Omega(t, t_0)\,dt + \gamma_{0i} + \gamma_{1i}\Omega(t_1, t_0)\right\}x(t_0) = -\int_{t_0}^{t_1} G_i(\tau, u(\tau))\,d\tau + \beta_i x(t_0), \tag{4}$$

where

$$G_i(\tau, u) = p_i(\tau)F(\tau, u) - f_i(\tau, u), \tag{5}$$

$$\beta_i = \gamma_{0i} - p_i(t_0), \tag{6}$$

and

$$p_i(\tau) = -\gamma_{1i}\Omega(t_1, \tau) - \int_{\tau}^{t_1} a_i(t)\Omega(t, \tau)\,dt, \quad i = 0, 1, \ldots, m. \tag{7}$$

Now problem (1) assumes the form

$$J_0(x(\cdot), u(\cdot)) = -\int_{\Delta} G_0(t, u(t))\,dt + \beta_0 x(t_0) \to \inf,$$

$$J_i(x(\cdot), u(\cdot)) = -\int_{\Delta} G_i(t, u(t))\,dt + \beta_i x(t_0) \leqslant c_i. \tag{8}$$

Problem (8) will be called a *Lyapunov-type problem*.

4.3.2. Lyapunov's Theorem. Lyapunov's theorem plays the key role in establishing the connection between optimal control problems and convex programming problems. Here it will be stated for the case of a vector function $p:\mathbf{R} \to \mathbf{R}^n$ and Lebesgue's measure m on the real line **R**. A slight perfection of the proof makes it possible to extend this theorem to the case of an arbitrary space with a σ-algebra $\mathfrak{S}$ defined in it and a finite continuous measure $\mu: \mathfrak{S} \longrightarrow \mathbf{R}^n$. In the formulation below $\mathfrak{S}$ is the σ-algebra of Lebesgue measurable subsets of **R** and $\mu(A) = \int_A p(t)\,dt$.

Here and in the next section, in contrast to the other parts of this book, we shall use some facts of functional analysis and the theory of measure which are not usually included in the traditional courses. We shall begin with the definitions. Here and henceforth the terms "measurability," "integrability," "almost everywhere," etc. are understood in Lebesgue's sense (for the measurability it is convenient to use the definition stated in Section 4.2.6). Two functions are said to be *equivalent* if they coincide almost everywhere.

Definition 1. Let Δ be an arbitrary interval on the real line. By the spaces $L_1(\Delta)$ and $L_\infty(\Delta)$ we shall mean the normed spaces whose elements are the classes of equivalent functions $x:\Delta \to \mathbf{R}$ possessing finite norms specified by the following equalities:

in the space $L_1(\Delta)$: $\|x(\cdot)\| = \int_\Delta |x(t)|\,dt;$ (1)

in the space $L_\infty(\Delta)$: $\|x(\cdot)\| = \operatorname{sup\,vrai}_\Delta |x(t)| = \inf_{N,\, m(N)=0} \sup_{t \in \Delta \setminus N} |x(t)|.$ (2)

Proposition 1. $L_1(\Delta)$ and $L_\infty(\Delta)$ are Banach spaces, the space $L_1(\Delta)$ being separable. The space $L_\infty(\Delta)$ is isometrically isomorphic to the conjugate space $L_1(\Delta)^*$ so that the closed balls are weakly* compact in this space.

Instead of proving these assertions we give the following references: for the completeness and the separability of the space $L_1(\Delta)$, see [KF, Chapter 7, Section 1] (the existence of a countable base of the Lebesgue measure on an infinite interval is an immediate consequence of the existence of such a base on a finite closed interval since every infinite interval is representable as a countable union of finite closed intervals); for the compactness of the closed balls in the weak* topology of the space conjugate to a separable space, see [KF, p. 202], and for the isometric isomorphism between $L_1(\Delta)^*$ and $L_\infty(\Delta)$ [which implies the completeness of $L_\infty(\Delta)$], see [4].

Definition 2. Let X be a linear space, and let $A \subset X$. A point x of the set A is said to be an *extremal point* of A if

$$x = \alpha x_1 + (1-\alpha) x_2, \quad 0 < \alpha < 1, \quad x_1, x_2 \in A \Rightarrow x_1 = x_2.$$

Exercise. Let $x_j \in X$, $j = 1,\dots,n$. Prove that at least one of the points $x_1,\dots,x_n$ is an extremal point of their convex hull $\operatorname{conv}\{x_1,\dots,x_n\}$.

Krein–Milman Theorem. A compact subset of a locally convex topological linear space X has at least one extremal point and coincides with the convex closure of its extremal points.

For the proof of this theorem, see [4].

Now we proceed to the basic theorem of the present section.

Theorem of A. A. Lyapunov. If $p:\Delta \to \mathbf{R}^n$, $p(\cdot) = (p_1(\cdot),\dots,p_n(\cdot))$, is an integrable vector function, then

$$M = \left\{ x \in \mathbf{R}^n \,\middle|\, x = \int_A p(t)\,dt,\ A \in \mathfrak{S} \right\}$$

is a convex compact set in $\mathbf{R}^n$. (We remind the reader that $\mathfrak{S}$ is the σ-algebra of Lebesgue measurable sets.)

Proof. A) Let us consider the mapping $\Lambda: L_\infty(\Delta) \to \mathbf{R}^n$ determined by the equality

$$\Lambda(\psi(\cdot)) = \int_\Delta \psi(t)\, p(t)\, dt = \left(\int_\Delta \psi(t)\, p_1(t)\, dt, \ldots, \int_\Delta \psi(t)\, p_n(t)\, dt\right).$$

This mapping is linear and, moreover, it is continuous in the weak* topology of $L_\infty(\Delta)$ [by definition, this topology is the weakest among those in which all the mappings $\psi(\cdot) \to \langle \psi(\cdot), p(\cdot)\rangle$, $p(\cdot) \in L_1(\Delta)$ are continuous].

The set

$$W = \{\psi(\cdot) \in L_\infty(\Delta) \mid 0 \leqslant \psi(t) \leqslant 1,\ t \in \Delta\}$$

is a closed ball in $L_\infty(\Delta)$ [of radius 1/2 and with center at the point $\hat{\psi}(\cdot)$, $\hat{\psi}(t) \equiv 1/2$]. Consequently, it is convex and, according to Proposition 1, it is compact in the weak* topology. Therefore, its image ΛW under the continuous linear mapping Λ is also convex and compact.

It is obvious that the characteristic function

$$\chi_A(t) = \begin{cases} 1, & t \in A, \\ 0, & t \notin A \end{cases}$$

of any set $A \in \mathfrak{S}$ belongs to W, and therefore

$$\int_A p(t)\, dt = \int_\Delta \chi_A(t)\, p(t)\, dt = \Lambda(\chi_A(\cdot)) \in \Lambda W,$$

and consequently $M \subset \Lambda W$. It remains to prove that $M = \Lambda W$.

B) Let us take a point $\xi \in \Lambda W$. Its inverse image $W_\xi = W \cap \Lambda^{-1}(\xi)$ is the intersection of the set W which is convex and compact in the weak* topology and the convex set (the affine manifold) $\Lambda^{-1}(\xi)$ which is closed in the same topology (because Λ is continuous with respect to that topology). Consequently, the set W_ξ is convex and compact, and, by the Krein–Milman theorem, it has an extremal point $x(\cdot) \in W_\xi$.

If we prove that $x(\cdot)$ is the characteristic function of a set $A \in \mathfrak{S}$, it will follow that

$$\xi = \int_\Delta x(t)\, p(t)\, dt = \int_A p(t)\, dt \in M,$$

and the theorem will be proved.

C) If $0 \leqslant x(t) \leqslant 1$ and $x(\cdot)$ is not a characteristic function, then the set $B_\varepsilon = \{t \mid \varepsilon \leqslant x(t) \leqslant 1 - \varepsilon\}$ has a positive measure for some $\varepsilon > 0$. The function $\alpha \to m(\alpha) = m[(t_0, \alpha) \cap B_\varepsilon]$ is continuous and varies from $0 = m(t_0)$ to $m(B_\varepsilon) = m(t_1)$ [$m(B)$ is the Lebesgue measure of the set B]. Let us take an arbitrary N and choose α_k, $k = 1, \ldots, N-1$, such that $m(\alpha_k) = k/NB_\varepsilon$. This specifies the partition $B_\varepsilon = B_1 \cup \ldots \cup B_N$ of the set B_ε into N pairwise disjoint subsets of positive measure:

$$B_1 = (-\infty, \alpha_1] \cap B_\varepsilon, \ldots, B_k = (\alpha_{k-1}, \alpha_k] \cap B_\varepsilon, \ldots, B_N = (\alpha_{N-1}, \infty) \cap B_\varepsilon.$$

Now we put

$$y(t) = \begin{cases} y_k, & t \in B_k, \\ 0, & t \notin B_\varepsilon. \end{cases} \tag{3}$$

The homogeneous linear system

$$\int_\Delta y(t)\, p(t)\, dt = \sum_k y_k \int_{B_k} p_i(t)\, dt = 0$$

possesses a nonzero solution $(\hat{y}_1,\ldots,\hat{y}_N)$ for $N > n$. We shall denote the corresponding function (3) by $\hat{y}(\cdot)$. If $0 < \lambda < \varepsilon \max\{|\hat{y}_k|\}^{-1}$, then

$$|\lambda y(t)| \begin{cases} \leqslant \varepsilon, & t \in B_\varepsilon, \\ = 0, & t \notin B_\varepsilon, \end{cases}$$

and therefore $x(\cdot) \pm \lambda\hat{y}(\cdot) \in W$. Moreover,

$$\Lambda(x(\cdot) \pm \lambda\hat{y}(\cdot)) = \Lambda(x(\cdot)) \pm \lambda\Lambda\hat{y}(\cdot) = \xi \pm \lambda \int_\Delta \hat{y}(t)\, p(t)\, dt = \xi.$$

Consequently, $x(\cdot) \pm \lambda y(\cdot) \in W_\xi$, and since $\lambda \neq 0$, we see that $x(\cdot)$ is not an extremal point.

Hence, $x(\cdot)$ is a characteristic function, and the theorem is proved. ■

4.3.3. Lagrange Principle for the Lyapunov-Type Problems. In this section Δ is a fixed (finite or infinite) interval on the real line and $\mathfrak{U}$ is a topological space.

Lemma on the Superpositional Measurability. If a function $f: \Delta \times \mathfrak{U} \to \mathbf{R}$ is continuous and a function $u: \Delta \to \mathfrak{U}$ is measurable (in the sense of the definition of Section 4.2.6), then the function $t \to f(t, u(t))$ is also measurable.

Proof. Let $A_n \subset \Delta$ be chosen so that $m\left(\Delta \setminus \bigcup_n A_n\right) = 0$ (where m is the Lebesgue measure) and the restrictions $u(\cdot)|A_n$ can be extended to continuous functions on $\bar{A}_n$. Then the functions $t \to f(t, u(t))$, $t \in A_n$, can also be extended to continuous functions on $\bar{A}_n$. ■

Now let us suppose that continuous functions $f_i: \Delta \times \mathfrak{U} \to \mathbf{R}$ are given. Let us denote by $\mathcal{U}$ the set of measurable mappings $u: \Delta \to \mathfrak{U}$ for which the compositions $t \to f_i(t, u(t))$ are not only measurable but also integrable on Δ; then functions $u(\cdot) \to \int_\Delta f_i(t, u(t))\, dt$ are defined on $\mathcal{U}$.

Let X be a linear space, let functions $g_i: X \to \bar{\mathbf{R}}$ be convex for $i = 0, 1, \ldots, m'$ and affine with finite values for $i = m' + 1, \ldots, m$, and let $A \subset X$ be a convex subset.

The extremal problem

$$\mathcal{F}_0(u(\cdot)) + g_0(x) = \int_\Delta f_0(t, u(t))\, dt + g_0(x) \to \inf,$$

$$\mathcal{F}_i(u(\cdot)) + g_i(x) = \int_\Delta f_i(t, u(t))\, dt + g_i(x) \begin{cases} \leqslant 0, & i = 1, \ldots, m', \\ = 0, & i = m' + 1, \ldots, m, \end{cases} \tag{1}$$

$$x \in A, \quad u(\cdot) \in \mathcal{U}$$

will be called a *Lyapunov-type problem*. The Lyapunov-type problems include as a special case the standard convex programming problem studied in Section 1.3.3 (for $m' = m$ and $f_i = 0$), and, as will be seen later, Lyapunov's theorem makes it possible to extend the argument with the aid of which the Kuhn–Tucker theorem was proved to the more general situation considered here.

The function

$$\mathcal{L}(u(\cdot), x, \lambda, \lambda_0) = \sum_{i=0}^{m} \lambda_i (\mathcal{F}_i(u(\cdot)) + g_i(x)) \tag{2}$$

is called the *Lagrange function* of problem (1).

THEOREM (Lagrange Principle for the Lyapunov-Type Problem). Let functions $f_i: \Delta \times \mathcal{U} \to \mathbf{R}$ be continuous, let functions $g_i: X \to \mathbf{R}$ be convex for

$i = 0, 1, \ldots, m'$ and affine with finite values for $i = m' + 1, \ldots, m$, and let $A \subset X$ be convex.

1. If a pair $(\hat{u}(\cdot), \hat{x})$ is a solution of problem (1), then there is a vector $\hat{\lambda}=(\hat{\lambda}_1, \ldots, \hat{\lambda}_m) \in \mathbf{R}^{m*}$ and a number λ_0, not vanishing simultaneously, such that:

$$\text{a)}\ \min_{u \in U} \sum_{i=0}^{m} \hat{\lambda}_i f_i(t, u) = \sum_{i=0}^{m} \hat{\lambda}_i (t, \hat{u}(t)) \text{ almost everywhere,} \tag{3}$$

$$\min_{x \in A} \sum_{i=0}^{m} \hat{\lambda}_i g_i(x) = \sum_{i=0}^{m} \hat{\lambda}_i g_i(\hat{x}) \tag{4}$$

(this is the minimum principle);

b)
$$\hat{\lambda}_i \geqslant 0, \quad i=0, 1, \ldots, m' \tag{5}$$

(this is the condition of concordance of signs);

c)
$$\hat{\lambda}_j \{\mathcal{F}_j(\hat{u}(\cdot)) + g_j(\hat{x})\} = 0, \quad j=1, \ldots, m \tag{6}$$

(these are the conditions of complementary slackness).

2. If $(\hat{u}(\cdot), \hat{x}) \in \mathcal{U} \times A$ and there exist $\hat{\lambda} \in \mathbf{R}^{m*}$ and $\hat{\lambda}_0 > 0$ such that conditions (3)-(6) hold, then $(\hat{u}(\cdot), \hat{x})$ is a solution of problem (1).

Proof. A) Lemma on the Convexity of the Image.

The image

$$\operatorname{im} \mathcal{F} = \{\xi = (\xi_0, \ldots, \xi_m) \mid \xi_i = \mathcal{F}_i(u(\cdot)),\ u(\cdot) \in \mathcal{U}\}$$

of the mapping $u(\cdot) \mapsto \mathcal{F}(u(\cdot)) = (\mathcal{F}_0(u(\cdot), \ldots, \mathcal{F}_m(u(\cdot))$ is a convex set in $\mathbf{R}^{m+1}$.

Proof. Let

$$\xi^k = (\xi_0^k, \xi_1^k, \ldots, \xi_m^k) = \mathcal{F}(u^{(k)}(\cdot)), \quad k=1, 2.$$

The function $p(\cdot) = (p_0(\cdot), \ldots, p_m(\cdot)) \colon \mathbf{R} \to \mathbf{R}^{m+1}$ determined by the equalities

$$p_i(t) = \begin{cases} f_i(t, u^{(1)}(t)) - f_i(t, u^{(2)}(t)), & t \in \Delta, \\ 0, & t \notin \Delta \end{cases}$$

is integrable since $u^{(k)}(\cdot) \in \mathcal{U}$. By Lyapunov's theorem, the set

$$M = \left\{\xi \mid \xi = \int_A p(t)\,dt, \quad A \in \mathfrak{S}\right\}$$

is convex, and since this set contains the points 0 ($A = \emptyset$) and $\xi^1 - \xi^2$ ($A = \Delta$), for any $\alpha \in [0, 1]$ there exists $A_\alpha \in \mathfrak{S}$ (i.e., A_α is Lebesgue measurable) such that

$$\alpha(\xi_i^1 - \xi_i^2) = \int_{A_\alpha} p_i(t)\,dt = \int_{A_\alpha \cap \Delta} [f_i(t, u^{(1)}(t)) - f_i(t, u^{(2)}(t))]\,dt =$$

$$= \int_{A_\alpha \cap \Delta} f_i(t, u^{(1)}(t))\,dt + \int_{\Delta \setminus A_\alpha} f_i(t, u^{(2)}(t))\,dt - \xi_i^2,$$

whence

$$\int_{A_\alpha \cap \Delta} f_i(t, u^{(1)}(t))\,dt + \int_{\Delta \setminus A_\alpha} f_i(t, u^{(2)}(t))\,dt = \alpha\xi_i^1 + (1-\alpha)\xi_i^2, \quad i=0, 1, \ldots, m. \tag{7}$$

Now let us put

$$u_\alpha(t) = \begin{cases} u^{(1)}(t), & t \in A_\alpha \cap \Delta, \\ u^{(2)}(t), & t \in \Delta \setminus A_\alpha. \end{cases}$$

Then, by virtue of (7),

$$\mathcal{F}_i(u_\alpha(\cdot)) = \alpha\xi_i^1 + (1-\alpha)\xi_i^2,$$

and hence $\mathcal{F}(u_\alpha(\cdot))=\alpha\xi^1+(1-\alpha)\xi^2$. ■

B) Existence of the Lagrange Multipliers. As was already mentioned, our argument will be similar to the proof of the Kuhn–Tucker theorem of Section 1.3.3.

By the hypothesis, the pair $(\hat{u}(\cdot), \hat{x})$ is a solution of the problem (1). We can assume, without loss of generality, that $\mathcal{F}_0(\hat{u}(\cdot))+g_0(\hat{x})=0$ (if otherwise, we can subtract the corresponding constant from $\mathcal{F}_0+g_0$). Let us prove that the set

$$\begin{gathered}C=\{\alpha=(\alpha_0, \ldots, \alpha_m) \mid \exists(\bar{u}(\cdot), \bar{x}) \in \mathcal{U}\times A, \\ \mathcal{F}_0(\bar{u}(\cdot))+g_0(\bar{x})<\alpha_0,\ \mathcal{F}_i(\bar{u}(\cdot))+g_i(\bar{x})\leqslant\alpha_i, \\ i=1, \ldots, m',\ \mathcal{F}_i(\bar{u}(\cdot))+g_i(\bar{x})=\alpha_i,\quad i=m'+1, \ldots, m\}\end{gathered} \tag{8}$$

is convex. Indeed, let $\alpha^k=(\alpha_0^k, \ldots, \alpha_m^k)\in C$, $k = 1, 2$, $0 < \theta < 1$, and let $(u^{(k)}(\cdot), x^{(k)})$ be elements of $\mathcal{U}\times A$ such that

$$\mathcal{F}_0(u^{(k)}(\cdot))+g_0(x^{(k)})<\alpha_0^k,$$
$$\mathcal{F}_i(u^{(k)}(\cdot))+g_i(x^{(k)})\begin{cases}\leqslant\alpha_i^k, & i=1, \ldots, m', \\ =\alpha_i^k, & i=m'+1, \ldots, m.\end{cases}$$

By the lemma on the convexity of the image, there exists a function $u_\theta(\cdot)\in\mathcal{U}$ such that

$$\mathcal{F}_i(u_\theta(\cdot))=\theta\mathcal{F}_i(u^{(1)}(\cdot))+(1-\theta)\mathcal{F}_i(u^{(2)}(\cdot)),\quad i=0, 1, \ldots, m.$$

Moreover, $x_\theta=\theta x^{(1)}+(1-\theta)x^{(2)}\in A$ since A is convex, and

$$g_i(x_\theta)=g_i(\theta x^{(1)}+(1-\theta)x^{(2)})=\theta g_i(x^{(1)})+(1-\theta)g_i(x^{(2)})$$

for $i = m' + 1,\ldots,m$ since these functions are affine. Consequently,

$$\begin{aligned}\mathcal{F}_i(u_\theta(\cdot))+g_i(x_\theta)&=\theta[\mathcal{F}_i(u^{(1)}(\cdot))+g_i(x^{(1)})]+(1-\theta)[\mathcal{F}_i(u^{(2)}(\cdot))+g_i(x^{(2)})]= \\ &=\theta\alpha_i^1+(1-\theta)\alpha_i^2,\quad i=m'+1, \ldots, m.\end{aligned}$$

Further, since g_0 is a convex function, we have

$$\begin{aligned}\mathcal{F}_0(u_\theta(\cdot))+g_0(x_\theta)&=\theta\mathcal{F}_0(u^{(1)}(\cdot))+(1-\theta)\mathcal{F}_0(u^{(2)}(\cdot))+g_0(\theta x^{(1)}+(1-\theta)x^{(2)})\leqslant \\ &\leqslant\theta(\mathcal{F}_0(u^{(1)}(\cdot))+g_0(x^{(1)}))+(1-\theta)(\mathcal{F}_0(u^{(2)}(\cdot))+g_0(x^{(2)}))<\theta\alpha_0^1+(1-\theta)\alpha_0^2.\end{aligned}$$

It is proved in a similar way that

$$\mathcal{F}_j(u_\theta(\cdot))+g_j(x_\theta)\leqslant\theta\alpha_j^1+(1-\theta)\alpha_j^2,\quad j=1, \ldots, m'.$$

Therefore, from the definition of the set C, it follows that $\theta\alpha^1+(1-\theta)\alpha^2\in C$, i.e., C is convex.

Putting $(\bar{u}(\cdot), \bar{x}) = (\hat{u}(\cdot), \hat{x})$ in (8), we obtain the inclusion

$$\begin{aligned}C\supset\{\alpha=(\alpha_0, \ldots, \alpha_m)\mid\alpha_0>0,\ \alpha_i\geqslant0,\ i=1, \ldots, m', \\ \alpha_i=0,\ i=m'+1, \ldots, m\}.\end{aligned} \tag{9}$$

In particular, the set C is nonempty. Moreover, $0\notin C$ because, if otherwise, then, according to (8), we have

$$\begin{gathered}\mathcal{F}_0(\bar{u}(\cdot))+g_0(\bar{x})<0,\quad \mathcal{F}_i(\bar{u}(\cdot))+g_i(\bar{x})\leqslant0, \\ i=1, \ldots, m', \\ \mathcal{F}_i(\bar{u}(\cdot))+g_i(\bar{x})=0,\quad i=m'+1, \ldots, m,\end{gathered}$$

for a pair $(\bar{u}(\cdot), \bar{x})$, and consequently $(\hat{u}(\cdot), \hat{x})$ is not a solution of problem (1).

Applying the finite-dimensional separation theorem we find $\hat{\lambda}_i$, $i = 0, 1,\ldots,m$, not all zero, such that the inequality $\sum_{i=0}^{m}\hat{\lambda}_i\alpha_i\geqslant0$ holds for all $\alpha\in C$.

We have $(\varepsilon, 0, \ldots, 0, \alpha_j = 1, 0, \ldots, 0) \in C$, where $1 \leqslant j \leqslant m'$ and $\alpha_i = 0$ for $i \neq 0, j$, and therefore the inequality $\varepsilon\hat{\lambda}_0 + \hat{\lambda}_j \geqslant 0$ holds. Passing to the limit for $\varepsilon \downarrow 0$ we obtain $\hat{\lambda}_j \geqslant 0$. In just the same way, we obtain

$$(\tilde{\varepsilon}, 0, \ldots, 0) \in C \Rightarrow \hat{\lambda}_0 \geqslant 0,$$

and hence (5) holds.

For any $\varepsilon > 0$ we have

$$(\varepsilon, 0, \ldots, 0, \alpha_j = \mathcal{F}_j(\hat{u}(\cdot)) + g_j(\hat{x}), 0, \ldots, 0) \in$$

$$\in C \Rightarrow \hat{\lambda}_0\varepsilon + \hat{\lambda}_j[\mathcal{F}_j(\hat{u}(\cdot)) + g_j(x)] \geqslant 0 \Rightarrow \hat{\lambda}_j[\mathcal{F}_j(\hat{u}(\cdot)) + g_j(\hat{x})] \geqslant 0. \tag{10}$$

Since $(\hat{u}(\cdot), \hat{x})$ is an admissible pair,

$$\mathcal{F}_j(\hat{u}(\cdot)) + g_j(\hat{x}) \begin{cases} \leqslant 0, & 1 \leqslant j \leqslant m', \\ 0, & j = m'+1, \ldots, m. \end{cases} \tag{11}$$

Relations (11) and (10) and the inequalities $\hat{\lambda}_j \geqslant 0$, $1 \leqslant j \leqslant m'$, which we have already proved yield (6).

Further, according to (8), for any $\varepsilon > 0$, $(u(\cdot), x) \in \mathcal{U} \times A$, we have

$$(\mathcal{F}_0(u(\cdot)) + g_0(x) + \varepsilon,\ \mathcal{F}_1(x(\cdot)) + g_1(x), \ldots, \mathcal{F}_m(u(\cdot)) + g_m(x)) \in$$

$$\in C \Rightarrow \sum_{i=0}^{m} \hat{\lambda}_i(\mathcal{F}_i(u(\cdot)) + g_i(x)) + \varepsilon\hat{\lambda}_0 =$$

$$= \mathcal{L}(u(\cdot), x, \hat{\lambda}, \hat{\lambda}_0) + \varepsilon\hat{\lambda}_0 \geqslant 0 \Rightarrow \mathcal{L}(u(\cdot), x, \hat{\lambda}, \hat{\lambda}_0) \geqslant 0.$$

However, according to (6), (11), the inequality $j \geqslant m' + 1$, and the above assumption about $\mathcal{F}_0 + g_0$, there must be $\mathcal{L}(\hat{u}(\cdot), \hat{x}, \hat{\lambda}, \hat{\lambda}_0) = 0$, and therefore the minimum principle holds for the Lagrange function (2):

$$\min_{\mathcal{U} \times A} \mathcal{L}(u(\cdot), x, \hat{\lambda}, \hat{\lambda}_0) = \mathcal{L}(\hat{u}(\cdot), \hat{x}, \hat{\lambda}, \hat{\lambda}_0). \tag{12}$$

If $\hat{\lambda}_0 > 0$, relation (12) takes place, and the conditions (5) and (6) hold, then

$$\hat{\lambda}_0(\mathcal{F}_0(u(\cdot)) + g_0(x)) \overset{(5)}{\geqslant} \sum_{i=0}^{m'} \hat{\lambda}_i(\mathcal{F}_i(u(\cdot)) + g_i(x)) =$$

$$= \sum_{i=0}^{m} \hat{\lambda}_i(\mathcal{F}_i(u(\cdot)) + g_i(x)) = \mathcal{L}(u(\cdot), x, \hat{\lambda}, \hat{\lambda}_0) \overset{(12)}{\geqslant}$$

$$\geqslant \mathcal{L}(\hat{u}(\cdot), \hat{x}, \hat{\lambda}, \hat{\lambda}_0) = \sum_{i=0}^{m} \hat{\lambda}_i(\mathcal{F}_i(\hat{u}(\cdot)) + g_i(\hat{x})) \overset{(6),\,(11)}{=}$$

$$= \hat{\lambda}_0(\mathcal{F}_0(\hat{u}(\cdot)) + g_0(\hat{x})) \Rightarrow \mathcal{F}_0(u(\cdot)) + g_0(x) \geqslant \mathcal{F}_0(\hat{u}(\cdot) + g_0(\hat{x}))$$

for every admissible pair $(u(\cdot), x)$, and hence $(\hat{u}(\cdot), \hat{x})$ is a solution of problem (1).

C) Reduction to the Elementary Problem. It can easily be seen that the fulfillment of relation (12) is equivalent to the simultaneous fulfillment of relations (4) and

$$\min_{u(\cdot) \in \mathcal{U}} \sum_{i=0}^{m} \hat{\lambda}_i \mathcal{F}_i(u(\cdot)) =$$

$$= \min_{u(\cdot) \in \mathcal{U}} \int_{\Delta} \sum_{i=0}^{m} \hat{\lambda}_i f_i(t, u(t))\, dt = \int_{\Delta} \sum_{i=0}^{m} \hat{\lambda}_i f_i(t, \hat{u}(t))\, dt = \sum_{i=0}^{m} \hat{\lambda}_i \mathcal{F}_i(\hat{u}(\cdot)). \tag{13}$$

Indeed, if (4) and (13) hold, then

$$\mathcal{L}(u(\cdot), x, \hat{\lambda}, \hat{\lambda}_0) = \sum_{i=0}^{m} \hat{\lambda}_i \mathcal{F}_i(u(\cdot)) + \sum_{i=0}^{m} \hat{\lambda}_i g_i(x) \geqslant \sum_{i=0}^{m} \hat{\lambda}_i \mathcal{F}_i(\hat{u}(\cdot)) + \sum_{i=0}^{m} \hat{\lambda}_i g_i(\hat{x}) = \mathcal{L}(\hat{u}(\cdot), \hat{x}, \hat{\lambda}, \hat{\lambda}_0)$$

for all $u(\cdot)\in\mathcal{U}$ and $x\in A$, i.e., (12) takes place. However, if, for instance, $\sum_{i=0}^{m}\hat{\lambda}_i g_i(\bar{x})<\sum_{i=0}^{m}\hat{\lambda}_i g_i(\hat{x})$ for some $\bar{x}\in A$, then

$$\mathcal{L}(\hat{u}(\cdot),\bar{x},\hat{\lambda},\hat{\lambda}_0)=\sum_{i=0}^{m}\hat{\lambda}_i\mathcal{F}_i(\hat{u}(\cdot))+\sum_{i=0}^{m}\hat{\lambda}_i g_i(\bar{x})<\sum_{i=0}^{m}\hat{\lambda}_i\mathcal{F}_i(\hat{u}(\cdot))+\sum_{i=0}^{m}\hat{\lambda}_i g_i(\hat{x}),$$

and hence (12) cannot hold.

Now in order to pass from (4) and (13) to the maximum principle, i.e., to (3) and (4), we have to show that it is legitimate to "insert min under the integral sign," i.e., to show that $\hat{u}(\cdot)$ is a solution of the auxiliary optimal control problem

$$\int_\Delta\sum_{i=0}^{m}\hat{\lambda}_i f_i(t,u(\cdot))\,dt\to\inf,\quad u(\cdot)\in\mathcal{U},\tag{14}$$

if and only if relation (3) holds.

A similar assertion was already proved in Section 3.1.4. However, it was assumed there that function $\hat{u}(\cdot)$ was piecewise continuous, whereas in the present case it is only measurable, and therefore we shall have to use a more subtle argument.

D) Elementary Problem with Measurable Controls. The Lemma on the Minimum Conditions for the Elementary Optimal Control Problem. Let Δ be an interval in $\mathbf{R}$, let $\mathfrak{U}$ be a topological space, let $f\colon\Delta\times\mathfrak{U}\to\mathbf{R}$ be a continuous function, and let $\mathcal{U}$ be the set of measurable mappings $u\colon\Delta\to\mathfrak{U}$ such that the functions $t\to f(t,u(t))$ are integrable on Δ.

For $\hat{u}(\cdot)\in\mathcal{U}$ to be a solution of the problem

$$\mathcal{F}(u(\cdot))=\int_\Delta f(t,u(t))\,dt\to\inf,\quad u(\cdot)\in\mathcal{U},\tag{15}$$

it is necessary and sufficient that the equality

$$f(t,\hat{u}(t))=\min_{u\in\mathfrak{U}}f(t,u)\tag{16}$$

hold almost everywhere in Δ.

Proof. "Sufficient." According to (16), we have $f(t,u(t))\geqslant f(t,\hat{u}(t))$ almost everywhere for any $u(\cdot)\in\mathcal{U}$, whence $\mathcal{F}(u(\cdot))\geqslant\mathcal{F}(\hat{u}(\cdot))$.

"Necessary." Let $\hat{u}(\cdot)$ be a solution of problem (15). Since $\hat{u}(\cdot)$ is measurable, there exist measurable sets $A_n\subset\Delta$ such that $m\left(\Delta\setminus\bigcup_n A_n\right)=0$ and $\hat{u}|A_n$ can be extended to a continuous function on A_n (see Section 4.2.6). A point $\tau\in A_n$ will be called *essential* if $m((\tau-\delta,\tau+\delta)\cap A_n)>0$ for any $\delta>0$ (and *inessential* if otherwise), and we shall show that equality (16) holds for all essential points.

Indeed, if the inequality $f(\tau,v)<f(\tau,\hat{u}(\tau))$ held for an essential point $\tau\in A_n$ and some $v\in\mathfrak{U}$ then, by virtue of the continuity of the functions $\hat{u}(\cdot)|A_n$ and $t\to f(t,v)$, the same equality would hold in a δ-neighborhood of the point τ in the set A_n, i.e.,

$$f(t,v)<f(t,\hat{u}(t)),\ \forall t\in(\tau-\delta,\tau+\delta)\cap A_n.$$

It is evident that the function

$$u(t)=\begin{cases}\hat{u}(t), & t\notin(\tau-\delta,\tau+\delta)\cap A_n,\\ v, & t\in(\tau-\delta,\tau+\delta)\cap A_n,\end{cases}$$

belongs to $\mathcal{U}$, and

$$\mathscr{F}(u(\cdot))-\mathscr{F}(\hat{u}(\cdot))=\int\limits_{(\tau-\delta,\,\tau+\delta)\cap A_n}[f(t,v)-f(t,\hat{u}(t))]dt<0$$

since $f(t, v) < f(t, \hat{u}(t))$ and $m((\tau-\delta,\tau+\delta)\cap A_n)>0$.

It remains to prove that the set of the inessential points is of measure zero. If $\tau\in A_n$ is inessential then, according to the definition, there is δ_τ such that $\tau \in (\alpha_\tau, \beta_\tau)$ and $m((\tau-\delta_\tau, \tau+\delta_\tau)\cap A_n)=0$. Narrowing the interval, we find rational $(\alpha_\tau, \beta_\tau)$ such that $m((\alpha_\tau,\beta_\tau)\cap A_n)=0$. The set of all the intervals with rational end points is countable, and therefore the set of those of them which are of the form $(\alpha_\tau, \beta_\tau)$ for an inessential point $\tau\in A_n$ is not more than countable. Let these intervals be $I_1, I_2, \ldots$. Then $A_n\cap\bigcup_k I_k$ contains all inessential points of A_n and $m\left(A_n\cap\bigcup_k I_k\right)\leqslant\sum_k m(A_n\cap I_k)=0$.

Since there are not more than countably many sets A_n, the union of the sets of the inessential points of each of them is also of measure zero. As was proved, relation (16) can be violated only on $\Delta\setminus\bigcup_n A_n$ and at inessential points; i.e., it holds almost everywhere.

E) Completion of the Proof. According to C) and D), for condition (12) to hold it is necessary and sufficient that (3) and (4) hold simultaneously. Replacing in the last paragraph of Section B) relation (12) by (3) and (4), we complete the proof of all the assertions of the theorem. ■

4.3.4. Duality Theorem. The notation in this section is the same as in the foregoing one. We remind the reader that the concept of duality for convex programming problems was introduced in Section 3.3.2. This was done in connection with the perturbation method, i.e., with the inclusion of an individual extremal problem in a family of problems of the same type. Here we shall carry out exactly the same procedure; for the sake of simplicity, we shall confine ourselves to integral functionals. We shall assume (and this assumption will be essential for the proof) that Δ is a closed interval of the real line and that the topological space $\mathfrak{U}$ to which the values of the various controls $u(\cdot)$ belong is separable, i.e., it contains a countable subset $\{u_n\}$ everywhere dense in $\mathfrak{U}$. For instance, $\mathfrak{U}$ can be an arbitrary subset of $\mathbf{R}^r$, $C(\Delta, \mathbf{R}^n)$, $C^1(\Delta, \mathbf{R}^n)$, or $L_1(\Delta)$; as to the space $L_\infty(\Delta)$ (see Section 4.3.2), it is not separable.

Exercise 1. Prove these assertions.

We shall consider the Lyapunov-type problem

$$\mathscr{F}_0(u(\cdot))=\int\limits_\Delta f_0(t,u(t))\,dt\to\inf,$$

$$\mathscr{F}_i(u(\cdot))=\int\limits_\Delta f_i(t,u(t))\,dt\leqslant 0,\quad u(t)\in\mathfrak{U},\ i=1,2,\ldots,m. \tag{$\mathfrak{z}$}$$

Let us fix f_i and include the problem ($\mathfrak{z}$) in the family of problems

$$\mathscr{F}_0(u(\cdot))\to\inf,\quad \mathscr{F}_i(u(\cdot))+\alpha_i\leqslant 0. \tag{$\mathfrak{z}(\alpha)$}$$

As in Section 3.3.2, the S-function of the family $\mathfrak{z}(\alpha)$ is determined by the equality

$$S(\alpha)=\inf\{\mathscr{F}_0(u(\cdot))\,|\,\mathscr{F}_i(u(\cdot))+\alpha_i\leqslant 0,\ u(\cdot)\in\mathscr{U}\}\ \alpha\in\mathbf{R}^m. \tag{1}$$

The analogy between Lyapunov-type problems and the convex programming problem is retained here too, and although the functions $\mathscr{F}_i$ are not convex, there holds the following:

LEMMA 1. Function $\alpha\to S(\alpha)$ is convex.

Proof. According to the lemma of Section 4.3.3, the set

$$\operatorname{im}\mathcal{F}=\{\xi=(\xi_0,\ \ldots,\ \xi_m)\mid \xi_i=\mathcal{F}_i(u(\cdot)),\ u(\cdot)\in\mathcal{U}\}$$

is convex in $\mathbf{R}^{m+1}$. Rewriting (1) in the form

$$S(\alpha)=\inf\{\xi_0\mid \xi_i+\alpha_i\leqslant 0,\ \xi\in\operatorname{im}\mathcal{F}\}$$

and putting $g_i(\xi) = g_i(\xi_0,\ldots,\xi_m) = \xi_i$, we find that $S(\alpha)$ is the value of the convex programming problem

$$g_0(\xi)\to\inf,\quad g_i(\xi)+\alpha_i\leqslant 0,\quad \xi\in\operatorname{im}\mathcal{F}.$$

The last problem is a special case of the problem considered in Section 3.3.2 (here there are no equality constraints and therefore Y, Λ, etc. should be dropped), and hence the convexity of S is guaranteed by Corollary 1 of Section 3.3.2.

Duality Theorem for the Lyapunov-Type Problems. Let Δ be a closed interval in $\mathbf{R}$, let $\mathfrak{U}$ be a separable topological space, let functions f_i: $\Delta\times\mathfrak{U}\to\mathbf{R}$, $i = 0, 1,\ldots,m$, be continuous, and let $\mathcal{U}$ be the set of measurable mappings $u\colon \Delta\to\mathfrak{U}$ such that the functions $t \to f_i(t, u(t))$ are integrable on Δ.

If the S-function of the family $\mathfrak{z}(\alpha)$ is continuous at the point $\alpha = 0$, then

$$S(\alpha)=\sup_{\lambda\geqslant 0}\left(\lambda\alpha+\int_\Delta \Phi(t,\ \lambda)\,dt\right) \tag{2}$$

for any $\alpha\in\operatorname{int}(\operatorname{dom}S)$, where

$$\Phi(t,\ \lambda)=\inf_{u\in\mathfrak{U}}\left(f_0(t,\ u)+\sum_{i=1}^m \lambda_i f_i(t,\ u)\right). \tag{3}$$

Proof. A) By Lemma 1, the function $\alpha \to S(\alpha)$ is convex, and the continuity of the function at one point implies its continuity for all $\alpha \in \operatorname{int}(\operatorname{dom} S)$ (Proposition 3 of Section 2.6.2) and the equality $S(\alpha) = (\overline{\operatorname{conv}} S)(\alpha)$, $\alpha \in \operatorname{int}(\operatorname{dom} S)$. By the Fenchel–Moreau theorem (Section 2.6.3), $\overline{\operatorname{conv}}\, S = S^{**}$, and hence

$$S(\alpha)=S^{**}(\alpha)=\sup_{\lambda\in\mathbf{R}^{m*}}\{\lambda\alpha-S^*(\lambda)\} \tag{4}$$

for the same α.

Now let us compute $S^*(\lambda)$. According to the definition,

$$S^*(\lambda)=\sup_\alpha(\lambda\alpha-S(\alpha))=\sup_\alpha\left\{\lambda\alpha-\inf\left(\int_\Delta f_0(t,\ u(t))\,dt\,\Big|\int_\Delta f_i(t,\ u(t))\,dt+\alpha_i\leqslant 0,\right.\right.$$

$$\left.\left.u(\cdot)\in\mathcal{U}\right)\right\}=\sup_{u(\cdot)\in\mathcal{U}}\ \sup_{\alpha_i\leqslant-\int_\Delta f_i(t,\ u(t))\,dt}\left\{\lambda\alpha-\int_\Delta f_0(t,\ u(t))\,dt\right\}=$$

$$=\begin{cases}+\infty, & \lambda\notin\mathbf{R}^m_+,\\ -\inf\limits_{u(\cdot)\in\mathcal{U}}\int_\Delta\left[f_0(t,\ u(t))\,dt+\sum\limits_{i=1}^m\lambda_i f_i(t,\ u(t))\right]dt.\end{cases}$$

Substituting the expression we have found into (4), we see that equality (2) and, along with it, the whole theorem will be proved if we verify that

$$\inf_{u(\cdot)\in\mathcal{U}}\int_\Delta\left[f_0(t,\ u(t))+\sum_{i=1}^m\lambda_i f_i(t,\ u(t))\right]dt=\int_\Delta\Phi(t,\ \lambda)\,dt, \tag{5}$$

where $\Phi(t, \lambda)$ is determined by equality (3).

B) LEMMA 2. Let Δ be a closed interval in R, let $\mathfrak{U}$ be a separable topological space, let a function $f: \Delta \times \mathfrak{U} \to \mathbf{R}$ be continuous, and let $\mathcal{U}$ be the set of measurable mappings $u: \Delta \to \mathfrak{U}$ such that the function $t \mapsto f(t, u(t))$ is integrable on Δ.

Then the function

$$\varphi(t) = \inf_{v \in \mathfrak{U}} \{f(t, v)\} \tag{6}$$

is measurable on Δ, the integral $\int_\Delta \varphi(t)\,dt$ (finite or equal to $-\infty$) exists, and

$$\int_\Delta \varphi(t)\,dt = \int_\Delta \inf_{\mathfrak{U}} f(t, u)\,dt = \inf_{u(\cdot) \in \mathcal{U}} \int_\Delta f(t, u(t))\,dt. \tag{7}$$

Proof. Let $\{v_k | k \geqslant 1\}$ be a countable subset everywhere dense in $\mathfrak{U}$. By virtue of the continuity of f, the functions

$$\varphi_n(t) = \min_{1 \leqslant k \leqslant n} f(t, v_k) \tag{8}$$

are continuous, and for $n \to \infty$ we have

$$\varphi_n(t) \downarrow \lim_{n \to \infty} \inf_{1 \leqslant k \leqslant n} f(t, v_k) = \inf_k f(t, v_k) = \inf_{v \in \mathfrak{U}} f(t, v) = \varphi(t). \tag{9}$$

Moreover, there exist functions $u_n(\cdot) \in \mathcal{U}$ such that $\varphi_n(t) = f(t, u_n(t))$. Indeed, this is true for n = 1, and if $\varphi_{n-1}(t) = f(t, u_{n-1}(t))$, then

$$\varphi_n(t) = \min_{1 \leqslant k \leqslant n} [f(t, v_k)] = \min[\varphi_{n-1}(t), f(t, v_n)] =$$
$$= \min[f(t, u_{n-1}(t)), f(t, v_n)] = f(t, u_n(t)),$$

where the function

$$u_n(t) = \begin{cases} u_{n-1}(t) & \text{for } \varphi_{n-1}(t) \leqslant f(t, v_n), \\ v_n & \text{for } \varphi_{n-1}(t) > f(t, v_n) \end{cases}$$

belongs to $\mathcal{U}$ [since the functions are continuous, the inequality $\varphi_{n-1}(t) > f(t, v_n)$ holds on an open set, and therefore the function $u_n(\cdot)$ is measurable; the integrability of the function $t \mapsto f(t, u_n(t)) = \varphi_n(t)$ follows from the continuity of $\varphi_n(t)$].

By virtue of (9), function $\varphi(\cdot)$ is a pointwise limit of continuous (and hence measurable) functions and therefore it is measurable [KF, p. 284]. Representing this function in the form of the difference

$$\varphi(t) = f(t, v) - [\varphi(t) - f(t, v)]$$

of a continuous function and a nonnegative function, we see that the integral $\int_\Delta \varphi(t)\,dt$ exists, its value being finite or equal to $-\infty$ [the function f(t, v) is continuous on the closed interval Δ and therefore the integral of f is finite while the integral of the nonnegative measurable function $\varphi(t) - f(t, v)$ exists but its value may be equal to $+\infty$].

Since $f(t, u(t)) \geqslant \varphi(t)$ for any $u(\cdot) \in \mathcal{U}$, the inequality

$$\inf_{u(\cdot) \in \mathcal{U}} \int_\Delta f(t, u(t))\,dt \geqslant \int_\Delta \varphi(t)\,dt \tag{10}$$

is obvious. If the left-hand member of (10) is equal to $-\infty$, then equality (7) holds. In case the left-hand member is finite, the inequalities

$$\int_\Delta \varphi_1(t)\,dt \geqslant \int_\Delta \varphi_n(t)\,dt = \int_\Delta f(t, u_n(t))\,dt \geqslant \inf_{u(\cdot) \in \mathcal{U}} \int_\Delta f(t, u(t))\,dt$$

imply that the sequence of integrals $\int_\Delta \varphi_n(t)dt$ is bounded. Applying the theorem of B. Levi [KF, p. 303] to the nondecreasing sequence $-\varphi_n(\cdot)$, we find that the function $\varphi(\cdot)$ is integrable and

$$\int_\Delta \varphi(t)\,dt=\lim_{n\to\infty}\int_\Delta \varphi_n(t)\,dt\geqslant\inf_n\int_\Delta \varphi_n(t)\,dt=\inf_n\int_\Delta f(t,\ u_n(t))\,dt\geqslant\inf_{u(\cdot)\in\mathcal{U}}\int_\Delta f(t,\ u(t))\,dt.$$

The last relation and inequality (10) yield (7). ■

By Lemma 2, relation (5) holds, which, as was seen, proves the theorem.

Exercise 2. State and prove the duality theorem for the Lyapunov-type problem of the general form [see (1) in Section 4.3.3].

4.3.5. Maximum Principle for Optimal Control Problems Linear with Respect to the Phase Coordinates. Let us come back to the problem posed in Section 4.3.1:

$$J_0(x(\cdot),\ u(\cdot))=\int_\Delta [a_0(t)\,x(t)+f_0(t,\ u(t))]\,dt+\gamma_{00}x(t_0)+\gamma_{10}x(t_1)\to\inf, \tag{1}$$

$$\dot{x}=A(t)\,x+F(t,\ u(t)),\quad u(t)\in\mathfrak{U}, \tag{2}$$

$$J_i(x(\cdot),\ u(\cdot))=\int_\Delta [a_i(t)\,x(t)+f_i(t,\ u(t))]\,dt+ \\ +\gamma_{0i}x(t_0)+\gamma_{1i}x(t_1)\leqslant c_i,\quad i=1,\ 2,\ \ldots,\ m. \tag{3}$$

THEOREM (Maximum Principle for Problems Linear with Respect to the Phase Coordinates). Let functions $a_i\colon \Delta\to\mathbf{R}^{n*}$, $i=0,\ 1,\ \ldots,\ m$, and $A\colon \Delta\to\mathscr{L}(\mathbf{R}^n,\ \mathbf{R}^n)$ be integrable on a closed interval $\Delta=[t_0,\ t_1]\subset\mathbf{R}$, let $\mathfrak{U}$ be a topological space, let $f_i\colon \Delta\times\mathfrak{U}\to\mathbf{R}$, $i=0,\ 1,\ \ldots,\ m$, $F\colon \Delta\times\mathfrak{U}\to\mathbf{R}^n$ be continuous functions, and let $\mathcal{U}$ be the set of measurable mappings $u\colon \Delta\to\mathfrak{U}$ such that the functions $t\to f_i(t,\ u(t))$ and $t\to F(t,\ u(t))$ are integrable.

1. If $(\hat{x}(\cdot),\ \hat{u}(\cdot))$ is an optimal process for the problem (1)-(3), then there exist a number $\hat{\lambda}_0\geqslant 0$, a vector $\hat{\lambda}=(\hat{\lambda}_1,\ \ldots,\ \hat{\lambda}_m)\in\mathbf{R}^{m*}$, and an absolutely continuous function $\hat{p}\colon\Delta\to\mathbf{R}^{m*}$, not all zero, such that:

a) the Euler–Lagrange equation (the adjoint equation) holds:

$$\dot{\hat{p}}(t)=-\hat{p}(t)\,A(t)+\sum_{i=0}^m\hat{\lambda}_i a_i(t); \tag{4}$$

b) the transversality conditions hold:

$$\hat{p}(t_k)=(-1)^k\sum_{i=0}^m\hat{\lambda}_i\gamma_{ki};\quad k=0,\ 1; \tag{5}$$

c) the maximum principle holds almost everywhere:

$$\max_{v\in\mathfrak{U}}G(t,\ v)=G(t,\ \hat{u}(t)), \tag{6}$$

where

$$G(t,\ v)=\hat{p}(t)\,F(t,\ v)-\sum_{i=0}^m\hat{\lambda}_i f_i(t,\ v); \tag{7}$$

d) the condition of concordance of signs holds:

$$\hat{\lambda}_i\geqslant 0,\quad i=1,\ \ldots,\ m; \tag{8}$$

e) the conditions of complementary slackness hold:

$$\hat{\lambda}_i[J_i(\hat{x}(\cdot),\ \hat{u}(\cdot))-c_i]=0,\quad i=1,\ \ldots,\ m. \tag{9}$$

2. If for an admissible process $(\hat{x}(\cdot),\ \hat{u}(\cdot))$ there exist $\hat{\lambda}_0>0$, $\hat{\lambda}\in\mathbf{R}^{m*}$ and an absolutely continuous function $\hat{p}(\cdot)\colon\Delta\to\mathbf{R}^{m*}$ such that conditions

(4)-(9) hold, then $(\hat{x}(\cdot), \hat{u}(\cdot))$ is an optimal process for problem (1)-(3).

Formally, conditions (4)-(9) of the present section are the same as the corresponding conditions in Section 4.2.2 (the reader can easily carry out the transformations from one group of conditions to the other). However, we remind the reader that the meaning of the conditions is somewhat different: in the case under consideration we only assume that the controls $\hat{u}(\cdot)$ are measurable, and consequently the necessary conditions (4)-(9) do not follow from the results of Section 4.2.

Moreover, the second part of the theorem yields sufficient conditions for the optimality, which was not considered in Section 4.2.

Proof. A) Reduction to a Lyapunov-Type Problem. Problem (1)-(3) was already reduced to a Lyapunov-type extremal problem in Section 4.3.1 [see problem (8) in Section 4.3.1] which differs from the problems of Section 4.3.3 because it may include inequality constraints $\geqslant$. However, this is inessential because the endpoint terms in problem (8) of Section 4.3.1 are linear, and therefore, changing the signs, when necessary, we again obtain convex functions.

Let us renumber the indices so that the inequality constraints occupy the last places in (3), and then, as in Section 3.2, let us put $\varepsilon_i = 1$ for $i = 0$ and for those i to which the signs $\leqslant$ and $=$ correspond in (3) and $\varepsilon_i = -1$ for those i to which the sign $\geqslant$ corresponds in (3). Further, we put $g_0(x(\cdot)) = \beta_0 x(t_0)$ and $g_i(x(\cdot)) = \varepsilon_i(\beta_i x(t_0) - c_i)$, $i \geqslant 1$.

Now problem (8) of Section 4.3.1 takes the standard Lyapunov-type form:

$$\tilde{J}_0(x(\cdot),\ u(\cdot)) = -\varepsilon_0 \int_\Delta G_0(t,\ u(t))\,dt + g_0(x(\cdot)) \to \inf, \tag{10}$$
$$\tilde{J}_i(x(\cdot),\ u(\cdot)) = -\varepsilon_i \int_\Delta G_i(t,\ u(t))\,dt + g_i(x(\cdot)) \begin{cases} \leqslant 0, & i = 1,\ \ldots,\ m', \\ = 0, & i = m'+1,\ \ldots,\ m, \end{cases}$$

where G_i are determined by formulas (5) of Section 4.3.1.

B) "Necessary." If $(\hat{x}(\cdot), \hat{u}(\cdot))$ is an optimal process, then, by the theorem of Section 4.3.3, there exist $\tilde{\lambda}_i$, nonnegative for $0 \leqslant i \leqslant m'$ and not all zero, such that

$$-\sum_{i=0}^{m} \varepsilon_i \tilde{\lambda}_i G_i(t,\ \hat{u}(t)) = \min_v \left(-\sum_{i=0}^{m} \varepsilon_i \tilde{\lambda}_i G_i(t,\ v) \right) \tag{11}$$

almost everywhere,

$$\sum_{i=0}^{m} \tilde{\lambda}_i g_i(\hat{x}(\cdot)) = \min_{x(\cdot)} \sum_{i=0}^{m} \hat{\lambda}_i g_i(x(\cdot)) \tag{12}$$

and

$$\tilde{\lambda}_i \tilde{J}_i(\hat{x}(\cdot),\ \hat{u}(\cdot)) = 0, \quad 1 \leqslant i \leqslant m. \tag{13}$$

It can readily be seen that condition (8) is satisfied if we put $\hat{\lambda}_i = \varepsilon_i \tilde{\lambda}_i$, and (13) implies (9).

Further, let us put

$$\hat{p}(t) = \sum_{i=0}^{m} \hat{\lambda}_i p_i(t), \tag{14}$$

where $p_i(\cdot)$ are determined by formulas (7) of Section 4.3.1, whence, by virtue of relations (6) of Section 4.3.1, it follows that

$$p_i(t_1) = -\gamma_{1i}, \quad p_i(t_0) = \gamma_{0i} - \beta_i. \tag{15}$$

By means of direct differentiation [or using formula (14) of Section 2.5.4], we verify that $\hat{p}(\cdot)$ is a solution of Eq. (4) and (5) holds for $k = 1$.

From (7), (14), and from formula (5) of Section 4.3.1, we obtain

$$G(t, v)=\hat{p}(t)F(t, v)-\sum_{i=0}^{m}\hat{\lambda}_i f_i(t, v)=\sum_{i=0}^{m}\varepsilon_i\tilde{\lambda}_i[p_i(t)F(t, v)-f_i(t, v)]=\sum_{i=0}^{m}\varepsilon_i\tilde{\lambda}_i G_i(t, v),$$

after which (11) goes into (6).

Finally, substituting the expressions of g_i taken from the definition into (12), we obtain

$$\left(\sum_{i=0}^{m}\varepsilon_i\tilde{\lambda}_i\beta_i\right)\hat{x}(t_0)=\min_{\Gamma}\left(\sum\varepsilon_i\tilde{\lambda}_i\beta_i\right)x.$$

This equality is only possible if $\sum_{i=0}^{m}\hat{\lambda}_i\beta_i=\sum_{i=0}^{m}\varepsilon_i\tilde{\lambda}_i\beta_i=0$, after which (5) with $k = 0$ follows from (14) and (15).

C) "Sufficient." Let us suppose that for an admissible process $(\hat{x}(\cdot), \hat{u}(\cdot))$ there are $\hat{\lambda}_0 > 0$, $\hat{\lambda}_i$, $\hat{p}(\cdot)$ such that conditions (4)-(9) hold. Further, let ε_i be the same as before, and let, as before, $\tilde{\lambda}_i = \varepsilon_i\hat{\lambda}_i$ and $p_i(\cdot)$ be determined by equalities (7) of Section 4.3.1 so that (15) takes place. Functions $p(\cdot)=\sum_{i=0}^{m}\hat{\lambda}_i p_i(\cdot)$ and $\hat{p}(\cdot)$ satisfy Eq. (4), and, according to (5) with $k = 1$, they coincide for $t = t_1$. By the uniqueness theorem, $p(t) \equiv \hat{p}(t)$, and consequently

$$G(t, v)\overset{(7)}{=}\hat{p}(t)F(t, v)-\sum_{i=0}^{m}\hat{\lambda}_i f_i(t, v)=$$

$$=\sum_{i=0}^{m}\varepsilon_i\tilde{\lambda}_i[p_i(t)F(t, v)-f_i(t, v)]=\sum_{i=0}^{m}\varepsilon_i\tilde{\lambda}_i G_i(t, v) \tag{16}$$

and

$$\sum_{i=0}^{m}\hat{\lambda}_i\beta_i\overset{(15)}{=}\sum_{i=0}^{m}\hat{\lambda}_i(\gamma_{0i}-p_i(t_0))=\sum_{i=0}^{m}\hat{\lambda}_i\gamma_{0i}-\sum_{i=0}^{m}\hat{\lambda}_i p_i(t_0)\overset{(5)}{=}\hat{p}(t_0)-p(t_0)=0. \tag{17}$$

Relations (16) and (6) imply (11), (9) implies (13), and (8) together with the equalities $\hat{\lambda}_i = \varepsilon_i\tilde{\lambda}_i$ implies that $\tilde{\lambda}_i \geqslant 0$, $i = 1,\ldots,m$. Moreover, from (17) and the definition of $g_i(\cdot)$, it follows that

$$\sum_{i=0}^{m}\tilde{\lambda}_i g_i(x(\cdot))=\sum_{i=0}^{m}\tilde{\lambda}_i\varepsilon_i\beta_i x(t_0)-\sum_{i=1}^{m}\tilde{\lambda}_i\varepsilon_i c_i=\left(\sum_{i=0}^{m}\hat{\lambda}_i\beta_i\right)x(t_0)-\sum_{i=1}^{m}\hat{\lambda}_i c_i=-\sum_{i=1}^{m}\hat{\lambda}_i c_i,$$

and therefore (12) also takes place. By the theorem of Section 4.3.3, conditions (11)-(13) together with the inequalities $\tilde{\lambda}_0 > 0$ and $\tilde{\lambda}_i \geqslant 0$ are sufficient for the optimality of $(\hat{x}(\cdot), \hat{u}(\cdot))$. ■

4.4. Application of the General Theory to the Simplest Problem of the Classical Calculus of Variations

In this section the general theory is applied to the simplest problem of the classical calculus of variations upon which so much emphasis is laid in mathematical textbooks. In the same manner it is also possible to derive necessary and sufficient conditions for a weak extremum and necessary conditions for a strong extremum in the general Lagrange problem. However, this would require a rather lengthy exposition without introducing anything essentially new. Therefore we shall confine ourselves to a special case in which an exhaustive investigation can be completed and the relationship between the old and the new theories can be demonstrated.

4.4.1. Euler Equation. Weierstrass' Condition. Legendre's Condition.

By the *simplest (vector) problem of the classical calculus of variations* we shall mean (cf. Section 1.2.6) the following extremal problem:

$$\mathcal{J}(x(\cdot))=\int_{t_0}^{t_1} L(t,\ x(t),\ \dot{x}(t))\,dt \to \text{extr}, \tag{1}$$

$$x(t_0)=x_0,\quad x(t_1)=x_1. \tag{2}$$

Here the function $L: V \to \mathbf{R}$ is assumed to be at least continuous in an open set $V \subset \mathbf{R}\times\mathbf{R}^n\times\mathbf{R}^n$, and x_0, x_1, t_0, and t_1 are fixed.

Two versions of problem (1), (2) are investigated. In the first of them we suppose that $x(\cdot)$ belongs to the space $C^1([t_0, t_1], \mathbf{R}^n)$, and in this case a local extremum is said to be *weak* (cf. Section 1.4.1). In the second version the function $x(\cdot)$ belongs to the space $KC^1([t_0, t_1], \mathbf{R}^n)$ of functions with piecewise-smooth derivatives endowed with the norm $\|x(\cdot)\| = \sup_{t_0 \leqslant t \leqslant t_1} |x(t)|$, and in this case a local extremum is said to be *strong*.

Exercise 1. Verify that $KC^1([t_0, t_1], \mathbf{R}^n)$ is a normed space but not a Banach space (i.e., it is not complete).

In both versions a function $x(\cdot)$ is said to be *admissible* if its *extended graph* lies in V: $\{(t,\ x(t),\ \dot{x}(t))\,|\,t_0 \leqslant t \leqslant t_1\} \subset V$, and the boundary conditions (2) are satisfied. The set of the admissible functions $x(\cdot)$ will be denoted by $\mathcal{V}^\partial$. The notation and the terminology introduced here and henceforth are retained throughout the present section. For brevity, problem (1), (2) will be called the *simplest problem* without mentioning the other details.

Exercise 2. Prove that

$$\inf_{\mathcal{V}^\partial \cap C^1([t_0,\ t_1],\ \mathbf{R}^n)} \mathcal{J}(x(\cdot)) = \inf_{\mathcal{V}^\partial \cap KC^1([t_0,\ t_1],\ \mathbf{R}^n)} \mathcal{J}(x(\cdot)).$$

Hint. Make use of the lemma on rounding the corners (Section 1.4.3).

If $x(\cdot)$ yields a strong extremum in problem (1) and if function $\dot{x}(\cdot)$ is continuous, then $x(\cdot)$, obviously, yields a weak extremum as well. Therefore, necessary conditions for a weak extremum are also necessary for a strong extremum [in the case when $\dot{x}(\cdot)$ is continuous]. Some necessary conditions are already known. In Section 1.4 we used elementary methods to derive the *Euler equation* which is a necessary condition for a weak extremum and *Weierstrass' condition* (for $n = 1$) which is a necessary condition for a strong minimum. Now we shall consider the simplest problem from the general viewpoint. On the one hand, problem (1), (2) is reducible to the following Lagrange problem:

$$\mathcal{J}(x(\cdot),\ u(\cdot))=\int_{t_0}^{t_1} L(t,\ x(t),\ u(t))\,dt \to \text{extr}, \tag{3}$$

$$\dot{x}=u,\quad x(t_0)=x_0,\quad x(t_1)=x_1.$$

If we suppose that not only the function L but also its derivatives with respect to x and u are continuous in V, then the Euler—Lagrange theorem of Section 4.1 can be applied to (3). In this case the weak optimality of a process $(\hat{x}(\cdot),\ \hat{u}(\cdot))$ in problem (3) means exactly the same as the weak extremality of $\hat{x}(\cdot)$ in problem (1), (2).

If $\hat{x}(\cdot)$ yields a weak extremum in problem (1), (2) then, by the Euler—Lagrange theorem, there exist $\hat{\lambda}_0$ and functions $\hat{p}(\cdot) \in C^1([t_0,\ t_1],\ \mathbf{R}^{n*})$ such that

$$\dot{\hat{p}}(t)=\hat{\lambda}_0 L_x(t,\ \hat{x}(t),\ \dot{\hat{x}}(t)),\quad \hat{p}(t)=\hat{\lambda}_0 L_{\dot{x}}(t,\ \hat{x}(t),\ \dot{\hat{x}}(t)).$$

If $\hat{\lambda}_0 = 0$, then all the Lagrange multipliers are zero, and therefore we can put $\hat{\lambda}_0 = 1$. Eliminating $\hat{p}(\cdot)$, we arrive at the system of Euler's equations

$$\frac{d}{dt}\hat{L}_{\dot{x}}(t)=\hat{L}_x(t) \tag{4}$$

[as usual, here $\hat{L}_{\dot{x}}(t) = L_{\dot{x}}(t, \hat{x}(t), \dot{\hat{x}}(t))$, etc.].

On the other hand, problem (3) (for definiteness, we assume that this is a minimum problem; in a maximum problem L should be replaced by –L) can be regarded as an optimal control problem with $\mathfrak{U} = \mathbf{R}^n$. If L and L_x are assumed to be jointly continuous with respect to the arguments, then the maximum principle of Section 4.2 can be applied to (1), (2). In this case the optimality of a process $(\hat{x}(\cdot), \hat{u}(\cdot))$ means that

$$x(\cdot) \in KC^1([t_0, t_1], \mathbf{R}^n)$$

yields a strong minimum in problem (1), (2).

From the maximum principle it follows that there exist $\hat{\lambda}_0 \geqslant 0$ and a piecewise-differentiable function $\hat{p}(\cdot):[t_0, t_1] \to \mathbf{R}^{n*}$ such that

$$\dot{\hat{p}}(t) = \hat{\lambda}_0 \hat{L}_x(t), \tag{5}$$

$$\min_{u \in \mathbf{R}^n} \{\hat{\lambda}_0 L(t, \hat{x}(t), u) - \hat{p}(t) u\} = \hat{\lambda}_0 L(t, \hat{x}(t), \dot{\hat{x}}(t)) - \hat{p}(t)\dot{\hat{x}}(t). \tag{6}$$

Here the equality $\hat{\lambda}_0 = 0$ is impossible because from (5) it follows that $\hat{\lambda}_0 = 0 \Rightarrow \hat{p}(t) \equiv \text{const}$, and, moreover, $\hat{p}(t) \neq 0$ (because, if otherwise, then all the Lagrange multipliers are equal to zero). However, if $\hat{\lambda}_0 = 0$ and $\hat{p} \neq 0$, then (6) cannot hold (a nontrivial linear function has no minimum in $\mathbf{R}^n$). Thus $\hat{\lambda}_0 \neq 0$, and we can assume that $\hat{\lambda}_0 = 1$.

Condition (6) can be rewritten in the subdifferential form:

$$\hat{p}(t) \in (\partial_{\dot{x}} L)(t, \hat{x}(t), \dot{\hat{x}}(t)). \tag{7}$$

If L is differentiable with respect to $\dot{x}$ too, then (6) implies the equality $\hat{p}(t) = \hat{L}_{\dot{x}}(t)$, and, on the one hand, we again arrive at the Euler equation (4), and, on the other hand, from (6) we obtain *Weierstrass' necessary condition*

$$L(t, \hat{x}(t), u) - L(t, \hat{x}(t), \dot{\hat{x}}(t)) - L_{\dot{x}}(t, \hat{x}(t), \dot{\hat{x}}(t))(u - \dot{\hat{x}}(t)) \geqslant 0, \quad \forall u \in \mathbf{R}^n, \quad \forall t \in [t_0, t_1]. \tag{8}$$

Let us introduce the *Weierstrass function* $\mathscr{E}$ (the letter $\mathscr{E}$ symbolizes the Latin *excessus* – "excess"):

$$\mathscr{E}(t, x, \xi, u) = L(t, x, u) - L(t, x, \xi) - L_{\dot{x}}(t, x, \xi)(u - \xi). \tag{9}$$

Then Weierstrass' condition (8) can be written in the form

$$\mathscr{E}(t, \hat{x}(t), \dot{\hat{x}}(t), u) \geqslant 0, \quad \forall u \in \mathbf{R}^n, \ \forall t \in [t_0, t_1]. \tag{10}$$

If we assume that not only L and $L_{\dot{x}}$ but also $L_{\dot{x}\dot{x}}$ is continuous in V, then (8) implies the inequality

$$\hat{L}_{\dot{x}\dot{x}}(t) \geqslant 0, \quad t \in [t_0, t_1]. \tag{11}$$

Condition (11) is called *Legendre's condition*.

Thus, the Euler equation is a necessary condition for both a weak extremum and a strong extremum. Moreover, from the maximum principle we have derived two more necessary conditions for a strong minimum: Weierstrass' condition and Legendre's condition (in the next section we shall see that the last condition must hold for a weak minimum too).

4.4.2. Second-Order Conditions for a Weak Extremum. Legendre's and Jacobi's Conditions.

LEMMA. Let the function L be continuous in V, and, moreover, let its derivatives L_{x_i}, $L_{\dot{x}_i}$, $L_{x_i x_j}$, $L_{\dot{x}_i x_j}$, $L_{\dot{x}_i \dot{x}_j}$ exist and be continuous in V.

If $\hat{x}(\cdot) \in C^1([t_0, t_1], \mathbf{R}^n)$ satisfies the Euler equation, then the following asymptotic formula holds for $\|h(\cdot)\|_{C^1} \to 0$:

$$\mathcal{J}(\hat{x}(\cdot)+h(\cdot)) = \mathcal{J}(\hat{x}(\cdot)) + \int_{t_0}^{t_1} \{h(t)^T \hat{L}_{xx}(t) h(t) + $$

$$+ 2h^T(t) \hat{L}_{x\dot{x}}(t) \dot{h}(t) + \dot{h}(t)^T \hat{L}_{\dot{x}\dot{x}}(t) \dot{h}(t)\} dt + o\left(\int_{t_0}^{t_1} [|h(t)|^2 + |\dot{h}(t)|^2] dt \right), \tag{1}$$

$$\forall h(\cdot) \in C^1([t_0, t_1], \mathbf{R}^n), \quad h(t_0) = h(t_1) = 0.$$

Proof. As in Section 2.4.2, let us embed $\hat{x}(\cdot)$ in a compact set entirely contained in V and use the uniform continuity of the second derivatives of the function L on that set to estimate the remainder term in Taylor's formula. This yields the expansion

$$L(t, \hat{x}(t)+h(t), \dot{\hat{x}}(t)+\dot{h}(t)) = \hat{L}(t) + \hat{L}_x(t) h(t) + \hat{L}_{\dot{x}}(t) \dot{h}(t) +$$

$$+ \frac{1}{2}\{h^T(t) \hat{L}_{xx}(t) h(t) + 2h^T(t) \hat{L}_{x\dot{x}}(t) \dot{h}(t) +$$

$$+ \dot{h}^T(t) \hat{L}_{\dot{x}\dot{x}}(t) \dot{h}(t)\} + \varepsilon(t, h(\cdot)) [|h(t)|^2 + |\dot{h}(t)|^2], \tag{2}$$

where $\varepsilon(t; h(\cdot)) \to 0$ uniformly on Δ for $\|h(\cdot)\|_{C^1} \to 0$.

Integrating (2) from t_0 to t_1 and then, as usual, performing integration by parts in the linear term involving $\dot{h}(t)$ and taking into consideration the Euler equation, we obtain (1). ■

If $\hat{x}(\cdot)$ yields a weak minimum in problem (1), (2) of Section 4.4.1, then replacing $h(\cdot)$ in (1) by $\alpha h(\cdot)$ and computing the limit $\lim [\mathcal{J}(\hat{x}(\cdot)+\alpha h(\cdot)) - \mathcal{J}(\hat{x}(\cdot))]/\alpha^2$ for $\alpha \downarrow 0$, we obtain

$$Q(h(\cdot)) = \int_{t_0}^{t_1} \{h^T(t) \hat{L}_{xx}(t) h(t) + 2h^T(t) \hat{L}_{x\dot{x}}(t) \dot{h}(t) + \dot{h}^T(t) \hat{L}_{\dot{x}\dot{x}}(t) \dot{h}(t)\} dt \geqslant 0, \tag{3}$$

$$\forall h(\cdot) \in C^1([t_0, t_1], \mathbf{R}^n), \quad h(t_0) = h(t_1) = 0.$$

This means that the function $\hat{h}(\cdot) \equiv 0$ yields a weak minimum in the following secondary extremal problem for the quadratic functional $Q(h(\cdot))$:

$$Q(h(\cdot)) \to \inf, \quad h(t_0) = h(t_1) = 0. \tag{4}$$

By the lemma on rounding the corners (Section 1.4.3), the infimum of the functional Q on the set of admissible functions belonging to the space $C^1([t_0, t_1], \mathbf{R}^n)$ coincides with its infimum on the set of admissible functions belonging to the space $KC^1([t_0, t_1], \mathbf{R}^n)$. Therefore, $\hat{h}(\cdot) \equiv 0$ also yields a strong minimum in problem (4). (It should be noted that, by virtue of the homogeneity of Q, the local minima turn out to be global, and the distinction between the weak and the strong minima reduces to the change of the set of admissible functions.)

According to what was proved in Section 4.4.1, for problem (4) there must hold Legendre's condition which obviously has the same form as in the case of problem (1), (2):

$$\hat{L}_{\dot{x}\dot{x}}(t) \geqslant 0, \quad \forall t \in [t_0, t_1]. \tag{5}$$

Thus, Legendre's condition turns out to be necessary not only for a strong minimum, as in Section 4.4.1, but also for a weak minimum.

The Euler equation for the secondary problem (4) has the form

$$\frac{d}{dt}[\hat{L}_{\dot{x}\dot{x}}(t) \dot{h}(t) + \hat{L}_{\dot{x}x}(t) h(t)] = \hat{L}_{x\dot{x}}(t) \dot{h}(t) + \hat{L}_{xx}(t) h(t), \tag{6}$$

and is called the *Jacobi equation* of problem (1), (2) of Section 4.4.1.

Definition. A point τ is said to be *conjugate to the point* t_0 (with respect to the functional Q) if the Jacobi equation (6) possesses a solution $\bar{h}(\cdot)$ for which $\bar{h}(t_0) = \bar{h}(\tau) = 0$ while†

$$\hat{L}_{\dot{x}\dot{x}}(\tau)\dot{\bar{h}}(\tau) \neq 0. \tag{7}$$

THEOREM 1 (Necessary Conditions for a Weak Extremum in the Simplest Problem). Let the function L satisfy the conditions of the lemma.

If $\hat{x}(\cdot)$ yields a weak minimum in problem (1), (2) of Section 4.4.1, then:

1) $\hat{x}(\cdot)$ satisfies the Euler equation (4) of Section 4.4.1;

2) Legendre's condition (5) holds;

3) Jacobi's condition holds: in the interval (t_0, t_1) there are no points conjugate to t_0;

4) inequality (3) holds.

In the case when $\hat{x}(\cdot)$ yields a weak maximum, the sign $\geqslant$ in (3) and (5) should be replaced by $\leqslant$.

Proof. Assertions 1), 2), and 4) are already proved [the assertion 1) was proved in the foregoing section]. Therefore, it only remains to prove 3). Let us suppose that the contrary is true, i.e., let $\tau \in (t_0, t_1)$ be a conjugate point to t_0, and let $\bar{h}(\cdot)$ be the corresponding solution of the Jacobi equation: $\bar{h}(t_0) = \bar{h}(\tau) = 0$ and (7) takes place. Let us put

$$\tilde{h}\, t() = \begin{cases} \bar{h}(t), & t_0 \leqslant t \leqslant \tau, \\ 0, & \tau \leqslant t \leqslant t_1. \end{cases} \tag{8}$$

Then

$$Q(\tilde{h}(\cdot)) = \int_{t_0}^{\tau} \{\bar{h}(t)^{\mathrm{T}} \hat{L}_{xx}(t)\bar{h}(t) + 2\bar{h}(t)^{\mathrm{T}} \hat{L}_{x\dot{x}}(t)\dot{\bar{h}}(t) + \dot{\bar{h}}(t)^{\mathrm{T}} \hat{L}_{\dot{x}\dot{x}}(t)\dot{\bar{h}}(t)\}\, dt =$$

$$= \int_{t_0}^{\tau} \bar{h}(t)^{\mathrm{T}} [\hat{L}_{xx}(t)\bar{h}(t) + \hat{L}_{x\dot{x}}(t)\dot{\bar{h}}(t)]\, dt + \int_{t_0}^{\tau} \dot{\bar{h}}(t)^{\mathrm{T}} [\hat{L}_{\dot{x}\dot{x}}(t)\dot{\bar{h}}(t) + \hat{L}_{\dot{x}x}(t)\bar{h}(t)]\, dt$$

[here we have used the fact that $\bar{h}(t)^{\mathrm{T}}\hat{L}_{x\dot{x}}(t)\dot{\bar{h}}(t) = \dot{\bar{h}}(t)^{\mathrm{T}}\hat{L}_{\dot{x}x}(t)\bar{h}(t)$]. Now, recalling that $\bar{h}(\cdot)$ satisfies Eq. (6) we integrate by parts in the first integral [taking into account the equalities $\bar{h}(t_0) = \bar{h}(\tau) = 0$], after which the integrals mutually cancel. Hence, $Q(\tilde{h}(\cdot)) = 0$, which means that $\tilde{h}(\cdot)$ [along with the function $\hat{h}(\cdot) \equiv 0$] yields a strong minimum in problem (4). Let us again apply the maximum principle of Section 4.2. According to the principle, there is $\hat{\lambda}_0 \geqslant 0$ and a piecewise-differentiable function $\tilde{p}(\cdot)$: $[t_0, t_1] \to \mathbf{R}^{n*}$ such that

$$\begin{gathered} \dot{\tilde{p}}(t) = 2\hat{\lambda}_0 \hat{L}_{x\dot{x}}(t)\dot{\tilde{h}}(t) + 2\hat{\lambda}_0 \hat{L}_{xx}(t)\tilde{h}(t), \\ \min_{u \in \mathbf{R}^n} \{\hat{\lambda}_0 u^{\mathrm{T}} \hat{L}_{\dot{x}\dot{x}}(t) u + 2\hat{\lambda}_0 \tilde{h}(t)^{\mathrm{T}} \hat{L}_{x\dot{x}}(t) u - \tilde{p}(t) u\} = \\ = \hat{\lambda}_0 \dot{\tilde{h}}^{\mathrm{T}}(t) \hat{L}_{\dot{x}\dot{x}}(t)\dot{\tilde{h}}(t) + 2\hat{\lambda}_0 \tilde{h}(t)^{\mathrm{T}} \hat{L}_{x\dot{x}}(t)\dot{\tilde{h}}(t) - \tilde{p}(t)\dot{\tilde{h}}(t). \end{gathered} \tag{9}$$

---

†In courses in the calculus of variations conjugate points are usually defined under the assumption that the strengthened Legendre condition $L_{\dot{x}\dot{x}}(t) > 0$ holds. Under this assumption we have $L_{\dot{x}\dot{x}}(\tau)\dot{\bar{h}}(\tau) = 0 \Rightarrow \dot{\bar{h}}(\tau) = 0$, and since $\bar{h}(\tau) = 0$, the uniqueness theorem implies that $\bar{h}(\cdot) \equiv 0$. Therefore, condition (7) is equivalent to the condition of the nontriviality of $\bar{h}(\cdot)$ and is usually replaced by the latter.

Using the same argument as above, we prove that $\hat{\lambda}_0 \neq 0$, and therefore it can be assumed that $\hat{\lambda}_0 = 1/2$.

From relations (9) it follows that

$$\tilde{p}(t) = \hat{L}_{\dot{x}\dot{x}}(t)\dot{\tilde{h}}(t) + \hat{L}_{\dot{x}x}(t)\tilde{h}(t). \tag{10}$$

However, $\tilde{h}(\cdot) \equiv 0$ for $t \geqslant \tau$, and therefore from (10) it follows that $\tilde{p}(\tau + 0) = 0$. On the other hand, by virtue of (7), we have $\tilde{p}(\tau - 0) = \hat{L}_{\dot{x}\dot{x}}(\tau)\dot{\tilde{h}} \times (\tau - 0) = \hat{L}_{\dot{x}\dot{x}}(\tau)\dot{\bar{h}}(\tau) \neq 0$. Therefore, the function $\tilde{p}$ is discontinuous whereas, by virtue of the maximum principle, it must be continuous. This contradiction proves that Jacobi's condition is fulfilled. ■

Thus, we have shown that all the necessary conditions for an extremum in the simplest problem which are usually presented in courses of calculus of variations, namely, Euler's, Legendre's, Jacobi's, and Weierstrass' conditions are consequences of the Pontryagin maximum principle. Now we proceed to sufficient conditions for a weak extremum.

THEOREM 2 (Sufficient Conditions for a Weak Extremum in the Simplest Problem). Let the function L satisfy the conditions of the lemma, let $\hat{x}(\cdot) \in C^1([t_0, t_1], \mathbf{R}^n)$, and let the extended graph of the function $\hat{x}(\cdot)$ lie in V: $\{t, \hat{x}(t), \dot{\hat{x}}(t)) \mid t_0 \leqslant t \leqslant t_1\} \subset V$.

If:

1) the function $\hat{x}(\cdot)$ satisfies the Euler equation [Eq. (4) of Section 4.4.1];
2) the boundary conditions $\hat{x}(t_0) = x_0$, $\hat{x}(t_1) = x_1$ are satisfied;
3) the strengthened Legendre condition

$$L_{\dot{x}\dot{x}}(t, \hat{x}(t), \dot{\hat{x}}(t)) > 0, \quad \forall t \in [t_0, t_1], \tag{11}$$

holds;

4) the strengthened Jacobi condition holds: in the half-open interval $(t_0, t_1]$ there are no conjugate points to t_0, then $\hat{x}(\cdot)$ yields a strict weak minimum [or maximum if the sign > in (11) is replaced by <] in the simplest problem (1), (2) of Section 4.4.1.

The proof of this theorem will be presented in Section 4.4.5.

4.4.3. Hamiltonian Formalism. The Theorem on the Integral Invariant. We already mentioned in Section 4.1.1 that the Euler–Lagrange equations together with the equation of the differential constraint can be brought to a *Hamiltonian* (or *canonical*) *system* of the form

$$\dot{x} = \mathcal{H}_p(t, x, p), \quad \dot{p} = -\mathcal{H}_x(t, x, p), \tag{1}$$

where the function $\mathcal{H}$ is called the *Hamiltonian* (or the *Hamiltonian function*) of the original problem. Here we shall study this transformation in greater detail under the conditions of the simplest problem. The passage from the Euler equation of this problem [Eq. (4) of Section 4.4.1] to the canonical system (1) is carried out with the aid of the Legendre transformation. We shall begin with the study of this transformation.

Definition 1. Let a function $l: v \to \mathbf{R}$ belong to the class $C^1(v)$ in an open set v of a normed space X, and let its derivative l' specify a one-to-one mapping of v onto $d = \{p \mid p = l'(\xi), \ \xi \in v\} \subset X^*$; let $\Xi: d \to v$ be the inverse mapping to l'. The function $h: d \to \mathbf{R}$ determined by the equality

$$h(p) = \langle p, \Xi(p) \rangle - l(\Xi(p)) \tag{2}$$

is called the *classical Legendre transform* of the function l.

Proposition 1. Let $l: v \to \mathbf{R}$ belong to the class $C^2(v)$ in an open convex set of a normed space X. If $l''(\xi)$ is positive-definite, i.e.,

$$d^2 l(\xi;\ \eta) = l''(\xi)[\eta,\ \eta] > 0, \quad \forall \xi \in v, \quad \forall \eta \in X, \quad \eta \neq 0, \tag{3}$$

then the Legendre transform $h(\cdot)$ of the function $l(\cdot)$ exists and coincides in d with the Young–Fenchel transform (Section 2.6.3) of the function

$$\tilde{l} = \begin{cases} l(\xi), & \xi \in v, \\ +\infty, & \xi \notin v, \end{cases} \tag{4}$$

which is convex.

Proof. The convexity of $\tilde{l}$ was proved in Section 2.6.1 [the example 2)]; the fact that the mapping $\xi \to l'(\xi)$ is one-to-one can be proved by analogy with Section B) of the proof of Proposition 2 below.

According to the definition of the conjugate function (Section 2.6.3),

$$\tilde{l}^*(p) = \sup_{\xi} \{\langle p,\ \xi\rangle - \tilde{l}(\xi)\} = -\inf \{\tilde{l}(\xi) - \langle p,\ \xi\rangle\}.$$

If $p \in d$, then $p = l'(\bar{\xi}) = \tilde{l}'(\bar{\xi})$ for some $\bar{\xi} \in v$. Consequently, $0 = (\tilde{l}'(\bar{\xi}) - p) \in \partial[\tilde{l}(\cdot) - \langle p,\ \cdot\rangle](\bar{\xi})$, which implies that $\bar{\xi}$ is a point of minimum of the function $\tilde{l}(\cdot) - \langle p,\ \cdot\rangle$ and

$$\tilde{l}^*(p) = -(\tilde{l}(\bar{\xi}) - \langle p,\ \bar{\xi}\rangle) = \langle p,\ \bar{\xi}\rangle - \tilde{l}(\bar{\xi}).$$

Since $p = l'(\bar{\xi})$, we have $\bar{\xi} = \Xi(p)$, and, according to the definition, $\tilde{l}^*(p) = h(p)$. ∎

Recalling Young's inequality (Section 2.6.3), we arrive at

COROLLARY 1. Under the conditions of Proposition 1,

$$l(\xi) + h(p) \geqslant \langle p,\ \xi\rangle, \quad \forall p \in d, \quad \forall \xi \in v. \tag{5}$$

Given the Lagrangian $L(t, x, \dot{x})$ of the simplest problem, to obtain the Hamiltonian of the problem we have to perform the Legendre transformation with respect to the argument $\dot{x}$. Consequently,

$$\mathscr{H}(t,\ x,\ p) = p\dot{x} - L(t,\ x,\ \dot{x}), \tag{6}$$

where $\dot{x}$ must be eliminated using the relation

$$p = L_{\dot{x}}(t,\ x,\ \dot{x}), \tag{7}$$

and hence $\mathscr{H}$ is defined correctly by formula (6) if the mapping $\dot{x} \to L_{\dot{x}}(t, x, \dot{x})$ is one-to-one. [The quantity $\dot{x}$ in formulas (6) and (7) plays the role of an independent argument; to avoid the confusion of $\dot{x}$ with dx/dy we shall denote the former as $\dot{x} = \xi$ although for the corresponding derivatives the notation $L_{\dot{x}}$, $\partial L/\partial \dot{x}_i$, etc. will be retained.]

Proposition 2. Let the function $(t, x, \xi) \to L(t, x, \xi)$ belong to the class $C^r(V)$, $r \geqslant 2$, in an open set $V \subset \mathbf{R} \times \mathbf{R}^n \times \mathbf{R}^n$.

If:

a) the second derivative $L_{\dot{x}\dot{x}}$ is nondegenerate in V, i.e.,

$$\det\left(\frac{\partial^2 L}{\partial \dot{x}_i \partial \dot{x}_j}(t,\ x,\ \xi)\right) \neq 0, \quad \forall (t,\ x,\ \xi) \in V;$$

b) the mapping P specified by the equality $P(t, x, \xi) = (t, x, L_{\dot{x}}(t, x, \xi))$ maps V onto $D \subset \mathbf{R} \times \mathbf{R}^n \times \mathbf{R}^{n*}$ in a one-to-one manner, then D is open, $P^{-1} \in C^{r-1}(D)$, and the function $\mathscr{H}\colon D \to \mathbf{R}$ determined by equalities (6) and (7) belongs to the class $C^r(D)$.

2) If the section $V_{t,x} = \{\xi \mid (t,\ x,\ \xi) \in V\}$ is convex for any (t, x) and the matrix $L_{\dot{x}\dot{x}}(t, x, \xi)$ is positive-definite for any $(t,\ x,\ \xi) \in V$, then the

conditions a) and b) are fulfilled and the assertion of Section 1) holds.

Proof. The inverse mapping of P, which exists according to b), is obviously of the form $(t, x, p) \to (t, x, \Xi(t, x, p))$, and hence

$$p = L_{\dot{x}}(t, x, \xi) \Leftrightarrow \xi = \Xi(t, x, p). \tag{8}$$

Since $L \in C^r(V)$, $r \geqslant 2$, we have $P \in C^{r-1}(V) \subset C^1(V)$. The Jacobi matrix of the mapping P has the form

$$\begin{pmatrix} 1 & 0 & 0 \\ 0 & E & 0 \\ L_{\dot{x}t} & L_{\dot{x}x} & L_{\dot{x}\dot{x}} \end{pmatrix},$$

and, according to a), its determinant, equal to $\det L_{\dot{x}\dot{x}}$, is nonzero.

Let $(t_0, x_0, p_0) = P(t_0, x_0, \xi_0)$ be an arbitrary point of D. By the inverse mapping theorem, there exists a neighborhood $U \ni (t_0, x_0, \xi_0)$ whose image $P(U) \subset D$ is a neighborhood of (t_0, x_0, p_0). Consequently, D is open. By the same theorem, the mapping P^{-1} (which is determined uniquely) coincides on P(U) with a mapping of class C^1. Since $(t_0, x_0, p_0) \in D$ is an arbitrary point, we have $P^{-1} \in C^1(D)$. Further, from (8) it is seen that $\Xi(t, x, p)$, as an implicit function, is determined by the equation $F(t, x, p, \xi) = p - L_{\dot{x}}(t, x, \xi) = 0$. This equation being formed by functions of class C^{r-1}, the function Ξ and, together with it, the function P^{-1} belong to the class $C^{r-1}(d)$ (see the remark in Section 2.3.4 made after the implicit function theorem).

From (6), (7), and (8) we derive the identity

$$\mathscr{H}(t, x, p) = p\Xi(t, x, p) - L(t, x, \Xi(t, x, p)). \tag{9}$$

The differentiation of (9) results in

$$\mathscr{H}_t = p\Xi_t - L_t - L_{\dot{x}}\Xi_t = [p - L_{\dot{x}}(t, x, \Xi)]\Xi_t - L_t = -L_t(t, x, \Xi(t, x, p)), \tag{10}$$

$$\mathscr{H}_x = p\Xi_x - L_x - L_{\dot{x}}\Xi_x = -L_x(t, x, \Xi(t, x, p)), \tag{11}$$

$$\mathscr{H}_p = \Xi + p\Xi_p - L_{\dot{x}}\Xi_p = \Xi(t, x, p). \tag{12}$$

From formulas (10)-(12) it is seen that the first derivatives of the function $\mathscr{H}$ are compositions of functions of class C^{r-1} and consequently they themselves belong to that class. Therefore, $\mathscr{H} \in C^r(D)$.

B) If the matrix $L_{\dot{x}\dot{x}}$ is positive-definite, then, by Sylvester's theorem [7], $\det L_{\dot{x}\dot{x}} > 0$, and hence a) holds. Let the sections $V_{t,x}$ be convex. Then, given any pair of points $(t, x, \xi_k) \in V$, $k = 1, 2$, we have $(t, x, \alpha\xi_1 + (1-\alpha)\xi_2) \in V, \alpha \in [0, 1]$. Denoting $p_k = L_{\dot{x}}(t, x, \xi_k)$, we write

$$p_1 - p_2 = L_{\dot{x}}(t, x, \xi_1) - L_{\dot{x}}(t, x, \xi_2) = \int_0^1 \frac{d}{d\alpha} L_{\dot{x}}(t, x, \alpha\xi_1 + (1-\alpha)\xi_2)\, d\alpha =$$

$$= \int_0^1 L_{\dot{x}\dot{x}}(t, x, \alpha\xi_1 + (1-\alpha)\xi_2)(\xi_1 - \xi_2)\, d\alpha.$$

If $\xi_1 \neq \xi_2$ while $P(t, x, \xi_1) = (t, x, p_1) = (t, x, p_2) = P(t, x, \xi_2)$, then $p_1 = p_2$ and

$$0 = \langle p_1 - p_2, \xi_1 - \xi_2 \rangle = \sum_{i=1}^{n} (p_{1i} - p_{2i})(\xi_{1i} - \xi_{2i}) =$$

$$= \int_0^1 \sum_{i,j=1}^{n} \frac{\partial^2 L}{\partial \dot{x}_i \partial \dot{x}_j}(t, x, \alpha\xi_1 + (1-\alpha)\xi_2)(\xi_{1i} - \xi_{2i})(\xi_{1j} - \xi_{2j})\, d\alpha > 0$$

because, by the hypothesis, the integrand function is positive ($L_{\dot{x}\dot{x}}$ is positive-definite). This contradiction shows that $p_1 = p_2 \Rightarrow \xi_1 = \xi_2$, i.e., b) holds. ■

Recalling Corollary 1, we obtain

COROLLARY 2. Under the conditions of the second part of Proposition 2, for any $(t, x, \xi)\in V$, $(t, x, p)\in D$, the inequality

$$\mathscr{H}(t, x, p)+L(t, x, \xi)\geqslant\langle p, \xi\rangle$$

holds.

Now we proceed to the relationship between the Euler equation

$$\frac{d}{dt}L_{\dot{x}}(t, x, \dot{x})=L_x(t, x, \dot{x}) \tag{13}$$

and the Hamiltonian system (1). When saying that $x:\Delta\to\mathbf{R}^n$ is a solution of Eq. (3) [or that a pair $(x, p):\Delta\to\mathbf{R}^n\times\mathbf{R}^{n*}$ is a solution of system (1)] we shall mean, without any special stipulation, that $(t, x(t), \dot{x}(t))\in V$ for all $t\in\Delta$ [or, accordingly, $(t, x(t), p(t))\in D$]. For brevity, a solution $x:\Delta\to\mathbf{R}^n$ of the Euler equation (13) and also its graph $\{(t, x(t))\,|\,t\in\Delta\}$ will be called an *extremal*, and a solution $(x, p):\Delta\to\mathbf{R}^n\times\mathbf{R}^{n*}$ of the Hamiltonian system (1) and its graph $\{(t, x(t), p(t))\,|\,t\in\Delta\}$ will be called a *canonical extremal* (it will always be clear from the context whether the functions or their graphs are meant).

Proposition 3. Let $L\in C^2(V)$, and let the function L satisfy the conditions a) and b) of Proposition 2.

A pair $(x(\cdot), p(\cdot))$ is a canonical extremal if and only if $x(\cdot)$ is an extremal and

$$p(t)\equiv L_{\dot{x}}(t, x(t), \dot{x}(t)). \tag{14}$$

Proof. A) Let $x(\cdot)$ be an extremal. If $p(\cdot)$ is determined by equality (14), then, according to (8), we have $\dot{x}(t) = \Xi(t, x(t), p(t))$. We see now that the first of Eq. (1) follows from (12) and the second follows from (13).

B) Let $(x(\cdot), p(\cdot))$ be a canonical extremal. Then (12), (8), and the first of Eqs. (1) imply (14), and then, according to (11), the second equation (1) turns into (13). ■

Remark. The assumption that $L\in C^2(V)$ is stronger than it is in fact necessary here: as in Section 4.2.2, it would be sufficient to require the existence and the continuity of the first and the second derivatives only with respect to $\dot{x}_i$ and x_j. The reader can verify this remark by making the corresponding changes in the statements of the implicit function and the inverse function theorems of Section 2.3.4.

The Hamiltonian system (1) possess many remarkable properties (e.g., see [1]). Here we shall limit ourselves to one of them which is related to the derivation of sufficient conditions for an extremum.

Theorem on the Integral Invariant. Let the function $\mathscr{H}$ belong to the class C^2 in an open set $D\subset\mathbf{R}\times\mathbf{R}^n\times\mathbf{R}^{n*}$, and let ω be the differential form in D specified by the equality

$$\omega=p\,dx-\mathscr{H}\,dt=\sum_{i=1}^{n}p_i\,dx_i-\mathscr{H}(t, x, p)\,dt. \tag{15}$$

Let $\Gamma_k=\{(t_k(\alpha), x_k(\alpha), p_k(\alpha))\,|\,a\leqslant\alpha\leqslant b\}$, k = 1, 2, be two piecewise-smooth closed contours in D [such that $t_k(a) = t_k(b)$, $x_k(a) = x_k(b)$, $p_k(a) = p_k(b)$].

If the contour Γ_1 is obtained from the contour Γ_2 by translating the latter along canonical extremals, i.e., if the points $\mathscr{P}_0(\alpha)=(t_0(\alpha), x_0(\alpha), p_0(\alpha))$ and $\mathscr{P}_1(\alpha)=(t_1(\alpha), x_1(\alpha), p_1(\alpha))$ lie on one canonical extremal for every α, then

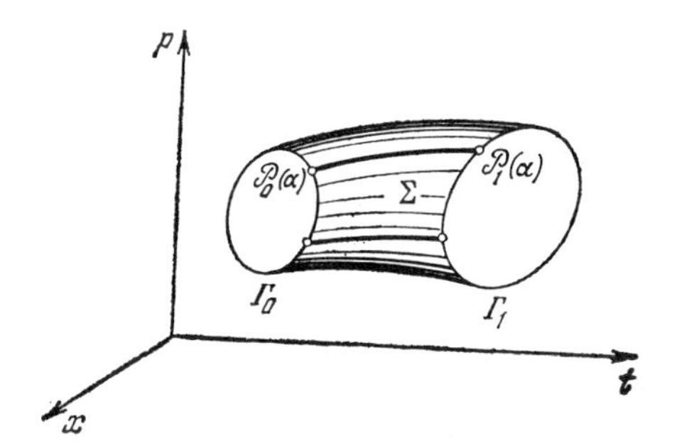

Fig. 38

$$\oint_{\Gamma_0} \omega = \oint_{\Gamma_1} \omega. \tag{16}$$

Proof. A) According to the classical theorem on the differentiability of the solutions with respect to the initial data (Section 2.2.7) and the composition theorem (Section 2.2.2), the solution of Cauchy's problem for system (1) with the initial conditions $(t_0(\alpha), x_0(\alpha), p_0(\alpha))$ [we shall denote this solution by $(x(t, \alpha), p(t, \alpha))$] is piecewise-continuously differentiable with respect to α and continuously differentiable with respect to t. The mapping σ: $\{(t, \alpha) | a \leqslant \alpha \leqslant b,\ t \in [t_0(\alpha), t_1(\alpha)]\} \to D$ (for $t_1(\alpha) < t_0(\alpha)$ we must write $[t_1(\alpha), t_0(\alpha)]$ here) determined by the equality $\sigma(t, \alpha) = (t, x(t, \alpha), p(t, \alpha))$ specifies a two-dimensional surface Σ in D (Fig. 38). Orienting Σ in an appropriate manner, we can write $\partial\Sigma = \Gamma_1 - \Gamma_0$ (the two arcs corresponding to $\alpha = a$ and $\alpha = b$ coincide geometrically but are oriented oppositely so that the integrals along these arcs mutually cancel). According to Stokes' theorem,

$$\oint_{\Gamma_1} \omega - \oint_{\Gamma_0} \omega = \int_{\partial\Sigma} \omega = \int_{\Sigma} d\omega = \int_a^b \int_{t_0(\alpha)}^{t_1(\alpha)} d\omega[\sigma_t, \sigma_\alpha]\, dt\, d\alpha,$$

and hence equality (16) will be proved if we verify that $d\omega[\sigma_t, \sigma_\alpha] \equiv 0$.

B) Differentiating form (15), we obtain

$$d\omega = \sum_{i=1}^n dp_i \wedge dx_i - \left[\mathcal{H}_t\, dt + \sum_{i=1}^n (\mathcal{H}_{x_i} dx_i + \mathcal{H}_{p_i} dp_i)\right] \wedge dt =$$
$$= \sum_{i=1}^n [dp_i \wedge dx_i - \mathcal{H}_{x_i} dx_i \wedge dt - \mathcal{H}_{p_i} dp_i \wedge dt] \tag{17}$$

(we remind the reader that for the exterior multiplication we have $dt \wedge dt = 0$).

If $\xi = (\xi_t, \xi_x, \xi_p)$ and $\eta = (\eta_t, \eta_x, \eta_p)$ are two tangent vectors, then (17) yields

$$d\omega[\xi, \eta] = \sum_{i=1}^n \left\{ \begin{vmatrix} \xi_{p_i} & \xi_{x_i} \\ \eta_{p_i} & \eta_{x_i} \end{vmatrix} - \mathcal{H}_{x_i} \begin{vmatrix} \xi_{x_i} & \xi_t \\ \eta_{x_i} & \eta_t \end{vmatrix} - \mathcal{H}_{p_i} \begin{vmatrix} \xi_{p_i} & \xi_t \\ \eta_{p_i} & \eta_t \end{vmatrix} \right\} = -\sum_i \begin{vmatrix} 1 & \mathcal{H}_{p_i} & -\mathcal{H}_{x_i} \\ \xi_t & \xi_{x_i} & \xi_{p_i} \\ \eta_t & \eta_{x_i} & \eta_{p_i} \end{vmatrix}.$$

Therefore

$$d\omega[\sigma_t, \sigma_\alpha] = -\sum_{i=1}^n \begin{vmatrix} 1 & \mathcal{H}_{p_i} & -\mathcal{H}_{x_i} \\ 1 & x_{it}(t, \alpha) & p_{it}(t, \alpha) \\ 0 & x_{i\alpha}(t, \alpha) & p_{i\alpha}(t, \alpha) \end{vmatrix} = 0$$

because the expression $\tilde{\sigma}(t, \alpha) = (x(t, \alpha), p(t, \alpha))$ is a solution of system (1) with respect to the argument t and hence the first and second rows of the determinant coincide. ■

The differential form (15) for which equality (16) holds is called the *Poincaré–Cartan integral invariant.*

Definition 2. We say that $A \subset D$ is a *Legendre set* if

$$\oint_\gamma \omega = 0 \tag{18}$$

for any piecewise-smooth contour $\gamma \subset A$. In other words, the restriction $\omega|A$ is a closed form.

[Here we need a more precise definition of the piecewise-smoothness: we shall suppose that A is the image of an open set $G \subset \mathbf{R}^k$ under a mapping $\varphi: G \to D$ of class $C^1(G)$ and that γ is the image of an ordinary piecewise-smooth closed contour in G under this mapping.]

The following three examples of Legendre sets will play an important role.

1) Let us fix a point $(t_0, x_0, p_0) \in D$ and consider the family of canonical extremals $\{(x(t, \lambda), p(t, \lambda)) | \alpha < t < \beta\}$ with the initial conditions

$$x(t_0, \lambda) = x_0, \quad p(t_0, \lambda) = \lambda. \tag{19}$$

The set

$$\Sigma = \{(t, x, p) \mid x = x(t, \lambda),\ p = p(t, \lambda),\ |p_0 - \lambda| < \varepsilon, \quad \alpha < t < \beta\}$$

is the desired Legendre set; $\varepsilon > 0$ and the interval $(\alpha, \beta) \ni t_0$ are chosen so that $\Sigma \subset D$.

Indeed, as was said above, a piecewise-smooth closed contour $\gamma \subset \Sigma$ is the image of a piecewise-smooth contour

$$\Gamma = \{(t(s), \lambda(s)) \mid a \leqslant s \leqslant b,\ t(a) = t(b),\ \lambda(a) = \lambda(b)\} \subset (\alpha, \beta) \times (p_0 - \varepsilon, p_0 + \varepsilon)$$

under the mapping $(t, \lambda) \to (t, x(t, \lambda), p(t, \lambda))$. Projecting Γ on the plane $t = t_0$, we obtain the contour $\Gamma_0 = \{(t_0, \lambda(s)) | a \leqslant s \leqslant b, \lambda(a) = \lambda(b)\}$, and, by virtue of (19), the image of Γ_0 is

$$\gamma_0 = \{(t_0, x(t_0, \lambda(s)), p(t_0, \lambda(s)) \mid a \leqslant s \leqslant b\} = \{(t_0, x_0, \lambda(s)) \mid a \leqslant s \leqslant b\}. \tag{20}$$

Since the contour γ_0 is obtained from the contour γ by means of the translation of the latter along canonical extremals, the theorem on the integral invariant implies

$$\oint_\gamma \omega = \oint_{\gamma_0} \omega = \oint_{\gamma_0} p\,dx - \mathcal{H}\,dt = 0$$

because, by virtue of (20), $dx = 0$ and $dt = 0$ on γ_0. Since (18) holds for any $\gamma \subset \Sigma$, we see that Σ is a Legendre set.

2) Let $x(\cdot)$ in problem (1), (2) be a scalar function, let us suppose that a family of extremals $x(\cdot, \lambda): (\alpha, \beta) \to \mathbf{R}$, $\lambda \in (\alpha_1, \beta_1)$, is given, and let $x(\cdot, \cdot) \in C^1((\alpha, \beta) \times (\alpha_1, \beta_1))$.

The set

$$\Sigma = \{(t, x, p) \mid x = x(t, \lambda), \quad p = p(t, \lambda) = L_{\dot x}(t, x(t, \lambda), \dot x(t, \lambda)),\ t \in (\alpha, \beta),\ \lambda \in (\alpha_1, \beta_1)\}$$

is the desired Legendre set. Indeed, let γ, γ_0, Γ, and Γ_0 be defined as above.

Then

$$\oint_\gamma \omega = \oint_{\gamma_0} \omega = \oint_{\gamma_0} p\,dx = \int_a^b p(t_0, \lambda(s)) \frac{\partial x}{\partial \lambda}(t_0, \lambda(s)) \lambda'(s)\,ds. \tag{21}$$

Denoting by $\Phi(\lambda)$ an antiderivative of the function $\lambda \to p(t_0, \lambda) x_\lambda(t_0, \lambda)$, we obtain from equalities (21) the relation

$$\oint_\gamma \omega = \int_a^b \Phi'(\lambda(s))\lambda'(s)\,ds = \Phi(\lambda(s))\Big|_{s=a}^{s=b} = 0$$

because $\lambda(a) = \lambda(b)$ (the contour is closed).

3) Proposition 4. Let a function $p\colon G \to \mathbf{R}^{n*}$, $p = (p_1, \ldots, p_n)$, be defined in an open set $G \subset \mathbf{R} \times \mathbf{R}^n$, the graph of p lying in D: $\Sigma = \{(t, x, p(t, x)) \mid (t, x) \in G\} \subset D$. For Σ to be a Legendre set:

a) under the assumption that p is continuous it is necessary and sufficient that there exist a function $S \in C^1(G)$ such that

$$p(t, x) = \frac{\partial S(t, x)}{\partial x}, \tag{22}$$

$$\frac{\partial S}{\partial t} + \mathcal{H}\left(t, x, \frac{\partial S}{\partial x}\right) \equiv 0; \tag{23}$$

b) under the assumption that $p \in C^1(G)$ it is necessary, and for a simply connected domain G it is also sufficient, that the relations

$$\frac{\partial p_i}{\partial x_k} = \frac{\partial p_k}{\partial x_i}, \tag{24}$$

$$\frac{\partial p_i}{\partial t} = -\frac{\partial \mathcal{H}}{\partial x_i} - \sum_{k=1}^{n} \frac{\partial \mathcal{H}}{\partial p_k}\frac{\partial p_k}{\partial x_i}, \quad i, k = 1, \ldots, n,$$

hold at each point $(t, x) \in G$.

Proof. Substituting $p = p(t, x)$ into (15), we obtain the form

$$\tilde{\omega} = p(t, x)\,dx - \mathcal{H}(t, x, p(t, x))\,dt.$$

The fact that Σ is a Legendre set means that $\oint_\Gamma \tilde{\omega} = 0$ for any $\Gamma \subset G$. According to the theorem of classical mathematical analysis, this is equivalent to the exactness of $\tilde{\omega}$, i.e., to the existence of a function S such that $\tilde{\omega} = dS$, whence follows a).

If $p \in C^1(G)$, then $\tilde{\omega} \in C^1(G)$ and $d\tilde{\omega} = d(dS) = 0$, which is equivalent to (24). The sufficiency of condition (24) in the case when G is simply connected follows from the same classical theorem which implied a). ■

Equation (23) is called the *Hamilton–Jacobi equation*. If a solution of this equation is known in the domain G, then (22) specifies a function whose graph is a Legendre set.

4.4.4. Sufficient Conditions for an Absolute Extremum in the Simplest Problem. We shall again consider the simplest problem of the classical calculus of variations:

$$\mathcal{J}(x(\cdot)) = \int_{t_0}^{t_1} L(t, x, \dot{x})\,dt \to \text{extr}, \tag{1}$$

$$x(t_0) = x_0, \quad x(t_1) = x_1. \tag{2}$$

Let us suppose that the Lagrangian L belongs to the class C^2 in an open set $V \subset \mathbf{R} \times \mathbf{R}^n \times \mathbf{R}^n$. By the admissible function for problem (1), (2) we shall mean the functions $x(\cdot) \in KC^1([t_0, t_1], \mathbf{R}^n)$ with piecewise-continuous derivatives whose extended graphs lie in V: $\{(t, x(t), \dot{x}(t) \mid t_0 \leqslant t \leqslant t_1\} \subset V$. For definiteness, the whole argument will be presented for the case of a minimum; the reader can easily make the corresponding changes in the statements for the case of a maximum by replacing L by $-L$.

As was already seen in Sections 4.4.1 and 4.4.2, the fulfillment of the Legendre condition $\hat{L}_{\dot{x}\dot{x}}(t) = L_{\dot{x}\dot{x}}(t, \hat{x}(t), \dot{\hat{x}}(t)) \geqslant 0$ is necessary for an admissible function $\hat{x}(\cdot)$ to yield a minimum in problem (1), (2). On the other hand, the strengthened Legendre condition $\hat{L}_{\dot{x}\dot{x}}(t) > 0$ is a part of the sufficient conditions for a minimum. This condition implies the convexity of the function $\dot{x} \to L(t, x, \dot{x})$ in the vicinity of the points $(t, \hat{x}(t), \dot{\hat{x}}(t))$, which suggests the idea that the convexity of the Lagrangian L with respect to $\dot{x}$ throughout its domain must be included in the sufficient conditions for an absolute minimum. Bogolyubov's theorem (see Section 1.4.3) confirms this idea. Therefore, we shall suppose that

$$L_{\dot{x}\dot{x}}(t, x, \xi) > 0, \quad \forall (t, x, \xi) \in V, \tag{3}$$

and that the section

$$V_{t,x} = \{\xi \mid (t, x, \xi) \in V\} \text{ is convex} \tag{4}$$

for any (t, x). Then from Proposition 1 of Section 4.4.3 it follows that the function

$$\bar{L}(t, x, \xi) = \begin{cases} L(t, x, \xi), & (t, x, \xi) \in V, \\ +\infty, & (t, x, \xi) \notin V \end{cases} \tag{5}$$

is convex with respect to ξ.

According to Proposition 2 of Section 4.4.3, conditions (3) and (4) guarantee the correctness of the definition of the Hamiltonian $\mathcal{H} \in C^2(D)$ by the relations

$$\begin{gathered} \mathcal{H}(t, x, p) = p\dot{x} - L(t, x, \dot{x}), \\ p = L_{\dot{x}}(t, x, \dot{x}) \end{gathered} \tag{6}$$

from which $\dot{x}$ must be eliminated. By Corollary 2 in the same section,

$$\begin{gathered} L(t, x, \xi) + \mathcal{H}(t, x, p) \geqslant p\xi, \\ \forall (t, x, \xi) \in V, \quad \forall (t, x, p) \in D. \end{gathered} \tag{7}$$

Finally, we shall again deal with the Poincaré–Cartan differential form

$$\omega = p\,dx - \mathcal{H}(t, x, p)\,dt. \tag{8}$$

THEOREM 1 (Sufficient Conditions for a Minimum). Let L belong to the class C^2 in an open set $V \subset \mathbf{R} \times \mathbf{R}^n \times \mathbf{R}^n$, and let conditions (3) and (4) hold. Further, let D be defined in the same way as in Proposition 2 of Section 4.4.3, and let G be an open set in $\mathbf{R} \times \mathbf{R}^n$.

If:

1) $\hat{x}(\cdot): [t_0, t_1] \to \mathbf{R}^n$ is an admissible extremal of problem (1), (2) and its graph lies in G:

$$\hat{\Gamma} = \{(t, \hat{x}(t)) \mid t_0 \leqslant t \leqslant t_1\} \subset G;$$

2) there exists a function $p: G \to \mathbf{R}^{n*}$ of class $C^1(G)$ whose graph lies in D:

$$\Sigma = \{(t, x, p(t, x)) \mid (t, x) \in G\} \subset D$$

and is a Legendre set;

3) $\hat{p}(t) \overset{\text{def}}{=} p(t, \hat{x}(t)) \equiv L_{\dot{x}}(t, \hat{x}(t), \dot{\hat{x}}(t))$;

then

$$\int_{t_0}^{t_1} L(t, x(t), \dot{x}(t))\,dt > \int_{t_0}^{t_1} L(t, \hat{x}(t), \dot{\hat{x}}(t))\,dt$$

for any admissible function $x(\cdot): [t_0, t_1] \to \mathbf{R}^n$ which satisfies boundary conditions (2) and whose graph lies in G:

$$\Gamma=\{(t, x(t)) \mid t_0 \leqslant t \leqslant t_1\} \subset G.$$

Remark. Proposition 3 of Section 4.4.3 implies the equality $\dot{\hat{x}}(t) = \mathscr{H}_p(t, \hat{x}(t), \hat{p}(t))$, and consequently the derivative $\dot{\hat{x}}(t)$ is continuous together with $\hat{x}(t)$ and $\hat{p}(t) = p(t, \hat{x}(t))$. Therefore, $\hat{x}(t)$ is continuously differentiable although in the case under consideration the admissible class consists of all piecewise-differentiable curves. However, this fact does not restrict the applicability of the theorem. Indeed, as was shown in Section 4.4.1, the existence of a continuous (and even piecewise-differentiable) function $\hat{p}(\cdot)$, such that $\hat{p}(t) = L_{\dot{x}}(t, \hat{x}(t), \dot{\hat{x}}(t))$, is a necessary condition for a (strong) minimum which is implied by the maximum principle. We see that under conditions (3) and (4) this means that the function $\dot{\hat{x}}(\cdot)$ must be continuous.

Proof. Let us denote $p(t) = p(t, x(t))$ and consider the curves

$$\gamma=\{(t, x(t), p(t)) \mid t_0 \leqslant t \leqslant t_1\} \subset \Sigma,$$
$$\hat{\gamma}=\{(t, \hat{x}(t), \hat{p}(t)) \mid t_0 \leqslant t \leqslant t_1\} \subset \Sigma.$$

Since $x(t_i) = \hat{x}(t_i)$, $i = 0, 1$, we have $p(t_i) = p(t_i, x(t_i)) = p(t_i, \hat{x}(t_i)) = \hat{p}(t_i)$. Therefore, the end points of the curve γ and $\hat{\gamma}$ coincide, and we can consider the closed contour $\gamma - \hat{\gamma} \subset \Sigma$ which is obtained if we first pass along γ from $(t_0, x(t_0), p(t_0))$ to $(t_1, x(t_1), p(t_1))$ and then pass in the opposite direction along $\hat{\gamma}$. Since Σ is a Legendre set,

$$0=\oint_{\gamma-\hat{\gamma}} \omega=\int_{\gamma} \omega-\int_{\hat{\gamma}} \omega. \tag{10}$$

Using equality (7), we obtain

$$\int_{\gamma} \omega=\int_{\gamma} p\,dx-\mathscr{H}\,dt=\int_{t_0}^{t_1}\{p(t)\dot{x}(t)-\mathscr{H}(t, x(t), p(t))\}\,dt \leqslant$$
$$\leqslant \int_{t_0}^{t_1} L(t, x(t), \dot{x}(t))\,dt=\mathscr{J}(x(\cdot)), \tag{11}$$

and the equality is only possible here if

$$p(t)\dot{x}(t)-\mathscr{H}(t, x(t), p(t)) \equiv L(t, x(t), \dot{x}(t)),$$

i.e. [we again use (7)], if

$$p(t)\dot{x}(t)-\mathscr{H}(t, x(t), p(t))=\max_{\{p \mid (t, x(t), p) \in D\}}\{p\dot{x}(t)-\mathscr{H}(t, x(t), p)\}.$$

By the Fermat theorem,

$$\dot{x}(t)=\mathscr{H}_p(t, x(t), p(t)). \tag{12}$$

Since the graph of p is a Legendre set, equality (24) of Section 4.4.3 must hold, whence, taking into consideration (12), we obtain

$$\frac{dp_i(t)}{dt}=\frac{d}{dt} p_i(t, x(t))=\frac{\partial p_i}{\partial t}+\sum_k \frac{\partial p_i}{\partial x_k} \frac{dx_k}{dt}=-\left(\frac{\partial \mathscr{H}}{\partial x_i}+\sum_k \frac{\partial \mathscr{H}}{\partial p_k} \frac{\partial p_k}{\partial x_i}\right)+$$
$$+\sum_k \frac{\partial p_k}{\partial x_i} \frac{\partial \mathscr{H}}{\partial p_k}=-\mathscr{H}_{x_i}(t, x(t), p(t)). \tag{13}$$

According to (12) and (13), the pair $(x(\cdot), p(\cdot))$ is a canonical extremal. Further, we have $(x(t_0), p(t_0)) = (\hat{x}(t_0), \hat{p}(t_0))$, and since $(\hat{x}(\cdot), \hat{p}(\cdot))$ is also a canonical extremal (Proposition 3 of Section 4.4.3), the uniqueness theorem implies that $x(t) \equiv \hat{x}(t)$, $p(t) \equiv \hat{p}(t)$.

Hence from (11) it follows that

$$\int_{\hat{\gamma}} \omega = \int_{t_0}^{t_1} L(t, \hat{x}(t), \dot{\hat{x}}(t))\, dt = \mathcal{J}(\hat{x}(\cdot)),$$

and for the other admissible functions $x(\cdot)$ whose graphs lie in G and which satisfy conditions (2) there must be

$$\int_{\gamma} \omega < \mathcal{J}(x(\cdot)).$$

By virtue of (10), we obtain $\mathcal{J}(x(\cdot)) > \mathcal{J}(\hat{x}(\cdot))$. ■

COROLLARY. If, under the conditions of the theorem, the set G coincides with the projection of D on $\mathbf{R} \times \mathbf{R}^n = \{(t, x)\}$, then $\hat{x}(\cdot)$ yields a strict absolute maximum in problem (1), (2).

A part of the conditions of the theorem we have just proved is usually presented in courses of the calculus of variations in a somewhat different form with the notion of a field of extremals.

Definition. Let Λ and G be open sets in $\mathbf{R}^n$ and $\mathbf{R} \times \mathbf{R}^n$, respectively. A family of functions $\{x(\cdot, \lambda): [t_0(\lambda), t_1(\lambda)] \to \mathbf{R}^n \mid \lambda \in \Lambda\}$ is said to form a *field of extremals covering* G if:

1) $x(\cdot, \lambda)$ is an extremal (i.e., a solution of the Euler equation) of problem (1), (2) for any $\lambda \in \Lambda$;

2) the mapping $(t, \lambda) \to (t, x(t, \lambda))$ is one-to-one and its image contains G:

$$\{(t, x) \mid x = x(t, \lambda),\ t_0(\lambda) \leqslant t \leqslant t_1(\lambda),\ \lambda \in \Lambda\} \supset G.$$

3) the function $p: G \to \mathbf{R}^{n*}$ determined by the equalities

$$p(t, x) = L_{\dot{x}}(t, x(t, \lambda), \dot{x}(t, \lambda)), \quad (t, x) = (t, x(t, \lambda)) \tag{14}$$

belongs to the class $C^1(G)$ and its graph is a Legendre set.

The function $u: G \to \mathbf{R}^n$ determined by the equalities

$$u(t, x) = \dot{x}(t, \lambda), \quad (t, x) = (t, x(t, \lambda)) \tag{15}$$

is called the *field slope function.* An extremal $\hat{x}(\cdot): [t_0, t_1] \to \mathbf{R}^n$ *is said to be embedded in the field* $x(\cdot, \cdot)$ if its graph is contained in G and $\hat{x}(t) \equiv x(t, \hat{\lambda})$ for some $\hat{\lambda} \in \Lambda$.

Substituting (14) and (15) into (8) and recalling the definition of a Legendre set (Section 4.4.3), we can restate condition (3) thus:

3') In the domain G the integral

$$\int_{(t_0, x_0)}^{(t_1, x_1)} \{L_{\dot{x}}(t, x, u(t, x))\, dx - [L_{\dot{x}}(t, x, u(t, x))\, u(t, x) - L(t, x, u(t, x))]\}\, dt \tag{16}$$

does not depend on the path of integration joining the points (t_0, x_0) and (t_1, x_1) (according to the theorem of classical mathematical analysis, this is equivalent to the property that the integral over every closed contour is equal to zero). Integral (16) is called *Hilbert's invariant integral.*

THEOREM 1'. Let function L satisfy the same conditions as in Theorem 1, and let $\hat{x}(\cdot): [t_0, t_1] \to \mathbf{R}^n$ be an admissible extremal whose graph is contained in G.

If $\hat{x}(\cdot)$ can be embedded in a field of extremals covering G, then $\hat{x}(\cdot)$ yields a minimum for functional (1) in the class of admissible functions which satisfy conditions (2) and whose graphs are contained in G.

The reader can easily see that this is a mere restatement of Theorem 1. It is particularly convenient for $n = 1$ because in this case condition (3) of the definition stated above follows from 1) and 2) (see the second of the examples of Legendre sets in Section 4.4.3). This remark can be used in the solution of many problems.

4.4.5. Conjugate Points. Sufficient Conditions for Strong and Weak Extrema. In this section we shall again suppose that $L\in C^2(V)$ and $\hat{x}(\cdot)\in C^1([t_0, t_1], \mathbf{R}^n)$ is an admissible extremal of problem (1), (2) of Section 4.4.4 along which the strengthened Legendre condition

$$L_{\dot{x}\dot{x}}(t, \hat{x}(t), \dot{\hat{x}}(t)) > 0, \quad \forall t\in[t_0, t_1], \tag{1}$$

holds. Since the matrix $L_{\dot{x}\dot{x}}$ is continuous, it remains positive-definite in a neighborhood $\hat{V}$ of the extended graph $\{(t, \hat{x}(t), \dot{\hat{x}}(t)) \mid t_0 \leqslant t \leqslant t_1\}$. Let us choose the neighborhood $\hat{V}$ so that its sections $\hat{V}_{t,x}=\{\xi \mid (t, x, \xi)\in\hat{V}\}$ are convex. Then the conditions of Proposition 2 of Section 4.4.3 are fulfilled, and, accordingly, we can define the domain $\hat{D}\subset\mathbf{R}\times\mathbf{R}^n\times\mathbf{R}^{n*}$ and the Hamiltonian $\mathscr{H}\colon \hat{D}\to\mathbf{R}$ of class C^2.

The right-hand sides of the canonical system

$$\dot{x}=\mathscr{H}_p(t, x, p), \quad \dot{p}=-\mathscr{H}_x(t, x, p) \tag{2}$$

are continuously differentiable in $\hat{D}$. The canonical extremal $(\hat{x}(\cdot), \hat{p}(\cdot))$ corresponding to the extremal $\hat{x}(\cdot)$, and hence $\hat{x}(\cdot)$ itself, can be extended to an open interval containing the closed interval $[t_0, t_1]$. After that, narrowing $\hat{V}$, if necessary, we can assume that V has the form

$$\hat{V}=\{(t, x, \xi) \mid |x-\hat{x}(t)|<\varepsilon,\ |\xi-\dot{\hat{x}}(t)|<\varepsilon,\ t_0-\varepsilon<t<t_1+\varepsilon\}.$$

Let $(t_0, x_0, p_0) = (t_0, \hat{x}(t_0), L_{\dot{x}}(t_0, \hat{x}(t_0), \dot{\hat{x}}(t_0)))$. Let us construct the family of canonical extremals $\{(x(\cdot, \lambda), p(\cdot, \lambda)) \mid |\lambda - p_0| < \delta\}$ determined by the initial conditions

$$x(t_0, \lambda)=x_0, \quad p(t_0, \lambda)=\lambda \tag{3}$$

(cf. the first example of a Legendre set in Section 4.4.3). For sufficiently small δ the extremals of this family, together with $(\hat{x}(\cdot), \hat{p}(\cdot)) = (x(\cdot, p_0), p(\cdot, p_0))$, are contained in $\hat{D}$ for $t\in[t_0, t_1]$; it is evident that the functions $x(\cdot, \cdot)$, $p(\cdot, \cdot)$ are continuously differentiable (Section 2.5.7).

Proposition 1. For a point $\tau\in(t_0, t_1]$ to be conjugate to t_0 it is necessary and sufficient that

$$\det\left(\left.\frac{\partial x(\tau, \lambda)}{\partial\lambda}\right|_{\lambda=p_0}\right)=0. \tag{4}$$

Proof. A) By the theorem of Section 2.5.7, the collection of the derivatives of the solutions of system (2) with respect to the initial data for $\lambda = p_0$ forms the principal matrix solution of the variational system of equations

$$\begin{aligned}\dot{\xi} &= \hat{\mathscr{H}}_{px}\xi+\hat{\mathscr{H}}_{pp}\eta,\\ \dot{\eta} &= -\hat{\mathscr{H}}_{xx}\xi-\hat{\mathscr{H}}_{xp}\eta.\end{aligned} \tag{5}$$

The derivatives $\partial x/\partial\lambda_i$, $\partial p/\partial\lambda_i$ we are interested in, which are arranged in columns, occupy half of the columns (n of the 2n columns) of this matrix, and, according to conditions (3),

$$\begin{pmatrix} \partial x/\partial \lambda_i \\ \partial p/\partial \lambda_i \end{pmatrix} \Big|_{t=t_0} = \begin{pmatrix} 0 \\ e_i \end{pmatrix}. \tag{6}$$

Equality (4) is equivalent to the existence of a nonzero vector $c \in \mathbf{R}^n$ such that

$$\frac{\partial x}{\partial \lambda}(\tau, p_0)\, c = \sum_i \frac{\partial x}{\partial \lambda_i}(\tau, p)\, c_i = 0.$$

Let us denote

$$\begin{pmatrix} \bar{\xi}(t) \\ \bar{\eta}(t) \end{pmatrix} = \begin{pmatrix} \sum_{i=1}^{n} \frac{\partial x}{\partial \lambda_i}(t, p_0)\, c_i, \\ \sum_{i=1}^{n} \frac{\partial p}{\partial \lambda_i}(t, p_0)\, c_i. \end{pmatrix}.$$

This expression is a linear combination of columns of the principal matrix solution and therefore it is a solution of the system (5), and, by virtue of (6), we have $\bar{\xi}(t_0) = 0$, $\bar{\eta}(t_0) = c \neq 0$. Moreover, $\bar{\xi}(\tau) = 0$.

Thus, equality (4) is equivalent to the existence of a nontrivial solution of system (5) for which $\bar{\xi}(t_0) = \bar{\xi}(\tau) = 0$.

B) LEMMA. A pair $\{\xi(\cdot), \eta(\cdot)\}$ is a solution of system (5) if and only if $\xi(\cdot)$ is a solution of the Jacobi equation

$$\frac{d}{dt}[\hat{L}_{\dot{x}\dot{x}}\dot{\xi} + \hat{L}_{\dot{x}x}\xi] = [\hat{L}_{x\dot{x}}\dot{\xi} + \hat{L}_{xx}\xi] \tag{7}$$

and

$$\eta(t) = \hat{L}_{\dot{x}\dot{x}}\dot{\xi}(t) + \hat{L}_{\dot{x}x}\xi(t). \tag{8}$$

Proof. According to the definition (see Section 4.4.2), Eq. (7) is the Euler equation for the secondary extremal problem with the Lagrangian

$$\mathfrak{L} = \frac{1}{2}\{\xi^{\mathrm{T}}\hat{L}_{xx}\xi + 2\xi^{\mathrm{T}}\hat{L}_{x\dot{x}}\dot{\xi} + \dot{\xi}^{\mathrm{T}}\hat{L}_{\dot{x}\dot{x}}\dot{\xi}\}. \tag{9}$$

The corresponding Hamiltonian $\mathfrak{H}$ is obtained by means of the Legendre transformation with respect to $\dot{\xi}$:

$$\begin{aligned} \mathfrak{H} &= \eta^{\mathrm{T}}\dot{\xi} - \mathfrak{L}, \\ \eta &= \mathfrak{L}_{\dot{\xi}} = \hat{L}_{\dot{x}\dot{x}}\dot{\xi} + \hat{L}_{\dot{x}x}\xi. \end{aligned} \tag{10}$$

To obtain the second of these equalities, we use the relation

$$\xi^{\mathrm{T}}\hat{L}_{x\dot{x}}\dot{\xi} = \sum_{j,k} \frac{\partial^2 \hat{L}}{\partial x_j \partial \dot{x}_k}\xi_j \dot{\xi}_k,$$

whence

$$\frac{\partial}{\partial \dot{\xi}_i}(\xi^{\mathrm{T}}\hat{L}_{x\dot{x}}\dot{\xi}) = \sum_j \frac{\partial^2 \hat{L}}{\partial x_j \partial \dot{x}_i}\xi_j = \sum_j \frac{\partial^2 \hat{L}}{\partial \dot{x}_i \partial x_j}\xi_j = (\hat{L}_{\dot{x}x}\xi)_i.$$

Similarly,

$$\frac{\partial}{\partial \dot{\xi}_i}(\dot{\xi}^{\mathrm{T}}\hat{L}_{\dot{x}\dot{x}}\dot{\xi}) = \frac{\partial}{\partial \dot{\xi}_i}\left(\sum_{j,k} \frac{\partial^2 \hat{L}}{\partial \dot{x}_j \partial \dot{x}_k}\dot{\xi}_j \dot{\xi}_k\right) = \sum_k \frac{\partial^2 \hat{L}}{\partial \dot{x}_i \partial \dot{x}_k}\dot{\xi}_k + \sum_j \frac{\partial^2 \hat{L}}{\partial \dot{x}_j \partial \dot{x}_i}\dot{\xi}_j = (2\hat{L}_{\dot{x}\dot{x}}\dot{\xi})_i.$$

Eliminating $\dot{\xi}$ from (10), we obtain

$$\mathfrak{H} = \frac{1}{2}\{\eta^{\mathrm{T}}\hat{L}_{\dot{x}\dot{x}}^{-1}\eta - 2\xi^{\mathrm{T}}\hat{L}_{x\dot{x}}\hat{L}_{\dot{x}\dot{x}}^{-1}\eta + \xi^{\mathrm{T}}[\hat{L}_{xx}\hat{L}_{\dot{x}x} - \hat{L}_{x\dot{x}}\hat{L}_{\dot{x}\dot{x}}^{-1}]\xi\}. \tag{11}$$

Now we make use of formulas (11), (12), and (8) of Section 4.4.3:

$$\mathscr{H}_x(t, x, p) = -L_x(t, x, \Xi(t, x, p)),$$
$$\mathscr{H}_p(t, x, p) = \Xi(t, x, p),$$
$$p \equiv L_{\dot{x}}(t, x, \Xi(t, x, p)).$$

The differentiation of these equalities with respect to x and p results in

$$\mathscr{H}_{xx} = -\hat{L}_{xx} - \hat{L}_{x\dot{x}}\hat{\Xi}_x = -\hat{L}_{xx} - \hat{L}_{x\dot{x}}\mathscr{H}_{px},$$
$$\mathscr{H}_{xp} = -\hat{L}_{x\dot{x}}\hat{\Xi}_p = -\hat{L}_{x\dot{x}}\mathscr{H}_{pp},$$
$$E = \frac{\partial p}{\partial p} = \hat{L}_{\dot{x}\dot{x}}\hat{\Xi}_p = \hat{L}_{\dot{x}\dot{x}}\mathscr{H}_{pp}.$$
$$0 = \frac{\partial p}{\partial x} = \hat{L}_{\dot{x}x} + \hat{L}_{\dot{x}\dot{x}}\hat{\Xi}_x = \hat{L}_{\dot{x}x} + \hat{L}_{\dot{x}\dot{x}}\mathscr{H}_{px}.$$

It follows that

$$\mathscr{H}_{pp} = \hat{L}_{\dot{x}\dot{x}}^{-1}, \qquad \mathscr{H}_{xp} = -\hat{L}_{x\dot{x}}\hat{L}_{\dot{x}\dot{x}}^{-1},$$
$$\mathscr{H}_{px} = -\hat{L}_{\dot{x}\dot{x}}^{-1}\hat{L}_{\dot{x}x}, \quad \mathscr{H}_{xx} = -\hat{L}_{xx} + \hat{L}_{x\dot{x}}\hat{L}_{\dot{x}\dot{x}}^{-1}\hat{L}_{\dot{x}x}. \tag{12}$$

Consequently, (11) can be rewritten thus:

$$\mathfrak{H} = \frac{1}{2}\{\eta^T\mathscr{H}_{pp}\eta + 2\xi^T\mathscr{H}_{xp}\eta + \xi^T\mathscr{H}_{xx}\xi\}.$$

According to Proposition 3 of Section 4.4.3, a pair $\{\xi(\cdot), \eta(\cdot)\}$ is a solution of the canonical system

$$\dot{\xi} = \mathfrak{H}_\eta, \quad \dot{\eta} = -\mathfrak{H}_\xi \tag{13}$$

if and only if $\xi(\cdot)$ is a solution of (7) and the second of relations (10) [coinciding with (8)] holds. It remains to note that system (13) is identical with (5). ■

<u>C) Completion of the Proof.</u> In the first part of the proof it was shown that equality (4) is equivalent to the existence of a nontrivial solution $\{\xi(\cdot), \eta(\cdot)\}$ of system (5) for which $\xi(t_0) = \xi(\tau) = 0$. By the lemma, $\xi(\cdot)$ is a solution of the Jacobi equation and (8) takes place, whence $\eta(\tau) = \hat{L}_{\dot{x}\dot{x}}(\tau)\dot{\xi}(\tau)$. The equality $\eta(\tau) = 0$ contradicts the nontriviality of the solution (by virtue of the uniqueness theorem), and therefore $\eta(\tau) \neq 0$. Recalling the definition stated in Section 4.4.2, we conclude that (4) is equivalent to the fact that the point τ is conjugate to the point t_0. ■

The proposition we have proved leads us to the following geometrical interpretation of the notion of a conjugate point. Let again $(\hat{x}(\cdot), \hat{p}(\cdot))$ be the canonical extremal corresponding to the extremal $\hat{x}(\cdot)$ under consideration. For a sufficiently small δ the canonical extremals determined by conditions (3) form a thin strip along the graph B_0B_1 (Fig. 39). For $t = t_0$ the edge of this strip is "vertical" and its projection coincides with the point $A_0 = (t_0, x_0)$. Condition (4) which, as was shown, is equivalent to the equality $(\partial x/\partial\lambda)c = 0$ means that for $t = \tau$ the strip is tangent to a "vertical" direction at the point $B_\tau = (\tau, \hat{x}(\tau), \hat{p}(\tau))$ (i.e., to a direction parallel to the planes $t = \text{const}$, $x = \text{const}$).

If we project the strip on the space (t, x), we obtain a pencil of extremals issued from the point A_0 (Fig. 39). In the vicinity of the point $A_\tau = (\tau, \hat{x}(\tau))$ the mapping $(t, \lambda) \to (t, x(t, \lambda))$ is not one-to-one ["the extremals lying infinitely close to the extremal $\hat{x}(\cdot)$ and having the same origin as $\hat{x}(\cdot)$ intersect at the point A_τ"]. The point A_τ is also said to be conjugate to A_0.

<u>Exercise.</u> The arcs of the great circles on the sphere $S^2 = \{(x, y, z) \mid x^2 + y^2 + z^2 = R^2\}$ are extremals of the distance functional. Draw a great circle through a point $A_0 \in S^2$ and find a point on it which is conjugate to A_0.

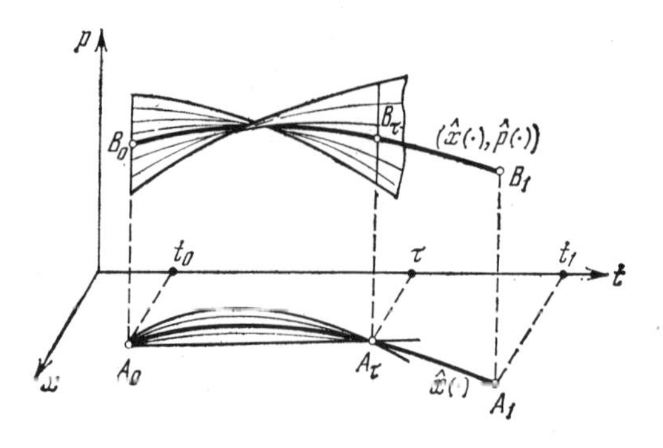

Fig. 39

Proposition 2. If on an extremal $\hat{x}(\cdot):[t_0, t_1] \to \mathbf{R}^n$ there hold the strengthened Legendre condition (1) and the strengthened Jacobi condition (i.e., the half-open interval $(t_0, t_1]$ contains no points conjugate to t_0), then this extremal can be embedded in a field of extremals covering a neighborhood G of its graph $\{(t, \hat{x}(t)) | t_0 \leqslant t \leqslant t_1\}$.

Proof. A) We shall first prove that for a sufficiently small $\delta > 0$ none of the points of the closed interval $[t_0, t_1]$ is conjugate to any of the points $t_0' \in (t_0-\delta, t_0)$. If otherwise, there exist $t_0^{(n)} \to t_0$, $t_1^{(n)} \in [t_0, t_1]$ and solutions $\xi^{(n)}(\cdot)$ of the Jacobi equation (7) such that $\xi^{(n)}(t_0^{(n)}) = \xi^{(n)}(t_1^{(n)}) = 0$, $\hat{L}_{\dot{x}\dot{x}}(t_1^{(n)})\dot{\xi}^{(n)}(t_1^{(n)}) \neq 0$. The equation being homogeneous, we can put $|\dot{\xi}^{(n)}(t_1^{(n)})| = 1$. Passing to a subsequence, if necessary, we can assume that $t_1^{(n)} \to t_1' \in [t_0, t_1]$, and

$$\dot{\xi}^{(n)}(t_1^{(n)}) \to \eta, \quad |\eta| = 1.$$

If condition (1) is fulfilled, the Jacobi equation is reducible to the linear system (5) whose coefficients are continuous on a closed interval $\Delta \supset [t_0, t_1]$, and Cauchy's problem for this system with the initial conditions

$$\xi(t_1') = 0,$$
$$\eta(t_1') = \hat{L}_{\dot{x}\dot{x}}(t_1')\dot{\xi}(t_1') + \hat{L}_{\dot{x}x}(t_1')\xi(t_1') = \hat{L}_{\dot{x}\dot{x}}(t_1')\eta$$

possesses a solution $(\xi(t), \eta(t))$ defined on Δ (Section 2.5.4). By the theorem on the continuity of the solution with respect to the initial data (Section 2.5.5), $\xi^{(n)}(t) \to \xi(t)$ and

$$\eta^{(n)}(t) = \hat{L}_{\dot{x}\dot{x}}(t)\dot{\xi}^{(n)}(t) + \hat{L}_{\dot{x}x}(t)\xi^{(n)}(t) \to \eta(t)$$

uniformly on Δ. Therefore, there must also be $\dot{\xi}^{(n)}(t) \to \dot{\xi}(t)$ uniformly on Δ.

Further, we have $\xi(t_0) = \lim\limits_n \xi(t_0^{(n)}) = \lim\limits_n \xi^{(n)}(t_0^{(n)}) = 0$, and the points $t_0 = \lim\limits_n t_0^{(n)}$ and $t_1' = \lim\limits_n t_1^{(n)}$ cannot coincide because, if otherwise, then $\eta = \dot{\xi}(t_1') = 0$ since

$$|\dot{\xi}(t_1')| = \left| \frac{1}{t_1^{(n)} - t_0^{(n)}} \int_{t_0^{(n)}}^{t_1^{(n)}} [\dot{\xi}^{(n)}(t) - \dot{\xi}(t_1')]\, dt \right| \leqslant \|\dot{\xi}^{(n)} - \dot{\xi}\|_C + \max_{[t_0', t_1']} |\dot{\xi}(t) - \dot{\xi}(t_1')| \to 0.$$

Since $\xi(t_0) = \xi(t_1') = 0$ and $\hat{L}_{\dot{x}\dot{x}}(t_1')\dot{\xi}(t_1') \neq 0$, the point $t_1' \in (t_0, t_1]$ is conjugate to t_0, which contradicts the hypothesis.

B) Let us choose $t_0' < t_0$ such that the closed interval $[t_0, t_1]$ does not contain conjugate points to t_0'. Let us construct a family of canonical extremals $(x(\cdot, \lambda), p(\cdot, \lambda))$ by replacing t_0 in (3) by t_0'. By virtue of Proposition 1, $\det \frac{\partial x}{\partial \lambda}(\tau, p_0) \neq 0$, $\forall \tau \in [t_0, t_1]$.

Let us define a mapping $\varphi: C_\delta \to \mathbf{R} \times \mathbf{R}^n$ of the cylinder $C_\delta = [t_0 - \delta, t_1 + \delta] \times B(p_0, \delta)$ by means of the equality $\varphi(t, \lambda) = (t, x(t, \lambda))$. At the points (τ, p_0), $\tau \in [t_0, t_1]$ the Jacobian of φ is nonzero:

$$\det \varphi' = \det \begin{pmatrix} 1 & 0 \\ x_t & x_\lambda \end{pmatrix} = \det x_\lambda \neq 0,$$

and we can choose δ so that this inequality is retained at all the points of C_δ. Moreover, by the inverse mapping theorem (Section 2.3.4), every point (τ, p_0) possesses a neighborhood U_τ which is mapped by φ onto $W_\tau = \varphi(U_\tau)$ in a one-to-one manner, and $\varphi^{-1} \in C^1(W_\tau)$.

We shall show that for a sufficiently small δ the mapping φ is one-to-one throughout C_δ. Indeed, if otherwise, there exist two sequences (t_n', λ_n'), (t_n'', λ_n'') such that $\lambda_n' \to p_0$, $\lambda_n'' \to p_0$ and $(t_n', x(t_n', \lambda_n')) = \varphi(t_n', \lambda_n') = \varphi(t_n'', \lambda_n'') = (t_n'', x(t_n'', \lambda_n''))$. It follows that $t_n' = t_n''$.

Passing to a subsequence we can assume that $t_n' = t_n'' \to \tau$. For sufficiently large n the points (t_n', λ_n') and (t_n'', λ_n'') belong to U_τ, and the equality $\varphi(t_n', \lambda_n') = \varphi(t_n'', \lambda_n'')$ contradicts the definition of U_τ.

Now we shall show that $G = \varphi(\operatorname{int} C_\delta)$ is a neighborhood of the graph $\hat{\Gamma} = \{(t, \hat{x}(t)) \mid t_0 \leqslant t \leqslant t_1\}$. Since $\hat{\Gamma} = \varphi(\{(t, p_0) \mid t_0 \leqslant t \leqslant t_1\}) \subset G$, it is necessary to prove that G is open. Indeed, if $(\bar{t}, \bar{x}) = \varphi(\bar{t}, \bar{\lambda}) \in G$, $(\bar{t}, \bar{\lambda}) \in C_\delta$, then $\det \varphi'(\bar{t}, \bar{\lambda}) \neq 0$, and, by the same theorem of Section 2.3.4, a neighborhood of the point $(\bar{t}, \bar{\lambda})$ is mapped on a neighborhood of the point $(\bar{t}, \bar{x})$. Therefore, $(\bar{t}, \bar{x})$ is an interior point of G, and hence G is an open set.

Thus, the family of extremals $\{x(\cdot, \lambda)\}$ covers G in a one-to-one manner. The fact that this family forms a field of extremals was proved in the first example of a Legendre set in Section 4.4.3. ■

The field we have constructed consists of extremals passing through one and the same point (t_0', x_0) and is called a *central field of extremals*.

Now we come back to sufficient conditions for a minimum. First of all we shall show how sufficient conditions for a weak minimum are derived from the general theorem of Section 4.4.4. The corresponding theorem was already stated in Section 4.4.2.

Proof of the Jacobi Theorem (Theorem 2 of Section 4.4.2). If the Legendre and the Jacobi strengthened conditions hold on an admissible extremal $\hat{x}(\cdot)$ of problem (1), (2) of Section 4.4.4, then, according to Proposition 2, it can be embedded in a field of extremals covering a neighborhood G of the graph of $\hat{x}(\cdot)$. By Theorem 1 (or Theorem 1') of Section 4.4.4, the inequality

$$\mathcal{J}(x(\cdot)) = \int_{t_0}^{t_1} L(t, x(t), \dot{x}(t))\, dt > \mathcal{J}(\hat{x}(\cdot))$$

holds for all $x(\cdot) \in C^1([t_0, t_1], \mathbf{R}^n)$, which satisfy the boundary conditions $x(t_0) = \hat{x}(t_0)$, $x(t_1) = \hat{x}(t_1)$ and whose graphs are contained in G while their extended graphs are contained in V. If the norm $\|x(\cdot) - \hat{x}(\cdot)\|_{C^1}$ is sufficiently small, the last two conditions are satisfied. Consequently, $\hat{x}(\cdot)$ yields a C^1-local, i.e., weak, minimum in problem (1), (2) of Section 4.4.4. ■

Now we proceed to sufficient conditions for a strong minimum. In this case we again assume that functions $x(\cdot)$ and $\hat{x}(\cdot)$ themselves are so close to one another that the graph of $x(\cdot)$ is contained in G; however,

$\dot{x}(\cdot)$ may be arbitrary, and the extended graph of $x(\cdot)$ is not necessarily contained in $\hat{V}$. Outside $\hat{V}$ condition (1) does not guarantee the convexity of the function $\xi \to L(t, x, \xi)$, and Young's inequality, which was used in the proof of Theorem 1 of Section 4.4.4 [see formula (7) of Section 4.4.4], does not necessarily hold. Therefore, in the formulation of the theorem we have to introduce an additional assumption (Weierstrass' condition) about the nonnegativity of the function

$$\mathscr{E}(t, x, u, \xi) = L(t, x, \xi) - L(t, x, u) - L_{\dot{x}}(t, x, u)(\xi - u).$$

Weierstrass' Theorem on Sufficient Conditions for a Strong Minimum. Let the function L be of class C^2 in an open set $V \subset \mathbf{R} \times \mathbf{R}^n \times \mathbf{R}^n$, let $\hat{x}(\cdot) \in C^1([t_0, t_1], \mathbf{R}^n)$,† and let the extended graph of function $\hat{x}(\cdot)$ lie in V:

$$\bar{\Gamma} = \{(t, \hat{x}(t), \dot{\hat{x}}(t)) \mid t_0 \leqslant t \leqslant t_1\} \subset V.$$

If:

1) $\hat{x}(\cdot)$ satisfies the Euler equation

$$\frac{d}{dt} \hat{L}_{\dot{x}}(t) = \hat{L}_x(t);$$

2) $\hat{x}(\cdot)$ satisfies the boundary conditions

$$\hat{x}(t_0) = x_0, \quad \hat{x}(t_1) = x_1;$$

3) the strengthened Legendre condition (1) holds along $\hat{x}(\cdot)$;

4) the strengthened Jacobi condition holds along $\hat{x}(\cdot)$: the half-open interval $(t_0, t_1]$ contains no conjugate points to t_0;

5) there exists a neighborhood $\bar{V} \supset \bar{\Gamma}$ for which Weierstrass' condition

$$\mathscr{E}(t, x, u, \xi) \geqslant 0$$

holds for all (t, x, u, ξ) such that $(t, x, u) \in \bar{V}$, $(t, x, \xi) \in V$, then $\hat{x}(\cdot)$ yields a strong minimum in the simplest problem (1), (2) of Section 4.4.4 (this minimum is strict if $\mathscr{E} > 0$ for $\xi \neq u$ in Weierstrass' condition).

Proof. Let us define a neighborhood $\hat{V} \supset \bar{\Gamma}$ in the same way as the beginning of the present section and construct a neighborhood $G \supset \hat{\Gamma} = \{(t, \hat{x}(t)) \mid t_0 \leqslant t \leqslant t_1\}$ and a field of extremals covering G; without loss of generality, we can put $\bar{V} = \hat{V}$. Let us denote by $u(t, x)$ the slope of the field at the point $(t, x) = (t, x(t, \lambda)) \in G$:

$$u(t, x) = \dot{x}(t, \lambda) \quad \text{for} \quad x = x(t, \lambda),$$

and let

$$p(t, x) = L_{\dot{x}}(t, x, u(t, x)).$$

According to Proposition 2 and the example 1) in Section 4.4.3, the graph of the function $p: G \to \mathbf{R}^{n*}$ is a Legendre set and lies in $\hat{D}$:

$$\Sigma = \{(t, x, p(t, x)) \mid (t, x) \in G\} \subset \hat{D}.$$

Now for an admissible extremal $x(\cdot)$ whose graph is contained in G, we write

$$\mathscr{J}(x(\cdot)) - \int_\gamma \omega = \int_{t_0}^{t_1} L(t, x(t), \dot{x}(t))\, dt - \int_{t_0}^{t_1} \{p(t, x(t))\dot{x}(t) - \mathscr{H}(t, x(t), p(t, x(t)))\}\, dt =$$

†We remind the reader that function $\hat{x}(\cdot)$, which yields a strong minimum when condition (1) holds, must possess a continuous derivative although in the present section the problem is considered in $KC^1([t_0, t_1], \mathbf{R}^n)$ (see the remark after the statement of Theorem 1 of Section 4.4.4).

$$=\int_{t_0}^{t_1}\{L(t, x(t), \dot{x}(t))-L_{\dot{x}}(t, x(t), u(t, x(t))\dot{x}(t)+$$
$$+L_{\dot{x}}(t, x(t), u(t, x(t))u(t, x(t))-L(t, x(t), u(t, x(t)))\}dt=$$
$$=\int_{t_0}^{t_1}\mathscr{E}(t, x(t), u(t, x(t)), \dot{x}(t))\,dt.$$

Since the extremal $\hat{x}(\cdot)$ is embedded in the field, we have $\dot{\hat{x}} = u(t, \hat{x}(t))$, and therefore

$$\mathcal{J}(\hat{x}(\cdot))-\int_{\hat{\gamma}}\omega=\int_{t_0}^{t_1}\mathscr{E}(t, \hat{x}(t), u(t, \hat{x}(t)), \dot{\hat{x}}(t))\,dt=0. \tag{14}$$

Consequently,

$$\mathcal{J}(x(\cdot))-\mathcal{J}(\hat{x}(\cdot))=\int_{t_0}^{t_1}\mathscr{E}(t, x(t), u(t, x(t)), \dot{x}(t))\,dt \tag{15}$$

(as in Theorem 1 of Section 4.4.4, we have $\left.\int_{\gamma}\omega=\int_{\hat{\gamma}}\omega\right)$.

By construction (see the proof of Proposition 2 of Section 4.4.5), we have $(t, x(t, \lambda), \dot{x}(t, \lambda))=(t, x, u(t, x))\in\hat{V}=\overline{V}$ for $(t, x)=(t, x(t, \lambda))\in G$ and $(t, x(t), \dot{x}(t))\in V$ since $x(\cdot)$ is an admissible function. Consequently, Weierstrass' condition is applicable, and (15) implies that $\mathcal{J}(x(\cdot))-\mathcal{J}(\hat{x}(\cdot))\geqslant 0$. Thus, if $x(\cdot)\in KC^1([t_0, t_1], \mathbf{R}^n)$ is admissible and satisfies the boundary conditions, and the norm $\|x(\cdot) - \hat{x}(\cdot)\|_C$ is so small that the graph of $x(\cdot)$ lies in G, then $\mathcal{J}(x(\cdot))\geqslant\mathcal{J}(\hat{x}(\cdot))$. Hence, $\hat{x}(\cdot)$ yields a strong minimum in problem (1)-(4) of Section 4.4.4.

Here the equality $\mathcal{J}(x(\cdot))=\mathcal{J}(\hat{x}(\cdot))$ can only hold when $\mathscr{E}(t, x(t), u(t, x(t)), \dot{x}(t)) = 0$ for all t except possibly the points of discontinuity of $\dot{x}(\cdot)$. If $u\neq\xi\Rightarrow\mathscr{E}>0$, then $\dot{x}(t) = u(t, x(t))$, and, in particular, $\dot{x}(\cdot)$ turns out to be continuous. Further, according to relations (8) and (12) of Section 4.4.3, we have

$$p(t, x)=L_{\dot{x}}(t, x, u(t, x))\Leftrightarrow u(t, x)=$$
$$=\Xi(t, x, p(t, x))\Leftrightarrow u(t, x)=\mathscr{H}_p(t, x, p(t, x))\Rightarrow\dot{x}(t)=\mathscr{H}_p(t, x, p(t, x(t))).$$

We have obtained a formula coinciding with formula (12) of Section 4.4.4, and using the same argument as in that section we find that $x(t) \equiv \hat{x}(t)$. ∎

Remark. Relation (15) allows us to show that the quadratic functional

$$Q(\xi(\cdot))=\int_{t_0}^{t_1}\mathscr{L}(t, \xi, \dot{\xi})\,dt$$

with Lagrangian (9) can be brought to a "perfect square." Computing the Weierstrass function [and putting $\hat{\xi}(\cdot) \equiv 0$], we obtain

$$Q(\xi(\cdot))=\int_{t_0}^{t_1}|P(t)\dot{\xi}+P(t)^{-1}[\hat{L}_{x\dot{x}}(t)-\Pi(t)\Xi^{-1}(t)]\xi|^2dt, \tag{16}$$

where $P(t)^2 = \hat{L}_{\dot{x}\dot{x}}(t)$, and $\Xi(\cdot)$ and $\Pi(\cdot)$ are the solutions of the matrix system

$$\begin{aligned}\dot{\Xi}(t)&=\hat{\mathscr{H}}_{px}(t)\Xi+\hat{\mathscr{H}}_{pp}(t)\Pi,\\ \dot{\Pi}(t)&=-\hat{\mathscr{H}}_{xx}(t)\Xi-\hat{\mathscr{H}}_{xp}(t)\Pi\end{aligned} \tag{17}$$

with the initial conditions

$$\Xi(t_0')=0, \quad \Pi(t_0')=E. \tag{18}$$

4.4.6. Theorem of A. E. Noether. In Section 1.4.1 we already demonstrated a number of situations in which the Euler equation possessed a first integral. For instance, if the Lagrangian does not involve t explicitly, i.e., if it is of the form $L(x, \dot{x})$, then $H = L_{\dot{x}}\dot{x} - L = \text{const}$ (the energy integral). In this section we shall consider a general method of constructing first integrals for systems of differential equations which are Euler equations of variational problems. This method is based on a simple and, at the same time, profound theorem proved by Amalie Emmy Noether (1882-1935), a distinguished German mathematician. The universal principle expressed by the Noether theorem is often formulated thus: "Every kind of symmetry in the world generates a conservation law" or thus: "The invariance of a system with respect to a transformation group implies the existence of a first integral of that system." For instance, the invariance of a system relative to translation with respect to time, which means that the time t is not involved in the Lagrangian, yields the energy integral.

Now we pass to strict definitions. Suppose that we are given a family of mappings

$$S_\alpha: \mathbf{R}\times\mathbf{R}^n \to \mathbf{R}\times\mathbf{R}^n, \quad |\alpha|<\varepsilon_0,$$
$$S_\alpha(t, x)=(\mathfrak{T}(t, x, \alpha), \mathfrak{X}(t, x, \alpha)).$$

We shall assume that this family possesses the following two properties:

1) the functions $\mathfrak{T}$ and $\mathfrak{X}$ belong to the class C^2;

2) for $\alpha \to 0$ the relations

$$\begin{aligned}\mathfrak{T}(t, x, \alpha)&=t+\alpha T(t, x)+o(\alpha),\\ \mathfrak{X}(t, x, \alpha)&=x+\alpha X(t, x)+o(\alpha)\end{aligned} \tag{1}$$

hold.

The vector field $(T(t, x), X(t, x))$ will be called the *tangent vector field of the family* $\{S_\alpha\}$. Under the action of the transformations S_α the point $P = (t, x)$ describes a curve in $\mathbf{R} \times \mathbf{R}^n$, and $\{T(t, x), X(t, x)\}$ is the tangent vector to this curve at the point P (i.e., for $\alpha = 0$; see Fig. 40).

LEMMA. Let $x(\cdot)\in C^2([t_0, t_1], \mathbf{R}^n)$. Then there exist $\delta>0$, $\varepsilon>0$, $x(\cdot, \cdot)\in C^2((t_0-\delta, t_1+\delta)\times(-\varepsilon, \varepsilon), \mathbf{R}^n)$, $t_k(\cdot)\in C^2((-\varepsilon, \varepsilon))$, $k = 0, 1$, such that for $\alpha \in (-\varepsilon, \varepsilon)$ the image of the graph of $x(\cdot)$ is the graph of the function $x(\cdot, \alpha)$: $[t_0(\alpha), t_1(\alpha)] \to \mathbf{R}^n$. Moreover,

$$x(t, 0)\equiv x(t), \quad x_\alpha(t, 0)\equiv X(t, x(t))-\dot{x}(t)\, T(t, x(t)), \tag{2}$$

$$t_i'(0)=T(t_i, x(t_i)). \tag{3}$$

Proof. The image of the graph of $x(\cdot)$ under the mapping S_α is specified by the parametric formulas

$$\begin{aligned}t&=\tau(s, \alpha)=\mathfrak{T}(s, x(s), \alpha), \quad t_0\leqslant s\leqslant t_1,\\ x&=\chi(s, \alpha)=\mathfrak{X}(s, x(s), \alpha), \quad |\alpha|\leqslant\varepsilon_0.\end{aligned} \tag{4}$$

The two functions τ and χ are compositions of functions of class C^2 and hence they themselves belong to this class.

According to (1),

$$\frac{\partial\mathfrak{T}}{\partial t}(t, x, 0)=1, \quad \frac{\partial\mathfrak{T}}{\partial x}(t, x, 0)=0,$$

and therefore

$$\frac{\partial\tau}{\partial s}(s, 0)=\frac{\partial\mathfrak{T}}{\partial t}\frac{dt}{ds}+\frac{\partial\mathfrak{T}}{\partial x}\dot{x}(s)\equiv 1.$$

Using the compactness of the closed interval $t_0 \leqslant s \leqslant t_1$ and the continuity of $\partial\tau/\partial s$, we find $\delta_1 > 0$ and $\varepsilon_1 > 0$ such that for $|\alpha| < \varepsilon_1$ and $t_0 - \delta_1 < s < t_1 + \delta_1$ the inequality $\frac{\partial\tau}{\partial s}(s, \alpha) > 1/2$ holds. Then for a fixed $\alpha\in(-\varepsilon, \varepsilon)$ the function $s \to \tau(s, \alpha)$ is monotone increasing and it maps $[t_0, t_1]$ onto

$$[t_0(\alpha), t_1(\alpha)]=[\tau(t_0, \alpha), \tau(t_1, \alpha)]=[\mathfrak{T}(t_0, x(t_0), \alpha), \mathfrak{T}(t_1, x(t_1), \alpha)]. \tag{5}$$

These equalities show that $t_i(\alpha)$, $i = 0, 1$, are functions of class C^2, and (1) implies (3).

By virtue of the same monotonicity, the mapping A determined by the equality $A(s, \alpha) = (\tau(s, \alpha), \alpha)$ maps the rectangle $\Pi = (t_0 - \delta_1, t_1 + \delta_1) \times (-\varepsilon_1, \varepsilon_1)$ onto its image $W = A(\Pi)$ in a one-to-one manner. Since we have

$$\det\begin{Bmatrix}\partial\tau/\partial s & \partial\tau/\partial\alpha\\ 0 & 1\end{Bmatrix}=\frac{\partial\tau}{\partial s}>\frac{1}{2}$$

for the Jacobian of A, the argument similar to the one in the proof of Proposition 2 of Section 4.4.5 [Section B) of the proof] shows that the set W is open and that the inverse mapping $A^{-1}:W \to \Pi$, which is obviously of the form $(t, \alpha) \to (\sigma(t, \alpha), \alpha)$, is continuously differentiable and $\sigma \in C^1(W)$.

From (4) and (1) we obtain $\tau(s, 0)=\mathfrak{T}(s, x(s), 0)\equiv s$. Therefore, the points $(s, 0)$ remain fixed under the mapping A, and A^{-1} possesses the same property, whence $\sigma(t, 0) = 0$.

Further, $\sigma(\cdot, \cdot)$ is found as an implicit function from the equation $F(\sigma, t, \alpha) = \tau(\sigma, \alpha) - t = 0$. It follows that, in the first place, according to (4) and (11), we have

$$\sigma_\alpha(t, 0)=-F_\sigma^{-1}F_\alpha=-(\tau_s(\sigma(t, 0), 0)^{-1}\tau_\alpha(\sigma(t, 0), 0)=$$
$$=-[\mathfrak{T}_t(t, x(t), 0)+\mathfrak{T}_x(t, x(t), 0)\dot{x}(t)]^{-1}\mathfrak{T}_\alpha(t, x(t), 0)=-T(t, x(t)), \tag{6}$$

and, in the second place, since F is of class C^2, the remark made in Section 2.3.4 implies that $\sigma\in C^2(W)$.

Further, A maps the closed interval $\{(s, 0) | t_0 \leqslant s \leqslant t_1\}$ into itself, and this interval is contained in W. Consequently, W contains the rectangle $(t_0 - \delta, t_1 + \delta) \times (-\varepsilon, \varepsilon)$ if $\delta > 0$ and $\varepsilon > 0$ are sufficiently small, and the function

$$x(t, \alpha)=\mathfrak{X}(\sigma(t, \alpha), x(\sigma(t, \alpha)), \alpha) \tag{7}$$

is of class C^2 on this rectangle. Let us choose ε so that the inequalities $|t_i(\alpha) - t_i| < \delta$ hold for $|\alpha| < \varepsilon$. Then recalling the definitions of functions $\sigma(\cdot, \cdot)$ and $t_i(\cdot)$ we conclude from (4) and (7) that the image of the graph of $x(\cdot)$ under the mapping S_α is the graph of the function $x(\cdot, \alpha)$: $[t_0(\alpha), t_1(\alpha)] \to \mathbf{R}^n$.

Finally, differentiating (7) and taking into account (1), (6) and the equality $\sigma(t, 0) = 0$, we find that

$$x_\alpha(t, 0)=[\mathfrak{X}_t(t, x(t), 0)+\mathfrak{X}_x(t, x(t), 0)\dot{x}(t)]\sigma_\alpha(t, 0)+$$
$$+\mathfrak{X}_\alpha(t, x(t), 0)=\dot{x}(t)[-T(t, x(t))]+X(t, x(t)),$$

i.e., (2) holds.

Definition. Let a function $L:V \to \mathbf{R}$ be at least continuous in an open set $V\subset\mathbf{R}\times\mathbf{R}^n\times\mathbf{R}^n$. The integral functional

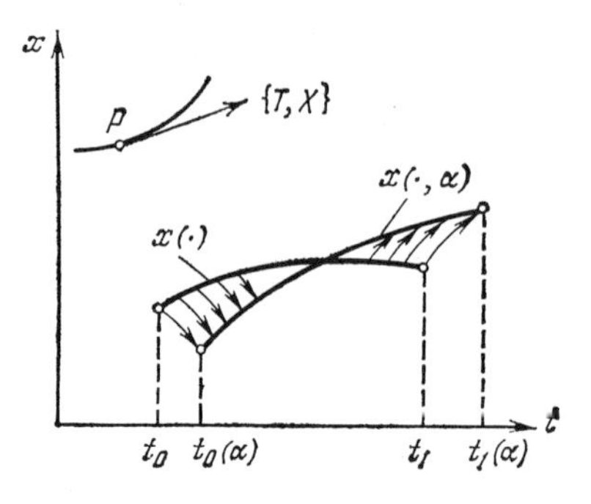

Fig. 40

$$\mathcal{J}(x(\cdot),\ t_0,\ t_1)=\int_{t_0}^{t_1} L(t,\ x(t),\ \dot{x}(t))\,dt \tag{8}$$

is said to be *invariant with respect to a family of mappings* $\{S_\alpha\}$ if for any function $x(\cdot)\in C^2([t_0,\ t_1],\ \mathbf{R}^n)$ such that $\{(t,\ x(t),\ \dot{x}(t))\,|\,t_0\leqslant t\leqslant t_1\}\subset V$ the relation

$$\mathcal{J}(x(\cdot,\ \alpha),\ t_0(\alpha),\ t_1(\alpha))\equiv\mathcal{J}(x(\cdot),\ t_0,\ t_1) \tag{9}$$

holds for all sufficiently small α.

[Here $x(\cdot,\ \alpha)$, $t_0(\alpha)$, $t_1(\alpha)$ are constructed from $x(\cdot)$ as in the lemma, and the interval where (9) holds may depend on $x(\cdot)$.]

Theorem of A. E. Noether. Let the function L, L_x, $L_{\dot{x}}$ be continuous in an open set $V\subset\mathbf{R}\times\mathbf{R}^n\times\mathbf{R}^n$, and let the integral functional (8) be invariant with respect to a family of transformations of class C^2 satisfying condition (1). Then the function

$$\varphi(t,\ x,\ \dot{x})=L_{\dot{x}}(t,\ x,\ \dot{x})X(t,\ x)-[L_{\dot{x}}(t,\ x,\ \dot{x})\dot{x}-L(t,\ x,\ \dot{x})]\,T(t,\ x) \tag{10}$$

is constant on every solution $x(\cdot)$ of the Euler equation

$$\frac{d}{dt}L_{\dot{x}}(t,\ x(t),\ \dot{x}(t))=L_x(t,\ x(t),\ \dot{x}(t)) \tag{11}$$

such that $x(\cdot)\in C^2([t_0,\ t_1],\ \mathbf{R}^n)$, $\{(t,\ x(t),\ \dot{x}(t))\,|\,t_0\leqslant t\leqslant t_1\}\subset V$ and $L_{\dot{x}}(\cdot,\ x(\cdot),\ \dot{x}(\cdot))\in C^1([t_0,\ t_1],\ \mathbf{R}^{n*})$.

Proof. A) Let $x(\cdot):[t_0,\ t_1]\to\mathbf{R}^n$ be the solution of Eq. (11) which is spoken of in the condition of the theorem. Using the lemma we construct a family $(x(\cdot,\ \alpha),\ t_0(\alpha),\ t_1(\alpha))$, where $x(\cdot,\ \cdot)$ is a function of class C^2 on $(t_0-\delta,\ t_1+\delta)\times(-\varepsilon,\ \varepsilon)$. Let us fix a closed interval $\Delta=[\beta,\ \gamma]$ such that the inequalities $t_0-\delta<\beta<t_0<t_1<\gamma<t_1+\delta$ hold and diminish ε, if necessary, for the inequalities $\beta<t_0(\alpha)<t_1(\alpha)<\gamma$ to hold for $|\alpha|<\varepsilon$.

Let us show that the mapping $\Phi:(-\varepsilon,\ \varepsilon)\to C^1(\Delta,\ \mathbf{R}^n)\times\mathbf{R}^2$, for which $\alpha\to\{x(\cdot,\ \alpha),\ t_0(\alpha),\ t_1(\alpha)\}$, is Fréchet differentiable at the point $\alpha=0$. It is clear that it suffices to verify this property for the first component of $\alpha\to x(\cdot,\ \alpha)$. Let us fix $\bar{\varepsilon}\in(0,\ \varepsilon)$ and apply the mean value theorem [to the differentiable mappings $\alpha\to x(t,\ \alpha)$; t is fixed]; then for $|\alpha|\leqslant\bar{\varepsilon}$, $\alpha\to 0$, we obtain

$$\sup_{t\in\Delta}\left|\frac{x(t,\ \alpha)-x(t,\ 0)-\alpha x_\alpha(t,\ 0)}{\alpha}\right|\leqslant\sup_{t\in\Delta}\ \sup_{c\in[0,\ \alpha]}|x_\alpha(t,\ c)-x_\alpha(t,\ 0)|\to 0,$$

$$\sup_{t\in\Delta}\left|\frac{x_t(t,\ \alpha)-x_t(t,\ 0)-\alpha x_{t\alpha}(t,\ 0)}{\alpha}\right|\leqslant\sup_{t\in\Delta}\ \sup_{c\in[0,\ \alpha]}|x_{t\alpha}(t,\ c)-x_{t\alpha}(t,\ 0)|\to 0,$$

since x_α and $x_{t\alpha}$ are uniformly continuous on the compact set $\Delta \times [-\bar{\varepsilon}, \bar{\varepsilon}]$. Consequently, in the space $C^1(\Delta, \mathbf{R}^n)$ we have

$$x(\cdot, \alpha) = x(\cdot, 0) + \alpha x_\alpha(\cdot, 0) + o(|\alpha|), \tag{12}$$

which means the Fréchet differentiability of Φ. Moreover, (12) implies that

$$\Phi'(0) = \{x_\alpha(\cdot, 0), t_0'(0), t_1'(0)\}. \tag{13}$$

B) Let us put $\mathcal{I}(\alpha) = (\mathcal{I} \circ \Phi)(\alpha) = \mathcal{I}(x(\cdot, \alpha), t_0(\alpha), t_1(\alpha))$. According to (9), we have $\mathcal{I}(\alpha) \equiv \mathcal{I}(0)$, and hence $\mathcal{I}'(0) = \mathcal{I}'(x(\cdot), t_0, t_1)[\Phi'(0)] = 0$. Using formula (9) of Section 2.4.2 for the derivative of an integral mapping, we derive from (13) the equality

$$0 = \mathcal{I}'(x(\cdot), t_0, t_1)[\Phi'(0)] = \int_{t_0}^{t_1} \{L_x(t, x(t), \dot{x}(t)) x_\alpha(t, 0) +$$

$$+ L_{\dot{x}}(t, x(t), \dot{x}(t)) x_{\alpha t}(t, 0)\} dt + L(t_i, x(t_i), \dot{x}(t_i)) t_i'(0) \Big|_{i=0}^{i=1}. \tag{14}$$

By the hypothesis, $L_{\dot{x}}(t, x(t), \dot{x}(t))$ is continuously differentiable; integrating by parts in (14) (as in Section 1.4.1) and taking into account (11), (2), (3), and (10), we obtain

$$0 = [L_{\dot{x}}(t_i, x(t_i), \dot{x}(t_i)) x_\alpha(t_i, 0) + L(t_i, x(t_i), \dot{x}(t_i)) t_i'(0)]_{i=0}^{i=1} =$$

$$= [L_{\dot{x}}(t_i, x(t_i), \dot{x}(t_i)) X(t_i, x(t_i)) + (L(t_i, x(t_i), \dot{x}(t_i)) -$$

$$- L_{\dot{x}}(t_i, x(t_i), \dot{x}(t_i)) \dot{x}(t_i)) T(t_i, x(t_i))]_{i=0}^{i=1} = \varphi(t_1, x(t_1), \dot{x}(t_1)) - \varphi(t_0, x(t_0), \dot{x}(t_0)).$$

Thus,

$$\varphi(t_1, x(t_1), \dot{x}_1(t_1)) = \varphi(t_0, x(t_0), \dot{x}(t_0)).$$

Repeating the same argument for an arbitrary closed interval $[t_0, t] \subset [t_0, t_1]$, we arrive at the equality

$$\varphi(t, x(t), \dot{x}(t)) \equiv \varphi(t_0, x(t_0), \dot{x}(t_0)) \equiv \text{const.} \ \blacksquare$$

COROLLARY. If, under the conditions of the Noether theorem, the assumptions a) and b) of Proposition 2 of Section 4.4.3 hold, then the function

$$\psi(t, x, p) = pX(t, x) - \mathcal{H}(t, x, p) T(t, x) \tag{15}$$

is constant on every solution $(x(t), p(t))$ of the Hamiltonian system (1) of Section 4.4.3.

The *proof* of the corollary is left to the reader. It is seen from formula (15) that the first integral constructed in the Noether theorem is the value of the Poincaré–Cartan differential form $\omega = p\,dx - \mathcal{H}\,dt$ on the tangent vector field of the family $\{S_\alpha\}$.

As an example, let us consider the derivation of the energy and the momentum integrals from the Noether theorem (Section 1.4.1). If L does not depend on t, functional (8) is invariant with respect to the transformations $(t, x) \to (t + \alpha, x)$ (check this!). Here $T = 1$, $X = 0$, and therefore $\varphi = L - L_{\dot{x}}\dot{x} = -\mathcal{H}$ is constant. In case L does not depend on one of the components of the vector x, say on x_k, then functional (8) is invariant with respect to the transformations $(t, x) \to (t, x + \alpha e_k)$ (check this!). In this case $T = 0$, $X = e_k$, and therefore $\varphi = L_{\dot{x}} e_k = L_{\dot{x}_k}$.

Some other examples of the application of the Noether theorem will be presented in the next section.

4.4.7. Lagrange Variational Principle and the Conservation Laws in Mechanics. Let us consider a system of n particles ("material points") with masses $m_1,\ldots,m_n$; let the radius vector of the i-th particle be $r_i = (x_i, y_i, z_i)$. For brevity, we put

$$r=(r_1, \ldots, r_n)\in \mathbf{R}^{3n}, \quad \frac{\partial}{\partial r}=\left(\frac{\partial}{\partial r_1}, \ldots, \frac{\partial}{\partial r_n}\right), \quad \frac{\partial}{\partial r_i}=\left(\frac{\partial}{\partial x_i}, \frac{\partial}{\partial y_i}, \frac{\partial}{\partial z_i}\right).$$

We shall suppose that the interaction between the particles and between the given system of particles and the surrounding medium is described by the potential energy $U(t, r)$ so that the equations of motion have the form

$$m_i\ddot{r}_i=F_i=-\partial U/\partial r_i. \tag{1}$$

The kinetic energy of the system is determined by the equality

$$K=\frac{1}{2}\sum_{i=1}^{n} m_i(\dot{r}_i\,|\,\dot{r}_i). \tag{2}$$

THEOREM (Lagrange Variational Principle). The equations of motion (1) of the mechanical system are the Euler equations for the variational problem

$$\int_{t_0}^{t_1} L(t, r, \dot{r})\,dt \to \text{extr}, \tag{3}$$

$$r(t_0)=r^0, \quad r(t_1)=r^1, \quad L=K-U.$$

Proof. From (1) and (2) we derive

$$m_i\ddot{r}_i=\frac{d}{dt}(m_i\dot{r}_i)=\frac{d}{dt}\frac{\partial K}{\partial \dot{r}_i}=\frac{d}{dt}L_{\dot{r}_i}=L_{r_i}=\frac{\partial U}{\partial r_i}. \quad \blacksquare$$

The significance of this theorem lies far beyond the simplest situation to which we are confining ourselves. In physics and in mechanics the properties of a system are often described by setting directly its Lagrangian L. In this case K, U, and $L = K - U$ are not necessarily expressed in terms of Cartesian coordinates and may involve some other coordinates. The Lagrangian may also involve certain terms describing magnetic or gyroscopic (nonpotential) forces, etc. If the Lagrangian is given, the Euler equations of the variational principle (3) determine the law of motion of the system under consideration.

In particular, the variational principle is convenient because it allows us to derive first integrals ("conservation laws") with the aid of the Noether theorem. We shall demonstrate this by the examples of the so-called classical integrals corresponding to the conservation laws for energy, linear momentum, and angular momentum.

a) A Conservative System. In this case U does not depend on t. The transformation group is $(t, r) \to (t + \alpha, r)$. The function $r(\cdot):[t_0, t_1] \to \mathbf{R}^{3n}$ goes into the function $r_\alpha(\cdot):[t_0 + \alpha, t_1 + \alpha] \to \mathbf{R}^{3n}$, $r_\alpha(t) = r(t - \alpha)$ under this transformation. The functional (3) is invariant:

$$\int_{t_0+\alpha}^{t_1+\alpha} L(r_\alpha(t), \dot{r}_\alpha(t))\,dt=\int_{t_0+\alpha}^{t_1+\alpha} L(r(t-\alpha), \dot{r}(t-\alpha))\,dt=\int_{t_0}^{t_1} L(r(t), \dot{r}(t))\,dt.$$

The tangent vector field is $T = 1$, $R = 0$. The first integral:

$$\varphi=\sum_{i=1}^{n} L_{\dot{r}_i}R_i-\left[\sum_{i=1}^{n} L_{\dot{r}_i}\dot{r}_i-L\right]T=-\sum_{i=1}^{n} m_i(\dot{r}_i\,|\,\dot{r}_i)+K-U=-K-U.$$

Consequently, the total energy

$$\mathcal{H}=K+U$$

is a first integral of the conservative system (this is the "*energy conservation law*").

b) A Free Particle. In this case U does not depend on the radius r_k of one of the particles. The transformation group is

$$(t,\ r_1,\ \ldots,\ r_k,\ \ldots,\ r_n)\mapsto(t,\ r_1,\ \ldots,\ r_k+\alpha l,\ \ldots,\ r_n),$$

where $l\in\mathbf{R}^3$ is arbitrary. The invariance of the functional (3) is obvious. The tangent vector field is

$$T=0,\ R=(R_1,\ \ldots,\ R_k,\ \ldots,\ R_n)=(0,\ \ldots,\ l,\ \ldots,\ 0).$$

The corresponding first integral is

$$\varphi=\sum_{i=1}^{n} L_{\dot r_i}R_i=\frac{\partial K}{\partial \dot r_k}l=m_k(\dot r_k\,|\,l).$$

Since l is arbitrary, $p_k = m_k\dot r_k$ must be constant. Hence, if there is no interaction between the k-th particle and the rest of the system (i.e., the k-th particle makes no contribution to the potential energy), then the "*conservation law for the linear momentum*" of that particle takes place.

c) There Are No External Forces. This means that U depends on the differences $r_i - r_j$ solely and L does not change if the system undergoes translation in space as a rigid body, i.e., if a transformation $(t, r_1,\ldots, r_n) \to (t, r_1 + \alpha l,\ldots,r_n + \alpha l)$ is performed where l is an arbitrary vector. Here $T = 0$, $R = (l,\ldots,l)$, and $\sum_{i=1}^{n}\left(\frac{\partial L}{\partial \dot r_i}\Big|\,l\right)=\left(\sum_{i=1}^{n} m_i\dot r_i\Big|\,l\right)$ is preserved. Since l is arbitrary,

$$P=\sum_{i=1}^{n} m_i\dot r_i=\text{const.}$$

Thus, if there are only internal forces, the "*law of conservation of the total linear momentum of the system*" holds.

d) Rotational Symmetry. In this case U depends on the pairwise distances $|r_i - r_j|$ solely. The Lagrangian does not change when the system undergoes an orthogonal transformation $\mathfrak{G}$ because

$$L(\mathfrak{G}r_i,\ \mathfrak{G}\dot r_i)=K-U=$$

$$=\frac{1}{2}\sum_{i=1}^{n} m_i(\mathfrak{G}\dot r_i\,|\,\mathfrak{G}\dot r_i)-U(|\,\mathfrak{G}r_i-\mathfrak{G}r_j\,|)=\frac{1}{2}\sum_{i=1}^{n} m_i(\dot r_i\,|\,\dot r_i)-U(|\,r_i-r_j\,|)$$

(the scalar products and the lengths do not change under an orthogonal transformation). Let us fix a vector ω and consider the transformation group corresponding to the uniform rotation of the system about the origin with the angular velocity ω. Here the tangent vector field is $T = 0$, $R = (\omega \times r_1,\ldots,\omega \times r_n)$. The first integral is

$$\varphi=\sum_{i=1}^{n}\frac{\partial L}{\partial \dot r_i}R_i=\sum_{i=1}^{n} m_i(\dot r_i\,|\,\omega\times r_i)=\sum_{i=1}^{n} m_i(r_i\times\dot r_i\,|\,\omega).$$

Since ω is arbitrary, the "*conservation law for the angular momentum* $M = \sum_{i=1}^{n} m_ir_i\times\dot r_i$ *of the system*" must hold:

$$M=\sum_{i=1}^{n} m_ir_i\times\dot r_i = \text{const.}$$

PROBLEMS

In this section we gather 100 problems. They are grouped by certain subjects irrespective of their difficulty. We begin with problems of geometrical character (Problems 1-14, 22, and 23) whose statement is evoked by the problems of Euclid, Kepler, Steiner, and Apollonius mentioned in Chapter 1. There is extensive literature devoted to extremal relations in geometry. The reader can find many interesting problems in the book by D. O. Shklyarskii, N. N. Chentsov, and I. M. Yaglom.‡ For most of them the standard solution (provided that the formalization is correct) is not more complicated than the "geometrical" one.

Problems 15-21 are related to the topic "Inequalities." Here too the reader can enlarge the list of the problems by resorting to the book by G. H. Hardy, G. E. Littlewood, and G. Polya.††

Problems 24-30 are taken from two topics of approximation theory: polynomials of least deviation from zero and inequalities for derivatives. Extremal problems play an important role in approximation theory, and most of the solved problems admit of the standard solution as well. A series of extremal problems in approximation theory is considered in the book by V. M. Tikhomirov [84].

Problems 31-100 are selected from the problems which were considered at the seminars in the course "Optimal Control" at the Department of Mathematics and Mechanics of Moscow State University. The most difficult problems are marked by stars.

We would like to draw the reader's attention to the fact that the application of the ideas of optimal control theory makes it possible to exhaustively analyze most of the well-known problems of the classical calculus of variations in a very simple way [for example, see the solution of Dido's problem (Problem 55), the problem on the harmonic oscillator (Problem 47), etc.].

---

†In the preparation of the material of this section the authors received valuable help from E. M. Galeev (Can. Phys. Math.) and from the students Le Chyong Tung, Yu. Aleksandrov, V. Sokolova, and K. Makhmutov. To all of them the authors express their gratitude.
‡D. O. Shklyarskii, N. N. Chentsov, and I. M. Yaglom, Geometrical Inequalities and Problems on Maximum and Minimum [in Russian], Nauka, Moscow (1970)
††G. H. Hardy, G. E. Littlewood, and G. Polya, Inequalities, Cambridge Univ. Press, Cambridge (1934).

The answers to all the problems and hints to many of them are given (or references are made to the literature where the problems are solved). For the key problems the solutions are presented.

1. Find the point in an n-dimensional Euclidean space for which the weighted sum of the squares of the distances (with positive weights) from that point to several fixed points is smallest.

2. Solve Problem 1 on condition that the sought-for point lies on the unit sphere.

3. Solve Problem 1 on condition that the sought-for point belongs to the unit ball.

4. Given three different points in a plane, find the point in the plane such that the weighted sum of the distances (with positive weights) from that point to the three given points is smallest (this is a generalization of Steiner's problem; cf. Section 1.1.2).

5. Find the point on a plane such that the sum of the distances from the point to four different points is smallest.

6. Inscribe in the unit circle the triangle with the maximum sum of the squares of its sides.

7. Inscribe in the unit circle the triangle with the maximum weighted sum (with positive weights) of the squares of its sides.

8. Inscribe in a sphere of unit radius in an n-dimensional Euclidean space the cylinder with maximum volume (this is a generalization of Kepler's problem; cf. Section 1.1.2).

9. Inscribe in a sphere of unit radius in an n-dimensional Euclidean space the cone with maximum volume.

10. Inscribe in a sphere of unit radius in an n-dimensional Euclidean space the rectangular parallelepiped with maximum volume.

11. Find the simplex with maximum volume whose one vertex is at the center of the unit sphere and the other vertices lie on that sphere.

12. Inscribe in a sphere of unit radius in an n-dimensional Euclidean space the simplex with maximum volume.

13. In a given triangle inscribe the triangle the sum of the squares of whose sides is smallest.

14. Find the shortest distance from a point lying outside an ellipsoid in a Hilbert space† to that ellipsoid (this is a generalization of the problem of Apollonius; cf. Section 1.1.2).

15. $\prod_{i=1}^{n} x_i^{\alpha_i} \to \sup; \quad \sum_{i=1}^{n} a_i x_i = \sigma, \; x_i \geqslant 0, \quad \alpha_i \geqslant 1, \quad i=1,\ldots,n$, where α_i are fixed numbers. From the solution of the problem derive the inequality for the arithmetic and the geometric means.

16. $\sum_{i=1}^{n} |x_i|^p \to \sup; \quad \sum_{i=1}^{n} |x_i|^q = 1$, where $p \geqslant 1$ and $q \geqslant 1$ are fixed numbers. From the solution of the problem derive the inequality for power means.‡

---

†By an ellipsoid in a Hilbert space H is meant the image of the unit ball under a continuous linear mapping Λ of the space into itself such that $\overline{\operatorname{Im} \Lambda} = H$.

‡By the power mean with an exponent $p \neq 0$ of nonnegative numbers $a_1, \ldots, a_n$ is meant the expression $\left(\sum_{i=1}^{n} a_i^p / n \right)^{1/p}$.

17. $\sum_{i=1}^{n} a_i x_i \to \sup$; $\sum_{i=1}^{n} x_i^2 = 1$, where a_i are fixed numbers.

18. $(\tilde{a}|x) \to \sup$; $(x|x) = 1$, $x \in H$, where H is a Hilbert space; a is a fixed element of M. From the solutions of Problems 17 and 18 derive the Cauchy—Bunyakovskii inequalities for $\mathbf{R}^n$ and for a Hilbert space.

19. $\sum_{i=1}^{n} a_i x_i \to \sup$; $\sum_{i=1}^{n} |x_i|^p = 1$, where p is a fixed number, $p \geqslant 1$. From the solution of the problem derive Hölder's inequality.

20. Among all discrete random variables assuming n values find the one with the maximum entropy.†

21. Among absolutely continuous probability distributions on the straight line $\mathbf{R}$ with a given mathematical expectation and a given dispersion find the random variable with the maximum differential entropy.†

22. Find the distance from a given point ξ in $\mathbf{R}^n$ to a given hyperplane $\sum_{i=1}^{n} a_i x_i = b$.

23. Find the distance from a point ξ in $\mathbf{R}^n$ to the straight line, with the direction vector ξ, passing through a point η.

24. Among the polynomials of the form $t^2 + \alpha_1 t + \alpha_2$ find the one of least deviation from zero in the space $Z_2([-1, 1])$.

25. Among the polynomials of the form $t^n + \alpha_1 t^{n+1} + \ldots + \alpha_n$ find the one of least deviation from zero in the space $Z_2([-1, 1])$.

26. Among the polynomials of the form $t^n + \alpha_1 t^{n-1} + \ldots + \alpha_n$ find the one of least deviation from zero in the space $C([-1, 1])$.

27. Among all nonnegative trigonometric polynomials of the form $x(t) = 1 + 2\rho_1 \cos t + \ldots + 2\rho_n \cos nt$ find the one for which the coefficient ρ_1 is greatest.

28. $x(0) \to \sup$; $\int_0^\infty (x^2 + (x^{(n)})^2)\, dt \leqslant 1$.

29. $\int_0^\infty \dot{x}^2\, dt \to \sup$; $\int_0^\infty (x^2 + \ddot{x}^2)\, dt \leqslant 1$.

30.* $\int_{-\pi}^{\pi} (x^{(k)})^2\, dt \to \sup$; $\int_{-\pi}^{\pi} x^2\, dt \leqslant \gamma_1$, $\int_{-\pi}^{\pi} (x^{(n)})^2\, dt = \gamma_2$, $x^{(j)}(-\pi) = x^{(j)}(\pi)$,

$j = 0, 1, \ldots, n-1$, $\gamma_1 < \gamma_2$, $0 \leqslant k < n$.

31.‡ $\int_{(0,0)}^{(T_0,0)} (\dot{x}^2 + x)\, dt \to \inf$.

---

†The entropy of a collection of positive numbers $p_1, \ldots, p_n$ whose sum is equal to unity is the number $H = \sum_{i=1}^{n} p_i \log(1/p_i)$. Similarly, by the (differential) entropy of a positive function $p(\cdot)$ whose integral is equal to unity is meant the number $-\int_{\mathbf{R}} p(x) \log p(x)\, dx$.

‡Here the limits of integration indicate the boundary conditions: $x(0) = 0$, $x(T_0) = 0$. In the next problem 32 the value $x(T_0)$ is free. We write T_0 when this instant is fixed and T when it is nonfixed (e.g., in Problems 71, 76, etc.).

32. $\int_{(0,0)}^{T_0} (\dot{x}^2+x)\,dt \to \inf.$

33. $\int_{(0,0)}^{T_0} (\dot{x}^2+x)\,dt \to \inf, \qquad |\dot{x}| \leq 1.$

34. $\int_{(0,0)}^{(T_0,0)} (\dot{x}^2+x)\,dt = \inf, \qquad |\dot{x}| \leq 1.$

35. $\int_{(1,3)}^{(2,1)} t^2\dot{x}^2\,dt \to \inf.$

36. $\int_{(0,0)}^{(1,1)} t^2\dot{x}^2\,dt \to \inf$ (Weierstrass' example)

37. $\int_{(0,0)}^{(1,1)} t^{2/3}\dot{x}^2\,dt \to \inf$ (Hilbert's example).

38.* $\int_{(0,0)}^{(1,\xi)} \frac{t^\alpha|\dot{x}|^\beta}{\beta}\,dt \to \inf, \qquad \alpha, \beta \in \mathbf{R}$ (the generalized example of Weierstrass and Hilbert).

39. $\int_{(0,0)}^{(1,1)} (\dot{x}^2-x)\,dt \to \inf.$

40. $\int_{(1,0)}^{(2,1)} (t\dot{x}^2-x)\,dt \to \inf.$

41. $\int_{(0,1)}^{(1,4)} (\dot{x}^2+x\dot{x}+12tx)\,dt \to \inf.$

42. $\int_{(0,0)}^{(T_0,\xi)} (\dot{x}^2+x^2)\,dt \to \inf.$

43. $\int_{(0,0)}^{(T_0,\xi)} (\dot{x}^2+x^2)\,dt \to \inf, \qquad |\dot{x}| \leq 1.$

44. $\int_{(0)}^{(1,\xi)} \left(\frac{\dot{x}^2+x^2}{2}+|\dot{x}|\right)dt \to \inf.$

45. $\int_{(-1,\xi)}^{(1,\xi)} ((|\dot{x}|-1)_+^2+x^2)\,dt \to \inf.$

46.* $\int_0^\infty x^2\,dt \to \inf, \qquad |\ddot{x}| \leq 1, \qquad x(0)=1.$

47. $\int\limits_{(0,0)}^{(T_0,\xi)} (\dot{x}^2 - x^2)\, dt \to \inf.$

48. $\int\limits_{(0,0)}^{T_0} (\dot{x}^2 - x^2)\, dt \to \inf.$

49.* $\int\limits_{(0,0)}^{(T_0,0)} (\dot{x}^2 - x^2)\, dt \to \inf, \qquad |\dot{x}| \leq 1.$

50.* $\int\limits_{(0,0)}^{(T_0,0)} ((|\dot{x}| - 1)_+^2 - x^2)\, dt \to \inf.$

51. $\int\limits_{(t_0,x_0)}^{(t_1,x_1)} \frac{x\, dt}{\dot{x}^2} \to \inf, \qquad x \geq 0.$

52. $\int\limits_0^{T_0} \dot{x}^2\, dt - \alpha x^2(T_0) \to \inf, \qquad x(0) = 0.$

53. $\int\limits_0^1 \dot{x}^2\, dt \to \inf; \qquad \int\limits_0^1 x\, dt = 3, \qquad x(0) = 1, \qquad x(1) = 6.$

54. $\int\limits_{(0,0)}^{(T_0,0)} \dot{x}^2\, dt \to \inf; \qquad \int\limits_0^{T_0} x^2\, dt \leq 1.$

55. $\int\limits_{-T_0}^{T_0} x\, dt \to \inf; \qquad \int\limits_{-T_0}^{T_0} \sqrt{1 + \dot{x}^2}\, dt \leq l, \qquad x(-T_0) = x(T_0)$ (Dido's problem).

56. $\int\limits_{(0,0)}^{(\pi,0)} \dot{x}^2\, dt \to \inf; \qquad \int\limits_0^{\pi} x \sin t\, dt = 1.$

57. $\int\limits_{(0,2)}^{(1,-14)} \dot{x}^2\, dt \to \inf; \qquad \int\limits_0^1 x\, dt = -1.5, \qquad \int\limits_0^1 tx\, dt = -2.$

58. $\int\limits_{(0,-4)}^{(1,4)} \dot{x}^2\, dt \to \inf; \qquad \int\limits_0^1 tx\, dt = 0.$

59. $\int\limits_0^1 \ddot{x}^2\, dt \to \inf; \qquad x(0) = \dot{x}(0) = 0, \qquad x(1) = 1.$

60. $\int\limits_0^1 \ddot{x}^2\, dt \to \inf; \qquad x(0) = 1, \qquad \dot{x}(0) = 0.$

61. $\int\limits_0^2 x\, dt \to \inf; \qquad -4 \leq \ddot{x} \leq 2, \qquad x(0) = 1, \qquad \dot{x}(0) = 0, \qquad \dot{x}(2) = -2.$

62. $\int_0^{T_0} x\,dt \to \inf; \quad |\ddot{x}| \leq 1, \quad x(0) = x(T_0) = 0.$

63. $\int_0^{T_0} x\,dt \to \inf; \quad |\ddot{x}| \leq 1, \quad x(0) = \dot{x}(0) = x(T_0) = 0.$

64. $\int_0^{T_0} x\,dt \to \inf; \quad |\ddot{x}| \leq 1, \quad x(0) = \dot{x}(0) = x(T_0) = \dot{x}(T_0) = 0.$

65.* $\int_{-1}^{1} x\,dt \to \inf; \quad |x^{(n)}| \leq 1, \quad x^{(k)}(\pm 1) = 0, \quad k = 0, 1, \ldots, n-1.$

66. $\int_0^1 x\,dt \to \inf; \quad \int_0^1 \ddot{x}^2\,dt = 144/5, \quad x(0) = 1, \quad \dot{x}(0) = -4.$

67. $x(2) = \sup; \quad |\dot{x}| \leq 2, \quad x(0) = 0, \quad \int_0^2 \dot{x}^2\,dt = 2.$

68. $T \to \inf; \quad -1 \leq \ddot{x} \leq 3, \quad x(0) = 1, \quad \dot{x}(0) = \dot{x}(T) = 0, \quad x(T) = -1.$

69. $T \to \inf; \quad |\ddot{x}| \leq 1, \quad x(0) = \xi_1, \quad \dot{x}(0) = \xi_2, \quad x(T) = 0.$

70. $T \to \inf; \quad |\ddot{x}| \leq 1, \quad x(0) = \xi_1, \quad \dot{x}(0) = \xi_2, \quad x(T) = \dot{x}(T) = 0$ (the simplest time-optimal problem).

71. $T \to \inf; \quad \int_0^T \ddot{x}^2\,dt = 4, \quad x(0) = 0, \quad \dot{x}(0) = 1, \quad \dot{x}(T) = -1.$

72. $T \to \inf; \quad |\ddot{x}| \leq 2, \quad x(-1) = -1, \quad \dot{x}(-1) = 0, \quad x(T) = 1, \quad \dot{x}(T) = 0.$

73. $T \to \inf; \quad \ddot{x} + x = u, \quad |u| \leq 1, \quad x(0) = \xi_1, \quad \dot{x}(0) = \xi_2, \quad x(T) = \dot{x}(T) = 0.$

74. $T \to \inf; \quad \ddot{x} + x = u, \quad |u| \leq 1, \quad x(0) = \xi_1, \quad \dot{x}(0) = \xi_2, \quad x^2(T) + \dot{x}^2(T) = R^2,$ $\xi_1^2 + \xi_2^2 > R^2.$

75. $T \to \inf; \quad \ddot{x} + (1 - \varepsilon u)x = 0, \quad x(0) = \xi_1, \quad \dot{x}(0) = \xi_2, \quad x^2(T) + \dot{x}(T) = 1,$ $0 \leq u \leq 1, \quad 0 < \varepsilon < 1.$

76.† $\int_0^T \dot{x}^2\,dt \to \text{extr}; \quad x(0) = 0, \quad T + x(T) + 1 = 0.$

77. $\int_0^1 x^2\,dt \to \sup; \quad |\dot{x}| \leq 1, \quad x(0) = 0$ (find all the extremals).

78. $\int_0^{T_0} (\ddot{x}^2 + x^2)\,dt \to \inf; \quad x(0) = \dot{x}(0), \quad x(T_0) = \xi_1, \quad \dot{x}(T_0) = \xi_2.$

79. $\int_0^{T_0} (\ddot{x}^2 - x^2)\,dt \to \inf; \quad x(0) = \dot{x}(0) = 0, \quad x(T_0) = 0, \quad \dot{x}(T_0) = 0.$

---

†Here T is nonfixed; see the footnote to Problem 31.

80. $\int_0^{T_0} (\ddot{x}^2 - x^2)\, dt \to \inf; \quad x(0) = \dot{x}(0) = 0, \quad x(T_0) = \xi_1, \quad \dot{x}(T_0) = \xi_2.$

81. $\int_0^1 \ddot{x}^2\, dt \to \inf; \quad |\ddot{x}| \leqslant 1, \quad x(0) = \dot{x}(0) = 0, \quad x(1) = -11/24.$

82. $\int_0^2 \ddot{x}^2\, dt = \inf; \quad \ddot{x} \geqslant 6, \quad x(0) = \dot{x}(0) = 0, \quad x(2) = 17.$

83. $\int_0^1 (\ddot{x}^2 - 48x)\, dt \to \inf; \quad x(0) = x(1) = 0.$

84. $\int_0^{T_0} \ddot{x}^2\, dt \to \inf; \quad \int_0^{T_0} x^2\, dt = 1, \quad x(0) = \dot{x}(0) = 0.$

85. $\int_0^2 u^2\, dt + x^2(0) = \inf; \quad \ddot{x} - x = u, \quad \dot{x}(2) = 1.$

86. $\int_0^2 u^2\, dt + x^2(0) \to \inf; \quad \ddot{x} - x = u, \quad \dot{x}(2) = 1, \quad x(2) = 0.$

87. $\int_0^2 u^2\, dt + 4x^2(2) \to \inf; \quad x(0) = 1, \quad \ddot{x} + 4x = u.$

88. $\int_0^2 u^2\, dt + 4x^2(2) \to \inf; \quad x(0) = 1, \quad \dot{x}(0) = 0, \quad \ddot{x} + 4x = u.$

89. $\int_0^{T_0} \left(\frac{\dot{x}_1^2 + \dot{x}_2^2}{2} + x_1 x_2\right) dt \to \inf; \quad x_1(0) = x_2(0) = x_1(1) = 0, \quad x_2(1) = 2 \sin 1 \sinh 1.$

90. $\int_0^1 (x_1 + x_2)\, dt \to \text{extr}; \quad \int_0^1 \dot{x}_1 \dot{x}_2\, dt = 0, \quad x_1(0) = x_2(0), \quad x_1(1) = 1, \quad x_2(1) = -3.$

91. $\int_0^1 |u|^p\, dt \to \inf; \quad \ddot{x} = u, \quad x(0) = 0, \quad \dot{x}(0) = 0, \quad x(1) = \xi_1, \quad \dot{x}(1) = \xi_2, \quad p \geqslant 1.$

92.* $\int_0^{T_0} |\ddot{x} - g|\, dt \to \inf; \quad x(0) = x_0, \quad \dot{x}(0) = v_0, \quad x(T_0) = x_1, \quad \dot{x}(T_0) = v_1, \quad x \in \mathbf{R}^n$

(Goddard's problem).

93. $\int_{(t_0, x_0)}^{(t_1, x_1)} x^\alpha \sqrt{1 + \dot{x}^2}\, dt \to \inf, \quad \alpha < 0.$

94. $\int_{(t_0, \xi)}^{(t_1, \xi)} x\sqrt{1 + \dot{x}^2}\, dt \to \inf.$

95. $\int_0^T \frac{\sqrt{1+\dot{x}^2}}{x}\,dt \to \inf;\quad x(0)=1,\quad T-x(T)=1.$

96.* $\int_0^T (1+\varepsilon|\ddot{x}|)\,dt \to \inf;\quad x(0)=\xi_1;\quad \dot{x}(0)=\xi_2,\quad x(T)=\dot{x}(T)=0,\quad |\ddot{x}|\leqslant 1.$

97.* $\int_0^1 |u|^p\,dt \to \inf;\quad \ddot{x}+ux=0,\quad x(0)=x(1)=0,\quad \dot{x}(0)=1,\quad p>1.$

98.* $\int_0^1 |u|\,dt \to \inf;\quad \ddot{x}+ux=0,\quad x(0)=x(1)=0,\quad \dot{x}(0)=1.$

99.* $\int_0^{T_0} (\sqrt{\dot{x}_1^2+\dot{x}_2^2}+\sqrt{(\dot{x}_1-\sin t)^2+(\dot{x}_2+\cos t)^2}\,dt \to \inf;\quad x_1(0)=x_2(0)=0,\quad x_1(T_0)=\xi_1,$
$x_2(T_0)=\xi_2$ (Ulam's problem).

100.* $\int_0^{T_0} \left(\frac{\cot x}{xg+\gamma t^2}+\frac{xt}{xg+\gamma t^2}\right)dt \to \sup;\quad x(0)=x_0,\quad x(T_0)=1,$ where c, g, and γ are positive constants (this is the problem on the maximum flight altitude of a rocket).

HINTS, SOLUTIONS, AND ANSWERS

Let us agree on some abbreviations: S will mean "solution," H will mean "hint," and F will mean "formalization." By $\hat{x}$ ($\hat{y}$, $\hat{z}$, etc.) the solution of the problem and by S its value will be denoted. If the problem depends on some parameters, then S is written as a function of the corresponding parameters; T_0 means that time is fixed, and if in the statement of the problem T is written, this means that time is nonfixed; $\hat{T}$ denotes the optimal time.

1. F: $\sum m_i|x-x_i|^2 \to \inf,\quad x, x_i \in \mathbf{R}^n$; this is an elementary smooth problem; $\hat{x}=\bar{x}=\sum m_i x_i/\sum m_i$ is the center of mass.

2. F: $\sum m_i|x-x_i|^2 \to \inf,\quad |x|=1,\quad x, x_i \in \mathbf{R}^n$; $\hat{x}=\bar{x}/|\bar{x}|$ if $\bar{x} \neq 0$, and $\hat{x}$ with $|\hat{x}| = 1$ is arbitrary if $\bar{x} = 0$; $\bar{x}$ is the center of mass.

3. $\hat{x} = \bar{x}$ if $|\bar{x}| \leqslant 1$, and $\hat{x} = \bar{x}/|\bar{x}|$ if $|\bar{x}| > 1$, $\bar{x}$ is the center of mass.

4. F: $f(x)=\sum m_i|x-x_i| \to \inf,\quad x, x_i \in \mathbf{R}^2,\quad i=1,2,3$; this is an elementary convex problem. S: $\hat{x}$ exists $\Rightarrow 0 \in \partial f(\hat{x})$. If $\hat{x} \notin \{x_1, x_2, x_3\}$, then $\partial f(\hat{x}) = \{\sum m_i(\hat{x}-x_i)/|\hat{x}-x_i|\}=\{0\}$, and this means that it is possible to construct a triangle with sides of lengths m_1, m_2, and m_3. Let α_{ij} be the angle of this triangle formed by the sides of lengths m_i and m_j. Then $\hat{x}$ is the point of intersection of the three arcs of circles, with the chords $[x_i, x_j]$, subtending the angles $\pi - \alpha_{ij}$. If the triangle with sides of lengths m_1, m_2, and m_3 does not exist or if the construction described above is impossible, the solution $\hat{x}$ coincides with one of the three given points.

5. We shall give the answer for the basic case when no three points among $\{x_1, x_2, x_3, x_4\}$ lie in one straight line. Then $\hat{x}$ is the point of intersection of the diagonals if the quadrilateral with the vertices x_i is convex and is the interior vertex of the convex hull if the quadrilateral is not convex.

6. $F: |x_1 - x_2|^2 + |x_1 - e|^2 + |x_2 - e|^2 \to \sup$, $|x_1|^2 = |x_2|^2 = 1$, $e = (1, 0)$, $x_i \in \mathbf{R}^2$, $i = 1, 2$. The problem possesses six stationary points; $(\hat{x}_1, \hat{x}_2, e)$ forms an equilateral triangle. For the detailed solution, see the book [54, p. 443].

7. $F: m_0|x_1 - x_2|^2 + m_1|x_1 - e|^2 + m_2|x_2 - e|^2 \to \sup$, $|x_1|^2 = |x_2|^2 = 1$. If it is possible to construct a triangle with sides of lengths m_1m_2, m_0m_1, and m_0m_2, and if α_1 and α_2 are the angles between the sides of lengths m_1m_2, m_0m_1 and m_1m_2, m_0m_2, respectively, then the angle between $\hat{x}_1$ and e is equal to $\pi - \alpha_1$, and the angle between $\hat{x}_2$ and e is equal to $\pi - \alpha_2$. If the construction is impossible, then $\hat{x}_1 = \hat{x}_2 = -e$ or $\hat{x}_1 = -e$, $\hat{x}_2 = e$.

Problems 8 and 9 admit of different formalizations connected with different interpretations of the terms "cylinder" and "cone." In what follows we shall understand a cylinder in $\mathbf{R}^n$ as a product of an $(n - 1)$-dimensional ball by a line segment orthogonal to it, and by a cone in $\mathbf{R}^n$ we shall mean the convex hull of an $(n - 1)$-dimensional ball and a line segment orthogonal to it.

8. $\hat{x}$ is half the altitude of the cylinder and is equal to $n^{-1/2}$.

9. $\hat{x}$ is the difference between the altitude of the cone and the radius of the ball and is equal to n^{-1}.

10. $\hat{x}$ is an n-dimensional cube.

11. $\hat{x}$ is a simplex whose n sides form an orthogonal basis, the lengths of the sides being equal to the radius of the sphere. The solution of this problem and the homogeneity imply *Hadamard's inequality*:

$$[\det(a_{ij})]^2 \leqslant \prod_{i=1}^{n} \left(\sum_{j=1}^{n} a_{ij}^2 \right).$$

For greater detail, see [54, p. 444].

12. $\hat{x}$ is a regular simplex.

13. $F: |x - y|^2 + |y - z|^2 + |z - x|^2 \to \inf$, $(x|a) \leqslant 0$, $(y|b) \leqslant 0$, $(z|c) + \alpha \leqslant 0$, $\alpha > 0$, $a + b + c = 0$, $a, b, c, x, y, z \in \mathbf{R}^2$. The origin coincides with the vertex C; a, b, and c are vectors orthogonal to [B, C], [C, A], and [A, B], respectively; $\hat{x}$ is the point of intersection of the straight lines $(x|a) = 0$ and $(x|b) = \beta(b - a|b)$; $\hat{y}$ is the point of intersection of $(y|b) = 0$ and $(y|a) = -\beta(b - a|a)$; $\hat{z} = \frac{(\hat{x} + \hat{y})}{2} + \frac{3}{2}\beta(a + b)$; $\beta = 2\alpha/(3|a + b|^2 + |b - a|^2)$.

14. $F: \|x_0 - \xi\| \to \inf$, $\xi = \Lambda y$, $\|y\| \leqslant 1$, $\hat{\xi} = \Lambda\hat{y}$, where $\hat{y} = (\Lambda^*\Lambda + \hat{\lambda}I)^{-1} \times \Lambda^* x_0$, and $\hat{\lambda}$ is the single positive solution of the equation $\|(\Lambda^*\Lambda + \lambda I) \times \Lambda^* x_0\| = 1$.

15. $\hat{x}_1 = \ldots = \hat{x}_n = \sigma/\Sigma\alpha_i$. The solution of the problem and the homogeneity imply the inequality

$$\prod x_i^{\alpha_i} \leqslant \left(\sum \alpha_i x_i / \sum \alpha_i\right)^{\Sigma\alpha_i} \qquad (x_i \geqslant 0, \alpha_i \geqslant 1).$$

For $\alpha_i = 1/n$ the *inequality for the arithmetic and the geometric means* is obtained: $\left(\prod_{i=1}^{n} x_i\right)^{1/n} \leqslant \left(\sum_i x_i\right)/n$.

16. $\hat{x} = \pm e_i$, where $e_1, \ldots, e_n$ is the standard basis in $\mathbf{R}^n$ if $q < p$; $\hat{x} = (\pm n^{-1/q}, \ldots, \pm n^{-1/q})$ if $q > p$; the case $p = q$ is trivial. The solution of this problem and the homogeneity imply the *inequality for power means*:

$$q \geqslant p \Rightarrow \sigma_p(x) \leqslant \sigma_q(x), \quad \sigma_p(x) = \left(\sum_{i=1}^{n} x_i^p / n\right)^{1/p}, \; x_i \geqslant 0.$$

17. $\hat{x}_i = a_i \Big/ \left(\sum_{i=1}^{n} a_i^2\right)^{1/2}$. The solution of this problem and the homogeneity imply the *Cauchy–Bunyakovskii inequality*:

$$|\sum a_i x_i| \leqslant (\sum a_i^2)^{1/2} (\sum x_i^2)^{1/2}.$$

18. $\hat{x} = a/|a|$. The solution of this problem and the homogeneity imply the Cauchy–Bunyakovskii inequality.

19. For $p > 1$: $\hat{x}_i = a_i^{(p'-1)}/(\sum |a_j|^{p'})^{1/p}$, $\frac{1}{p}+\frac{1}{p'}=1$. The solution of this problem and the homogeneity imply *Hölder's inequality*:

$$|\sum a_i x_i| \leqslant (\sum |a_i|^{p'})^{1/p'} (\sum |x_i|^p)^{1/p}, \qquad \frac{1}{p}+\frac{1}{p'}=1.$$

For $p = 1$:

$$\hat{x}_i = \begin{cases} \operatorname{sgn} a_i & \text{if } |a_i| = \|a\| = \max\{|a_1|, \ldots, |a_n|\}, \\ 0 & \text{if } |a_i| < \|a\|. \end{cases}$$

20. F: $\sum_{i=1}^{n} p_i \log p_i \to \inf$, $\sum p_i = 1$, $p_i \geqslant 0$, $\hat{p}_i = \frac{1}{n}$ (the uniform distribution).

21. F: $\int_{\mathbf{R}} p(x) \log p(x)\, dx \to \inf$; $\int_{\mathbf{R}} p(x)\, dx = 1$, $\int_{\mathbf{R}} x p(x)\, dx = a$; $\int_{\mathbf{R}} (x-a)^2 p(x)\, dx = \sigma^2$, $p(x) \geqslant 0$; this is a Lyapunov-type problem. S: $\mathscr{L} = \int_{\mathbf{R}} [\lambda_0 p(x) \log p(x) + \lambda_1 p(x) + \lambda_2 x p(x) + \lambda_3 (x-a)^2 p(x)]\, dx$. The extremum condition for $\hat{p}(\cdot)$: $\hat{\lambda}_0 p \log p + \hat{\lambda}_1 p + \hat{\lambda}_2 x p + \hat{\lambda}_3 \times (x-a)^2 p \geqslant \hat{\lambda}_0 \hat{p}(x) \log \hat{p}(x) + \hat{\lambda}_1 \hat{p}(x) + \hat{\lambda}_2 x \hat{p}(x) + \hat{\lambda}_3 (x-a)^2 \hat{p}(x)$, $\forall p \geqslant 0$. It follows that $\hat{\lambda}_0 \neq 0$ and $\hat{p}(x) = e^{-(a_1 x^2 + b_1 x + c_1)}$, where a_1, b_1, and c_1 are chosen from the constraints, whence

$$\hat{p}(x) = (2\pi\sigma)^{-1/2} e^{-((x-a)^2/2\sigma)}$$

(the Gauss distribution).

22. $S = |\sum a_i \xi_i - b| (\sum a_i^2)^{-1/2}$.

23. $S = (|\xi - \eta|^2 - (\xi - \eta | \zeta / |\zeta|^2))^{1/2}$.

24. $\hat{x}(t) = t^2 - 1/3$. H (to Problems 22-24): problems of this kind were investigated in Section 3.5.3 (Problem 24 can be interpreted as the problem on the shortest distance from the function t^2 to the subspace lin $\{1, t\}$).

25. F: $\int_{-1}^{1} x^2\, dt \to \inf$; $x^{(n)} = n!$. S: $\mathscr{L} = \int_{-1}^{1} \left[\frac{\lambda_0 x^2}{2} + p(x^{(n)} - n!)\right] dt$. The Euler–Poisson equation: $\hat{p}^{(n)}(t) + (-1)^n \hat{\lambda}_0 \hat{x} = 0$; the transversality conditions: $p^{(k)}(\pm 1) = 0$; $k = 0, 1, \ldots, n-1$; $\hat{\lambda}_0 = 0 \Rightarrow \hat{p} \equiv 0$; $\hat{\lambda}_0 = 1 \Rightarrow \hat{p}^{(2n)} = (-1)^{n+1} \hat{x}^{(n)} = (-1)^{n+1} n!$; $p^{(k)}(\pm 1) = 0 \Rightarrow \hat{p}(t) = C(t^2-1)^n$, $\hat{x}(t) = (-1)^{n+1}$, $\hat{p}^{(n)}(t) = t^n + \ldots = \frac{n!}{(2n)!}\left(\frac{d}{dt}\right)^n (t^2-1)^n$. When proving the sufficiency it is possible to refer to the convexity of the problem.

26. F: $f(x) = \max_{-1 \leqslant t \leqslant 1} \{f(t, x)\} \to \inf$; $f(t, x) = \left|t^n + \sum_{k=1}^{n} x_k t^{k-1}\right|$; this is an elementary convex problem whose solution exists. The maximum of the modulus of the

function $\hat{x}(t)=t^n+\sum_{k=1}^{n}\hat{x}_k t^{k-1}$ is attained at $s \leqslant n+1$ points because if $s > n+1$, then $\dot{\hat{x}}(\cdot)$ has not less than n zeros in the interval (−1, 1), which is impossible because $\deg \dot{\hat{x}}(\cdot) = n-1$. Let us denote these points by $\tau_1,\ldots,\tau_s$, $\tau_1 < \tau_2 < \ldots < \tau_s$. If $\hat{x}(\cdot)$ is a solution, then $0\in \delta f(\hat{x}) \overset{2.6.4}{=} \text{conv} \times (\delta f(\tau_1,\hat{x}),\ldots,\delta f(\tau_s,\hat{x}) \Leftrightarrow \exists \hat{\alpha}_1,\ldots,\hat{\alpha}_s;\ \hat{\alpha}_i \geqslant 0,\quad \sum_{i=1}^{s}\hat{\alpha}_i = 1,\quad \sum_{i=1}^{s}\hat{\alpha}_i \operatorname{sgn}\hat{x}(\tau_i)\tau_i^k = 0,\quad k=0,1,\ldots,$ $n-1$. The last relation implies that $s = n+1$, i.e., $\hat{x}(\cdot)$ assumes its maximum and minimum values at $n+1$ points. Therefore, these points must include ±1 [because, if otherwise, then $\dot{\hat{x}}(\cdot)$ again has not less than n zeros in the interval (−1, 1)]. It follows that $(1-t^2)\dot{\hat{x}}^2(t)$ and $f^2(\hat{x}) - \hat{x}^2(t)$ are polynomials of degree 2n having simple zeros at $\hat{\tau}_1 = 1$ and $\hat{\tau}_{n+1} = +1$ and twofold zeros at $\hat{\tau}_2,\ldots,\hat{\tau}_n$. Now, equating the coefficients in t^{2n}, we obtain the equation $(1-t^2)\dot{\hat{x}}^2 + n^2(f^2(\hat{x}) - \hat{x}^2(t))$; solving this equation we find $\hat{x}(t) = 2^{-(n-1)}\cos(n \arccos t)$.

Now we can easily compute $\hat{\alpha}_k$ (this result will be of use to us later):

$$\hat{\alpha}_1=\hat{\alpha}_{n+1}=(2n)^{-1},\qquad \hat{\alpha}_i=\frac{1}{n},\qquad i=1,\ldots,n-1.$$

From what has been proved we can derive the following important equality:

$$\int_{-1}^{1}\operatorname{sgn}\hat{x}(t)t^k\,dt=0,\qquad k=0,1,\ldots,n-1.$$

27. $\hat{\rho}_1 = \cos \pi/(n+2)$. H: By the well-known Riesz' theorem, the nonnegative trigonometric polynomial $x(\cdot)$ admits of the representation $x(t)=1+2\rho_1\cos t+\ldots+2\rho_n\cos nt=\left|\sum_{k=0}^{n}x_k e^{ikt}\right|^2$, where x_k are real numbers. It follows that

$$\sum_{k=0}^{n}x_k^2=1,\qquad \sum_{k=0}^{n-1}x_k x_{k+1}=\rho_1.$$

Further, the Lagrange multiplier rule should be applied. For greater detail, see [84, pp. 113 and 114].

28. $\hat{x}(t)=\sum_{j=1}^{n}\dfrac{k_j^n e^{k_j t}}{\left(\sum_{s=1}^{n}k_s\right)P'(k_j)}$, where k_j are the roots of the equation $x^{(2n)} + (-1)^n x = 0$ lying in the left half-plane, and $P(z) = (z-k_1)\ldots(z-k_n)$. For greater detail, see the book [84, pp. 121-123].

29. $\hat{x}(t) = Ae^{-t/2}\sin(t\sin\pi/3 - \pi/3)$, where A is chosen from the isoperimetric condition. For the solution of the problem, see the book by G. H. Hardy, G. E. Littlewood, and G. Polya.* For the solution of the general problem on inequalities for derivatives by the direct application of the Lagrange principle, see the work by A. P. Buslaev.†

30. $S = \gamma_1 p_1 + \gamma_2 p_2$, where p_1 and p_2 satisfy the system $p_1 + s^{2n}p_2 = s^{2k}$, $p_1 + (s+1)^{2n}p_2 = (s+1)^{2k}$, where $s=[(\gamma_2/\gamma_1)^{1/[2(n-1)]}]$ and [ ] denotes the integral part. H: the function $x(\cdot)$ should be expanded into Fourier's series to reduce the problem to a linear programming problem. For greater detail, see the article by Din Zung and V. M. Tikhomirov.‡

*Loc. cit. p. 227.
†A. P. Buslaev, "On an extremal problem connected with inequalities for derivatives," Vestn. Mosk. Gos. Univ., Ser. Mat., No. 3, 67-77 (1978).
‡Din Zung and V. M. Tikhomirov, "On inequalities for derivatives in the metric of L_2," Vestn. Mosk. Gos. Univ., Ser. Mat., No. 5, 3-11 (1979).

H (to Problems 31-100): in most of the problems that follow, the proof of the fact that the solution $\hat{x}(\cdot)$ yields absmin can be obtained either by the direct analysis of the values of the functional at the points $\hat{x}(\cdot)$ and $\hat{x}(\cdot) + h(\cdot)$ (in such a case we write "is verified directly") or by an independent proof of the existence of the solution. In the latter case one can make use of the following existence theorem for the simplest problem of the classical calculus of variations with a constraint on the derivative which can be derived from the general results presented in [88]†:

Let the integrand L in the problem

$$\int_{t_0}^{t_1} L(t, x(t), \dot{x}(t))\, dt \to \inf; \qquad x(t_0) = x_0, \qquad x(t_1) = x_1, \qquad |\dot{x}| \leqslant A, \tag{$*$}$$

be a continuous function on $[t_0, t_1] \times \mathbf{R} \times [-A, A]$, and, moreover, let the function $\xi \to L(t, x, \xi)$ be convex. Then there exists an absolutely continuous function $\hat{x}(\cdot)$ which yields a minimum in the problem (*).

The introduction of the "compulsory" constraint $|\dot{x}| \leqslant A$ (which makes it possible to guarantee the existence of the solution) followed by the passage to the limit for $A \to \infty$ will be called the "compulsory constraint method."

31. $\hat{x}(t) = (t^2 - T_0 t)/4$ (absmin).

32. $\hat{x}(t) = (t^2 - 2T_0 t)/4$ (absmin).

33. If $T_0 \leqslant 2$, then $\hat{x}(t) = (t^2 - 2T_0 t)/4$; if $T_0 > 2$, then $\hat{x}(t) = -t$ for $0 \leqslant t \leqslant T_0 - 2$ and $\hat{x}(t) = t(t - 2T_0)/4 + (T_0 - 2)^2/4$ for $T_0 - 2 \leqslant t \leqslant T_0$. In all the cases $\hat{x}(t)$ yields absmin.

34. If $T_0 \leqslant 4$, then $\hat{x}(t) = (t^2 - T_0 t)/4$. If $T_0 > 4$, then $\hat{x}(t) = -t$ for $0 \leqslant t \leqslant (T_0 - 4)/2$, $\hat{x}(t) = t(t - T_0)/4 + (T_0 - 4)^2/16$ for $(T_0 - 4)/2 \leqslant t \leqslant (T_0 + 4)/2$, and $\hat{x}(t) = t - T_0$ if $(T_0 + 4)/2 \leqslant t \leqslant T_0$. In all the cases $\hat{x}(\cdot)$ yields absmin.

H (to Problems 31-34): one should apply the Lagrange principle and find the single extremal $\hat{x}(\cdot)$. The fact that $\hat{x}(\cdot)$ yields absmin can be verified directly, and it is also possible to resort to the above-mentioned existence theorem.

35. $\hat{x}(t) = 4/t - 1$.

36. $S = 0$, and there is no solution. H: consider the minimizing sequence $\{x_n(t)\}$, where $x_n(t) = nt$ for $0 \leqslant t \leqslant 1/n$ and $x_n(t) = 1$ for $1/n \leqslant t \leqslant 1$. This example was put forward by Weierstrass as an argument against Riemann's proof of the Dirichlet principle.

37. $\hat{x}(t) = t^{1/3}$ (this example was considered in Section 1.4.3; the solution exists but does not belong to the class C^1).

38. This problem can be used to demonstrate the application of the duality methods. F: $\int_0^1 \frac{t^\alpha |u|^\beta}{\beta}\, dt \to \inf$; $\int_0^1 u\, dt = \xi, \alpha, \beta \in \mathbf{R}$; this is a Lyapunov-type problem. The conjugate function is

$$S^*(\eta) = \sup_\xi \left[\eta\xi - \inf\left\{ \int_0^1 t^\alpha \frac{|u|^\beta}{\beta}\, dt \,\middle|\, \int_0^1 u\, dt = \xi \right\} \right] = \sup_{u(\cdot)} \int_0^1 \left(u\eta - \frac{t^\alpha |u|^\beta}{\beta} \right) dt$$

†Also see W. H. Fleming and R. W. Rishel, Deterministic and Stochastic Control, Springer-Verlag (1975).

$$= \begin{cases} \delta\{0\} \text{ for } \beta < 1 \text{ or } \beta \geqslant 1, \quad \beta \leqslant \alpha + 1; \\ \delta\{[-1, 1]\} \text{ for } \beta = 1, \quad \alpha \leqslant 0; \\ \dfrac{\beta - 1}{\beta - \alpha - 1} \cdot \dfrac{|\eta|^{\beta'}}{\beta'} \text{ for } \beta > 1, \quad \beta > \alpha + 1 \left(\dfrac{1}{\beta} + \dfrac{1}{\beta'} = 1\right). \end{cases}$$

The value of the problem is

$$S(\xi, \alpha, \beta) = \begin{cases} \text{(i) } 0 \text{ for } \beta < 1 \text{ or } \beta \geqslant 1, \quad \beta \leqslant \alpha + 1; \\ \text{(ii) } |\xi| \text{ for } \beta = 1, \quad \alpha \leqslant 0; \\ \text{(iii) } \left(\dfrac{\beta - \alpha - 1}{\beta - 1}\right)^{\beta - 1} \dfrac{|\xi|^\beta}{\beta} \text{ for } \beta > 1, \quad \beta > \alpha + 1. \end{cases}$$

For (i) $\beta < 1$ there is no solution because the integrand is not convex (see Bogolyubov's theorem in Section 1.4.3); for (i) $\beta \geqslant 1$ there is a discontinuous solution (a "jump at zero"); for (ii) $\alpha = 0$ there are infinitely many solutions (any monotone function is a solution); for (ii) $\alpha < 0$ there is a generalized solution (a "jump at unity"); for (iii) the solution is $\hat{x}(t) = \xi t^{(\beta-\alpha-1)/\beta}$ (it belongs to $C^1([0, 1])$ for $\alpha \leqslant 0$).

39. $\hat{x}(t) = -t^2/4 + 5t/4$ (absmin).

40. $\hat{x}(t) = -t/2 + 3 \log t/2 \log 2 + 1/2$ (absmin).

41. $\hat{x}(t) = t^3 + 2t + 1$ (absmin).

H (to Problems 39-41): solve the corresponding Euler equations.

42. $\hat{x}(t, \xi) = \xi \sinh t/\sinh T_0$ (absmin).

43. If $|\xi| \cosh T_0 \leqslant 1$, then $\hat{x}(t, \xi) = \xi \sinh t/\sinh T_0$; if $|\xi| \cosh T_0 > 1$, then $\hat{x}(t, \xi) = \pm \sinh t/\cosh \tau$ for $0 \leqslant t \leqslant \tau$ and $\hat{x}(t, \xi) = \pm(\tanh \tau + (t - \tau))$ for $\tau \leqslant t \leqslant T_0$, where τ is the solution of the equation $T_0 - \xi = \tau - \tanh \tau$. In all the cases $\hat{x}(\cdot)$ yields absmin.

44. If $|\xi| \leqslant 1$, then $\hat{x}(t, \xi) = \xi$; if $|\xi| > 1$, then $\hat{x}(t, \xi) = \eta$ for $0 \leqslant t \leqslant |\eta|^{-1}$ and $\hat{x}(t, \xi) = \eta \cosh(t - |\eta|^{-1})$ for $|\eta|^{-1} \leqslant t \leqslant 1$, where η is the solution of the equation $\eta \cosh(1 - |\eta|^{-1}) = \xi$. In all the cases $\hat{x}(\cdot)$ yields absmin.

45. $\hat{x}(t, \xi) = e^{|t|} + (\xi - e)\cosh t/\cosh 1$. (This problem was discussed in Section 1.4.3.)

46. H: let us suppose that $x(\cdot)$ is a solution of the problem. Then, according to the Pontryagin maximum principle, there exists a function $p(\cdot)$ such that the pair $(x(\cdot), p(\cdot))$ satisfies the following conditions:

$$\ddot{x} = \operatorname{sgn} p, \quad -\ddot{p} = x, \quad x(0) = 1, \quad x(\infty) = p(\infty) = p(0) = 0. \tag{1}$$

If τ is the first positive zero of $p(\cdot)$ and $x(\tau) = a \neq 0$, then it can easily be verified that the pair $(x_1(\cdot), p_1(\cdot))$, where

$$x_1(t) = a^{-1}x(\sqrt{|a|}\, t + \tau), \quad p_1(t) = a^{-2} \operatorname{sgn} a p(\sqrt{|a|}\, t + \tau),$$

also satisfies (1). The solution is unique due to the strict convexity of the functional which must be minimized, and therefore

$$x(t) = a^{-1}(\sqrt{|a|}\, t + \tau), \quad p(t) = a^{-2} \operatorname{sgn} a p(\sqrt{|a|}\, t + \tau).$$

It readily follows that $x(\cdot)$ and $p(\cdot)$ are finite functions, i.e., functions of finite (compact) support, formed by "sticking together" a countable number of polynomials of the second degree and the fourth degree, respectively, and are completely determined by setting three parameters: $\dot{x}(0) = -\alpha$, $\dot{p}(0) = \beta$, and τ. Moreover, β and τ are expressed explicitly in terms of

α, and α itself [under the assumption that $\alpha \in (\sqrt{2}, 2)$] satisfies the equation

$$\int_0^{\varphi(\alpha)} (t^2/2 - \alpha t + 1)(\psi(\alpha) - t)\, dt = 0, \tag{2}$$

where $\varphi(\alpha) = \alpha\left(\sqrt{\dfrac{\alpha^3 - 2}{\alpha^2 + 2}} + 1\right)$, $\psi(\alpha) = \varphi(\alpha)\left(\left(\dfrac{\alpha^2 + 2}{\alpha^2 - 2}\right)^{3/2} + 1\right)^{-1}$.

It can be shown that (2) does in fact have a solution $\hat{\alpha}$ on $(\sqrt{2}, 2)$, and if $\hat{\beta}$ and $\hat{\tau}$ are the values of the parameters β and τ corresponding to this value of $\hat{\alpha}$, then we can easily construct a pair $(\hat{x}(\cdot), \hat{p}(\cdot))$ satisfying (1). This means that $\hat{x}(\cdot)$ is a solution of the original problem.

Moreover, the function $\hat{p}(\cdot)$ is a solution of a dual problem which is in fact equivalent to the problem investigated in the work by L. D. Berkovitz and H. Pollard.* The solutions of the two problems (considered for a somewhat more general situation) are contained in the article by G. G. Magaril-Ilyaev.†

47. $$S(T_0, \xi) = \begin{cases} -\infty & \text{for } T_0 > \pi \text{ or } T_0 = \pi,\ \xi \neq 0, \\ 0 & \text{for } T_0 = \pi,\ \xi = 0, \\ \xi^2 \cot T_0 & \text{for } 0 < T_0 < \pi, \end{cases}$$

$\hat{x}(t; T_0, \xi) = (\xi \sin t)/\sin T_0$ (absmin), $T_0 < \pi$; $\hat{x}(t; \pi, 0) = A \sin t$ (absmin).

48. $$S(T_0) = \begin{cases} -\infty & \text{for } T_0 > \pi/2 \\ 0 & \text{for } 0 < T_0 < \pi/2, \end{cases}$$

$\hat{x}(t; T_0) \equiv 0$, $0 < T_0 < \pi/2$ (absmin); $\hat{x}(t; \pi/2) = A \sin t$ (absmin).

49. $$\hat{x}(t; T_0) = \begin{cases} 0 \text{ for } T_0 < \pi, \\ A \sin t,\ |A| \leq 1, \text{ for } T_0 = \pi, \\ \begin{cases} t \text{ for } 0 \leq t_0 \leq T_0/2 - \theta, \\ \sqrt{1 + (T_0/2 - \theta)^2} \cos(T_0/2 - t) \text{ for } |T_0/2 - t| \leq \theta, \\ T_0 - t \text{ for } T_0/2 + \theta \leq t \leq T_0, \end{cases} \end{cases}$$

where $\theta = \theta(T_0)$ is the minimum positive solution of the equation $\cot z = T_0/2 - z$.

H (to Problems 47-49): there exists an extremal which yields a minimum in Problem 49. Further, one should apply the maximum principle to show that there is always a finite number of extremals. It can easily be verified that the extremal we have written yields the smallest value for the functional. Problems 47 and 48 can be investigated exhaustively by using the "compulsory constraint method"($|\dot{x}| \leqslant A$). After that it is necessary to solve some problems analogous to Problem 49 and then to pass to the limit for $A \to \infty$.

50. $S(T_0) = \begin{cases} 0 \text{ for } T_0 < \pi, \\ -\infty \text{ for } T_0 > \pi. \end{cases}$ For $T_0 < \pi$ there are countably many polygonal extremals; the first corner point lies on the curve $(\tau, \tan\tau)$, $0 \leqslant \tau \leqslant \pi/2$. The curve with one corner point has the following form:

$$x_1(t) = \begin{cases} (\sin t)/\cos(T_0/2) & \text{for } 0 \leq t \leq T_0/2, \\ \sin(T_0 - t)/\cos(T_0/2) & \text{for } T_0/2 \leq t \leq T_0, \end{cases}$$

---

*L. D. Berkovitz and H. Pollard, "A nonclassical variational problem arising from optimal filter problem. I," Arch. Rat. Mech. Analysis, 26, 281-304 (1967); also see [88].

†G. G. Magaril-Ilyaev, "On Kolmogorov's inequalities on a half line," Vestn. Mosk. Univ., 5, 33-41 (1976).

and the others are constructed in a similar way.

51. $S = 0$. H: apply Bogolyubov's theorem (see Section 1.4.3).

52. $\hat{x}(t) \equiv 0$ (absmin) for $\alpha \leqslant T_0^{-1}$; $S = -\infty$ for $\alpha > T_0^{-1}$.

53. $\hat{x}(t) = 3t^2 + 2t + 1$.

54. The stationary points of the problem are

$$\hat{x}_k(t) = (2/T_0)^{-1/2} \sin k(\pi/T_0)t, \quad k = 1, 2, \dots .$$

The solution is $\hat{x}(t) = x_1(t) = (2/T_0)^{-1/2} \sin (\pi/T_0)t$ (absmin).

The solution of this problem perfectly illustrates Hilbert's theorem (Section 3.5.4).

55. The solution of the problem does not always exist. Let us apply the "compulsory constraint method": $|\dot{x}| \leqslant A$. $\mathscr{L} = \int_{-T_0}^{T_0} (-\lambda_0 x + \lambda_1\sqrt{1+u^2} + p(\dot{x} - u))\, dt \Rightarrow \dot{\hat{p}} = -\hat{\lambda}_0$, $\hat{p} = \hat{\lambda}_1 \dfrac{\hat{u}}{\sqrt{1+\hat{u}^2}}$; $\hat{\lambda}_0 = 0 \Rightarrow \hat{u}(t) \equiv \text{const} \Rightarrow \hat{u}(t) \equiv 0$ (due to the boundary conditions) $\Rightarrow$ $l = 2T_0$. If $l > 2T_0$ then, $\hat{\lambda}_0 \neq 0$; i.e., we can assume that $\hat{\lambda}_0 = +1$. Then $\hat{p}(t) = -t$. Integrating we conclude that $\hat{x}(\cdot)$ is either an arc of a circle or a curve composed of the line segments $A(t + T_0)$ and $A(T_0 - t)$ joined by an arc of a circle or a broken line:

$$A(t + T_0) \text{ for } -T_0 \leqslant t \leqslant 0; \; A(T_0 - t) \text{ for } 0 \leqslant t \leqslant T_0.$$

Passing to the limit we obtain the following solution of the problem: for $2T_0 < l \leqslant \pi T_0$ the curve $\hat{x}(\cdot)$ is an arc of a circle with center on the straight line $t = 0$, and for $l > \pi T_0$ the curve $\hat{x}(\cdot)$ is the semicircle "lifted" by a distance of $(l - \pi T_0)/2$.

56. $\hat{x}(t) = 2 \sin t/\pi$.

57. $\hat{x}(t) = -10t^3 - 12t^2 + 6t + 2$.

58. $\hat{x}(t) = 5t^3 + 3t - 4$.

59. $\hat{x}(t) = (3t^2 - t^3)/2$.

H (to Problems 56-59): $\hat{x}(\cdot)$ yields absmin in all the cases, and this can be verified directly.

60. $\hat{x}(t) \equiv 1$ (this can be noticed at once without any calculations although the problem can also be solved using standard methods).

61. $\hat{x}(t) = -2t^2 + 1$ for $0 \leqslant t \leqslant 1$ and $\hat{x}(t) = t^2 - 6t + 4$ for $1 \leqslant t \leqslant 2$.

62. $\hat{x}(t) = t(t - T_0)/2$.

63. $\hat{x}(t) = -t^2/2$ if $0 \leqslant t \leqslant (1 - 2^{-1/2})T_0$, and $\hat{x}(t) = t^2/2 + (\sqrt{2} - 2)T_0 t + (\sqrt{2} - 3/2)T_0^2$ if $(1 - 2^{-1/2})T_0 \leqslant t \leqslant T_0$.

64. $\hat{x}(t) = \int_0^t (t - \tau)\hat{u}(\tau)\, d\tau, \qquad \hat{u}(t) = -\operatorname{sgn} \cos \dfrac{2\pi t}{T_0}.$

H (to Problems 61-64): one should prove the existence of the solution and apply the maximum principle. It turns out that there is a single stationary point which yields absmin.

Here the existence of the solution follows from the compactness in the space C of the set of the functions with bounded second derivatives (for the same boundary conditions as those involved in the statement of the problem) and the continuity of the functional $f(x(\cdot)) = \int_0^{T_0} x\,dt$.

65. $\hat{x}(t) = \frac{1}{(n-1)!}\int_{-1}^{t} (t-\tau)^{n-1}\hat{u}(\tau)\,d\tau$, $\quad \hat{u}(t) = \pm\operatorname{sgn}\cos[(n+1)\arccos t]$, where the sign is chosen so that $\hat{x}(t) \leqslant 0$, $t \in [-1, 1]$.

H: here the maximum principle leads to the following relations:

$$\hat{p}^{(n)} + (-1)^{n+1} = 0, \qquad \hat{x}^{(n)}(t) = \hat{u}(t) = \operatorname{sgn}\hat{p}(t),$$

i.e., $\hat{p}(\cdot)$ is a polynomial of degree $(n + 1)$ and

$$\hat{x}(t) = \frac{1}{(n-1)!}\int_{-1}^{1} (t-\tau)^{n-1}\operatorname{sgn}\hat{p}(\tau)\,d\tau.$$

If $\hat{p}(\cdot)$ is Chebyshev's $(n + 1)$-th polynomial then, as it follows from the solution of Problem 26,

$$\int_{-1}^{1} (1-\tau)^{n-k-1}\operatorname{sgn}\hat{p}(\tau)\,d\tau = 0, \qquad k = 0, 1, \ldots, n-1,$$

i.e., $\hat{x}^{(k)}(\pm 1) = 0$, $k = 0, 1, \ldots, n-1$. Hence $\hat{x}(\cdot)$ is a stationary point. The fact that this point is a solution follows from the convexity of the problem.

66. There are two extremals: $-t^4 + t^3 - 6t^2 - 4t + 1$ and $(t - 1)^4$.

67. $\ddot{\hat{x}}(t) = t$ (absmin).

H: the fact that this solution yields absmin follows from the relations

$$x(2) = \int_0^2 \dot{x}\,dt \leqslant \sqrt{2}\left(\int_0^2 \dot{x}^2\,dt\right)^{1/2} = 2.$$

As a rule, in the problems that follow we only indicate the stationary points. The verification of whether or not they yield the solutions of the problems is left to the reader.

68. $\hat{x}(t) = -t^2/2 + 1$ if $0 \leqslant t \leqslant \sqrt{3}$, and $\hat{x}(t) = 3(t - 4/\sqrt{3})^2/2 - 1$ if $\sqrt{3} \leqslant t \leqslant 4/\sqrt{3}$; $\hat{T} = 4/\sqrt{3}$.

69. $\hat{u}(t) \equiv +1$ or $\hat{u}(t) \equiv -1$.

70. For the solution, see Section 1.6.3.

71. $\hat{T} = 1$; $\hat{x}(t) = t - t^2$.

72. $\hat{T} = 1$, $\hat{x}(t) = (t + 1)^2 - 1$ for $-1 \leqslant t \leqslant 0$, and $\hat{x}(t) = 1 - (t - 1)^2$ for $0 \leqslant t \leqslant 1$.

For the solutions of Problems 73 and 74, see [12, pp. 34 and 63]; for the solution of Problem 75, see [54, pp. 436-439].

76. $\hat{x}(t) = -2t$.

77. The extremals are $x_k(t) \int_0^t \operatorname{sgn}\sin(2k+1)\frac{\pi\tau}{2}\,d\tau$, $k = 0, \pm 1, \ldots$; $\hat{x}(t) = \pm t$.

78. $\hat{x}(t) = \hat{C}_1 \sinh\frac{t}{\sqrt{2}}\sin\frac{t}{\sqrt{2}} + \hat{C}_2\left(\cosh\frac{t}{\sqrt{2}}\sin\frac{t}{\sqrt{2}} - \sinh\frac{t}{\sqrt{2}}\cos\frac{t}{\sqrt{2}}\right)$; $\hat{C}_1$ and $\hat{C}_2$ are found from the conditions $\hat{x}(T_0) = \xi_1$, $\dot{\hat{x}}(T_0) = \xi_2$.

79. $\hat{x}(t) \equiv 0$, $T_0 \leqslant \hat{T}$, where $\hat{T}$ is the first zero of the equation $\cos T \cosh T = 1$; $S = -\infty$, $T_0 > \hat{T}$.

80. $\hat{x}(t) = \hat{C}_1(\cosh t - \cos t) + \hat{C}_2(\sinh t - \sin t)$ for $T_0 < \hat{T}$; $\hat{T}$ is determined in Problem 79; $\hat{C}_1$ and $\hat{C}_2$ are found from the relations $\hat{x}(T_0) = \xi_1$, $\dot{\hat{x}}(T_0) = \xi_2$; $S = 0$ for $T_0 = \hat{T}$, $\xi_1 = \xi_2 = 0$; $S = -\infty$ for $T_0 > \hat{T}$.

81. $\hat{x}(t) = -t^2/2$ if $0 \leqslant t \leqslant 1/2$, and $\hat{x}(t) = t^3/3 - t^2 + t/4 - 1/24$ if $1/2 \leqslant t \leqslant 1$.

82. $\hat{x}(t) = -t^3 + 6t^2$ for $0 \leqslant t \leqslant 1$, and $\hat{x}(t) = 3t^2 + 3t - 1$ for $1 \leqslant t \leqslant 2$.

83. $\hat{x}(t) = t^4 - 2t^3 + t$.

84. The function $\hat{x}(t) = \hat{C}[(\sinh \omega t - \sin \omega t)(\cosh \omega T_0 - \cos \omega T_0) - (\cosh \omega t - \cos \omega t)(\sinh \omega T_0 + \sin \omega T_0)]$ yields the infimum which is equal to ω^4, where $\omega > 0$ is the minimum root of the equation $\cosh \omega T_0 \cos \omega T_0 = -1$, and C is found from the condition

$$\int_0^{T_0} \hat{x}^2\, dt = 1$$

85. $\hat{x}(t) = \sinh t/\cosh 2$.

86. $\hat{x}(t) = (t - 2)\cosh t/\cosh 2$.

87. $\hat{x}(t) = -\sin 2(t - 2)/\sin 4$.

88. $\hat{x}(t) = [2t \cos 2(t - 2) - \sin 2(t - 2)]/\sin 4$.

89. $\hat{x}_1(t) = \sin 1 \sinh t - \sinh 1 \sin t$, $\hat{x}_2(t) = \sinh 1 \sin t + \sin 1 \times \sinh t$.

90. $\begin{cases} \hat{x}_1(t) = 3t^2 - 2t, \\ \hat{x}_2(t) = 3t^2 - 6t, \end{cases}$ $\begin{cases} \hat{x}_1(t) = 4t - 3t^2, \\ \hat{x}_2(t) = -3t^2. \end{cases}$

91. $\hat{x}(t) = \dfrac{1}{a^2 p^{1/(p-1)}} |at+b|^{(2p-1)/(p-1)} \operatorname{sgn}(at+b) + \lambda_1 t + \lambda_2$ for $p > 1$, where $a = \alpha \operatorname{sgn} \alpha$, and

$$\alpha = \frac{\xi_1^{(p-1)} p}{(|\xi_2/\xi_1|^{P/(p-1)} - |\xi_2/\xi_1 - 1|^{P/(p-1)})}, \qquad b = a\frac{\xi_2 - \xi_1}{\xi_1},$$

$$\lambda_1 = \frac{b^{p/(p-1)}}{ap^{1/(p-1)}}, \qquad \lambda_2 = \frac{|b|^{(2p-1)/(p-1)}}{a^2 p^{1/(p-1)}}.$$

H: apply the Lagrange principle. For the case $p = 1$, see Problem 92.

92. Let us denote $\eta = v_0 - v_1 - gT_0$, $\xi = x_0 + v_0 T_0 - x_1 - v_1 + v_0 + gT_0$. Then $S = S(\eta, \xi) = |\xi/T_0| + |\eta - \xi/T_0|$.

H: putting $u = \ddot{x} - g$ we arrive at the following Lyapunov-type problem:

$$\int_0^{T_0} |u|\, dt \to \inf; \quad \dot{x} = y, \quad \dot{y} = u + g \Leftrightarrow \int_0^{T_0} |u|\, dt \to \inf; \quad \int_0^{T_0} u\, dt = \eta, \int_0^{T_0} tu(t)\, dt = \xi.$$

Further use the results of Section 4.3.

93. The extremals of the problem were found in Section 1.6.5. For $\alpha < 0$ the extremal joining the two points (t_0, x_0) and (t_1, x_1) is unique if $x_i > 0$ and can easily be embedded in a family of extremals covering the whole plane. Hence it yields an absolute minimum in the problem (see Section 4.4).

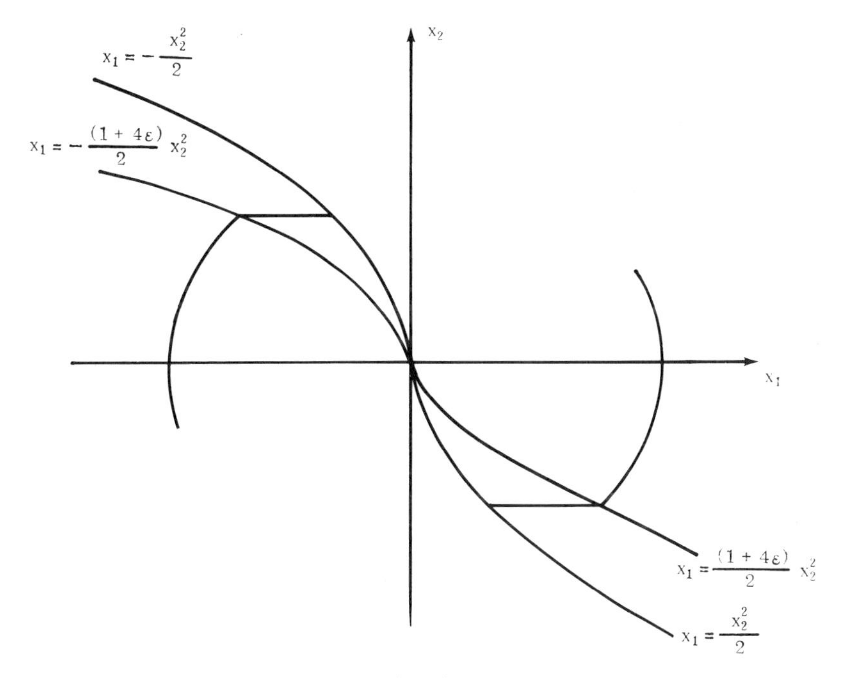

Fig. 41

94. This is the minimum-surface-of-revolution problem. The extremals were found in Section 1.6.5. For the complete solution of the problem, see the book [54, p. 427].

95. $\hat{x}(t) = \sqrt{2 - (t-1)^2}$. H: show that the Lagrange principle implies that for the problems of geometrical optics [in which the integrand has the form $\sqrt{1 + \dot{x}^2}/v(t, x)$] the transversality condition coincides with the orthogonality condition.

96. In Fig. 41 we present the phase diagram of the problem (cf. Fig. 28).

97. S: let us bring the problem to the standard form:

$$\int_0^1 |u|^p\,dt \to \inf; \qquad \dot{x}_1 = x_2, \qquad \dot{x}_2 = ux_1, \qquad x_1(0) = x_1(1) = 0, \qquad x_2(0) = 1.$$

In the last problem it is possible to prove the existence of the solution. The Lagrangian of the problem is

$$L = \lambda_0 |u|^p + p_1(\dot{x}_1 - x_2) + p_2(\dot{x}_2 + ux_1).$$

Show that $\lambda_0 \neq 0$. Then the Euler equations with respect to x and the transversality conditions take the following form:

$$-\dot{p}_1 + up_2 = 0,\ -\dot{p}_2 - p_1 = 0,\ p_2(1) = 0 \Leftrightarrow \ddot{p}_2 + up_2 = 0,\ p_2(1) = 0.$$

Thus, $x_1(\cdot)$ and $p_2(\cdot)$ satisfy one and the same second-order linear differential equation, and both the functions vanish at one and the same point $t = 1$. Consequently, $x_1(\cdot)$ and $p_2(\cdot)$ are proportional: $Cx_1(\cdot) = p_2(\cdot)$. From the Euler equation with respect to u ($L_u = 0$) we obtain

$$p_2(t)x_1(t) = -p|u(t)|^{p-1}\,\mathrm{sgn}\,u(t) \Rightarrow C_1x_1^2(t) = -p|u(t)|^{p-1}\,\mathrm{sgn}\,u(t).$$

Hence, the optimal $\hat{u}(\cdot)$ does not change sign. It can easily be seen that $u \geqslant 0$. Therefore, $C < 0$, and thus

$$\hat{u}(t) = C_1(\hat{x}_1(t))^{2/(p-1)}, \quad C_1 > 0,$$

i.e., $\hat{x}(\cdot)$ satisfies the equation $\ddot{x} + C_1 x^{(p+1)/(p-1)} = 0$. Denoting $(p + 1)/(p - 1) = q$ we derive the integral $\dot{x}^2/2 + (C_1/(q + 1))x^{q+1} = D$ from the equation with respect to x. From the conditions $x(0) = 0$, $\dot{x}(0) = 1$ we obtain $D = 1/2$ and $t = \int_0^x \frac{dz}{\sqrt{1+(2C_1/(q+1))z^{q+1}}}$. The constant C_1 is found from the relation

$$\tfrac{1}{2} - \int_0^{((q+1)/2C_1)^{1/(q+1)}} \frac{dz}{\sqrt{1+(2C_1/(q+1))z^{q+1}}} = ((q+1)/2C_1)^{1/(q+1)} \times \int_0^1 \frac{d\eta}{\sqrt{1-\eta^{q+1}}} = C_1^{-1/(q+1)}\Gamma_q,$$

where Γ_q is a known quantity. Thus, the extremal $\hat{x}(\cdot)$ is a function symmetric with respect to the straight line $t = 1/2$, and on the interval $[0, 1/2]$ this function is inverse to the function $t = \int_0^x \frac{dz}{\sqrt{1-(2\hat{C}_1/(q+1))z^{q+1}}}$, where $\hat{C}_1 = (2\Gamma_q)^{q+1}$.

98. $\hat{x}(t) = \begin{cases} t \text{ for } 0 \leqslant t \leqslant 1/2, \\ 1 - t \text{ for } 1/2 \leqslant t \leqslant 1. \end{cases}$

H: one should apply the compulsory constraint method ($|u| \leqslant A$), apply the maximum principle, make use of the relation $Cx_1(\cdot) = p_2(\cdot)$ analogous to the one obtained in the solution of Problem 97, and then pass to the limit for $A \to \infty$.

Problems 97 and 98 were considered in stability theory.*

99. H: Problem 99 is one of the formalizations of the well-known Ulam problem of making line segments coincide. In his book† Ulam writes: "Suppose two segments are given in the plane, each of length one. One is asked to move the first segment continuously without changing its length to make it coincide at the end of the motion with the second interval in such a way that the sum of the lengths of the two paths described by the end points should be a minimum."

In the formalization the role of the parameter t is played by the angle between the moving line segment and its initial position.

The problem can be reduced to a Lyapunov-type problem (because the integrand does not depend on x_1 and x_2) and investigated with the aid of the duality methods of Section 4.3.‡ However, the simplest way is to apply the maximum principle. In Fig. 42 we present the optimal solution for line segments lying in one straight line (here we require that the point x_1 go to x_1' and the point x_2 go to x_2'). The arrows indicate the trajectories of the points x_1 and x_2. The point O is determined uniquely from the conditions of the symmetry, of the parallelism of Ox_1' to $x_2\xi$ and of x_1O to $\xi' x_2'$, and of the perpendicularity of $O\xi$ to $x_2\xi$ and of $O\xi'$ to $x_2'\xi$.

*G. Borg, "Über die Stabilität gewisser Klassen von linearen Differentialgleichungen," Arch. Mat. Astron. Phys., Bd 31A, No. 41, 460-482 (1944). Yu. A. Levin, "On the zeroth stability zone," Dokl. Akad. Nauk SSSR, 145, No. 6, 1021-1023 (1962). Yu. A. Levin, "On a stability criterion," Usp. Mat. Nauk, 17, No. 3, 211-212 (1962).

†S. M. Ulam, A Collection of Mathematical Problems, Interscience, New York (1960), p. 79.

‡See M. A. Rvachev, On Ulam's Problem of Making Line Segments Coincide, Trudy Seminara "Kombinatsionnaya Geometriya i Optimalnye Razmeshcheniya, Kiev (1973), pp. 42-52.

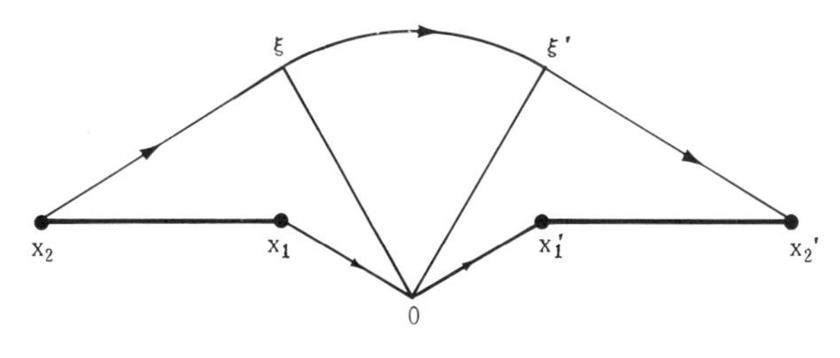

Fig. 42

100. H: Problem 100 is a formalization of one of the problems on the maximum flight altitude of a rocket.* The problem can easily be reduced to an elementary control problem (since the Lagrangian depends linearly on $\dot{x}$).

$$\hat{x}(t)=\begin{cases} x_0 & \text{for } 0\leq t\leq t_1, \\ (\gamma/cg)t^3+(\gamma/g)t^2 & \text{for } t_1\leq t\leq t_2, \\ 1 & \text{for } t>t_2, \end{cases}$$

where t_1 is the positive root of the equation $(\gamma/cg)t^3 + (\gamma/g)t^2 = x_0$ and t_2 is the positive root of the equation $(\gamma/cg)t^3 + (\gamma/g)t^2 = 1$.†

*A. A. Kosmodemyanskii, Konstantin Eduardovich Tsiolkovskii [in Russian], Nauka, Moscow (1976), pp. 263-265.

†For greater detail, see Yu. L. Aleksandrov, "The problem on the maximum flight altitude of a meteorological rocket in a homogeneous atmosphere and a homogeneous gravitation field," Vestn. Mosk. Gos. Univ., Ser. 7, Mat. Mekh., No. 5, 70-74 (1981).

NOTES AND GUIDE TO THE LITERATURE

There is a huge literature on the theory of extremal problems. Neither the bibliography nor the notes and the guide to the literature presented here pretend to be complete: we confine ourselves to a number of fundamental works, basic monographs, textbooks, and review articles.

To Chapter 1. Section 1.1 . The history of the appearance of the first maximum and minimum problems is presented in [86] (besides the book by Van der Waerden quoted in Section 1.1.2); for the initial stage of the classical calculus of variations, see [83] and [87]. The extremal properties of a circle and a ball are dealt with in the monograph [21]. The transportation problem was already investigated in the work of L. V. Kantorovich [55], the first work on linear programming. The simplest automatic control problems were first investigated by Bushaw [96]. The time-optimal problem and a number of others were thoroughly considered in the fundamental monograph by L. S. Pontryagin, V. G. Boltyanskii, R. V. Gamkrelidze, and E. F. Mishchenko [12] and in many other books on optimal control ([23], [27], etc.).

Sections 1.2-1.5. For the history of the Lagrange principle, see the article [45]. The same topic but in the framework of the classical calculus of variations is considered in the works [43], [44].

We also point out some other books and articles devoted to the topics touched upon in these sections and meant for a broad audience: [71], [74], [76], [77], and [115]. The fundamentals of the classical calculus of variations are presented in the textbooks [2], [3], and [8].

To Chapter 2. Section 2.1. Besides the textbooks on functional analysis already mentioned, we point out the books [92] and [107] specially meant for the "needs" of the theory of extremal problems.

Section 2.2. The differential calculus in normed linear spaces is presented (besides [KF]) in the textbooks [5], [56], and [69] and in the monographs [28], [54], and [59].

Section 2.3. Finite-dimensional versions of the implicit function theorem are presented in the textbooks [5], [56], and [69].

The theorem of Section 2.3.1 is sufficiently convenient for constructing the general theory since it contains the classical implicit function theorem and Lyusternik's theorem on the tangent space and yields an estimate for the deviation from the kernel of a mapping. The constructions used in the proof were in fact contained in the initial work [68] (also see [69]). For other proofs and modifications, see [54] and [65].

Section 2.4. For the differentiability of other important concrete functionals, see [28] and [59].

Section 2.6. The fundamentals of finite-dimensional convex analysis were laid down by Minkowski [113] and [114] and Fenchel [101] and [102]. Convex analysis in finite-dimensional spaces was constructed in the Sixties in the works of Brφnsted [95], Dubovitskii and Milyutin [46], Moreau, Rockafellar, et al. The most complete review of the finite-dimensional theory is contained in the monograph by Rockafellar [81] and also in the monograph by Bonnesen and Fenchel [94] (in the aspect related only to convex sets). The infinite-dimensional theory is presented in [16], [52], [54], [62], [79], and [84].

Lately some attempts have been made to create a synthetic ("smooth-convex") calculus (see [70], [99], and [118]).

To Chapter 3. Section 3.2. It is difficult to say who was the first to prove the Lagrange multiplier rule for smooth finite-dimensional problems with equality and inequality constraints. Some American works contain references to the work of Valentine [120] and the thesis of Karush [109]. The multiplier rule for the case of infinitely many inequalities was proved by John [108]. The work of Kuhn and Tucker [111] played a very important role in these problems. Infinite-dimensional versions of the multiplier rule are presented in [37], [39], [46], [54], [79], and [84].

Section 3.3. Linear and convex programming is widely represented in Russian textbooks and monographs: [38], [40], [41], [50], [57], [58], [67], [75], [76], [81], [89], and [90].

Section 3.4. The work [100] was one of the first works devoted to necessary conditions for problems with inequality constraints. The finite-dimensional theory is perfectly presented in the textbook by Hestenes [106]; also see [51] and [85]. Recently Levitin, Milyutin, and Osmolovskii have elaborated the complete theory of second-order conditions for problems with constraints [64] and [65]. In the last of these works there is an extensive bibliography. We used some of the constructions of [65] in the presentation of the material in this book. Also see [105] and [117].

To Chapter 4. Sections 4.1 and 4.4. There are many textbooks and monographs devoted to the classical calculus of variations. Besides those mentioned above, see [20], [63], [91], [93], [97], and [103]. The theory of necessary and sufficient conditions in the Lagrange problem is elaborated most completely in the monograph by Bliss [20]. The history of the problem is also presented in that monograph. For the relationship between the calculus of variations and classical mechanics, see [1].

Section 4.2. The initial outline of optimal control theory was presented in [26] and in the review article by Pontryagin [78]. After that Pontryagin, Boltyanskii, Gamkrelidze, and Mishchenko devoted their monograph [12] to optimal control theory, which gave rise to the rapid development of the whole branch of mathematics connected with the investigation of extremal problems. The proof of the maximum principle presented in Section 4.2 is a modification of the original proof given by V. G. Boltyanskii although we used no other means than those of classical mathematical analysis. The method of centered systems was used in the works of Dubovitskii and Milyutin. At present there are many proofs of the maximum principle, e.g., see [24], [31], [46]-[49], [54], [104], and [112].

Optimal control theory is dealt with in the textbooks [23], [31], and [35] and also in the monographs [25], [27], [32], [61], [66], [73], etc. In these monographs much emphasis is laid upon the solution of various concrete applied problems.

Section 4.3. The maximum principle for linear systems was first proved by Gamkrelidze [36]. The complete theory of linear systems was elaborated by Krasovskii [60]. The article [17] is devoted to the review of the subjects connected with Lyapunov-type problems; this article contains a generalization of the results of Section 4.3 and an extensive bibliography.

The development of the theory of Lyapunov-type problems has led to the investigation of convex integral functionals in the works [54], [98], [119], etc.

In conclusion, we mention several monographs which may help the reader to get acquainted with some fundamental questions of the theory of extremal problems which are not touched upon in the present book:

Many-dimensional calculus of variations: [110] and [116].

Existence theorems: [35], [54], and [116].

Sufficient conditions in optimal control problems: [22], [23], and [61].

Extension of extremal problems: [53] and [88].

Duality methods in the theory of extremal problems: [52] and [88].

Numerical methods: [29], [42], [72], [80], and [82].

Sliding regimes: [34].

Dynamic programming: [18] and [19].

Phase constraints and mixed-type constraints: [33], [47], and [48].

Also see the review [30].

REFERENCES

KF. A. N. Kolmogorov and S. V. Fomin, Elements of the Theory of Functions and Functional Analysis [in Russian], Nauka, Moscow (1976).

1. V. I. Arnold, Mathematical Methods of Classical Mechanics, Springer-Verlag, New York (1978).
2. N. I. Akhiezer, Lectures on the Calculus of Variations [in Russian], Gostekhizdat, Moscow (1955).
3. I. M. Gelfand and S. V. Fomin, Calculus of Variations, Prentice-Hall, Englewood Cliffs, N.J. (1963).
4. N. Dunford and J. I. Schwartz, Linear Operators, Part 7: General Theory, Interscience, New York (1958).
5. H. Cartan, Calcul Différentiel. Formes Différentielles [in French], Hermann, Paris (1967).
6. E. A. Coddington and N. Levinson, Theory of Ordinary Differential Equations, McGraw-Hill, New York (1955).
7. A. G. Kurosh, Higher Algebra [Russian translation], Mir, Moscow (1972).
8. M. A. Lavrent'ev and L. A. Lyusternik, A Course in the Calculus of Variations [in Russian], Gostekhizdat, Moscow (1950).
9. S. M. Nikol'skii, A Course of Mathematical Analysis [Russian translation], Vols. 1-2, Mir, Moscow (1977).
10. I. G. Petrovskii, Lectures on the Theory of Ordinary Differential Equations [in Russian], Gostekhizdat, Moscow (1952).
11. L. S. Pontryagin, Ordinary Differential Equations [in Russian], Nauka, Moscow (1965).
12. L. S. Pontryagin, V. G. Boltyanskii, R. V. Gamkrelidze, and E. F. Mishchenko, The Mathematical Theory of Optimal Processes, Interscience, New York (1962).
13. I. I. Privalov, Introduction to the Theory of Functions of a Complex Variable [in Russian], Nauka, Moscow (1977).
14. G. M. Fikhtengol'ts, A Course in Differential and Integral Calculus [in Russian], Vols. 1 and 2, Nauka, Moscow (1969).
15. P. Hartman, Ordinary Differential Equations, Wiley, New York (1964).
16. G. P. Akilov and S. S. Kutateladze, Ordered Vector Spaces [in Russian], Nauka, Novosibirsk (1978).
17. V. I. Arkin and V. L. Levin, "The convexity of the values of vector integrals, the measurable selection theorem and variational problems," Usp. Mat. Nauk, 27, No. 3, 21-77 (1972).
18. R. Bellman, Dynamic Programming, Princeton Univ. Press, Princeton (1957).

19. R. Bellman and R. Kalaba, Dynamic Programming and Modern Control Theory, Academic Press, New York (1965).
20. G. A. Bliss, Lectures on the Calculus of Variations, Univ. of Chicago, Chicago (1946).
21. V. B. Blyashke, The Circle and the Ball [in Russian], Nauka, Moscow (1967).
22. V. G. Boltyanskii, "Sufficient conditions for optimality and the justification of the dynamic programming methods," Izv. Akad. Nauk SSSR, Ser. Mat., 28, No. 3, 481-514 (1964).
23. V. G. Boltyanskii, Mathematical Methods of Optimal Control, Holt, Rinehart, and Winston, New York (1971).
24. V. G. Boltyanskii, "The tent method in the theory of extremal problems," Usp. Mat. Nauk, 30, No. 3, 3-55 (1975).
25. V. G. Boltyanskii, Optimal Control of Discrete Systems [in Russian], Nauka, Moscow (1973).
26. V. G. Boltyanskii, R. V. Gamkrelidze, and L. S. Pontryagin, "On the theory of optimal processes," Dokl. Akad. Nauk SSSR, 110, No. 1, 7-10 (1956).
27. A. E. Bryson and Y. C. Ho, Applied Optimal Control, Blaisdell, New York (1969).
28. M. M. Vainberg, The Variational Method and the Monotone Operator Method in the Theory of Nonlinear Equations [in Russian], Nauka, Moscow (1972).
29. F. P. Vasil'ev, Lectures on the Methods of Solving Extremal Problems [in Russian], Moscow State Univ. (1974).
30. R. Gabasov and F. M. Kirillova, Optimal Control Methods [in Russian], Itogi Nauki i Tekhniki. Sovremennye Problemy Matematiki, Vol. 6, Moscow, pp. 133-206.
31. R. Gabasov and F. M. Kirillova, Optimization Methods [in Russian], Belorussian State Univ., Minsk (1975).
32. R. Gabasov and F. M. Kirillova, Singular Optimal Control [in Russian], Nauka, Moscow (1973).
33. R. V. Gamkrelidze, "Optimal control processes with constrained phase coordinates," Izv. Akad. Nauk SSSR, Ser. Mat., 24, No. 3, 315-356 (1960).
34. R. V. Gamkrelidze, "On sliding optimal regimes," Dokl. Akad. Nauk SSSR, 143, No. 6, 1243-1245 (1962).
35. R. V. Gamkrelidze, Fundamentals of Optimal Control [in Russian], Tomsk State Univ. (1977).
36. R. V. Gamkrelidze, "The theory of time-optimal processes in linear systems," Izv. Akad. Nauk SSSR, Ser. Matem., 22, No. 4, 449-474 (1958).
37. R. V. Gamkrelidze and G. L. Kharatishvili, First-Order Necessary Conditions in Extremal Problems [in Russian], Nauka, Moscow (1972).
38. S. I. Gass, Linear Programming, McGraw-Hill, New York (1958).
39. I. V. Girsanov, Lectures on the Mathematical Theory of Extremal Problems [in Russian], Moscow State Univ. (1970).
40. E. G. Gol'shtein, Duality Theory in Mathematical Programming [in Russian], Nauka, Moscow (1971).
41. G. B. Dantsig, Linear Programming and Extensions, Princeton Univ. Press, Princeton, New Jersey (1963).
42. V. F. Demyanov and A. M. Rubinov, Approximate Methods of Solving Extremal Problems [in Russian], Leningrad State Univ. (1968).
43. A. V. Dorofeeva, "Calculus of variations in the second half of the 19th century," Istoriko-Matematicheskiye Issledovaniya, XV, 99-128 (1963).
44. A. V. Dorofeeva, "The development of the variational calculus as a calculus of variations," Istoriko-Matematicheskie Issledovaniya, XIV, 101-180 (1961).

45. A. V. Dorofeeva and V. M. Tikhomirov, "From the Lagrange multiplier rule to the Pontryagin maximum principle," Istoriko-Matematicheskie Issledovaniya, XXV (1979).
46. A. Ya. Dubovitskii and A. A. Milyutin, "Extremum problems in the presence of constraints," Zh. Vychisl. Mat. Mat. Fiz., No. 3, 395-453 (1965).
47. A. Ya. Dubovitskii and A. A. Milyutin, Necessary Conditions for Weak Extremum in the General Optimal Control Problem [in Russian], Nauka, Moscow (1971).
48. A. Ya. Dubovitskii and A. A. Milyutin, "Necessary conditions for weak extremum in optimal control problems with mixed-type inequality constraints," Zh. Vychisl. Mat. Mat. Fiz., No. 4, 725-770 (1968).
49. A. Ya. Dubovitskii and A. A. Milyutin, "Translation of Euler's equation," Zh. Vychisl. Mat. Mat. Fiz., No. 6, 1263-1284 (1969).
50. I. I. Eremin and N. N. Astaf'ev, Introduction to the Theory of Linear and Convex Programming [in Russian], Nauka, Moscow (1976).
51. W. Zangwill, Nonlinear Programming — Unified Approach, Prentice-Hall, Englewood Cliffs, New Jersey (1969).
52. A. D. Ioffe and V. M. Tikhomirov, "Duality of convex functions and extremal problems," Usp. Mat. Nauk, No. 6, 51-116 (1968).
53. A. D. Ioffe and V. M. Tikhomirov, "Extension of variational problems," Trudy MMO, 18, 187-246 (1968).
54. A. D. Ioffe and V. M. Tikhomirov, Theory of Extremal Problems, North-Holland, Amsterdam—New York—Oxford (1979).
55. L. V. Kantorovich, Mathematical Methods of Organizing and Planning Production [in Russian], Leningrad State Univ. (1939).
56. L. V. Kantorovich and G. P. Akilov, Functional Analysis [in Russian], Nauka, Moscow (1977).
57. S. Karlin, Mathematical Methods and Theory in Games, Programming and Economics, Addison-Wesley, Reading (1959).
58. V. G. Karmanov, Mathematical Programming [in Russian], Nauka, Moscow (1975).
59. M. A. Krasnosel'skii, P. P. Zabreiko, E. I. Pustyl'nik, and P. E. Sobolevskii, Integral Operators in the Space of Summable Functions [in Russian], Nauka, Moscow (1966).
60. N. N. Krasovskii, Theory of Motion Control. Linear Systems [in Russian], Nauka, Moscow (1968).
61. V. F. Krotov and V. I. Gurman, Methods and Problems in Optimal Control [in Russian], Nauka, Moscow (1973).
62. S. S. Kutateladze and A. M. Rubinov, The Minkowski Duality and Its Applications [in Russian], Nauka, Novosibirsk (1976).
63. M. A. Lavrent'ev and L. A. Lyusternik, Foundations of the Calculus of Variations [in Russian], Vols. I and II, ONTI, Moscow—Leningrad (1935).
64. E. S. Levitin, A. A. Milyutin, and N. P. Osmolovskii, "Conditions for local minimum in problems with constraints," in: Mathematical Economics and Functional Analysis [in Russian], Nauka, Moscow (1974), pp. 139-202.
65. E. S. Levitin, A. A. Milyutin, and N. P. Osmolovskii, "Higher-order conditions for local minimum in problems with constraints," Usp. Mat. Nauk., 33, No. 6, 85-148 (1978).
66. E. B. Lee and L. Markus, Foundations of Optimal Control Theory, Wiley, New York (1967).
67. H. W. Kuhn and A. W. Tucker (eds.), "Linear inequalities and related systems," Ann. Math. Stud., No. 38 (1956).
68. L. A. Lyusternik, "On conditional extrema of functions," Mat. Sb., 41, No. 3, 390-401 (1934).
69. L. A. Lyusternik and V. I. Sobolev, Elements of Functional Analysis, Ungar, New York (1961).

70. G. G. Magaril-Il'yaev, "Implicit function theorem for the Lipschitzian mappings," Usp. Mat. Nauk, 33, No. 1, 221-222 (1978).
71. N. Kh. Rozov (ed.), Mathematics for Engineers [in Russian], Znaniye, Moscow (1973).
72. N. N. Moiseev, Numerical Methods in the Theory of Optimal Systems [in Russian], Nauka, Moscow (1971).
73. N. N. Moiseev, Elements of the Theory of Optimal Systems [in Russian], Nauka, Moscow (1971).
74. A. D. Myskis, Advanced Mathematics for Engineers, Special Courses [Russian translation], Mir, Moscow (1975).
75. H. Nikaido, Convex Structures and Economic Theory, Academic Press, New York—London (1968).
76. I. V. Nit, Linear Programming with Discussion of Some Nonlinear Problems [in Russian], Moscow State Univ. (1978).
77. Optimal Control [in Russian], Znaniye, Moscow (1978).
78. L. S. Pontryagin, "Optimal control processes," Usp. Mat. Nauk, 14, No. 1, 3-20 (1959).
79. N. N. Pshenichnyi, Necessary Conditions for Extremum [in Russian], Nauka, Moscow (1969).
80. B. N. Pshenichnyi and Yu. M. Danilin, Numerical Methods in Extremal Problems [in Russian], Nauka, Moscow (1975).
81. R. T. Rockafellar, Convex Analysis, Princeton Univ. Press, Princeton, New Jersey (1970).
82. I. V. Romanovskii, Algorithms for Solving Extremal Problems [in Russian], Nauka, Moscow (1977).
83. K. A. Rybnikov, "The first stages of the calculus of variations," Istoriko-Matematicheskie Issledovaniya, No. 2, GITTL, Moscow—Leningrad (1949).
84. V. M. Tikhomirov, Some Problems in the Approximation Theory [in Russian], Moscow State Univ. (1976).
85. A. V. Fiacco and G. P. McCormick, Nonlinear Programming. Sequential Unconstrained Technique, Wiley (1968).
86. H. G. Zeuthen, Histoir des Mathématique dans L'Antiquité et la Moyen Age [in French], Paris (1902).
87. H. G. Zeuthen, Geschichte der Mathematik in XVI and XVIII Jahrhundert [in German], Leipzig (1903).
88. I. Ekeland and R. Temam, Analyse Convexe et Problèmes Variationelles [in French], Hermann, Paris (1974).
89. K. J. Arrow, L. Hurwicz, and H. Uzawa, Studies in Linear and Nonlinear Programming, Stanford Univ. Press, Stanford (1958).
90. D. B. Yudin and E. G. Gol'shtein, Linear Programming (Theory, Methods, and Applications) [in Russian], Nauka, Moscow (1969).
91. L. C. Young, Lectures on the Calculus of Variations and Optimal Control Theory, W. B. Saunders, Philadelphia—London—Toronto (1969).
92. A. V. Balakrishnan, Applied Functional Analysis, Springer-Verlag, New York (1976).
93. O. Bolza, Vorlesungen über Variationsrechnung [in German], Leipzig (1949).
94. T. Bonnesen and W. Fenchel, Theorie der konvexen Körper [in German], Springer-Verlag, Berlin (1934).
95. A. Brøndsted, "Conjugate convex functions in topological vector spaces," Mat. Fys. Medd. Dansk. Vid. Selsk., 34, No. 2, 1-26 (1964).
96. D. W. Bushaw, Differential Equations with a Discontinuous Forcing Term, Princeton: Dept. of Math. Princeton Univ. (1952).
97. C. Carathéodory, Variationsrechnung und partielle Differentialgleichungen erster Ordnung [in German], Teubner, Leipzig—Berlin (1935).
98. C. Castaing and M. Valadier, "Convex analysis and measurable multifunctions," Lect. Notes Math., No. 580, Springer-Verlag, Berlin (1977).

99. F. H. Clarke, "Generalized gradients and applications," Trans. Am. Math. Soc., 205, 247-262 (1975).
100. M. J. Cox, "On necessary conditions for relative minima," Am. J. Math., 66, No. 2, 170-198 (1944).
101. W. Fenchel, Convex Cones, Sets and Functions, Princeton Univ. Press, Princeton, New Jersey (1951).
102. W. Fenchel, "On conjugate convex functions," Can. J. Math., 1, 73-77 (1949).
103. J. Hadamard, Leçon sur le Calcul des Variations [in French], Hermann, Paris (1910).
104. H. Halkin, "On necessary conditions for optimal control of nonlinear systems," J. Analyse Math., 12, 1-82 (1964).
105. M. R. Hestenes, Calculus of Variations and Optimal Control Theory, Wiley, New York (1966).
106. M. R. Hestenes, Optimization Theory. The Finite-Dimensional Case, Wiley, New York (1975).
107. R. B. Holmes, Geometric Functional Analysis and Its Applications, Springer-Verlag, New York (1975).
108. F. John, "Extremum problems with inequalities as subsidiary conditions," in: Studies and Essays. Courant Anniversary Volume, Interscience, New York (1948), pp. 187-204.
109. W. E. Karush, Minima of Functions of Several Variables with Inequalities as Side Conditions, Univ. of Chicago Press, Chicago (1939).
110. R. Klötzler, Mehrdimensionale Variationsrechnung, Berlin: VEB Deutscher Verlag (1971).
111. H. W. Kuhn and A. W. Tucker, "Nonlinear programming," in: Proc. of Second Berkeley Symp., Univ. of California Press, Berkeley (1951), pp. 481-492.
112. P. Michel, "Une démonstration élémentaire du principe du maximum de Pontriaguine," Bull. Math. Économiques, 14 (1977).
113. H. Minkowski, Geometrie der Zahlen [in German], Teubner, Leipzig (1910).
114. H. Minkowski, "Theorie der konvexen Körper," in: Gesammelte Abhandlungen [in German], Vol. II, Teubner, Leipzig–Berlin (1911).
115. N. N. Moiseev and V. M. Tikhomirov, "Optimization," in: Mathematics Applied to Physics, Springer-Verlag, New York (1970), pp. 402-467.
116. Ch. B. Morrey, Multiple Integrals in the Calculus of Variations, Springer-Verlag, New York (1966).
117. L. W. Neustadt, Optimization. A Theory of Necessary Conditions, Princeton Univ. Press, Princeton, New Jersey (1976).
118. R. T. Rockafellar, "The theory of subgradients and its application to problems of optimization," Lect. Notes Univ. Montreal (1978).
119. R. T. Rockafellar, "Convex–integral functionals and duality," in: Contributions to Nonlinear Functional Analysis, Academic Press, New York (1971), pp. 215-236.
120. F. A. Valentine, "The problem of Lagrange with differential inequalities as added side conditions," in: Contributions to the Calculus of Variations, Univ. of Chicago Press, Chicago (1933-1937).

BASIC NOTATION

$\forall$ universal quantifier, "for every"

$\exists$ existential quantifier, "there exists"

$\Rightarrow$ sign of implication "...implies..."

$\Leftrightarrow$ sign of equivalence

$\overset{\text{def}}{=}$ by definition, is equal to

$\overset{(4)}{=}$ by virtue of (4), is equal to

$x \in A$ the element x belongs to the set A

$x \notin A$ the element x does not belong to the set A

$\varnothing$ empty set

$A \cup B$ the union of the sets A and B

$A \cap B$ the intersection of the sets A and B

$A \setminus B$ the difference of the sets A and B, i.e., the set of elements that belong to the set A but do not belong to the set B

$A \subset B$ the set A is contained in the set B

$A \times B$ the Cartesian product of the sets A and B

$A + B$ the arithmetical sum of the sets A and B

$\{x | P(x)\}$ the set of those elements x that possess the property $P(\cdot)$

$\{x_1, \ldots, x_n, \ldots\}$ the set which consists of the elements $x_1, \ldots, x_n, \ldots$

$F: X \to Y$ the mapping F of the set X into the set Y; the function F with domain X whose values belong to the set Y

$x \to F(x)$ the mapping (function) F assigns the element F(x) to an element x; the notation of the mapping (function) F in the case when it is desirable to indicate the notation of its argument

$F(\cdot)$ the notation which stresses that F is a mapping (function)

$F(A)$ the image of the set A under the mapping F

$\operatorname{im} F = \{y | y = F(x),\ x \in X\}$ the image of the mapping $F: X \to Y$

$F^{-1}(A)$ the inverse image of the set A under the mapping F

$F|A$ the restriction of the mapping F to the set A

$F \circ G$ the composition of the mappings G and F: $(F \circ G)(x) = F(G(x))$

$\mathbf{R}$ the set of all real numbers; the number line (the real line)

$\overline{\mathbf{R}} = \mathbf{R} \cup \{-\infty, +\infty\}$ the extended number line

$\inf A$ ($\sup A$) the infimum (supremum) of the numbers which belong to the set $A \subset \mathbf{R}$

$\mathbf{R}^n$ the arithmetical n-dimensional space endowed with the standard Euclidean structure; the elements of $\mathbf{R}^n$ should be thought of as column vectors even when they are printed in rows

$\mathbf{R}^n_+ = \{x = (x_1, \ldots, x_n) \in \mathbf{R}^n \mid x_i \geqslant 0\}$ the nonnegative orthant of $\mathbf{R}^n$

$e_1, \ldots, e_n$ the vectors of the standard orthonormal basis in $\mathbf{R}^n$; $e_1 = (1, 0, \ldots, 0), \ldots, e_n = (0, 1, \ldots, 1)$

$\mathbf{R}^{n*}$ the arithmetical n-dimensional space conjugate to $\mathbf{R}^n$, the elements of $\mathbf{R}^{n*}$ should be thought of as row vectors

$px = \langle p, x \rangle = \sum_{i=1}^{n} p_i x_i$ for any $x \in \mathbf{R}^n$, $p \in \mathbf{R}^{n*}$

x^T the row vector which is the transpose of the column vector x

$|x|$ the Euclidean norm in $\mathbf{R}^n$; $|x|^2 = (x|x) = x^T x$

$(x|y) = \sum_{i=1}^{n} x_i y_i = x^T y$ the scalar product in $\mathbf{R}^n$

$\|x\|$ the norm of an element x in a normed space

$\rho(x, y)$ the distance from an element x to an element y

$\rho(x, A) = \inf\{\rho(x, y) \mid y \in A\}$ the distance from an element x to a set A

$B(x, r) = \{y \mid \rho(y, x) \leqslant r\}$ the closed ball with center at x and radius r

$\mathring{B}(x, r) = \{y \mid \rho(y, x) < r\}$ the open ball with center at x and radius r

$\overline{A}$ the closure of the set A

$\operatorname{int} A$ the interior of the set A

$T_x M$ ($T_x^+ M$) the set of the tangent vectors (one-sided tangent vectors) to the set A at the point x

$\mathscr{L}_a f = \{x \mid f(x) \leqslant a\}$ an a-level set of the function f

X^* the conjugate space of X

x^* an element of the conjugate space X^*

$\langle x^*, x \rangle$ the value of the linear functional $x^* \in X^*$ on the element $x \in X$

$(x|y)$ the scalar product of elements x and y of a Hilbert space

$A^\perp = \{x^* \mid \langle x^*, x \rangle = 0, x \in A\}$ the annihilator of the set A

$\dim L$ the dimension of the space L

X/L the factor space of the space X generated by the subspace L

$\mathscr{L}(X, Y)$ the space of continuous linear mappings of the space X into the space Y; the mappings belonging to the space $\mathscr{L}(\mathbf{R}^n, \mathbf{R}^m)$ can be identified with their matrices relative to the standard bases in $\mathbf{R}^n$ and $\mathbf{R}^m$

$\mathscr{L}^n(X, Y)$ the space of continuous multilinear mappings of the space $\underbrace{X \times \ldots \times X}_{n \text{ times}}$ into the space Y

I the unit operator; E or E_n is the matrix of the unit operator of $\mathscr{L}(\mathbf{R}^n, \mathbf{R}^n)$, i.e., the unit n-th-order matrix (sometimes denoted by the same symbol I)

Λ^* the operator adjoint to the operator Λ; $\langle \Lambda^* y^*, x \rangle = \langle y^*, \Lambda x \rangle$

$\operatorname{Ker} \Lambda = \{x \mid \Lambda x = 0\}$ the kernel of the linear operator Λ

$\operatorname{Im} \Lambda = \{y \mid y = \Lambda x\}$ the image of the linear operator Λ

$x^T A x (y^T A x)$ the notation for the quadratic (bilinear) form with a matrix $A = (a_{ij})$: $y^T A x = \sum_{i, j=1}^{n} A_{ij} y_i x_j$.

$A > 0$ the matrix A is positive-definite

$A \geqslant 0$ the matrix A is positive semidefinite (nonnegative definite)

$C(K, Y)$ the space of continuous mappings from K into Y; if Y is normed, then $\|x(\cdot)\| = \|x(\cdot)\|_0 = \sup_{t \in K} \|x(t)\|$ for $x(\cdot) \in C(K, Y)$; $C(K) = C(K, Y)$ if it is clear what the space Y is or if Y = R

$C([t_0, t_1])$ the space of continuous functions on the closed interval $[t_0, t_1]$

$C^r(U)$ the set of mappings having in U continuous derivatives up to the order r inclusive; the notation is usually used in expressions of the type "the function f belongs to the class $C^r(U)$" or "F is a mapping of class $C^r(U)$" ($F \in C^r(U)$)

$C^r(\Delta, Y)$ the space of mappings of the closed interval $\Delta \subset \mathbf{R}$ into the space Y (usually $Y = \mathbf{R}^n$ or $Y = \mathbf{R}^{n*}$) possessing continuous derivatives up to the order r inclusive; for $x(\cdot) \in C^r(\Delta, Y)$ we have $\|x(\cdot)\| = \|x(\cdot)\|_r = \max\{\|x(\cdot)\|_0, \|\dot{x}(\cdot)\|_0, \ldots, \|x^{(r)}(\cdot)\|_0\}$

$KC^1(\Delta, Y)$ the space of mappings of the closed interval $\Delta \subset \mathbf{R}$ into the space Y (usually $Y = \mathbf{R}^n$ or $Y = \mathbf{R}^{n*}$) possessing piecewise-continuous derivatives. It is equipped with a norm as a subspace of the space $C(\Delta, Y)$

$W_\infty^1(\Delta, Y)$ the space of mappings of the closed interval $\Delta \subset \mathbf{R}$ into the space Y (usually $Y = \mathbf{R}^n$ or $Y = \mathbf{R}^{n*}$) each of which satisfies the Lipschitz condition with some constant. It is equipped with a norm as a subspace of the space $C(\Delta, Y)$

$F'(x; h)$ the derivative of the function F at the point x in the direction of the vector h

$\delta_+ F(x; \cdot)$ the first variation of the mapping F at the point x

$\delta F(x; \cdot)$ the first Lagrange variation of the mapping F at the point x

$\delta^n F(x; \cdot)$ the n-th Lagrange variation of the mapping F at the point x

$F'_\Gamma(x)$ the Gateaux derivative of the mapping F at the point x

$F'(x)$ the Fréchet derivative of the mapping F at the point x

$F'(x)[h]$ or $F'(x)h$ the value of the derivative of the mapping F at the point x on the vector h

$F \in D^1(x)$ the mapping F is Fréchet differentiable at the point x

$F \in SD^1(x)$ the mapping F is strictly differentiable at the point x

$F''(x)$ the second Fréchet derivative of the mapping F at the point x

$F''(x)[h_1, h_2]$ the value of the second derivative (regarded as a bilinear function) on the pair of vectors h_1 and h_2

$h \to d^2F(x; h) = F''(x)[h, h]$ the second differential of the mapping F at the point x

$F_{x_1}(x_1, x_2)$ ($F_{x_2}(x_1, x_2)$) the partial derivative with respect to x_1 (with respect to x_2) of the mapping F, i.e., the derivative of the mapping $x_1 \to F(x_1, x_2)$ ($x_2 \to F(x_1, x_2)$)

$[x_1, x_2] = \{x \mid x = \alpha x_1 + (1-\alpha)x_2, \ 0 \leqslant \alpha \leqslant 1\}$ the line segment joining the points x_1 and x_2

$\mathrm{dom}\, f = \{x \in X \mid f(x) < +\infty\}$ the effective domain of the function $f: X \to \overline{\mathbf{R}}$

$\mathrm{epi}\, f = \{(x, \alpha) \in X \times \mathbf{R} \mid f(x) \leqslant \alpha, \ x \in \mathrm{dom}\, f\}$ the epigraph of the function $f: X \to \overline{\mathbf{R}}$

δA the indicator function of the set A

sA the support function of the set A

μA the Minkowski function of the set A

lin A the linear hull of the set A

aff A the affine hull of the set A

cone A the conic hull of the set A

conv A the convex hull of the set A

$\overline{\mathrm{conv}}\, A$ the convex closure, i.e., the closure of the convex bull, of the set A

$\partial f(x)$ the subdifferential of the function f at the point x

$f^*(\cdot)$ the function conjugate to the function $f(\cdot)$, or the Young–Fenchel transform of the function f

$f(x) \to \inf$ (sup, extr), $x \in C$ the notation of an extremal problem

$\hat{x}$ a solution of an extremal problem

absmin (absmax, absextr) absolute minimum (maximum, extremum)

$\hat{x} \in \mathrm{absmin}(\mathfrak{z})$ ($\mathrm{absmax}(\mathfrak{z})$, $\mathrm{absextr}(\mathfrak{z})$) $\hat{x}$ yields an absolute minimum (maximum, extremum) in the problem $\mathfrak{z}$

locmin (locmax, locextr) a local minimum (maximum, extremum)

$\hat{x} \in \mathrm{locmin}(\mathfrak{z})$, ($\mathrm{locmax}(\mathfrak{z})$, $\mathrm{locextr}(\mathfrak{z})$) $\hat{x}$ yields a minimum (maximum, extremum) in the problem $\mathfrak{z}$

$\mathcal{J}, \mathcal{G}, \mathcal{B}, \mathcal{F}, \mathcal{P}$ functionals in problems of the classical calculus of variations and optimal control problems

Q the quadratic functional of a secondary problem of the calculus of variations

$L = L(t, x, \dot{x})$ or $L = L(t, x, \dot{x}, u)$ the Lagrangian

$\mathcal{L}$ the Lagrange function

$\mathcal{H}$ the Hamiltonian

H the Pontryagin function

$\mathcal{E}$ the Weierstrass function